24 0150993 4

AF598537

*Mechanics of fracture*

VOLUME 1

# Methods of analysis and solutions of crack problems

*Mechanics of fracture*

edited by GEORGE C. SIH

VOLUME 1

*Methods of analysis and solutions of crack problems*

# *Mechanics of fracture* 1

# *Methods of analysis and solutions of crack problems*

Recent developments in fracture mechanics
Theory and methods of solving crack problems

Edited by

G. C. SIH
*Professor of Mechanics and*
*Director of the Institute of*
*Fracture and Solid Mechanics*

*Lehigh University*
*Bethlehem, Pennsylvania*

NOORDHOFF INTERNATIONAL PUBLISHING
LEYDEN

ISBN: 90 01 79860 8

Library of Congress Catalog Card Number: 72-76784



PRINTED IN THE NETHERLANDS BY
NEDERLANDSE BOEKDRUK INDUSTRIE N.V.,–'S-HERTOGENBOSCH

# Contents

Chapter 5

## Field singularities and related integral expressions
*H. F. Bueckner*



Chapter 6

## Integral transform methods *I. N. Sneddon*



Chapter 7

## Numerical solution of singular integral equations
*F. Erdogan, G. D. Gupta and T. S. Cook*

Chapter 8

## Application of the finite element method to the calculation of stress intensity factors *P. D. Hilton and G. C. Sih*



Chapter 9

## Finite element methods for elastic bodies containing cracks *W. K. Wilson*

# *Introduction to a series on mechanics of fracture*

The main objective of this series is to present a unified view of the many different branches of fracture mechanics with emphasis placed on the application of the theory to engineering problems. With the large quantity of accumulated research results and the still larger quantity being generated, it has become necessary to develop a coherent presentation of these research findings. Each volume will concentrate on a subject of fundamental nature and contain several chapters written by experts in their own areas of specialization. In this way the research materials on timely subjects are carefully synthesized and made available to the newcomers and practitioners of fracture mechanics in a relatively short period of time.

The subject matter in this series will cover the experimental and theoretical, the static and dynamic, and the microscopic and macroscopic. Attempts will be made to integrate the atomistic and continuum approaches as well as to bridge the gap between laboratory measurements and structural design. It is generally agreed that the vast amount of fracture data collected on metals, polymers, composites, etc., still remains as laboratory information. The rational procedure for incorporating them in the design of structures such as bridges and aircraft remains to be worked out. Because of the interdisciplinary nature of this field which includes chemists, physicists, metallurgists, mechanical engineers, civil engineers, and others, misconceptions of the basic idea and inconsistencies in the applications of the theory cannot be avoided. Conference proceedings and journal publications are not the best vehicles for settling these differences since the opinions of the individual contributors are often diversified and many readers are not made aware of the discredited speculations. In this series, I intend to present a refreshed outlook on the theory of fracture mechanics. To accomplish this ambitious goal, I will need a great deal of help as well as understanding from my colleagues.

I will now attempt to show the desirability of a careful review of the present-day version of fracture mechanics. There is no doubt that the theoretical foundation of the fracture theory based on continuum mechanics should be credited to A. A. Griffith. In his second paper (1924), Griffith raised a question concerning his calculation (1921) of the change in strain energy $\Delta W$ of an elastic plate due to the presence of a crack. He claimed that the 1921 version of

$$\Delta W = \frac{\pi \sigma_g^2 a^2 \nu}{E}, \quad (1921) \tag{1}$$

is in error since according to his own words

> *... the expression used for the stresses gave values at infinity differing from the postulated uniform stress at infinity by an amount which, though infinitesimal, yet made a finite contribution to the energy when integrated round the infinite boundary. This difficulty has been overcome by slightly modifying the expressions for the stresses so as to make this contribution to the energy vanish.*

Griffith modified his original result to

$$\Delta W = \frac{\pi \sigma_g^2 a^2}{E}, \quad (1924) \tag{2}$$

where $E$ is the Young's modulus, $\sigma_g$ the uniform stress and $a$ the half crack length. No mathematical details were given by Griffith. While the difference between equations (1) and (2) lies only in the Poisson's ratio $\nu$, it has far reaching physical implications. This discrepancy has caused a great deal of confusion in the earlier and recent publications on the Griffith theory. A detailed explanation of this apparent discrepancy can be found in the work of Sih and Liebowitz (1967) where it is shown that the correct version is given by equation (2). From an energy analysis the stress corresponding to an unstable crack is found to be

$$\sigma_g = \left(\frac{2E\gamma_g}{\pi a}\right)^{\frac{1}{2}} \tag{3}$$

in which $\gamma_g$ is the energy required to create additional crack surface.

Some twenty-five years later, the validity of the Griffith equation (3) was questioned by Irwin (1948) and Orowan (1952) when applied to materials that deform plastically near the crack tips. In a later paper, Orowan (1955) reported that his X-ray photographic measurements of apparently brittle fractures of low carbon steels had detected a thin layer of material at the surface that contains significant plastic deformation. A quantity $\gamma_p$ called "plastic work" was introduced and estimated to be on the order of $10^3$ times greater than the Griffith surface energy $\gamma_g$. It was suggested that since $\gamma_p \gg \gamma_g$, the Griffith equation can be extended to include plasticity by stating that

$$\sigma_g = \left\{\frac{2E(\gamma_g + \gamma_p)}{\pi a}\right\}^{\frac{1}{2}} \simeq \left(\frac{2E}{\pi a}\right)^{\frac{1}{2}} \gamma_p^{\frac{1}{2}} . \tag{4}$$

Surprisingly enough, the meaning of this revision has never been carefully studied. First of all, it should be remembered that the Griffith energy principle is based on the global potential energy in the system reaching a maximum for unstable equilibrium. Hence, the quantities $a$, $\sigma_g$ and $\gamma_g$ are measured on the macroscale. The arbitrary insertion of $\gamma_p$ into the Griffith equation becomes questionable should it be interpretated as a microquantity. Moreover, the association of $\gamma_p$ with continuum plasticity, if any, is also not clear.

Briefly, it can be said that $\gamma_p$ in equation (4) should not be confused with $\gamma_c$ which is associated with the theoretical cleavage strength of the material. The open literature result

$$\frac{\gamma_c}{\gamma_g} \simeq \frac{\rho}{a_0} \tag{5}$$

which predicts the correct order of magnitude with $\rho/a_0 \simeq 10^3$ or $10^4$ was derived by using an incorrect approach. In equation (5), $\rho$ is the macroscopic crack tip radius and $a_0$ the microscopic equilibrium spacing between atomic planes. I have used a linear stress-displacement relationship for the atomic model and found that

$$\frac{\gamma_c}{\gamma_g} = \frac{\pi}{2}\left(\frac{\rho}{a_0}\right)\left(\frac{e_0}{E}\right) \tag{6}$$

where $e_0$ is the elastic modulus of the atomistic stress-strain curve. The Griffith stress at fracture becomes

$$\sigma_g = \left\{\frac{2}{\pi}\left(\frac{a_0}{\rho}\right)\left(\frac{E}{e_0}\right)\right\}^{\frac{1}{2}} \left(\frac{2E}{\pi a}\right)^{\frac{1}{2}} \gamma_c^{\frac{1}{2}} . \tag{7}$$

The quantity

$$\left\{\frac{2}{\pi}\left(\frac{a_0}{\rho}\right)\left(\frac{E}{e_0}\right)\right\}^{\frac{1}{2}} \simeq 10^{-2} \tag{8}$$

is simply a scale shifting factor.

In the first volume of this open series on the "Mechanics of Fracture", a more general theory of fracture mechanics is presented. The new discipline introduces the concept of the field strength of the crack tip strain-energy-density, marking a fundamental departure from the classical and current trends. The theory is a significant advance in that it can resolve all mixed mode crack extension problems for the first time in history. Unlike the Griffith–Irwin theory which accounts through $k$ or $G$ only for the amplitude of the local stresses, the fundamental parameter in the new theory, the "strain-energy-density" factor, $S$, is also directionally sensitive. The difference between $k$ (or $G$) and $S$ is analogous to the difference between a scalar and vector quantity. Loosely speaking, the stress-intensity factors $k_1$ and $k_2$ in a two-dimensional problem are both represented in the magnitude of $S$. Hence, the stress-intensity factors still play an important role in the fracture process.

This first volume is devoted to the evaluation of stress-intensity factors using analytical and numerical methods presented by some of the outstanding people in the field. The methods include conformal mapping, collocation, Laurent series, asymptotic expansion, integral transforms, alternating procedure and finite elements. Collections of stress-intensity factor solutions are given, many of which are presented for the first time. This comprehensive survey of the methods of analysis and solutions of crack problems should be most useful to research and practicing engineers.

Fundamental misconceptions and inconsistencies are scattered throughout the literature on fracture mechanics, and the practitioner must always be on guard against unquestioning acceptance of them. Additional examples of these miscues can be found in the Preface of the first volume.

## References

Griffith, A. A., *Phil. Trans. Roy. Soc. London*, Ser. A, Vol. A221, p. 163 (1921).

Griffith, A. A., *Proceedings of the* 1st *International Congress for Applied Mechanics*, Delft (1924) p. 55.

Irwin, G. R., *Fracturing of Metals*, ASM, Cleveland, Ohio, (1948).

Orowan, E., *Proceedings of the Symposium on Fatigue and Fracture of Metals*, Wiley, New York, (1952) p. 139.

Orowan, E., *Energy Criterion of Fracture*, Welding Research Supplement, (1955) p. 157.

Sih, G. C. and Liebowitz, H., *Intern. J. Solids and Structures*, (1967) Vol. 3, p. 1.

Sih, G. C. (forthcoming). *A New Outlook on Fracture Mechanics.*

# *Editor's preface*

It is well known that the traditional failure criteria cannot adequately explain failures which occur at a nominal stress level considerably lower than the ultimate strength of the material. The current procedure for predicting the safe loads or safe useful life of a structural member has been evolved around the discipline of linear fracture mechanics. This approach introduces the concept of a crack extension force which can be used to rank materials in some order of fracture resistance. The idea is to determine the largest crack that a material will tolerate without failure. Laboratory methods for characterizing the fracture toughness of many engineering materials are now available. While these test data are useful for providing some rough guidance in the choice of materials, it is not clear how they could be used in the design of a structure. The understanding of the relationship between laboratory tests and fracture design of structures is, to say the least, deficient. Fracture mechanics is presently at a standstill until the basic problems of scaling from laboratory models to full size structures and mixed mode crack propagation are resolved. The answers to these questions require some basic understanding of the theory and will not be found by testing more specimens.

The current theory of fracture is inadequate for many reasons. First of all it can only treat idealized problems where the applied load must be directed normal to the crack plane. In this case, values of the Mode I critical stress-intensity factors, $k_{1c}$ or crack extension force, $G_{1c}$, can be used to predict the applied loads that cause failure or, conversely, to find the allowable crack size that will not cause premature failure. However, in a structural member the crack is seldom in a plane normal to the principal stress. As a rule, the crack follows a curved path. For a two-dimensional problem, the influence of Mode II crack propagation is not always negligible. Complications arise here because the crack does not spread in a self-similar manner and the classical treatment of the strain-energy release rate concept breaks down. Mode III crack propagation has also been mistreated in the literature.

In this first volume of an open series on the "Mechanics of Fracture", I shall only touch on the subject of mixed mode crack extension. The problem of fracture design will be treated in the future. To define the fracture toughness of a material under mixed mode conditions, I find that the direction of crack propagation must first be determined. To accomplish this, it was necessary to cast away the classical concept of Griffith–Irwin which turns out to be a special case of a more general theory based on the intensity of the strain-energy-density field reaching a critical value at incipient fracture. This theory is still being developed and marks a fundamental departure from classical and current trends. I offer this simplified version as a first step toward predicting the critical loads of a cracked member under combined loading. The discrepancy on load predictions between the mixed mode theory and that of Griffith and Irwin can be very significant depending on the relative position of the crack to the applied load.

In dealing with the crack instability problem for a structure, it is essential to consider a slight perturbation of the load from the ideal condition of having it placed perfectly normal to the crack plane. This is analogous to the introduction of a small eccentricity in column studies. An instability phenomenon is always associated with imperfections that exist in the system. Thus, the mixed mode crack extension problem cannot be dismissed as being academic. It is real and must be accounted for in structural design. The material parameter in this new theory is referred to as the "strain-energy-density factor", $S$, which is a function of the stress-intensity factors $k_1$ and $k_2$ as given by

$$S = a_{11}k_1^2 + 2a_{12}k_1k_2 + a_{22}k_2^2$$

where $a_{ij}$ depend on the elastic properties of the material. The critical value $S_{cr}$ is an intrinsic material property. As can be seen, the stress-intensity factors $k_1$ and $k_2$ still play an important role in the fracture process. Hence, the correct determination of these factors is a necessary step toward the safe design of a structure.

The purpose of this volume is to present a comprehensive presentation of the methods of analysis and solutions to crack problems with emphasis placed on finding the stress-intensity factors from which the strain-energy density factor can be obtained. This is the only area in fracture mechanics that is relatively free from misconceptions and inconsistencies. All the chapters are written by outstanding people who have already made many important contributions in the past. Their views and experience are most valuable. It should be emphasized that the accuracy of any crack problem solution can be achieved only through the sound reasoning of the analyst who has a clear understanding of the physical problem. Brute force techniques are not the answer. In solving crack problems, the stress singularities that prevail at the crack tips must be treated with respect. When handled properly, all methods such as integral transforms, collocation, conformal mapping, finite elements, etc. should lead to the same answer. One of the least understood problems at the present is the analysis of surface flaws where the crack front intersects with a free boundary. The published results on this class of problems are unsatisfactory, both numerically and conceptually.

I shall briefly comment on the contents of this book without referring to each chapter and the individual author. The *conformal mapping* technique in the complex variable formulation is a powerful tool for solving crack problems where the geometries do not conform to coordinate systems. Problems such as cracks emanating from straight boundaries and holes can be easily handled by this method provided that the problem is two-dimensional. Another version of the complex variable technique is to expand the Goursat functions in *series form.* For a direct evaluation of the stress-intensity factor, it suffices to determine a finite number of the coefficients in the series. The computations are tedious but the method solves some most difficult boundary value problems involving cracks oriented at different positions in stiffened or unstiffened plates and cracks near inclusions. One of the most ingenious concepts that has been employed in all branches of applied mathematics is the *asymptotic expansion.* For some reason, the researchers in fracture mechanics have not been able to take advantage of this concept to its fullest extent, particularly in obtaining approximate but effective solutions to crack problems. The idea behind this approach is to evade the difficult problem of obtaining a complete solution. By knowing the asymptotic solutions for both ends of the range of the parameter, the solution for intermediate values of the parameter are found by interpolation to complete the picture.

The method of *integral transforms* has also played a key role in solving many linear fracture mechanics crack problems. The "Fourier" and "Mellin" are two commonly employed transforms depending upon the geometry under consideration. The problem involves the specification of tractions and displacements along single or multiple boundaries. A cardinal rule for solving mixed boundary-value problems is first to determine the nature of the stress singularity correctly and then to isolate it from any numerical calculations. The resulting integral equations usually come under the names of Cauchy, Fredholm, etc. This procedure has been used to solve many interesting crack problems in the area of layered composites where the crack-tip stress singularities are not necessarily of the inverse square root type. One of the salient features of this method is that it can be applied in a straightforward manner for solving the class of linear partial differential equations that arise in plate and shell theories. A departure from the conventional approach can be made by introducing the concept of fundamental singularities and fundamental fields. The stress-intensity factors can be given by integral representations involving a *weight function* and for a given crack geometry the result covers the loading condition in an arbitrary fashion. This is similar to using the concentrated load solution of a particular crack problem as the Green's function and expressing the stress-intensity factors in integral form that applies to arbitrary tractions on the crack surface. In many cases, these weight functions can be identified as solutions to homogeneous linear and singular integral equations. Axially symmetric crack problems can also be treated.

When the problem becomes three-dimensional most of the conventional methods can no longer be applied directly and only a handful of effective solutions are available. The common trend is to use numerical procedures. Among the popular ones are finite element, boundary collocation, successive superposition, etc. Generally speaking, if the three-dimensional crack is completely embedded in the solid and the stress singularity is known analytically everywhere around the crack border, any one of the aforementioned methods is satisfactory. Should the crack front intersect with a free boundary, serious problems can arise for a number of reasons. First of all, the specification of the stress state near the point where the crack penetrates through the surface is no longer obvious. Hence, the accuracy of any numerical solution becomes questionable. It should be cautioned that the errors committed near a free surface are not necessarily confined locally but they can affect the interior solution as well. Numerical methods can only estimate the average stress in a finite region and by definition they cannot describe local phenomena. For instance, the method of *finite elements* as originally designed for solving large scale structures is not effective unless special elements are placed near the crack tip region. For a surface flaw, the problem can never be resolved by resorting to a larger number of smaller elements near the surface. The answer lies in the inadequacy of the continuum mechanics model to explain a surface phenomenon rather than the inaccuracies of the numerical procedure. It becomes necessary to introduce the concept of a surface boundary layer which has never been advocated in solid mechanics. It is within this layer estimated to be approximately equal to the crack tip radius that the continuum definition of stress becomes inadequate. The physically meaningful quantity on a solid boundary is surface tension and not stress. Stress-intensity factors simply cannot be properly defined on a free surface. In other words, the Griffith–Irwin concept of $k$ (or $G$) is basically not adequate for treating the surface flaw problem. As I mentioned earlier, the classical theory is strictly limited to simple loading conditions and crack geometries.

For many years, my colleagues and I at Lehigh have pondered the surface flaw problem. Numerous analytical and numerical methods were tried before the procedure of successive superposition referred to as the *alternating method* in this book was adopted. Such an approach controlled the accuracy of numerical results through analytical separation of the singularities. Careful analysis of opposing contributions to the singularities enabled the numerical instability to be minimized. Considerable judgement using known qualitative characteristics of the solution was required to balance the numerical and analytical components of the approach. The stress-intensity factor solution was obtained for a semi-circular surface crack opened by uniform pressure. The $k$-factor starts out with a finite value at the utmost interior point of the crack border and increases first slowly and then more sharply as the free boundary is approached. After the peak, it drops very rapidly toward zero stopping short at a boundary layer distance from the free surface. This trend of the stress-intensity factor solution is believed to be correct.

Because of space and time limitations, many other useful methods such as the Riemann–Hilbert formulation, eigenfunction expansions, etc. are not included in this volume. The most important accomplishment of this work may not necessarily be the comprehensive coverage of the basic methods of analyzing crack problems, but the primary attention should be focused on the ingenuity and imagination of the individual authors who have a full command of the physical nature as well as a mathematical understanding of the problem. To the authors I am deeply indebted for their contributions which have made this volume possible.

*Lehigh University* G. C. SIH
*Bethlehem, Pa.*

February 1972

# *Contributing authors*

J. P. BENTHEM
Laboratorium voor Technische Mechanica, Technische Hogeschool, Delft, Holland

O. L. BOWIE
Army Materials and Mechanics Research Center, Watertown, Massachusetts

H. F. BUECKNER
Turbine Engineering, General Electric Company, Schenectady, New York

Adjunct Professor of Mechanics, Rensselaer Polytechnic Institute, Troy, N.Y.

T. S. COOK
Department of Mechanical Engineering and Mechanics, Lehigh University, Bethlehem, Pennsylvania

F. ERDOGAN
Department of Mechanical Engineering and Mechanics, Lehigh University, Bethlehem, Pennsylvania

G. D. GUPTA
Department of Mechanical Engineering and Mechanics, Lehigh University, Bethlehem, Pennsylvania

R. J. HARTRANFT
Department of Mechanical Engineering and Mechanics, Lehigh University, Bethlehem, Pennsylvania

P. D. HILTON
Department of Mechanical Engineering and Mechanics, Lehigh University, Bethlehem, Pennsylvania

M. Isida
National Aerospace Laboratory, Tokyo, Japan

W. T. Koiter
Laboratorium voor Technische Mechanica, Technische Hogeschool, Delft, Holland

G. C. Sih
Department of Mechanical Engineering and Mechanics, Lehigh University, Bethlehem, Pennsylvania

I. N. Sneddon
Department of Mathematics, University of Glasgow, Scotland

W. K. Wilson
Research and Development Center, Westinghouse Electric Company, Pittsburgh, Pennsylvania

G. C. Sih*

# Introductory chapter: A special theory of crack propagation

## I. Historical remarks

The Griffith concept [1, 2] of imperfection instability in a solid was the first step toward predicting the fracture strength of solids. The basic idea behind his theory is that a crack will begin to propagate if the elastic energy released by its growth is greater than the energy required to create the fractured surfaces. As a model, Griffith considered the problem of a crack of length $2a$ in a plate under tension $\sigma$ as in Figure 1a. He then found that the critical stress $\sigma_{cr}$ required for crack growth is

$$\sigma_{cr} a^{\frac{1}{2}} = \left(\frac{2E\gamma}{\pi}\right)^{\frac{1}{2}} \tag{1}$$

where $E$ is the Young's modulus and $\gamma$ the specific surface energy. Since the quantity $(2E\gamma/\pi)^{\frac{1}{2}}$ contains only material constants, the factor $\sigma_{cr} a^{\frac{1}{2}}$ should be an intrinsic material parameter. Twice the specific surface energy $\gamma$ is equal to the critical elastic energy release rate $G_{1c}$, i.e., $G_{1c} = 2\gamma$. The experiments Griffith performed on glass show that the values of $\sigma_{cr} a^{\frac{1}{2}}$ were indeed the same over a wide range of crack lengths.

The concept of crack energy release leads to serious drawbacks in carrying out the mathematical details for cracks in a combined stress field. The energy release concept assumes the direction of crack propagation to be known an *a priori*. Hence, the Griffith theory can only treat problems with the crack lying in a plane normal to the applied stress as in Figure 1a. A simple question, such as what will be the direction of crack propagation if the crack was

* Professor of Mechanics and Director of the Institute of Fracture and Solid Mechanics, Lehigh University, Bethlehem, Pa. 18015, USA.

inclined at an angle $\beta$ to the loading axis, has not yet been answered satisfactorily. In such a case, equation (1) is obviously no longer valid. Moreover, a reversal of loading (Figure 1b) will produce crack propagation along a different path. A realistic theory of fracture mechanics should be able to explain the fracture phenomena for both types of loading on the inclined crack in Figures 1a and 1b.

Irwin [3] in applying the Griffith's concept to solve fracture problems recognized the importance of the intensity of the local stress field. He proposed three modes of crack extension which are identified by their respective stress-intensity factors $k_1$, $k_2$ and $k_3$. The Mode I intensity factor $k_1$ is in fact related to the Griffith energy release rate $G_1$ as

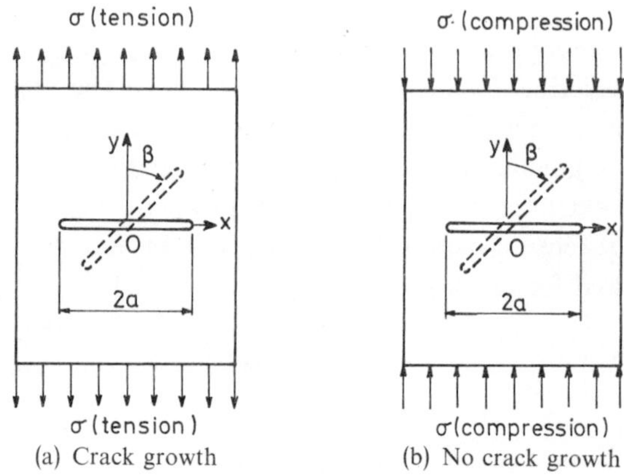

(a) Crack growth (b) No crack growth

Figure 1. A central crack in a plate.

$$G_1 = \frac{\pi k_1^2}{E} \text{ (generalized plane stress)}. \tag{2}$$

Similar relations for $G_2 = \pi k_2^2/E$ and $G_3 = k_3^2/2\mu$ with $\mu$ being the shear modulus of elasticity were formally computed by assuming that the crack extends in a plane collinear with the original crack as in Mode I. In fact, the physical meaning of Mode II and Mode III crack extension was never questioned and only a few attempts [4, 5] have been made to study the direction of Mode II and Mode III crack propagation. To the author's best knowledge, experimental values of $G_{2c}$ and $G_{3c}$ were never recorded and hence the present day understanding of crack propagation is restricted to Mode I problems.

The crack model of Barenblatt [6] should also be mentioned. In his theory, Barenblatt chooses to remove the $r^{-\frac{1}{2}}$ stress singularity by considering a cohesive zone along a line ahead of the crack. The criterion of fracture is based on the concept of a critical modulus of cohesion. Subsequently, Dugdale [7] used a mathematically similar but physically different model to study the crack tip plastic zone sizes in thin sheets. A generalization of some of these physical concepts was later made by Rice [8] through the application of a path independent integral which is identical in form to a component of the energy momentum tensor introduced by Eshelby [9] to characterize generalized forces on dislocations in elastic fields. The above crack models which put more emphasis on the physical aspects of the material, are all inherently restricted to cracks that extend along the line of load symmetry. Although this limitation is not essential to the experimental studies of fracture toughness, it presents a major set-back to the prediction of applied stress for crack initiation in structural members. In the design of bridges and aircraft, the stress-state around the crack tip is in most cases of the mixed type where the assumption of Mode I fracture would be unrealistic.

## II. The strain-energy-density concept

In a series of recent papers, Sih [10–12] has proposed a theory of fracture based on the field strength of the local strain-energy-density which marks a fundamental departure from the classical and current concepts. The theory requires no calculation on the energy release rate and thus possesses the inherent advantage of being able to treat all mixed mode crack extension problems for the first time. Unlike the conventional theory of $G$ and $k$ which measures only the amplitude of the local stresses, the fundamental parameter in the new theory, the "strain-energy-density" factor $S$, is also direction sensitive. The difference between $k$ (or $G$) and $S$ is analogous to the difference between a scalar and vector.

Referring to Figures 2a and 2b, the Griffith–Irwin theory can be viewed as a scalar theory in that it specifies only the critical value of a scalar $G_{1c}$ (or $k_{1c}$) at incipient fracture. The direction of crack propagation is always preassumed to be normal to the load. Moreover, the crack front must be straight such that $G$ or $k$ does not vary along the leading edge of the crack. In addition, a scalar theory cannot yield the correct material parameter if two or more stress-intensity factors are present along the crack border. The $S$-factor in the Sih theory behaves like a director. It senses the direction of least resistance by attaining a stationary value with respect to the angle $\theta$ as indicated in Figure 2b. As it will be shown, the stationary value of $S_{min}$ can be used as an intrinsic material parameter whose value at the point of crack

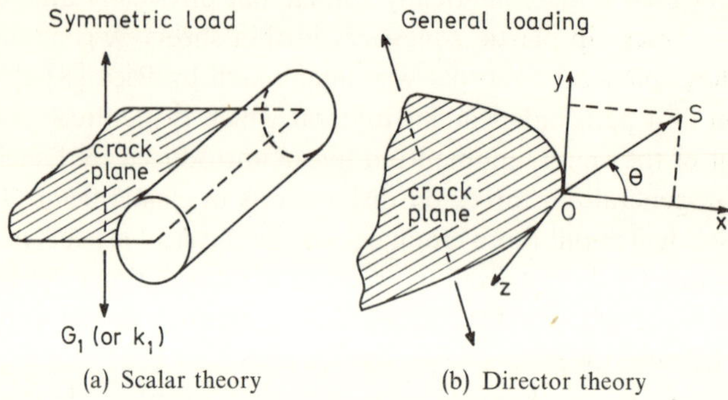


(a) Scalar theory (b) Director theory

Figure 2. Comparison of scalar and director theory.

instability $S_{cr}$ is independent of the crack geometry and loading. In the general context, the Griffith–Irwin theory is the special case when $\theta=0$ and the director $S$ coincides with the $x$-axis.

*Special form of local energy density.* Consider the three-dimensional case of a crack in a combined stress field and focus attention on a coordinate system $(x, y, z)$ in Figure 2 with the $x$-axis normal to the crack, the $y$-axis perpendicular to the crack plane and the $z$-axis tangent to the crack border. While the origin 0 traces the crack periphery, the functional form of the stress components in $r$, $\theta$ [13] remains unchanged, i.e.,

$$\sigma_x = \frac{k_1}{(2r)^{\frac{1}{2}}} \cos(\theta/2)[1-\sin(\theta/2)\sin(3\theta/2)] - \frac{k_2}{(2r)^{\frac{1}{2}}} \sin(\theta/2)[2+\cos(\theta/2)\cos(3\theta/2)] + \ldots$$

$$\sigma_y = \frac{k_1}{(2r)^{\frac{1}{2}}} \cos(\theta/2)\,[1+\sin(\theta/2)\sin(3\theta/2)] + \frac{k_2}{(2r)^{\frac{1}{2}}} \sin(\theta/2)\cos(\theta/2)\cos(3\theta/2) + \ldots$$

$$\tau_{xy} = \frac{k_1}{(2r)^{\frac{1}{2}}} \cos(\theta/2)\sin(\theta/2)\cos(3\theta/2)$$

$$+ \frac{k_2}{(2r)^{\frac{1}{2}}} \cos(\theta/2)[1 - \sin(\theta/2)\sin(3\theta/2)] + \dots$$

$$\sigma_z = 2\nu \frac{k_1}{(2r)^{\frac{1}{2}}} \cos(\theta/2) - 2\nu \frac{k_2}{(2r)^{\frac{1}{2}}} \sin(\theta/2) + \dots$$

$$\tau_{xz} = -\frac{k_3}{(2r)^{\frac{1}{2}}} \sin(\theta/2) + \dots$$

$$\tau_{yz} = \frac{k_3}{(2r)^{\frac{1}{2}}} \cos(\theta/2) + \dots \tag{3}$$

where the non-singular terms have been dropped and $r, \theta$ are the polar components in the $yz$-plane (Figure 3). For an elastic material, the strain

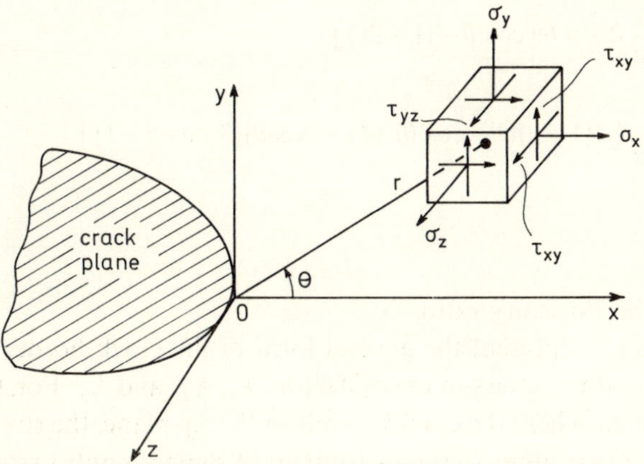


Figure 3. Stress components near crack border.

energy stored in the element $dV = dx\,dy\,dz$ under a general three-dimensional stress system is

$$dW = \left[\frac{1}{2E}(\sigma_x^2 + \sigma_y^2 + \sigma_z^2) - \frac{\nu}{E}(\sigma_x\sigma_y + \sigma_y\sigma_z + \sigma_z\sigma_x)\right.$$

$$\left. + \frac{1}{2\mu}(\tau_{xy}^2 + \tau_{xz}^2 + \tau_{yz}^2)\right] dV. \tag{4}$$

Substituting equations (3) into (4) yields the quadratic form for the strain-energy-density function

$$\frac{dW}{dV} = \frac{1}{r}(a_{11}k_1^2 + 2a_{12}k_1k_2 + a_{22}k_2^2 + a_{33}k_3^2) + \ldots \tag{5}$$

Note that the higher order terms in $r$ have been neglected and that the strain-energy-density function near the crack possesses a $1/r$ energy singularity. Hence, the quadratic

$$S = a_{11}k_1^2 + 2a_{12}k_1k_2 + a_{22}k_2^2 + a_{33}k_3^2 \tag{6}$$

represents the amplitude or the intensity of the strain-energy-density field and it varies with the polar angle $\theta$ in Figure 3. The coefficients $a_{ij}$ $(i, j = 1, 2, 3)$ are given by

$$a_{11} = \frac{1}{16\mu}[(3-4\nu-\cos\theta)(1+\cos\theta)]$$

$$a_{12} = \frac{1}{16\mu}2\sin\theta[\cos\theta-(1-2\nu)]$$

$$a_{22} = \frac{1}{16\mu}[4(1-\nu)(1-\cos\theta)+(1+\cos\theta)(3\cos\theta-1)]$$

$$a_{33} = \frac{1}{4\mu} \tag{7}$$

where $\nu$ is the Poisson's ratio.

Equations (3) represent the general form of the crack border stress field involving the three stress-intensity factors $k_1$, $k_2$, and $k_3$. For two-dimensional problems where the crack extends in the $xy$-plane, the stress-intensity factors do not vary along the crack front and $S$ depends only on one variable, namely the angle $\theta$. In three dimensions, $k_1$, $k_2$, and $k_3$ may occur simultaneously and they can also vary from point to point on the crack border. An example of this is given in [12].

*Invariant property of density factor.* Before proceeding with the application of the field strength concept to the fracture problem, it is worthwhile to examine the invariant property of the local strain-energy-density field. The important point to be made here is that the quadratic $S$ in equations (6) has a value that is independent of the choice of the stress-intensity factors $k_1$, $k_2$

and $k_3$. In other words, $S$ is a form "invariant" with reference to one system of stress-intensity factors to another. This implies that

$$S(k_1, k_2, k_3) = S(k'_1, k'_2, k'_3) \tag{8}$$

where $k'_i$ and $k_i$ are the stress-intensity factors of two different systems and $S$ possesses the inherent property of being a constant. Equation (8) defines a quadric surface for each state of stress-intensity factors and the discussion of this quadric parallels closely that of the strain or stress quadric of Cauchy in the theory of elasticity. Briefly, equation (6) may be solved as an eigenvalue problem. For a nontrivial solution, the three eigenvalues are found to be

$$\lambda^2 - (a_{11} + a_{22})\lambda + (a_{11}a_{22} - a_{12}^2) = 0 \tag{9}$$
$$\lambda_3 = a_{33}\,.$$

Using relations of $a_{ij}$ in equation (7), the first of equations (9) gives

$$\lambda_1 = \frac{1}{16\mu}\,[4(1-2\nu) + (1-\cos\theta)^2]$$

$$\lambda_2 = \frac{1}{16\mu}\,(1+\cos\theta)^2 \tag{10}$$

Hence, the quadric surface in eigenform is

$$\lambda_1 K_1^2 + \lambda_2 K_2^2 + \lambda_3 K_3^2 = \text{constant} \tag{11}$$

and the associated transformation

$$\begin{bmatrix} \cos(\theta/2) & \sin(\theta/2) & 0 \\ -\sin(\theta/2) & \cos(\theta/2) & 0 \\ 0 & 0 & 1 \end{bmatrix} \begin{bmatrix} k_1 \\ k_2 \\ k_3 \end{bmatrix} = \begin{bmatrix} K_1 \\ K_2 \\ K_3 \end{bmatrix}$$

The solution determines the points on a central quadric surface for which the distance from the origin is stationary relative to neighboring points. With this background, the property of the quadric $S$ will be shown to play an important role in the theory of crack propagation.

## III. Fundamental hypotheses on crack initiation and direction

Since the strain-energy-density factor $S$ in equation (6) has some attributes of the intensity of a force field associated with a type of potential, it is natural to inquire on the relationship between $S$ and the potential energy $\Sigma$ in the system. If the cracked body is subjected to tractions only, then the potential energy is equal to the negative of the strain energy*. Now let $P$ stand for the potential energy per unit volume of the element located at a distance $r$ from the crack border as shown in Figure 3, i.e., $P = \mathrm{d}\Sigma/\mathrm{d}V$, and similarly let $U$ be the strain energy per unit volume given by $U = \mathrm{d}W/\mathrm{d}V$. Making use of the relation $P = -U$ and $U = S/r$, the potential energy per unit volume becomes

$$P = -\frac{S}{r}. \tag{12}$$

*Basic assumptions.* Two fundamental hypotheses** of crack extension will now be laid down:

Hypothesis (1): *The crack will spread in the direction of maximum potential energy density.*

Hypothesis (2): *The critical intensity $S_{cr}$ of this potential field governs the onset of crack propagation.*

Note that for crack propagation to take place in the $xy$-plane the direction of maximum potential energy density must be found. In two-dimensional problems, the direction of crack propagation can be determined by a single variable $\theta$ and hence hypothesis (1) can be satisfied by the application of the calculus of variations. A necessary condition for the potential energy density $P$ to have a stationary value is that

$$\frac{\partial P}{\partial \theta} = 0\,, \quad \text{at} \quad \theta = \theta_0\,. \tag{13}$$

The value of $\theta_0$, which makes $P$ a maximum, determines the angle of the plane along which the crack spreads and can be found by further requiring that

$$\frac{\partial^2 P}{\partial \theta^2} < 0\,, \quad \text{at} \quad \theta = \theta_0 \tag{14}$$

which is a position of unstable equilibrium. From the canonical form of $S$ in equation (11), it can be concluded that $S$ is positive definite and thus $P$ is negative.

* The opposite holds for displacement loading conditions.

** These hypotheses will be further discussed in a general theory of crack propagation being developed by Sih [14].

Rewriting the conditions in equations (13) and (14) in terms of the strain-energy-density function renders

$$\frac{\partial S}{\partial \theta} = 0\,, \quad \frac{\partial^2 S}{\partial \theta^2} > 0 \text{ at } \theta = \theta_0 \tag{15}$$

which are the necessary and sufficient conditions for $S$ to be a minimum. Hypothesis (1) is thus equivalent to the assumption that

*crack initiation will start in a radial direction along which the strain energy density is a minimum.*

An essential point to be made here is that the above criterion is based on the local density of the energy field in the crack tip region and requires no special assumption on the direction in which the energy released by the separating crack surfaces is computed as in the Griffith theory [1, 2] and others. This removes the fundamental difficulties involved in the past for computing energy release rate in mixed mode problems. It is now clear that any fracture criterion based on a single stress parameter such as $k_1$ alone will not be sufficient to describe the problem of mixed mode fracture. Furthermore, the Mode II energy release rate commonly computed as $G_2 = \pi k_2^2/E$, where $k_2$ is the stress-intensity factor for skew-symmetric loadings, is not valid since in Mode II the crack does not run straight ahead as in Mode I. Hence the crack energy release for a mixed mode problem cannot be obtained by simply adding $G_1$ and $G_2$. It should also be emphasized that the so-called "crack closure" method though conceptually correct has misled a number of previous investigators to solve the branch crack or kinked crack problem, analytically as well as numerically. This type of boundary value problems is extremely difficult to solve and thus far no effective solutions have been found.

## IV. Prediction of crack growth direction

One of the least understood problems in fracture mechanics has been on the estimate of applied loads at crack instability in mixed mode situations where a crack can spread in any direction depending on the relative orientation of the load and crack position. Strangely enough, this problem has never been examined seriously. Admittedly, this is not an important problem in fracture toughness testing studies since all experiments could be carried out under Mode I crack extension. However, the omission of the mixed mode effect in predicting the applied stress to trigger fracture in structures could be dangerously optimistic. For this reason, the classical fracture mechanics approach is basically a theory for characterizing materials and has limited use in structural design.

A major shortcoming of the classical theory is that it cannot predict the direction of crack propagation. It is basically a theory for cracks that propagate in a self-similar manner, although Griffith did postulate that the crack will open up in the plane normal to the direction of maximum stress. At first sight, this statement appears to be plausible and has received a certain degree of acceptance in the literature. The validity and completeness of the Griffith's postulate has never been clearly understood and is not a simple matter to settle. It will be shown that the predictions based on the maximum stress criterion do deviate from those of the strain-energy-density criterion and can be questioned simply on physical grounds. Additional discussions on this subject will be brought up in the general theory of crack propagation in a future communication.

At the present time, it is more pertinent to illustrate the method of determining the direction of crack propagation by considering a few simple examples. All of the crack problems treated by the energy density theory are taken to be in a state of plane strain.

*Mode I crack extension.* In his classical paper, Griffith examined the problem of an infinite body containing a central crack of length $2a$ subjected to applied stress $\sigma$ normal to the crack plane as in Figure 1a or 4a. Because of load symmetry, the direction of crack propagation never entered into the problem. Suppose now that this is an unknown in the problem, then it is necessary to apply equations (13) and (14) or (15).

For the Griffith crack

$$k_1 = \sigma a^{\frac{1}{2}}, \quad k_2 = 0 \tag{16}$$

and hence equation (6) takes the simple form

$$S = \frac{\sigma^2 a}{16\mu}\left[(3-4\nu-\cos\theta)(1+\cos\theta)\right]. \tag{17}$$

Differentiating $S$ with respect to $\theta$ and setting $\partial S/\partial\theta = 0$, two possible solutions are found

$$\theta = 0 \quad \text{and} \quad \cos\theta = 1-2\nu\,. \tag{18}$$

The second derivative of $S$ with $\theta$ gives

$$\frac{\partial^2 S}{\partial \theta^2} = \frac{\sigma^2 a}{8\mu} [\cos 2\theta - (1-2\nu) \cos \theta] . \tag{19}$$

Inserting the results in equation (18) into (19), it is found that the solution $\theta_0 = 0$ yields $\partial^2 S/\partial \theta^2 > 0$ and thus $S$ is a minimum:

$$S_{\min} = \frac{(1-2\nu)\sigma^2 a}{4\mu} . \tag{20}$$

The plane $\theta_0 = 0$ corresponds to the direction of maximum potential energy, a position of unstable equilibrium. The critical applied stress to initiate crack growth becomes

$$\sigma_{\rm cr} = [4\mu S_{\rm cr}/(1-2\nu)a]^{\frac{1}{2}} \tag{21}$$

where the parameter $\sigma_{\rm cr} a^{\frac{1}{2}}$ is a material constant as in the classical case. Thus for cracks propagating in a self-similar fashion $S_{\rm cr}$ can be related to $\gamma$ in the Griffith theory and to $k_{1c}$ or $G_{1c}$ as follows:

$$\sigma_{\rm cr}^2 a = [4\mu/(1-2\nu)] S_{\rm cr} = [2E/\pi(1-\nu^2)]\gamma = k_{1c}^2 . \tag{22}$$

*Mode II crack extension.* Let an infinite solid with a through crack of length $2a$ in Figure 4b be subjected to shear stresses $\tau$ such that the crack occupies a plane of skew-symmetry. For this problem

$$k_1 = 0 , \quad k_2 = \tau a^{\frac{1}{2}} \tag{23}$$

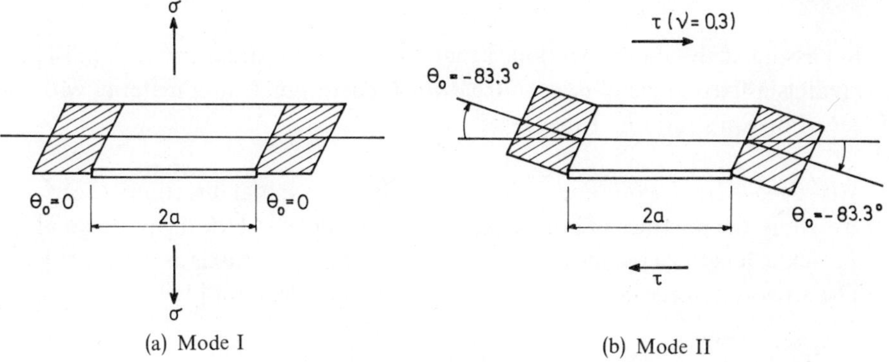


(a) Mode I (b) Mode II

Figure 4. Mode I and Mode II crack extension.

and the direction of crack growth is no longer obvious. Applying the conditions stated in equation (15) to

$$S = \frac{\tau^2 a}{16\mu}[4(1-\nu)(1-\cos\theta)+(1+\cos\theta)(3\cos\theta-1)] \tag{24}$$

it is found that

$$\cos\theta_0 = \frac{1-2\nu}{3} \tag{25}$$

which gives $S$ a minimum or $P$ a maximum. The other solution $\sin\theta=0$ is not of interest in discussing crack propagation. The predicted fracture angle $\theta_0$ in equation (25) depends on the Poisson's ratio $\nu$ (see Table I).

A point to be recognized is that the energy-density theory predicts the crack to initiate in a plane which is no longer collinear with the crack itself. Furthermore, this direction depends on the material property of the solid.

TABLE I

*Mode II fracture angle*

| $\nu$ | 0.0 | 0.1 | 0.2 | 0.3 | 0.4 | 0.5 |
|---|---|---|---|---|---|---|
| $\theta_0$ | −70.5° | −75.6° | −79.3° | −83.3° | −87.2° | −90.0° |

This marks a fundamental departure from the classical concept of fracture mechanics. The minimum value of $S$ is found to be

$$S_{\min} = \frac{[8(1-\nu)-4\nu^2]\tau^2 a}{48\mu}. \tag{26}$$

In passing, it should be mentioned that the maximum stress criterion in [4] predicts a fixed angle of $\theta_0=-70.5°$ which corresponds to a material with zero Poisson's ratio in Table I.

*Mixed mode crack extension.* The first study of the initial direction of crack growth in the presence of both $k_1$ and $k_2$ was made in [4] for the problem of a crack of length $2a$ inclined at an angle $\beta$ with the loading axis as in Figure 1. The stress-intensity factors $k_1$ and $k_2$ for this problem are [15]

$$\begin{aligned} k_1 &= \sigma a^{\frac{1}{2}} \sin^2\beta \\ k_2 &= \sigma a^{\frac{1}{2}} \sin\beta\cos\beta \end{aligned} \tag{27}$$

where $\sigma$ is the applied stress. It was assumed that the crack will start to extend in the plane which is normal to the maximum circumferential stress $\sigma_\theta$ in accordance with the condition

$$k_1 \sin \theta_0 + k_2 (3 \cos \theta_0 - 1) = 0 \tag{28}$$

for determining the initial angle of crack growth $\theta_0$. Using equation (27), $k_1$ and $k_2$ may be eliminated. This renders

$$\sin \theta_0 + (3 \cos \theta_0 - 1) \cot \beta = 0\,, \qquad \beta \neq 0 \tag{29}$$

which contains no elastic constants. This result implies that for the crack configuration and loading condition given in Figure 1 the initial angle of crack growth is independent of the material properties.

Turning now to the strain-energy-density theory which states that crack propagates in the direction of

$$S = \sigma^2 a(a_{11} \sin^2 \beta + 2a_{12} \sin \beta \cos \beta + a_{22} \cos^2 \beta) \sin^2 \beta \tag{30}$$

being a minimum. The coefficients $a_{ij}$ in equation (30) as derived from equations (6) and (27) are those given in equations (7). Differentiating equation (30) with respect to $\theta$ and setting the result to zero, the fracture angle $\theta_0$ for a given position of the crack specified by $\beta$ can be calculated from

$$2(1-2\nu) \sin (\theta_0 - 2\beta) - 2 \sin [2(\theta_0 - \beta)] - \sin 2\theta_0 = 0\,, \qquad \beta \neq 0 \tag{31}$$

(a) *Uniaxial tension.* The numerical results of equation (31) for negative values of $\theta_0$ and positive $\sigma$ are shown in Figure 5 which is a plot of the fracture angle $\theta_0$ *versus* the crack angle $\beta$ from 0° to 90°. The curve based on the maximum stress criterion which is dotted agrees well with equation (31) for large values of $\beta$ and represents a lower bound for small values of $\beta$. In general, it can be taken as an average curve.

The validity of these predictions can be checked with the results of a series of experiments [4] performed on the specimen in Figure 5. Plexiglass sheets of approximately $9'' \times 18'' \times \frac{3}{16}''$ were used with a central crack of approximately 2″ in length positioned at angles of $\beta$ from 30° to 80° in increments of 10°. The initial fracture angles at both ends of the crack were measured. The experimental data for four sets of tests are given in Table II with $(\theta_0)_{avg}$ being the average fracture angle of all the measured values. The last two rows give the theoretical calculations of equation (28) for the

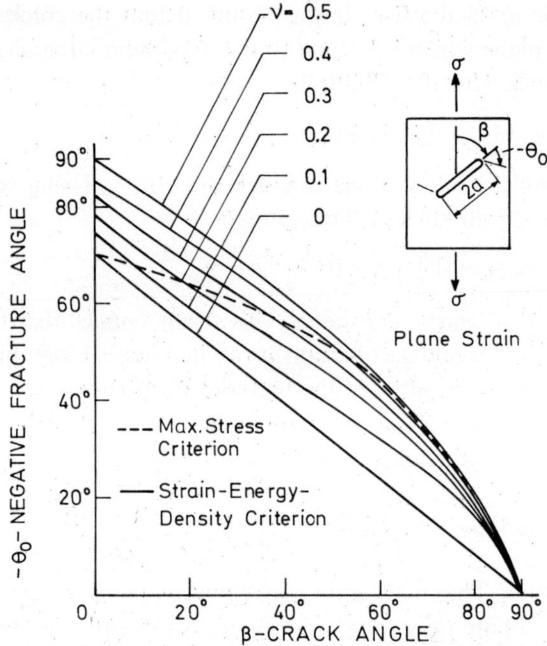


Figure 5. Crack angle *versus* fracture angle in tension.

TABLE II

*Measured and calculated values of the fracture angle*

| $\beta$ | | 30° | 40° | 50° | 60° | 70° | 80° |
|---|---|---|---|---|---|---|---|
| | 1 | −64° | −55.5° | −50° | −40° | −29° | −17° |
| $\theta_0$ | 2 | −60° | −52° | −50° | −43.5° | −30.5° | −18° |
| @ right | 3 | −63° | −57° | −53° | −44.5° | — | −15.5° |
| | 4 | — | −57° | −52° | −43.5° | — | — |
| | 1 | −65° | −58° | −50.5° | −44° | −31.5° | −18.5° |
| $\theta_0$ | 2 | — | −53° | −52° | −40° | −31° | −17.5° |
| @ left | 3 | −60° | −55° | −51.5° | −46° | −31.5° | −17° |
| | 4 | — | −57° | −50° | −43° | — | — |
| $(\theta_0)_{avg.}$ | | −62.4° | −55.6° | −51.1° | −43.1° | −30.7° | −17.3° |
| Eq. (28) | | −60.2° | −55.7° | −50.2° | −43.2° | −33.2° | −19.3° |
| Eq. (31) | | −63.5° | −56.7° | −49.5° | −41.5° | −31.8° | −18.5° |

maximum stress criterion and equation (31) based on the strain-energy-density criterion with $\nu=\frac{1}{3}$. As it can be seen the agreement between theory and experiment is good. Additional experiments are needed to verify the influence of Poisson's ratio on the fracture angle $\theta_0$ as predicted by the solid curves in Figure 5.

(b) *Uniaxial compression.* There exists another set of solutions of positive $\theta_0$ to equation (31) which will also yield a minimum on *S*. Physically, the positive values of $\theta_0$ correspond to the problem of Figure 1b in which the inclined crack is now under uniaxial compression. Since *S* depends on $\sigma^2$,

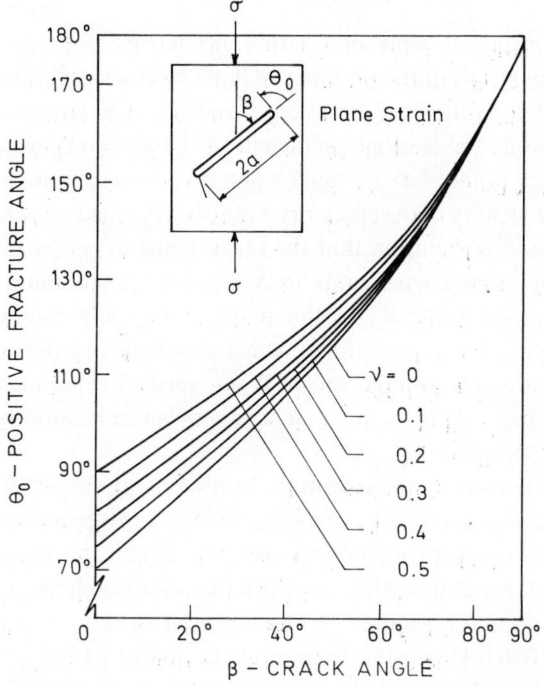


Figure 6. Crack angle *versus* fracture angle in compression.

equation (30) contains both the solutions of uniaxial tension $(+\sigma)$ and compression $(-\sigma)$. A plot of the positive fracture angle $\theta_0$ against $\beta$ is given in Figure 6 for different values of the Poisson's ratio. In contrast to tensile loading, where the crack tends to become horizontal, the crack path under uniaxial compression is towards the direction of loading. Such a phenomenon has indeed been observed by Hoek and Bieniawski [17] who have tested a number of glass plates with an inclined crack under uniaxial compression. Unfortunately, they did not report the initial angle of crack extension so that a comparison of the theoretical results in Figure 6 with experiments cannot be made at this time. Nevertheless, they did publish the values of the applied stress $\sigma_{cr}$ to initiate crack growth for different positions of the crack. It will be shown that their results on the variations of $\sigma_{cr} a^{\frac{1}{2}}$ with $\beta$ are indeed predictable from the present theory.

## V. Intrinsic property of strain-energy-density factor

One of the principal aims of fracture mechanics is to characterize the behavior of materials in the presence of flaws or cracks. This requires a clear distinction of the difference between $S$ and $S_{cr}$. The strain-energy-density factor $S$ is simply the leading coefficient of the series expansion of $\mathrm{d}W/\mathrm{d}V$ about the crack point $r=0$ in Figure 3 and it varies as a function of the polar angle $\theta$. In the vicinity of the crack tip, $S$ may be regarded as a crack resistance force with the interpretation that the crack tends to run in the direction of least resistance that corresponds to $S$ reaching a minimum. Once $S$ has attained a critical value $S_{cr}$ at the point of incipient fracture it may be regarded as a crack extension force which should be independent of loading conditions and crack configuration. When verified experimentally, $S_{cr}$ can be used as a material constant that serves as an indication of the fracture toughness of the material.

In order to become familiar with the strain-energy-density factor $S$, which can be viewed as a vector-like quantity, further considerations will be given to the inclined crack problem. For the case of uniaxial tension, the crack will spread in the negative $\theta$-direction in a plane for which the crack resistance force $S$ is a minimum. Figure 7 shows a plot of $16\mu S_{min}/\sigma^2 a$ against $\beta$ for $v$ varying from 0 to 0.4 inclusive. In general, the quantity $16\mu S_{min}/\sigma^2 a$ increases with the crack angle $\beta$ reaching a maximum along the axis of Mode I crack extension. As $S_{min}$ will be used as a material constant, the above statement implies that the lowest value of the applied stress $\sigma_{cr}$ to initiate crack propagation occurs at $\beta=\pi/2$ for a material with low Poisson's ratio. A similar graph for uniaxial compression is shown in Figure 8. An interesting point to be observed here is that the quantity $16\mu S_{min}/\sigma^2 a$ first increases with the crack angle $\beta$ reaching a peak and then decreases in magnitude. The peak

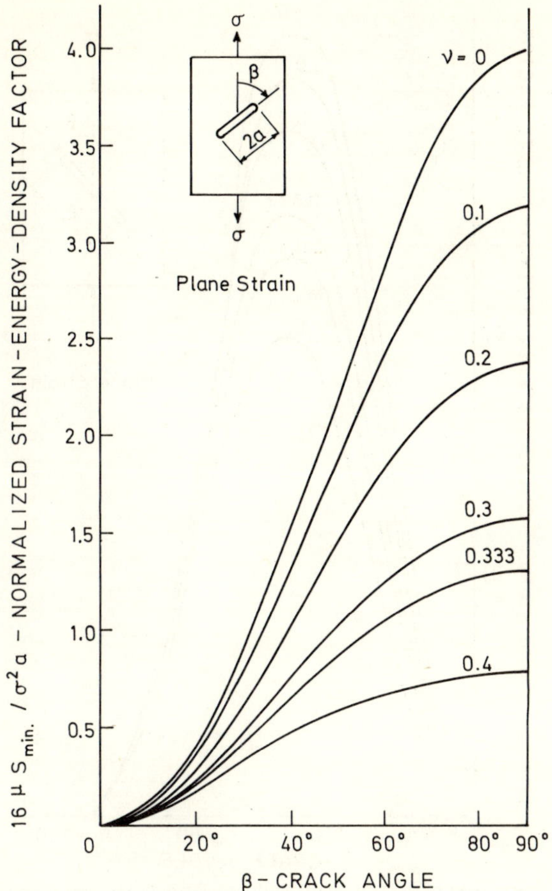


Figure 7. Variations of density factor with crack angle for tensile loading.

value is a function of the Poisson's ratio. This suggests that given $16\mu S_{min}/\sigma^2 a$=constant there exists a critical angle $\beta_0$ at which the critical applied compressive stress is a minimum.

Having completed the preliminaries, the theoretical results will now be compared with the experimental data obtained on DTD 5050-5½% Zn aluminum alloy in [18]. All the tests were carried out in specimens containing an inclined crack. The measured values of $\sigma_{cr} a^{\frac{1}{2}}$ for different crack sizes and failure loads are plotted against the crack angle $\beta$ in Figure 9. The solid and dotted curves represent the predicted values for the aluminum alloy with $k_{1c}$ equal to 28.2 ksi in$^{\frac{1}{2}}$ and 29.2 ksi in$^{\frac{1}{2}}$, respectively. The agreement is good.

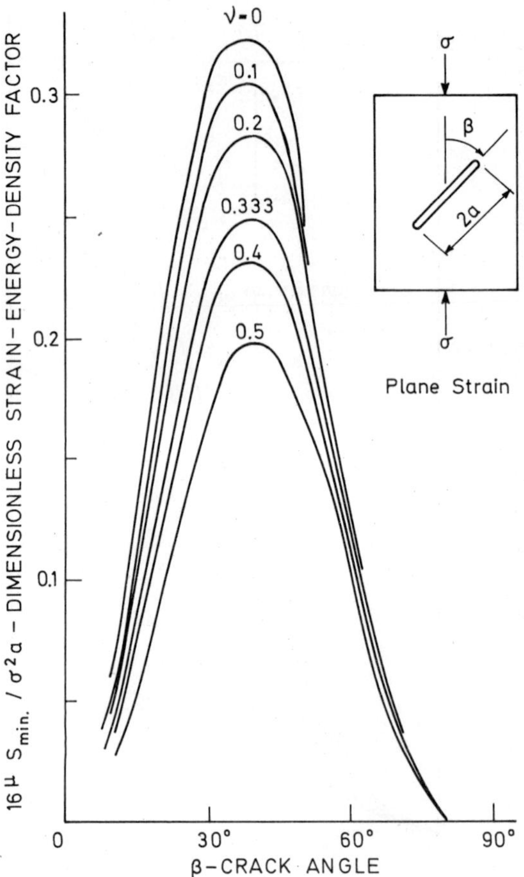


Figure 8. Variations of density factor with crack angle for compressive loading.

The same data is given in Figure 10 with the critical stress-density factor $S_{cr}$ normalized with respect to its value $(S_{cr})_{\pi/2}$ corresponding to Mode I crack extension and $S_{cr}$ remained essentially constant. Although corrections for plasticity ahead of the crack can be made, it is not considered to be essential in this discussion.

The fracture mechanics of an inclined crack under compression is basically different from that of extension. For glass with a Poisson's ratio of $\nu = 0.25$, the theoretical curve in Figure 11 predicts a critical angle of $\beta_0 \simeq 37°$ at which the applied stress to initiate fracture is a minimum. In [17], compression tests on $6'' \times 6''$ precracked glass plates were performed. The critical

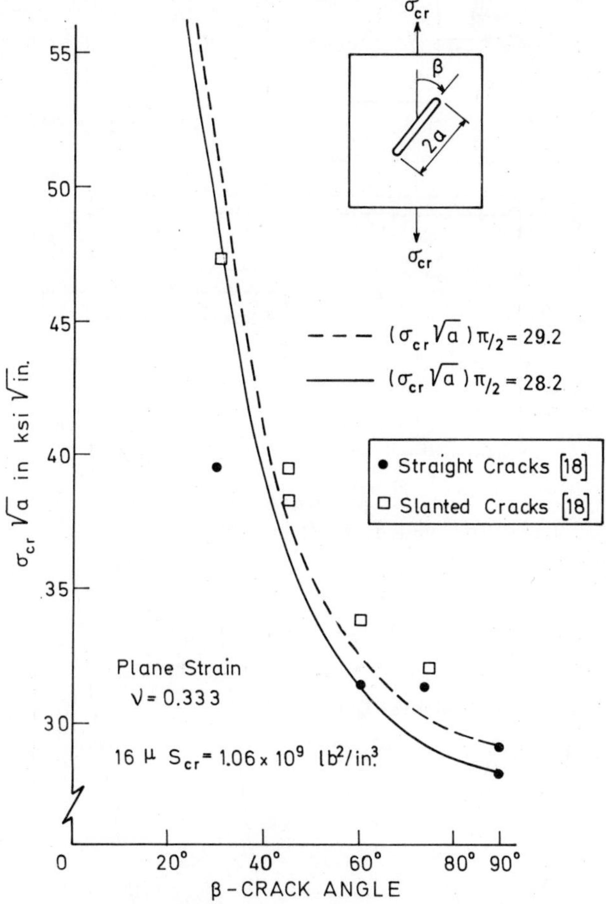

Figure 9. Critical tensile stress *versus* crack angle.

loads were then measured for cracks inclined at various different angles with respect to the axis of loading. The experimental curve is dotted and indicates a critical angle $\beta_0 \simeq 31°$. For cracks loaded under remote compression, there is a tendency for the crack surfaces to come into contact and to rub against one another. Thus, the important point to be made here may not be in the quantative agreement of theory and experiment but in the trend of the failure stress variations with crack angle for fracture under compression which has been predicted by a theory based on the concept of a strain-energy-density factor.

In addition, the new theory can explain that the apparent compressive

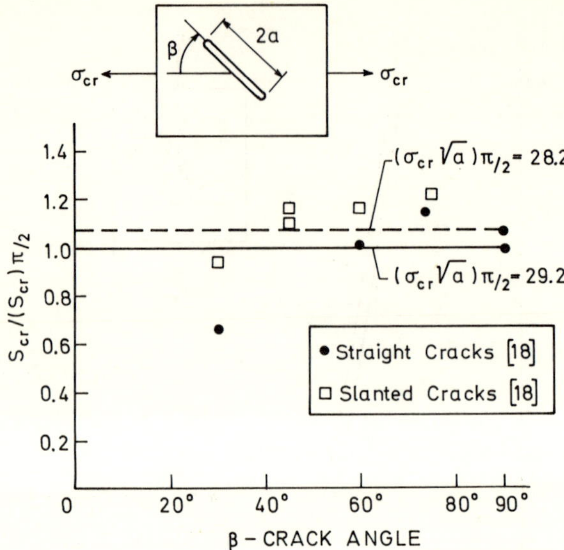


Figure 10. Critical density factor as a material constant.

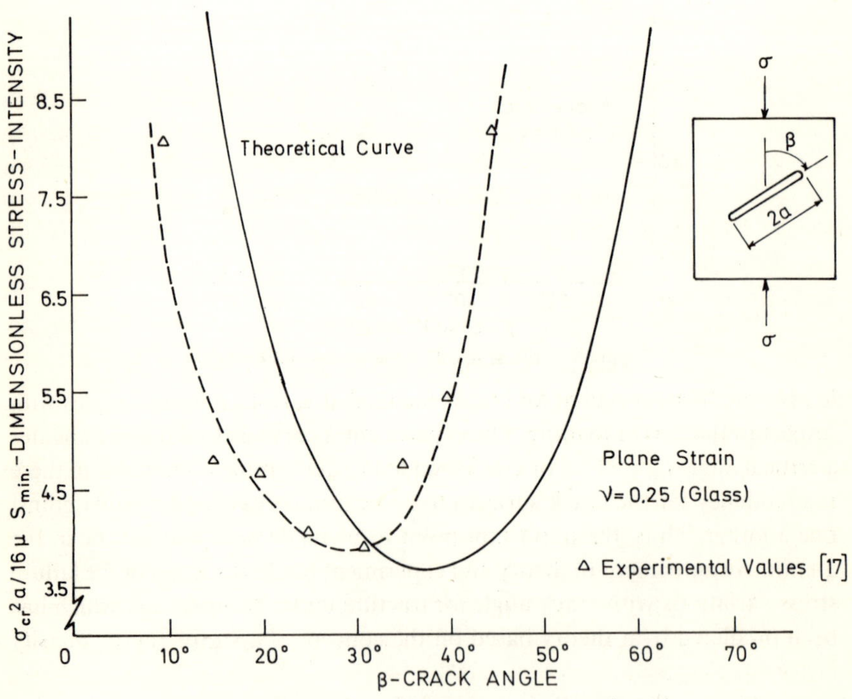


Figure 11. Critical compressive stress *versus* crack angle.

strength of brittle materials can be many times greater than their tensile strength depending upon the geometry and loading conditions. Griffith has attempted to use the maximum stress criterion to explain this apparent strength* difference. It is now well known that the Griffith concept must be defective since it predicts that the compressive fracture stress of a material is exactly eight times the tensile stress. This obviously cannot be true in general, particularly in rocks, where compressive fracture stress in excess of one hundred times the tensile stress have been reported. McClintock and Walsh [16] pointed out this defect in the Griffith conception and modified the Griffith's model by assuming that the cracks close up under compression, developing friction on the sliding crack surfaces. However, as they point out, the coefficient of surface friction would have to be unrealistically high in order to explain compressive fracture stress in excess of ten times the tensile stress. The deficiency lies in using the maximum stress theory as the criterion of fracture and the failure to realize that the fracture stress does not represent the strength of the material.

The answer to the problem posed by Griffith is given by the curves in Figures 7 and 8 or Figures 9 and 11. On a qualitative basis, it is easily seen that for sufficiently large values of $\beta$, say 60° or 70°, the ratio of $\sigma_{cr}$(compression) to $\sigma_{cr}$(tension) can be very large. This ratio depends on the Poisson's ratio and the position of the crack relative to the applied load. It should be emphasized that the material possesses only one strength characterized by the critical value of the strain-energy-density factor $S_{cr}$ regardless of the nature of loading. The fact that Griffith chooses to distinguish between compressive strength and tensile strength of the same material is in itself a weakness of the theory.

## VI. Mixed mode fracture criterion

Having shown that $S_{cr}$ can be used as a material constant, a mixed mode fracture criterion can be stated. The critical values of $k_1$ and $k_2$, i.e., $k_{1c}$ and $k_{2c}$ in a given problem will lie on a curve in the $k_1$, $k_2$-plane determined by the hypotheses stated earlier. The theoretical values of $k_1$ and $k_2$ may be determined from equations (30) and (31) for a given material, i.e., a specific value of $S_{cr}$. From the reported values of $k_{1c}$ in [18] on the aluminum alloy, the

* The strength or the intrinsic property of a material and the maximum stress at failure are two different quantities. The failure to observe such a distinction has caused numerous misconceptions in the open literature.

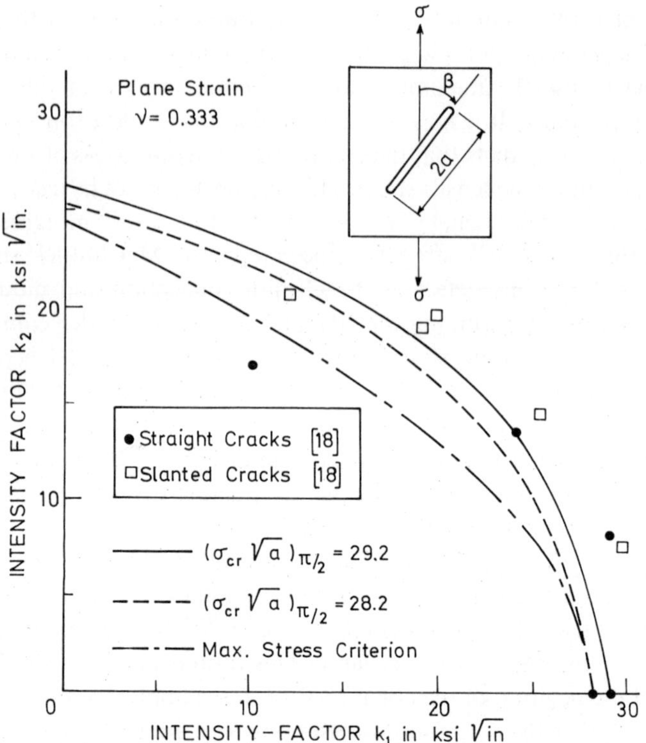


Figure 12. Mixed mode fracture criterion of $k_{1c}$ *versus* $k_{2c}$.

values of $4.8\,(\mu S_{cr})^{\frac{1}{2}} = 28.2$ psi in$^{\frac{1}{2}}$ and 29.2 psi in$^{\frac{1}{2}}$ are used and the theoretical plots of $k_2$ *versus* $k_1$ are given in Figure 12. The third curve represents the prediction based on a criterion of maximum stress [4]. It is evident that the strain-energy-density theory is closer to the experimental results. The same observation has been made on plexiglass plates tested in [4]; i.e., the measured points of $(k_{1c}, k_{2c})$ lie outside of the $k_1k_2$-curve of the maximum stress criterion.

The $k_1k_2$-curve governing the mixed mode fracture of cracks under remote compression is basically different from that of tension. First, the curve does not intersect the $k_1$-axis which implies the obvious fact that Mode I crack extension does not exist in compression. This can be easily verified by solving equations (30) and (31) for $k_1$ and $k_2$ with the constraint that the crack angle and fracture angle satisfies the relations dictated by the curves given in Figure 6. For a glass with $\nu = 0.25$, the theoretical prediction gives a slanted

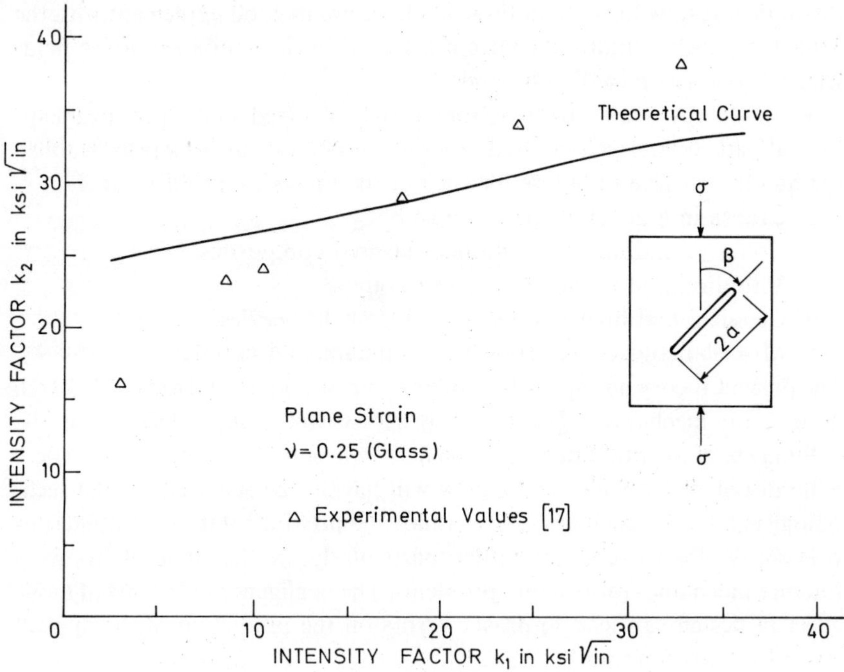


Figure 13. Mixed mode criterion for cracks under compression.

curve in the $k_1$, $k_2$-plane as shown in Figure 13. Again the qualitative feature of the solution is in agreement.

## VII. Concluding remarks

A theory based on the concept of a strain-energy-density factor $S$ has been presented. The stationary values of this density factor can predict the direction of crack growth under mixed mode conditions, whereas it is well known that the classical theory of Griffith lacks this basic feature of predicting the direction of crack propagation. In addition, the critical value $S_{cr}$ has been shown to be independent of the crack geometry and loading and hence it can be used as a material parameter for measuring the resistance against fracture. Using the example of an inclined crack, the theory predicts two basic solutions, one for tensile loading and the other for compressive loading. The theoretical results for both of these loadings are in good agreement with the available experimental data on crack extension in combined stress fields where Mode I is mixed with Mode II.

Strain-energy-density factors $S$ for a variety of mixed mode crack problems have already been developed and experimentations are underway to establish the quadric surface of failure for each of the following problem areas:

1. Cracks in a generally anisotropic body.
2. Cohesive and adhesive failure of layered composites.
3. Vibration and impact of cracked bodies.
4. Classical and higher order plate and shell theories.
5. Miscellaneous crack problems of fundamental nature.

The present theory has opened the door to a new and fruitful area of research in fracture mechanics. The future progress will depend largely on the willingness of the practitioners in the field to accept this new concept. There is no doubt that the classical theory will have to be replaced so that technology in fracture mechanics can advance and provide solutions to numerous previously unanswered questions, particularly in the area of applying fracture mechanics to structure problems. The negligence of the mixed mode effect in design can lead to drastic errors on the prediction of the applied stress to cause fracture.

This work represents a simplified version of a more general theory of fracture [14] which has already been developed far beyond the basic concept presented here. The special theory being the first deviation from the classical thought can be easily understood and immediately applied for resolving many practical problems in the field.

## References

[1] Griffith, A. A., *Phil. Trans. Royal Society*, A221, p. 163 (1921).
[2] Griffith, A. A., *Proc. 1st. Int. Congr. Applied Mech.*, Delft, p. 55 (1924).
[3] Irwin, G. R., *Structural Mechanics*, Pergamon Press, London, England, p. 560 (1960).
[4] Erdogan, F. and Sih, G. C., *J. of Basic Engrg.*, 85, p. 519 (1963).
[5] Knauss, W. G., *Int. Journal of Fracture Mech.*, 6, p. 183 (1970).
[6] Barenblatt, G. I., *Prikl. Mat. Mech.*, 23, p. 434 (1959).
[7] Dugdale, D. S., *J. Mech. Phys. Solids*, 8, p. 100 (1960).
[8] Rice, J. R., *J. of Appl. Mech.*, 35, p. 379 (1968).
[9] Eshelby, J. D., *Proc. of Roy. Soc. London*, Series A, 241 (1957).
[10] Sih, G. C., *Strain-Energy-Density Factor Applied to Mixed Mode Crack Problems*, Institute of Fracture and Solid Mechanics Technical Report, Lehigh University (1972).

[11] Sih, G. C., Some Basic Problems in Fracture Mechanics and New Concepts, *Journal of Engineering Fracture Mechanics*, in press.

[12] Sih, G. C., *Application of the Strain-Energy-Density Theory to Fundamental Fracture Problems*, Institute of Fracture and Solid Mechanics Technical Report, Lehigh University (1972).

[13] Sih, G. C. and Liebowitz, H., *Mathematical Fundamentals of Fracture*, Academic Press, New York, p. 67 (1968).

[14] Sih, G. C., *A New Outlook on Fracture Mechanics* (forthcoming).

[15] Sih, G. C., Paris, P. C. and Erdogan, F., *J. of Appl. Mech.*, 29, p. 306 (1962).

[16] McClintock, F. A. and Walsh, J. B., *Proc. of the 4th U.S. Nat. Congress of Appl. Mech.*, p. 1015 (1962).

[17] Hoek, E. and Bieniawski, Z. T., *Fracture Propagation Mechanics in Hard Rock*, Technical Report—Rock Mech. Div., South African Council for Scientific and Industrial Research (1965).

[18] Pook, L. P., *J. of Engrg. Frac. Mech.*, 3, p. 205 (1966).

*O. L. Bowie*

# 1 *Solutions of plane crack problems by mapping technique*

## 1.1 Introduction

The analysis of crack problems in plane elasticity has intrigued mathematicians for nearly sixty years. Inglis [1] found the solution for a single crack in an infinite sheet with the use of elliptic coordinates. Since then, many mathematical approaches with wide ranges of sophistication have been applied to a variety of crack configurations and loading conditions. It is easy to appreciate the mathematical interest in a problem area in which solution techniques span such diverse topics as analytic function theory, integral equations, transform methods, conformal mapping, boundary collocation, finite differences, finite elements, asymptotic methods, etc.

Muskhelishvili's work on the complex form of the two-dimensional equations due to Kolosoff [2] has undoubtedly been the major development and influence on analytical techniques for solving plane crack problems during this era. As a result of his work, particularly the formulation of the problems of "linear relationship" using analytic continuation arguments, the analyst is provided with considerable insight into the mathematical character of the solution in terms of analytic function theory. With this insight, mathematical singularities due to geometry or loading can usually be anticipated and frequently the structure of the singularity can be predicted. This feature is invaluable in the choice of solution representation for bypassing numerical convergence difficulties.

Many numerical solutions and techniques have appeared in the last decade following the great surge of interest in crack problems due to Irwin [3]. Irwin's concept of stress intensity related the stress distribution local to the crack tip with the earlier Griffith [4] energy concept and thus stimulated the growth of the recent field of fracture mechanics. Considerable credit for many of the solutions must also be given to the substantial growth of computer technology in this same period. Few, if any, of the effective numerical techniques in use today would have been considered feasible fifteen years ago in the age of the desk calculator.

The great variety of numerical techniques for computing crack stresses has played a healthy role in ensuring reliability of numerical results. This has been essential in a problem area where all techniques are prone to convergence difficulties. Most key problems have been duplicated by several techniques and a common accuracy requirement is an error toleration of one or two percent. This has minimized gross errors from appearing in the literature and forced careful refinements of such approaches as finite elements, etc.

This chapter will be concerned primarily with techniques involving conformal mapping combined with the Muskhelishvili formulation for plane elasticity. The basic formulations for the isotropic and orthotropic theory will first be summarized and illustrated briefly in the following two sections. Then a series of techniques drawn from the author's methodology will be described and illustrated. These techniques range from the method of polynomial approximation, e.g., Bowie [5], to a recent method, the modified mapping-collocation technique, Bowie and Neal [6]. The extension of the latter technique to plane crack problems in orthotropic materials, Bowie and Freese [7] will be described. The several numerical solutions included to illustrate the techniques have not been previously published in the open literature.

The first section is introductory in nature and is intended to serve as a guide for the reader uninitiated to Muskhelishvili methods of analysis. On the other hand, it was considered worthwhile to include this section for the more experienced analyst for the sake of continuity and efficiency in presentation of the techniques in the later sections.

## 1.2 Complex variable formulation

Several results of the complex variable formulation of two-dimensional elasticity for both isotropic and orthotropic materials will be summarized briefly in this section. Only those areas primarily relating to the mapping techniques of this chapter will be considered. For complete accounts of the isotropic and anisotropic theories, the reader is referred to the classical treatments of Muskhelishvili [8] and Lekhnitskii [9], respectively. Brief introductions to the respective theories may be found in Timoshenko and Goodier [10] and Savin [11].

*Isotropic formulation.* In the two-dimensional theory of elasticity for homogeneous isotropic bodies, a well-known device is the description of the stresses in terms of suitable derivatives of the Airy stress function, $U(x, y)$. In rectangular coordinates, if $\sigma_x = \partial^2 U/\partial y^2$, $\sigma_y = \partial^2 U/\partial x^2$, $\tau_{xy} = -\partial^2 U/\partial x\,\partial y$, then the two-dimensional equations of equilibrium and compatibility are satisfied provided

$$\nabla^4 U(x, y) = 0 \tag{1.1}$$

where $\nabla^4$ denotes the biharmonic operator.

To describe the complex variable formulation of Muskhelishvili, introduce the complex coordinate system, $z = x + \mathrm{i}y$, and, henceforth, refer to the $z$-plane as the physical plane. The formulation depends upon the representation of $U(x, y)$ in terms of two analytic functions of the complex variable, $z$, namely, $\phi(z)$ and $\psi(z)$, where

$$U(x, y) = \mathrm{Re}\left[\bar{z}\phi(z) + \int^z \psi(z)\,\mathrm{d}z\right]. \tag{1.2}$$

In equation (1.2), Re denotes "real part" and bars denote complex conjugates.

With this representation and the conventional use of primes to denote differentiation, the stresses and displacements in rectangular coordinates can be written as

$$\begin{aligned} \sigma_y + \sigma_x &= 2[\phi'(z) + \overline{\phi'(z)}] = 4\,\mathrm{Re}\,[\phi'(z)] \\ \sigma_y - \sigma_x + 2\mathrm{i}\tau_{xy} &= 2[\bar{z}\phi''(z) + \psi'(z)] \end{aligned} \tag{1.3}$$

and

$$2\mu(u + \mathrm{i}v) = \kappa\phi(z) - z\overline{\phi'(z)} - \overline{\psi(z)} \tag{1.4}$$

where $\mu = E/2(1+\nu)$ and $\kappa = 3 - 4\nu$ (plane strain) and $\kappa = (3-\nu)/(1+\nu)$ (generalized plane stress). $E$ and $\nu$ are Young's modulus and Poisson's ratio, respectively.

An additional relation which will be used frequently throughout this chapter is provided by the stress resultant acting on an arbitrary arc of the material. Consider the arc AB (Figure 1.1) with element of length d$s$ and normal $n$ drawn to the right of the arc when looking along it in the positive direction. Then, if $X_n$d$s$ and $Y_n$d$s$ denote the horizontal and vertical components, respectively, of the force on the elements d$s$ (exerted on the side of the positive normal), it can be shown that

$$(X_n + \mathrm{i}\, Y_n)\,\mathrm{d}s = -\mathrm{i}\,\mathrm{d}\,\{\phi(z) + z\overline{\phi'(z)} + \overline{\psi(z)}\}\,. \tag{1.5}$$

Thus,

$$\begin{aligned}\phi(z) + z\overline{\phi'(z)} + \overline{\psi(z)} &= \mathrm{i}\int^{s} (X_n + \mathrm{i}\, Y_n)\,\mathrm{d}s \\ &= f_1(s) + \mathrm{i} f_2(s)\,.\end{aligned} \tag{1.6}$$

When the arc AB is chosen as the boundary of the physical region, $f_1(s) + \mathrm{i} f_2(s)$ is known to within a constant on those intervals where the boundary loads

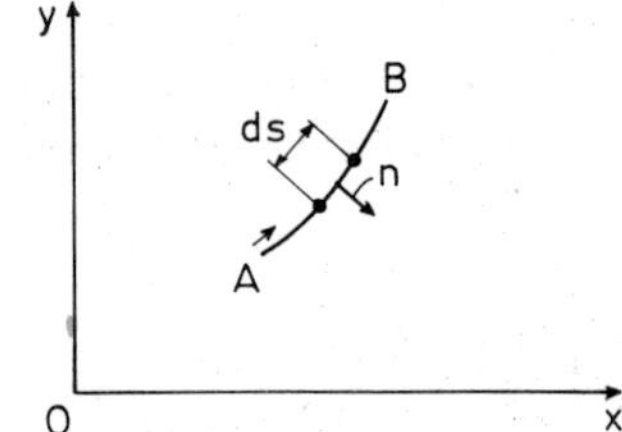


Fig. 1.1. Sign convention for stress resultant.

are specified. In a similar calculation, the resultant moment, $M_0$, (with respect to the origin of the coordinate system) can be written

$$\begin{aligned}M_0 &= \int_{\mathrm{AB}} (x Y_n - y X_n)\,\mathrm{d}s \\ &= \mathrm{Re}\,\{\chi(z) - z\psi(z) - z\bar{z}\phi'(z)\}_{\mathrm{A}}^{\mathrm{B}}\end{aligned} \tag{1.7}$$

where $\chi'(z) = \psi(z)$.

There are, of course, many additional details of the formulation which for the sake of brevity must be left to the reader. Important properties of the analytic functions $\phi(z)$ and $\psi(z)$ are elegantly deduced by Muskhelishvili from physical considerations. For example, when rigid body motions are not precluded, the analytic functions are obviously not unique. The extent of arbitrariness in their definitions can be systematically considered for the several boundary-value problems of elasticity. Overall equilibrium requirements and the single-valuedness of the displacements lead to additional relations. For example, if AB in equation (1.7) is a closed curve and the boundary of a simple connected region of the material, then the integral over the total loop must vanish (equilibrium). Such a condition clearly places specific requirements on multi-valued functions which might be used for stress function representation. When the physical region is infinite or multiply-connected further properties of the stress functions can be deduced. These refinements of the formulation should be familiar to the reader before he undertakes actual problem solving.

The formulation of the two-dimensional problem in terms of analytic functions lends itself to the application of conformal mapping. Conformal mapping, in turn, provides the analyst with a powerful tool for the systematic introduction of curvilinear coordinates compatible with the geometry of the physical region. The formulation in terms of mapping functions will be summarized as follows:

Let $S_z$ denote an open region in the physical plane and let us introduce a complex parameter plane, the $\zeta$-plane along with the transformation.

$$z=\omega(\zeta)\,. \tag{1.8}$$

It will be assumed that there exists an open region $S_\zeta$ in the $\zeta$-plane such that equation (1.8) provides a one-to-one correspondance between the points $z \in S_z$ and $\zeta \in S_\zeta$. Such a transformation is said to be conformal provided the mapping function $\omega(\zeta)$ is analytic and $\omega'(\zeta) \neq 0$ for $\zeta \in S_\zeta$. Singularities in $\omega(\zeta)$ can exist on the boundary of $S_\zeta$ for corner-describing properties of the boundary of $S_z$.

The significant feature of the analytic function formulation is the carry-over of analyticity to the parameter region. From the well-known property of analytic functions, if $\phi(z)$ is analytic for $z \in S_z$ and $\omega(\zeta)$ is the conformal mapping function in equation (1.8), then $\phi[\omega(\zeta)]$ is an analytic function of $\zeta$ for $\zeta \in S_\zeta$. Thus the explicit determination of $\phi(z)$ and $\psi(z)$ can be by-passed and the analyst can seek directly the analytic functions $\phi[\omega(\zeta)]$ and $\psi[\omega(\zeta)]$ as functions of $\zeta$.

The necessity for introducing considerable new notation can be avoided by designating $\phi[\omega(\zeta)]$ as $\phi(\zeta)$, etc., which leads to such relationships as $\phi'(z)=\phi'(\zeta)/\omega'(\zeta)$, etc. Thus, the stresses, displacements, etc. can now be written

$$\sigma_y+\sigma_x=4\,\mathrm{Re}\,\{\phi'(\zeta)/\omega'(\zeta)\} \tag{1.9}$$

$$\sigma_y-\sigma_x+2i\,\tau_{xy}=2\,\{\overline{\omega(\zeta)}[\phi'(\zeta)/\omega'(\zeta)]'+\psi'(\zeta)\}/\omega'(\zeta) \tag{1.10}$$

$$2\mu(u+\mathrm{i}v) = \kappa\phi(\zeta) - \omega(\zeta)\overline{\phi'(\zeta)}/\overline{\omega'(\zeta)} - \overline{\psi(\zeta)} \tag{1.11}$$

$$\phi(\zeta) + \omega(\zeta)\overline{\phi'(\zeta)}/\overline{\omega'(\zeta)} + \overline{\psi(\zeta)} = \mathrm{i}\int^{s}(X_n + \mathrm{i}\,Y_n)\,\mathrm{d}s \tag{1.12}$$

etc.

It is important intuitively to keep in mind the parametric nature of the $\zeta$-plane. The original physical problem always remains defined in the $z$-plane. The solution in terms of $\zeta$ is parametric in nature and must be related back to the physical problem through the mapping function. In section 1.6, the parametric nature of mapping by using an unconventional version of the procedure will be considered. (The above remarks are in contrast to mapping by inversion, Timoshenko [12], where several solutions are obtained in which both geometry and loading are considered in the image region.)

In the conventional use of conformal mapping, the parameter region $S_\zeta$ is often chosen as the interior (exterior) of a circle or the upper (lower) half plane for simply connected regions $S_z$. In general, the parameter region must enjoy the same connectivety as the physical region. (Exceptions to this remark will be encountered later when certain analytic continuation arguments are made for the stress functions.) The choice of a circular region for $S_\zeta$ is particularly convenient for handling boundary conditions with a power series representation of the stress functions. The choice of a half plane is more natural for an integral representation.

The most remarkable of Muskhelishvili's contributions to the two-dimensional theory are the elegant analytic continuation concepts he introduced which lead to a formulation in terms of a Hilbert problem. In addition to leading to explicit solutions in terms of Cauchy integrals for the wide class of problems of "linear relationship", such a formulation provides valuable insight into the mathematical character of the solution for all two-dimensional problems. Due to the frequent use of these concepts in this chapter, the basic argument will be summarized.

Let $S_z^-$ and $S_z^+$ denote the region $(y<0)$ and $(y>0)$, respectively, and let us assume the elastic body occupies the lower half plane, $S_z^-$. Consider $\phi(z)$ as defined in the unoccupied region, $S_z^+$, by

$$\phi(z) = -z\bar{\phi}'(z) - \bar{\psi}(z)\,, \qquad z\in S_z^+ \tag{1.13}$$

where the bar notation is defined by

$$\bar{f}(z) = \overline{f(\bar{z})} \tag{1.14}$$

The function $\psi(z)$ can now be expressed as

$$\psi(z) = -\bar{\phi}(z) - z\phi'(z)\,, \qquad z \in S_z^- \,. \tag{1.15}$$

Replacing $\psi(z)$ in the force condition (1.6) by its definition (1.15), we find

$$\phi(z) - \phi(\bar{z}) + (z-\bar{z})\overline{\phi'(z)} = f_1(s) + \mathrm{i}f_2(s)\,. \tag{1.16}$$

For most practical loading systems acting on the real axis, $z = x$, one can correctly assume that

$$|\phi'(z)| < A/|z-x|^{\alpha}\,, \qquad 0 \leqslant \alpha < 1 \tag{1.17}$$

for $z$ near $x$. Thus the force condition (1.16) on the boundary $z = x$ reduces to

$$\phi^-(x) - \phi^+(x) = f_1(x) + \mathrm{i}f_2(x) \tag{1.18}$$

where $\phi^-(x)$ and $\phi^+(x)$ denote the values $\phi(x)$ as approached through $S_z^-$ and $S_z^+$, respectively.

The consequences which can now be deduced are remarkable. Consider an interval $S_x$ of the real axis as traction-free. Then, by adjustment of the constant of integration in equation (1.6),

$$\phi^-(x) - \phi^+(x) = 0\,, \qquad x \in S_x\,. \tag{1.19}$$

Thus, $\phi(z)$ is analytic for $z \in S_x$. The problem of determining two analytic functions $\phi(z)$ and $\psi(z)$ in $S_z^-$ is reduced to finding a single analytic function $\phi(z)$ defined in $S_z^- + S_z^+$ which continues analytically through the unloaded intervals of the real axis. The corresponding Hilbert problem can often be solved effectively in terms of Cauchy integrals. For example, if the stresses at infinity are assumed to vanish and if loads are prescribed on the real axis such that $f_1(x)$ and $f_2(x)$ satisfy the Hölder condition, then the solution can be written directly as

$$\phi(z) = \frac{1}{2\pi \mathrm{i}} \int_L \frac{f_1(x) + \mathrm{i}f_2(x)}{x-z}\,\mathrm{d}x \tag{1.20}$$

where $L$ denotes the real axis.

Similar continuation arguments can be made for other regions with the use of Schwarz's symmetry principle. For example, frequent use is made of Kartzivadze's continuation form [13] for mapping onto the exterior (interior) of the unit circle. Now consider equation (1.8) as a conformal mapping of the experior of the unit circle $S_\zeta^-$ into the region of the elastic material. Points on the unit circle, $\zeta=\sigma$, will be assumed to map into the finite boundary of the physical region. The function $\phi(\zeta)$ is extended by definition into the interior of the unit circle, $S_\zeta^+$, as

$$\phi(\zeta) = -\omega(\zeta)\bar{\phi}'(1/\zeta)/\bar{\omega}'(1/\zeta) - \bar{\psi}(1/\zeta)\,, \qquad \zeta\in S_\zeta^+ \tag{1.21}$$

where the bar notation is now defined by

$$\bar{f}(1/\zeta) = \overline{f(1/\bar{\zeta})}\,. \tag{1.22}$$

The function $\psi(\zeta)$ becomes

$$\psi(\zeta) = -\bar{\phi}(1/\zeta) - \bar{\omega}(1/\zeta)\,\phi'(\zeta)/\omega'(\zeta)\,, \qquad \zeta\in S_\zeta^-\,. \tag{1.23}$$

The resultant force condition (1.12) when evaluated on the unit circle can be written with obvious notation as

$$\phi^-(\sigma) - \phi^+(\sigma) + [\omega^-(\sigma) - \omega^+(\sigma)]\,\overline{\phi'^-(\sigma)}/\overline{\omega'^-(\sigma)} = f_1(s) + \mathrm{i}f_2(s)\,. \tag{1.24}$$

If the mapping function $\omega(\zeta)$ is continuous across the unit circle, then again, in general, (1.24) reduces to

$$\phi^-(\sigma) - \phi^+(\sigma) = f_1(s) + \mathrm{i}f_2(s)\,. \tag{1.25}$$

Again the problem has been reduced to the determination of a single analytic function $\phi(\zeta)$ which is defined in $S_\zeta^-$ and $S_\zeta^+$ and continues analytically across the intervals of the unit circle corresponding to unloaded intervals of the physical boundary.

One of the useful applications of the continuation arguments which is made later in crack analysis is the generation of stress functions ensuring traction-free conditions on the crack. If, for example, the unit circle above maps into a crack and the continuation argument is used, then analyticity of $\phi(\zeta)$ across the unit circle ensures traction-free conditions on the crack.

*Anisotropic formulation.* The formulation for the plane problem for rectilinearly anisotropic bodies will now be briefly summarized. For a state of generalized plane stress, the inplane stress-strain relations are

$$\begin{aligned} \varepsilon_x &= a_{11}\sigma_x + a_{12}\sigma_y + a_{16}\tau_{xy} \\ \varepsilon_y &= a_{12}\sigma_x + a_{22}\sigma_y + a_{26}\tau_{xy} \\ \gamma_{xy} &= a_{16}\sigma_x + a_{26}\sigma_y + a_{66}\tau_{xy}\,. \end{aligned} \tag{1.26}$$

A similar set of relations can be written for plane strain. Included in these considerations is the familiar case of orthotropy which is defined when the body possesses three planes of elastic symmetry. In the case of orthotropy, the additional symmetries imply that $a_{16}=a_{26}=0$ in equation (1.26).

When the Airy stress function, $U(x, y)$, is substituted into the compatibility equation, there results the condition

$$a_{22}\partial^4 U/\partial x^4 - 2a_{26}\partial^4 U/\partial x^3\partial y + (2a_{12}+a_{66})\partial^4 U/\partial x^2\partial y^2 - 2a_{16}\partial^4 U/\partial x\partial y^3 + a_{11}\partial^4 U/\partial y^4 = 0\,. \tag{1.27}$$

If the roots $\mu_j$ of the characteristic equation

$$a_{11}\mu_j^4 - 2a_{16}\mu_j^3 + (2a_{12}+a_{66})\mu_j^2 - 2a_{26}\mu_j + a_{22} = 0 \tag{1.28}$$

are distinct, the solution of equation (1.27) is

$$U(x, y) = U_1(x+\mu_1 y) + U_2(x+\mu_2 y) + U_3(x+\mu_3 y) + U_4(x+\mu_4 y)\,. \tag{1.29}$$

Leknitskii [9] showed that the characteristic roots are either complex or purely imaginary and occur in conjugate pairs. When $\mu_1=\mu_2=\mathrm{i}$, the situation corresponds to the isotropic case and equation (1.29) can no longer be considered the general solution of (1.27). The stress function, for distinct roots $\mu_1, \mu_2, \bar{\mu}_1, \bar{\mu}_2$ (with $\mu_1 \neq \mu_2$) can be expressed as

$$U(x, y) = 2\,\mathrm{Re}\,[U_1(z_1) + U_2(z_2)] \tag{1.30}$$

where $U_1(z_1)$ and $U_2(z_2)$ are analytic functions of the complex variables $z_1 = x + s_1 y$ and $z_2 = x + s_2 y$ where

$$s_1 = \mu_1 = \alpha_1 + \mathrm{i}\beta_1\,, \quad s_2 = \mu_2 = \alpha_2 + \mathrm{i}\beta_2 \tag{1.31}$$

and $\alpha_j$, $\beta_j$ are real constants computed from the material properties. Without loss of generality, assume $\beta_1 > 0$, $\beta_2 > 0$; $\beta_1 \neq \beta_2$.

The stresses and displacements can now be expressed in terms of the Airy stress function in the usual manner. To simplify notation, the new functions

$$\phi(z_1) = \mathrm{d}U_1/\mathrm{d}z_1\,, \quad \psi(z_2) = \mathrm{d}U_2/\mathrm{d}z_2 \tag{1.32}$$

are introduced. Then

$$\begin{aligned} \sigma_x &= 2\,\mathrm{Re}\,[s_1^2\phi'(z_1) + s_2^2\psi'(z_2)] \\ \sigma_y &= 2\,\mathrm{Re}\,[\phi'(z_1) + \psi'(z_2)] \\ \tau_{xy} &= -2\,\mathrm{Re}\,[s_1\phi'(z_1) + s_2\psi'(z_2)] \end{aligned} \tag{1.33}$$

and

$$\begin{aligned} u &= 2\,\mathrm{Re}\,[p_1\phi(z_1)+p_2\psi(z_2)] \\ v &= 2\,\mathrm{Re}\,[q_1\phi(z_1)+q_2\psi(z_2)] \end{aligned} \tag{1.34}$$

where

$$\begin{aligned} p_i &= a_{11}s_i^2+a_{12}-a_{16}s_i \\ q_i &= a_{12}s_i+(a_{22}/s_i)-a_{26}\,. \end{aligned} \tag{1.35}$$

The resultant force condition (corresponding to equation (1.6)) becomes

$$(1+\mathrm{i}s_1)\phi(z_1)+(1+\mathrm{i}s_2)\psi(z_2)+(1+\mathrm{i}\bar{s}_1)\overline{\phi(z_1)} + (1+\mathrm{i}\bar{s}_2)\overline{\psi(z_2)} = \mathrm{i}\int^s (X_n+\mathrm{i}Y_n)\mathrm{d}s \tag{1.36}$$

The conformal mapping arguments summarized previously for the isotropic case carry over with some conceptual modifications to the formulation above. Clearly the preservation of analyticity affected by conformal transformation must be considered with regard to the $z_1$ and $z_2$ coordinate planes rather than the physical $z$-plane in the present case.

It will be convenient to consider three complex planes, $z$, $z_1$, and $z_2$ where

$$\begin{aligned} z\ &= x+\mathrm{i}y \\ z_1 &= x+s_1y = x+\alpha_1y+\mathrm{i}\beta_1y \\ z_2 &= x+s_2y = x+\alpha_2y+\mathrm{i}\beta_2y\,. \end{aligned} \tag{1.37}$$

Equation (1.37) can be considered as a mapping of the physical region $S_z$ into two image regions $S_{z1}$ and $S_{z2}$. The image sets are interrelated since, for fixed $s_1$ and $s_2$, a given value of $z$ determines uniquely a corresponding set of coordinates $z_1$ and $z_2$. These mappings are not conformal, thus conformal mapping when applied directly to the physical geometry is not feasible in this case.

Now introduce two additional parameter planes, the $\zeta_1$-plane and the $\zeta_2$-plane, along with a pair of conformal transformations

$$\begin{aligned} z_1 &= \omega_1(\zeta_1) \\ z_2 &= \omega_2(\zeta_2)\,. \end{aligned} \tag{1.38}$$

Two parameter regions $S_{\zeta 1}$ and $S_{\zeta 2}$ will be assumed to map conformally into $S_{z1}$ and $S_{z2}$, respectively. Again, in conventional mapping procedures, the regions $S_{\zeta 1}$ and $S_{\zeta 2}$ are taken as half-planes, circles, etc. When boundary conditions are considered in terms of the parameter regions, it is clearly necessary to retain the identity of the pairs $\zeta_1$ and $\zeta_2$ associated with a given point $z$.

Formally, the formulation can now be written in parallel to the isotropic case. We set $\phi(z_1)=\phi[\omega_1(\zeta_1)]=\phi(\zeta_1)$ and $\psi(z_2)=\psi[\omega_2(\zeta_2)]=\psi(\zeta_2)$; thus, $\phi'(z_1)=\phi'(\zeta_1)/\omega'(\zeta_1)$, etc. Then,

$$\sigma_x = 2\,\mathrm{Re}\,\{s_1^2\phi'(\zeta_1)/\omega_1'(\zeta_1)+s_2^2\psi'(\zeta_2)/\omega_2'(\zeta_2)\}$$
etc. (1.39)

Similarly, the resultant force condition becomes

$$(1+\mathrm{i}s_1)\phi(\zeta_1)+(1+\mathrm{i}s_2)\psi(\zeta_2)+(1+\mathrm{i}\bar{s}_1)\overline{\phi(\zeta_1)} + (1+\mathrm{i}\bar{s}_2)\overline{\psi(\zeta_2)} = f_1(s)+\mathrm{i}f_2(s)\,. \tag{1.40}$$

The requirements above are somewhat deceiving in their simplicity. In practice, it is very difficult to find suitable mapping functions (1.38) and, until recently, only a few simple cases have been solved. In section 1.6, however, a practical utilization of this formalization in the "modified mapping-collocation" technique will be described.

The analytic continuation arguments described previously for the isotropic case can be similarly carried out for rectilinear anisotropy, e.g. Bowie and Freese [7].

First, consider the case when the material occupies the lower half-plane $S_z^-$, $(z<0)$. It is clear from equation (1.37) that the image regions $S_{z1}^-$ and $S_{z2}^-$ will also correspond to lower half planes in the $z_1$- and $z_2$-planes, respectively. It will be convenient, conceptually, to introduce an additional complex plane, the $W$-plane, and to consider the two image planes as mapped into this single plane. Thus, $S_{z1}^-$ and $S_{z2}^-$ will be considered as two overlapping coverings of the lower half of the $W$-plane, i.e., on $S_W^-$.

Introduce a function $F(W)$, analytic in $S_W^-$ and extend it by definition into the upper half-plane $S_W^+$ such that

$$\bar{B}F(W)=\bar{\psi}(W)-\bar{C}\bar{F}(W)\,, \qquad W\in S_W^+ \tag{1.41}$$

where $B$ and $C$ are complex constants and the functional bar notation is again defined as

$$\bar{f}(W)=\overline{f(\overline{W})} \tag{1.42}$$

Since $\bar{z}_2\in S_W^+$, it follows from equation (1.41) that

$$\psi(z_2) = B\bar{F}(z_2) + CF(z_2), \qquad z_2 \in S_W^- . \tag{1.43}$$

Now, if

$$\phi(z_1) = F(z_1), \qquad z_1 \in S_W^- \tag{1.44}$$

then $\phi(z_1)$ and $\psi(z_2)$ have been defined in terms of $F(W)$ and its extension across the real axis.

Now choose

$$\begin{aligned} B &= (\bar{s}_2 - \bar{s}_1)/(s_2 - \bar{s}_2) \\ C &= (\bar{s}_2 - s_1)/(s_2 - \bar{s}_2) \end{aligned} \tag{1.45}$$

and insert equations (1.43)–(1.45) into (1.40) and evaluate on the real axis $W = \xi = x$. This gives

$$\begin{aligned} &(1 + \mathrm{i}\bar{s}_2)(s_2 - s_1)(\bar{s}_2 - s_2)^{-1}[F^-(\xi) - F^+(\xi)] \\ &\quad + (1 + \mathrm{i}s_2)(\bar{s}_2 - \bar{s}_1)(s_2 - \bar{s}_2)^{-1}[\overline{F^-(\xi)} - \overline{F^+(\xi)}] = -\mathrm{i}\int^s (X_n + \mathrm{i}\,Y_n)\mathrm{d}s \end{aligned} \tag{1.46}$$

where $F^-(\xi)$ and $F^+(\xi)$ denote $F(\xi)$ approached through $S_W^-$ and $S_W^+$, respectively. Again, a traction-free interval of the real axis implies

$$F^-(\xi) = F^+(\xi) \tag{1.47}$$

on this interval by adjusting the constant of integration in equation (1.46). As in the isotropic case, $F(W)$ continues analytically across an unloaded interval of the real axis.

The extension argument for the case when the image regions are mapped onto the exteriors (interiors) of the unit circles of respective parameter planes is similar. The parameter regions, $S_{\zeta 1}^-$ and $S_{\zeta 2}^-$, (exteriors of unit circles) will again be considered as overlapping regions on a common $\zeta$-plane (in the sense of the previous $W$-plane). Now assume that it is possible to determine the mappings (1.38) such that $\zeta_1 = \zeta_2 = \sigma$ on the unit circle, i.e., on the common boundary in the $\zeta$-plane. We then define $F(\zeta)$ as analytic in $S_\zeta^-$ and extend its definition to the interior of the unit circle $S_\zeta^+$ as

$$\bar{B}F(\zeta) = \bar{\psi}(1/\zeta) - \bar{C}\bar{F}(1/\zeta) \tag{1.48}$$

where

$$\bar{F}(1/\zeta) = \overline{F(1/\bar{\zeta})} . \tag{1.49}$$

The remaining arguments are essentially identical to equations (1.43)–(1.46) with $\xi$ replaced by $\sigma$.

## 1.3 Crack problems and conformal mapping

The earliest application of conformal mapping to two-dimensional crack problems perhaps can be considered as Kolosoff's solution [14] for an elliptic hole in an infinite region using the mapping $z = c \cosh \zeta$. Next, Muskhelishvili [15] used mapping of a circular region to solve this problem in terms of Cauchy integrals. In spite of the considerable interest in crack problems generated by Griffith's theory [4] and the subsequent refinements of the complex variable theory by Muskhelishvili, remarkably few solutions were available prior to the late 1950's. Griffith [16] and Neuber [17] found solutions for hyperbolic notches in a region of infinite extent. Outstanding crack solutions in this period include Sneddon and Elliot's solution [18] for varying pressure inside a crack (by transforms) and Willmore's solution [19] for the case of two equal collinear cracks in an infinite region. An interesting account of the evolution of several of these earlier solutions is contained in a survey by Sneddon [20].

This section will relate the previous formulation specifically to crack analysis. First, a simple example of mapping combined with analytic continuation will be described to illustrate basic procedures and the nature of singularities at crack tips. Then, the nature of the stress distribution and the stress intensity concept of Irwin [3] is summarized. Finally, the difficulties introduced by branch points in the mapping function are examined in terms of the analytic continuation arguments.

*A simple illustration of conformal mapping.* Several aspects of mapping, analytic continuation, etc. will now be illustrated by the following solution of the classical problem of an infinite region under simple tension, $T$, containing an internal crack (Figure 1.2).

First consider the isotropic case for which the crack is traction-free. Utilization will be made of the mapping function,

$$z = \omega(\zeta) = (L/2)(\zeta + \zeta^{-1}) \tag{1.50}$$

which maps the unit circle and its exterior $S_\zeta^-$ in the $\zeta$-plane into a crack of length $2L$ and its exterior, respectively, in the $z$-plane. From Muskhelishvili's formulation for infinite regions, it follows from the loading in Figure 1.2 that, for large $|z|$, $\phi(z) \to Tz/4$ and $\psi(z) \to Tz/2$. Thus, for large $|\zeta|$, $\phi(\zeta) \to TL\zeta/8$ and $\psi(\zeta) \to TL\zeta/4$. Except for these simple poles at infinity, both $\phi(\zeta)$ and $\psi(\zeta)$ are analytic in $S_\zeta^-$.

Traction-free conditions on the crack can be ensured by using the continuation arguments of the previous section. A tentative form of the extended definition of $\phi(\zeta)$ for the entire $\zeta$-plane, $S_\zeta$, can be written as

$$\phi(\zeta) = TL\zeta/8 + \sum_{n=1}^{\infty} \alpha_n \zeta^{1-2n}, \qquad \zeta \in S_\zeta \tag{1.51}$$

where the $\alpha_n$'s are real coefficients. The form of equation (1.51) is consistent with the properties of $\phi(\zeta)$ for $\zeta \in S_\zeta^-$ and utilization of the obvious stress

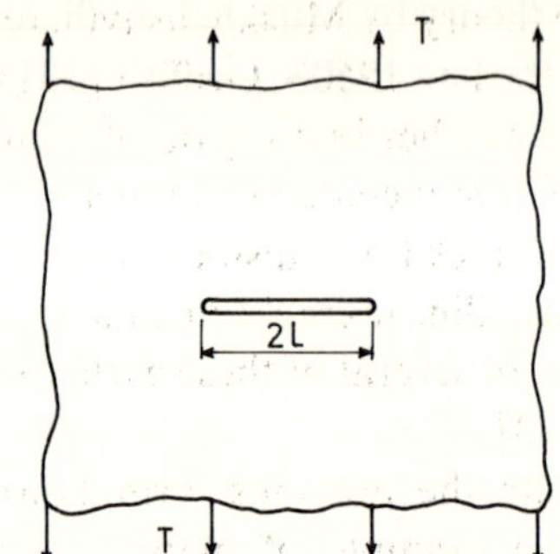


Figure 1.2. Crack in an infinite sheet under tension.

symmetries of the physical problem has been made. The extension of $\phi(\zeta)$ into the interior of the unit circle, $S_\zeta^+$, by equation (1.21) requires consistency of the properties of $\psi(\zeta)$ in (1.23). In this case $\psi(\zeta)$ must be analytic in $S_\zeta^-$ except for a simple pole at infinity with residue $TL/4$. After substituting equations (1.50) and (1.51) into (1.23), one finds this condition requires $\alpha_1 = -3TL/8$, $\alpha_2 = \alpha_3 = \ldots = 0$. Thus,

$$\phi(\zeta) = (TL/8)(\zeta - 3\zeta^{-1}), \qquad \zeta \in S_\zeta \tag{1.52}$$

and this completes the solution. This remarkably simple solution illustrates the effectiveness of the analytic continuation argument. Since the crack was traction-free, a solution in terms of a single analytic function defined in $S_\zeta$ is anticipated.

For the problem of a pressurized crack, $\sigma_n = -P$, with stresses vanishing at infinity, it is equally clear that with extension, $\phi(\zeta)$ will not continue across the unit circle. This problem from equation (1.25) reduces to the solution of the Hilbert problem for the determination of the analytic function $\phi(\zeta)$, vanishing at infinity, and satisfying the condition

$$\phi^-(\sigma) - \phi^+(\sigma) = -P\omega(\sigma). \tag{1.53}$$

The solution can be expressed as

$$\phi(\zeta) = \frac{-P}{2\pi \mathrm{i}} \int \frac{\omega(\sigma)\mathrm{d}\sigma}{\sigma - \zeta}, \tag{1.54}$$

or

$$\begin{aligned} \phi(\zeta) &= -PL/2\zeta, \qquad \zeta \in S_\zeta^- \\ &= \quad PL\zeta/2, \qquad \zeta \in S_\zeta^+ \end{aligned} \tag{1.55}$$

Thus, $\phi(\zeta)$ is sectionally analytic in $S_\zeta^-$ and $S_\zeta^+$ but, as anticipated, does not continue across the unit circle.

In the solution (1.52), it will be noted that $\phi(\zeta)$ is well-behaved at the crack tips, i.e., at $\zeta = \pm 1$. On the other hand, from equations (1.23) and (1.52),

$$\psi(\zeta) = TL[\zeta^3 - 4\zeta - \zeta^{-1}]/4(\zeta^2 - 1). \tag{1.56}$$

The stress function $\psi(\zeta)$ has simple poles in the parameter plane at the crack tips. This property is similar to a variety of crack problems and must be taken into account in the stress function representation.

The problem in Figure 1.2 has been solved for plane rectilinear anisotropy, e.g., Savin [11]. The solution can be found more elegantly by using the continuation arguments of the previous section. From equations (1.37), it is evident that if the crack is oriented along the real axis of the $z$-plane then $z_1 = z_2$ on the crack images in the $z_1$- and $z_2$-planes. The mappings (1.38) can be written as

$$\begin{aligned} z_1 &= (L/2)(\zeta_1 + \zeta_1^{-1}) \\ z_2 &= (L/2)(\zeta_2 + \zeta_2^{-1}). \end{aligned} \tag{1.57}$$

The structures of the stress functions for large $z$ are again determined by the applied loads, e.g., Savin [11]. For $\sigma_y^{(\infty)} = T$,

$$\begin{aligned} &\phi(z_1) \to B^* z_1 \\ &\psi(z_2) \to (C^* + \mathrm{i}D^*) z_2 \end{aligned} \tag{1.58}$$

where

$$\begin{aligned} (s_1^2 + \bar{s}_1^2)B^* + (s_2^2 + \bar{s}_2^2)C^* + \mathrm{i}(s_2^2 - \bar{s}_2^2)D^* &= 0 \\ 2B^* \qquad\quad + 2C^* \qquad\qquad\qquad &= T \\ (s_1 + \bar{s}_1)B^* + (s_2 + \bar{s}_2)C^* + \mathrm{i}(s_2 - \bar{s}_2)D^* &= 0 \end{aligned} \tag{1.59}$$

Therefore, for large $|\zeta_1|$ and $|\zeta_2|$,

$$\phi(\zeta_1)\to(LB^*/2)\zeta_1$$
$$\psi(\zeta_2)\to(L/2)(C^*+\mathrm{i}D^*)\zeta_2\,. \tag{1.60}$$

Now apply the continuation argument of the previous section and choose the function $F(\zeta)$ of equation (1.48) so that

$$F(\zeta)=a\zeta+b\zeta^{-1} \tag{1.61}$$

Since $\phi(\zeta_1)=F(\zeta_1)$ for $\zeta_1\in S_\zeta^-$, it follows from equation (1.60) that $a=LB^*/2$. Furthermore, the extension (1.48) implies that

$$\psi(\zeta_2)=B\bar{F}(1/\zeta_2)+CF(\zeta_2) \tag{1.62}$$

where $B$ and $C$ are defined by equations (1.45).

$$\psi(\zeta_2)=B[\bar{a}\zeta_2^{-1}+\bar{b}\zeta_2]+C[a\zeta_2+b\zeta_2^{-1}] \tag{1.63}$$

From equations (1.60) and (1.63)

$$B\bar{b}+aC=(L/2)(C^*+\mathrm{i}D^*) \tag{1.64}$$

Thus,

$$F(\zeta)=(LB^*/2)\zeta+L[C^*-\mathrm{i}D^*-B^*\bar{C}]\zeta^{-1}/2\bar{B} \tag{1.65}$$

and the solution is complete.

*Crack-tip stresses, stress intensity factors.* It was Sneddon [21] who was the first to give stress field expansion around a crack tip. When one considers the time interval between this expansion and the Inglis solution [1], the reluctance of the earlier elasticians to accept even qualitatively the local stress infinities at crack tips is evident. Furthermore, there was the matter of selection of a useful class of loadings involved in a formal classification of stress infinities. Clearly any number of applied loads, e.g., hydrostatic tension, can be found for which the crack-tip stresses are bounded and well-behaved. Williams [22] introduced polar coordinates $(r, \theta)$ at the apex of a wedge with traction-free sides and examined the non-trivial solutions. In this manner, he established the general $r^{-\frac{1}{2}}$ character of the stresses in the vicinity of a traction-free crack tip for two-dimensional isotropic crack problems. Irwin's characterization [23] of the stress fields into three basic types with corresponding stress intensity factors, $k_i$ $(i=1, 2, 3)$ followed.

If the crack lies along the $x$-axis in the $z$-plane and $(r, \theta)$ are the local polar coordinates measured from a crack tip, then, for the isotropic case of Mode I,

$$\begin{aligned} \sigma_x &= k_1(2r)^{-\frac{1}{2}} \cos(\theta/2)[1-\sin(\theta/2)\sin(3\theta/2)]+\ldots \\ \sigma_y &= k_1(2r)^{-\frac{1}{2}} \cos(\theta/2)[1+\sin(\theta/2)\sin(3\theta/2)]+\ldots \\ \tau_{xy} &= k_1(2r)^{-\frac{1}{2}} \cos(\theta/2)\sin(\theta/2)\cos(3\theta/2)+\ldots \end{aligned} \tag{1.66}$$

and, for Mode II,

$$\begin{aligned} \sigma_x &= -k_2(2r)^{-\frac{1}{2}} \sin(\theta/2)[2+\cos(\theta/2)\cos(3\theta/2)]+\ldots \\ \sigma_y &= \quad k_2(2r)^{-\frac{1}{2}} \sin(\theta/2)\cos(\theta/2)\cos(3\theta/2)+\ldots \\ \tau_{xy} &= \quad k_2(2r)^{-\frac{1}{2}} \cos(\theta/2)[1-\sin(\theta/2)\sin(3\theta/2)]+\ldots . \end{aligned} \tag{1.67}$$

For Mode I, $\sigma_x$ and $\sigma_y$ are even functions of $\theta$ and $\tau_{xy}$ is odd in $\theta$, whereas, in Mode II the situation is reversed. In isotropic analysis, symmetry conditions of the loading and geometry are usually sufficiently apparent to the analyst for the prediction as to whether the problem is one of Mode I or II or, frequently, a combination of both.

Sih, Paris and Irwin [24] extended several of the basic concepts of fracture mechanics to the case of plane rectilinear anisotropy. For Mode I,

$$\sigma_x = \frac{k_1}{(2r)^{\frac{1}{2}}} \operatorname{Re}\left[\frac{s_1 s_2}{s_1 - s_2}\left(\frac{s_2}{(\cos\theta + s_2 \sin\theta)^{\frac{1}{2}}} - \frac{s_1}{(\cos\theta + s_1 \sin\theta)^{\frac{1}{2}}}\right)\right] + \ldots$$

etc. (1.68)

and, for Mode II,

$$\sigma_x = \frac{k_2}{(2r)^{\frac{1}{2}}} \operatorname{Re}\left[\frac{1}{s_1 - s_2}\left(\frac{s_2^2}{(\cos\theta + s_2 \sin\theta)^{\frac{1}{2}}} - \frac{s_1^2}{(\cos\theta + s_1 \sin\theta)^{\frac{1}{2}}}\right)\right] + \ldots$$

etc. (1.69)

The definitions of $k_1$ and $k_2$ in equations (1.68) and (1.69) are consistent with the case of isotropy. In fact, in several problems, including the problem in Figure 1.2, the stress intensity factors for the isotropic and rectilinearly anisotropic cases can be shown to be identical. This property, however, is not true in general. Furthermore, in the case of anisotropy, prediction of the type of mode requires the consideration of symmetries of the material properties as well as the symmetries in loading and geometry.

The computation of stress intensity factors in terms of the Muskhelishvili analysis can be reduced to a simple statement. For the isotropic case, Sih, Paris and Erdogan [25] observed that

$$k = k_1 - ik_2 = 2 \cdot 2^{\frac{1}{2}} \lim_{z \to z_1} (z - z_1)^{\frac{1}{2}} \phi'(z) \tag{1.70}$$

where $z_1$ corresponds to a crack tip. In terms of mapping, therefore,

$$k = 2 \cdot 2^{\frac{1}{2}} \lim_{\zeta \to \zeta_1} [\omega(\zeta) - \omega(\zeta_1)]^{\frac{1}{2}} \phi'(\zeta)/\omega'(\zeta) \tag{1.71}$$

where $\zeta_1$ corresponds to the parametric coordinate identifying the crack tip.

This matter was considered independently by the author [26], using the extension arguments in section 1.2. It was shown for mappings on the exterior of the unit circle and the subsequent continuation arguments, if $\zeta = \sigma_0$ is a point on the unit circle corresponding to a crack tip, then $\phi(\zeta)$ can be considered as analytic in the neighborhood of $\sigma_0$ provided traction-free conditions are assumed on the crack in the neighborhood of the tip. By expanding $\phi(\zeta)$ and $\omega(\zeta)$ about $\zeta = \sigma_0$ and utilizing the property $\omega'(\sigma_0) = 0$, it can be shown by series reversion and the definition of $k$, that

$$k = 2\phi'(\sigma_0)/[\omega''(\sigma_0)]^{\frac{1}{2}} . \tag{1.72}$$

The equivalence of equations (1.71) and (1.72) can easily be shown.

Similar arguments can be made for the case of plane rectilinear anisotropy. In fact,

$$k_1 + k_2/s_2 = 2 \cdot 2^{\frac{1}{2}} \left( \frac{s_2 - s_1}{s_2} \right) \lim_{z_1 \to z_0} (z_1 - z_0)^{\frac{1}{2}} \phi'(z_1)$$

$$\text{etc.} \tag{1.73}$$

where $z_0$ is the image point in the $z_1$-plane corresponding to the crack tip.

*Multi-valued mapping functions.* Corner points in the boundary geometry occur in many practical problems involving cracks. The local character of mapping functions with corner-describing properties is well-known, in fact, the exact mapping functions for a large class of useful geometries can be found by using the Schwartz–Christoffel transformation for polygonal regions. Such mapping functions are multi-valued with branch points occurring at points corresponding to the vertices of the corners.

It will be shown that the knowledge of the mapping function does not always lead to a simple solution for the stress analysis. The class of multi-valued mapping functions in general does not lead to simple closed solutions. It is usually necessary to introduce specialized techniques to analyze the corresponding geometries.

To illustrate the difficulties introduced by multi-valued mapping functions, consider

$$dz/z=(1-\zeta^{-K})d\zeta/\zeta(1+2\varepsilon\zeta^{-K}+\zeta^{-2K})^{\frac{1}{2}} \qquad 0\leqslant|\varepsilon|\leqslant 1 \tag{1.74}$$

which was introduced by Bowie [5] to map the exterior of the unit circle in the $\zeta$-plane into the exterior of a circle with $k$ radial cracks emanating from its boundary. This mapping function has branch points occurring on the unit circle at the roots of

$$\zeta^{-2k}+2\varepsilon\zeta^{-k}+1=0\,. \tag{1.75}$$

These branch points describe the corner points corresponding to the intersections of the radial cracks and the circle in the $z$-plane. For the case, $k=1$,

$$z=C[\zeta+\zeta^{-1}+(1+\varepsilon)+(1+\zeta^{-1})(\zeta^2+2\varepsilon\zeta+1)^{\frac{1}{2}}] \tag{1.76}$$

and the two branch points, $\zeta=e^{\pm i\lambda}$, lie on the unit circle with

$$\cos\lambda=-\varepsilon\,,\ \ \sin\lambda=(1-\varepsilon^2)^{\frac{1}{2}}\,. \tag{1.77}$$

In order to render the mapping function single-valued, it is necessary to choose an appropriate branch cut. Although this cut can be chosen in a variety of ways, the simplest and most natural choice is along the arc of the

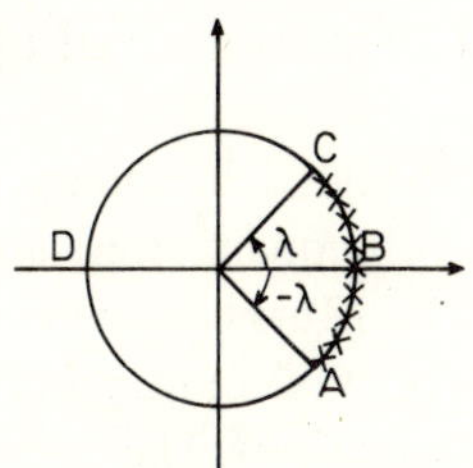


Figure 1.3. Branch cut in $\zeta$-plane.

unit circle joining the two branch points, e.g., Figure 1.3, $\widehat{ABC}$ (or $\widehat{CDA}$). In either choice, the mapping function must be considered as discontinuous across the cut. Therefore, there is an interval on the unit circle for which

$$\omega^-(\sigma)-\omega^+(\sigma)\neq 0\,. \tag{1.78}$$

It is the property (1.78) that causes the difficulty in solution. In the previous continuation arguments, the property of linear relationship depended on the reduction of equation (1.24) to (1.25). In the present case, such a reduction is no longer valid on the interval defined by equation (1.78) and the problem is no longer one of linear relationship. In general, therefore, the relatively direct solutions outlined in the previous section are no longer valid. Alternative approaches are considered later on.

There are interesting exceptions of the solution difficulties described above. Recently, Andersson [27] found the solution for a star-shaped contour in an infinite tensile sheet. This solution is of considerable practical interest since it provides a basis for the study of crack branching observed frequently in glass and other very brittle materials.

In Andersson's solution, a mapping function carrying the unit circle and its exterior into a branching crack contour and its exterior, respectively, is described. Such a mapping function has $k$ branch points corresponding to the $k$ cracks emanating from a common point, and the solution difficulties would seem consistent with our discussion above. However, for this particular class of mapping, an unusual property exists, namely,

$$\omega(\sigma)/\overline{\omega'(\sigma)} = g(\sigma) \tag{1.79}$$

where $g(\sigma)$ is a rational function of $\sigma$. When equation (1.79) is introduced in (1.12), the stress functions can be found in closed form by the application of standard Cauchy integral arguments.

## 1.4 Method of polynomial approximation

The approximation of complicated boundary geometries by polynomial mapping functions and the subsequent stress analysis has been utilized for many years by authors of the Muskhelishvili school. The method gained much of its initial impetus in this country from a solution by Morkovin [28] for an ovaloid in an infinite sheet. The main attraction of the method is the directness by which a stress analysis can be carried out for a given polynomial approximation. The analysis clearly falls into the category of linear relationship and the stress functions can be found by the arguments of section 1.2. In fact, Muskhelishvili [8] provides a systematic treatment of the mathematical analysis for most practical boundary-value problems in terms of polynomial mapping functions.

*Application to crack problems.* The application of the method of polynomial mapping approximation to complicated configurations involving cracks was first carried out by Bowie [5] for the solution for radial cracks emanating from a circular hole in an infinite sheet. Although the exact mapping function for this problem could be found, e.g. equation (1.74), the stress analysis was complicated by the multi-valued character of the mapping function.

In this earlier solution, it was recognized that in mapping approximations of crack configurations it was important to preserve the accuracy of the crack tip geometry. This was accomplished by ensuring the existence of a cusp at each approximate crack tip, i.e., $\omega'(\sigma_i)=0$ where $\sigma_i$ are the points on the unit circle corresponding to the crack tips.

The determination of suitable polynomial approximations was accomplished as follows: First, a series representation of equation (1.74) was found in the form

$$z=\omega(\zeta)=C\left[\zeta+\sum_{n=1}^{\infty} A_n\zeta^{1-kn}\right] \tag{1.80}$$

where the $A_n$'s are real. The $A_n$'s were obtained numerically from simple recursive formulae determined by expanding both sides of equation (1.74) in series form, squaring to remove the radical, then equating coefficients of equal powers of $\zeta$. The existence of cusps at locations corresponding to the crack roots implies

$$\omega'(\zeta)=(1-\zeta^{-K})g(\zeta) \tag{1.81}$$

where $g(\zeta)$ is a polynomial with coefficients chosen so that the roots of $g(\zeta)=0$ fall inside the unit circle (conformality should be preserved exterior to the unit circle). Suitable polynomial approximations

$$\omega(\zeta)=C\left[\zeta+\sum_{n=1}^{N} \varepsilon_n\zeta^{1-kn}\right] \tag{1.82}$$

were then obtained by setting $\varepsilon_n\approx A_n$ with small perturbations made to satisfy equation (1.81). Due to the rapid convergence of (1.80) for all but the very short crack lengths, there was no need at that time to devise a more systematic truncation process. Convergence to reliable values of the energy release rate was found by comparison of the solutions for several values of $N$ for each crack depth.

*Systematic truncation plan.* When the series representation of the exact mapping is slowly convergent, the selection of polynomial approximations which preserve crack tip geometry without disturbing the overall configuration requires a systematic truncation plan. Such a plan was proposed by

Bowie [26] in the solution for rectangular tensile sheet with symmetric edge cracks. Incorporated in this plan is the additional requirement of the accuracy of the second derivative of the mapping function at the crack tip. This latter requirement was motivated by the appearance of $\omega''(\sigma_0)$ in the calculation of the stress intensity factor, equation (1.72).

The truncation plan is simple, yet, it has proved to be remarkably effective. Let the expansion of the exact mapping function be given, for example, by the following form:

$$\omega(\zeta) = \sum_{n=1}^{\infty} A_n \zeta^{2n-1} \tag{1.83}$$

where it is presumed that the $A_n$'s can be determined numerically from the exact mapping function. Furthermore, assume that $\omega''(\sigma_0) = Q$ (where $\sigma_0$ corresponds to the crack tip) can be calculated directly from the exact mapping function. Now, the partial sums of the expansions for $\omega'(\zeta)$ and $\omega''(\zeta)$ are given by

$$\begin{aligned} \omega'(\zeta) &= \sum_{n=1}^{M} (2n-1) A_n \zeta^{2n-2} \\ \omega''(\zeta) &= \sum_{n=1}^{M} (2n-1)(2n-2) A_n \zeta^{2n-3} . \end{aligned} \tag{1.84}$$

The numerical values of $\omega'(\sigma_0)$ and $\omega''(\sigma_0)$ in (1.84) as a function of the truncation index $M$ will reveal, in general, preferred values, $M^*$, for which $\omega'(\sigma_0) \approx 0$ and $\omega''(\sigma_0) \approx Q$, simultaneously. Effective truncations can now be constructed as follows: Let

$$\omega_T(\zeta) = \sum_{n=1}^{M^*+2} \varepsilon_n \zeta^{2n-1} \tag{1.85}$$

where,

$$\varepsilon_n = A_n , \quad n = 1, 2, \ldots, M^* ; \qquad \varepsilon_{M^*+1} = +R ; \; \varepsilon_{M^*+2} = +S . \tag{1.86}$$

The constants $R$ and $S$ can be simply calculated to ensure that $\omega_T'(\sigma_0) = 0$ and $\omega_T''(\sigma_0) = Q$, exactly. Thus, preservation of the geometry at the crack tip is made with little disturbance of the overall configuration.

In practice, one should consider approximately three distinct values of $M^*$, preferably in the ranges, (0, 30), (30, 60), (60, 90) to ensure convergence. Most solutions exhibit remarkably rapid convergence with respect to $M^*$. In contrast, if truncations of equation (1.83) are based only on index number, experience has shown that, in many such problems, the corresponding solutions often exhibit very slow convergence.

*An example.* To illustrate the truncation process and the ensuing stress analysis, consider the solution for periodic edge cracks in a semi-infinite tensile sheet (as illustrated in the Appendix, Figure 1.13). This solution was found by Bowie and Freese [29] but has not been reported previously.

The mapping function, described by

$$dz/d\zeta = 2^{\frac{1}{2}}\, W \cos(\zeta/2)/2\pi\,(\cos\zeta+\alpha)^{\frac{1}{2}} \tag{1.87}$$

maps the real axis and the lower half of the $\zeta$-plane into the region shown in Figure 1.13. The parameter $\alpha$ can be varied to control the ratio $L/W$, in fact,

$$2\pi L/W = \ln\left\{\frac{2^{\frac{1}{2}}+(1-\alpha)^{\frac{1}{2}}}{2^{\frac{1}{2}}-(1-\alpha)^{\frac{1}{2}}}\right\}. \tag{1.88}$$

The mapping function can be expressed in series form as

$$z=\omega(\zeta)=(W/2\pi)\left[\zeta+\mathrm{i}\sum_{n=1}^{\infty}(A_n/n)(e^{-in\zeta}-1)\right] \tag{1.89}$$

where the $A_n$'s can be computed from simple recursive relations derived from equations (1.87) and (1.89).

Since the mapping function is periodic, crack tip accuracy can be specified by considering a typical crack tip, e.g., $\zeta=\pi$. Differentiating equation (1.87) and evaluating at $\zeta=\pi$ give

$$\omega''(\pi)=Q=\mathrm{i}\,W/2\pi(2-2\alpha)^{\frac{1}{2}}. \tag{1.90}$$

The choices of $M^*$ as described previously are made from an examination of the conditions

$$\begin{aligned}\omega_T'(\pi)&=(W/2\pi)\left(1+\sum_{n=1}^{M}A_n(-1)^n\right)\approx 0\\ \omega_T''(\pi)&=-\mathrm{i}(W/2\pi)\sum_{n=1}^{M}nA_n(-1)^n\approx Q.\end{aligned} \tag{1.91}$$

Effective mapping approximations can then be chosen in the form

$$\omega_T(\zeta)=(W/2\pi)\left[\zeta+\mathrm{i}\sum_{n=1}^{M^*+2}(\varepsilon_n/n)(e^{-in\zeta}-1)\right] \tag{1.92}$$

where $\varepsilon_n = A_n$ $(n=1, 2, \ldots, M^*)$ and the last two coefficients are adjusted so that equations (1.91) are satisfied exactly.

To complete the stress analysis, set

$$\phi(\zeta) = (TW/2\pi)\left[\zeta/4 - \mathrm{i}\sum_{n=1}^{M^*+2} \alpha_n \mathrm{e}^{-\mathrm{i}n\zeta}\right] \tag{1.93}$$

where the $\alpha_n$'s are real. The structure of equation (1.93) is consistent with the obvious symmetry of the stresses and loading conditions at infinity. Traction-free conditions on the finite boundary can be ensured by the previous analytic continuation arguments. Here use is made of the reflection across the real axis in the $\zeta$-plane. Thus,

$$\psi(\zeta) = -\bar{\phi}(\zeta) - \bar{\omega}(\zeta)\,\phi'(\zeta)/\omega'(\zeta) \tag{1.94}$$

with the notation of equation (1.14). It is necessary to ensure that $\psi(\zeta)$ as defined by equation (1.94) satisfies analyticity requirements at infinity. From the loading conditions at infinity, it can easily be shown that $\psi(\zeta)$ must approach $-TW\zeta/4\pi$ for large $|\zeta|$. From equations (1.92)–(1.94), this requires

$$\alpha_k + \sum_{n=1}^{M^*+2-k} \{\varepsilon_n \alpha_{k+n} + n\varepsilon_{n+k}\alpha_n/(n+k)\} = \varepsilon_k/4k \qquad k = 1, 2, \ldots, M^*+1 \tag{1.95}$$

and

$$\alpha_{M^*+2} = \varepsilon_{M^*+2}/4(M^*+2)\,. \tag{1.96}$$

This linear system of simultaneous equations must be solved for the $\alpha_n$'s to complete the solution.

In the calculation of the stress intensity factor, it will be recalled that in equations (1.66) the crack was assumed to lie along the $x$-axis. To account for the crack(s) lying along the imaginary axis in the present problem, equation (1.72) must be modified to

$$k_1 = 2\phi'(\pi)/[-\mathrm{i}\omega''(\pi)]^{\frac{1}{2}}\,. \tag{1.97}$$

It then follows that

$$k_1/T(L^{\frac{1}{2}}) = \left(\frac{2W}{\pi L}\right)^{\frac{1}{2}} (2-2\alpha)^{\frac{1}{4}} \left[\left(\tfrac{1}{4}\right) - \sum_{n=1}^{M^*+2} n\alpha_n(-1)^n\right]. \tag{1.98}$$

Numerical results are listed in the Appendix, Table I. For very short crack lengths, the numerical results approach the well-known limit for a single edge crack (Koiter [30]), $k_1 \rightarrow 1.12T(L^{\frac{1}{2}})$.

*Observations.* The method of polynomial mapping approximation for crack problems has been applied successfully to several problems where the exact mapping function is known. In addition to the previous reference, Bowie [31] considered clamped end conditions for finite rectangular plates with symmetric edge cracks. This work was extended by J. M. Bloom [32] to single edge cracks in rectangular sheet subjected to end rotations. Bowie and Neal [33] used the method to verify the accuracy of the collocation solution of Gross et al. [34] for the single edge crack in finite tensile sheet. Akas and Kobayashi [35] analyzed short edge-notched specimens subjected to three-point loading. Brown and Kobayashi [36] applied the method to solve the Dugdale model for estimating the extent of the plastic zone in single edge-cracked tensile specimens. Rich and Roberts [37] found the solution for out-of-plane bending of an infinite sheet with radial cracks emanating from an interior circular hole. Wieselmann [38] automated the method in a computer program and found the solution for several cases of oblique edge cracks.

The method of polynomial mapping approximation as described in this section is seriously limited by the relatively small class of geometries for which exact mapping functions can be found. Although the Schwartz–Christoffel transformation has proved useful for a limited class of edge-crack geometries, direct methods for obtaining mapping functions for most geometries are still unavailable. Additional complications are evident when the physical region is multiply connected. Methods for obtaining mapping functions for parameter regions with corresponding multiple connectivity are virtually non-existent.

Some consideration has been given to the matter of introducing iterative schemes to find suitable polynomial mapping approximations. In principle, the crack tip geometry can be preserved to the extent of specifying $\omega'(\sigma_0)=0$ and the polynomial coefficients determined numerically on the basis of the remaining geometry by any of several well-known iterative schemes. It is believed, that such a process would be non-competitive with other methods, e.g., the modified mapping collocation technique which will be described later.

## 1.5 An unconventional application of mapping to edge cracks and notches in semi-infinite regions

The previous sections considered mapping techniques from a conventional viewpoint. Mapping functions were sought which mapped either exactly or approximately a rigidly prescribed parameter region into the physical geometry. The prescribed parameter regions were chosen as the upper (lower) half plane or the exterior (interior) of the unit circle. In this and the following sections, departure from the conventional choice of parameter regions will be made. Although the procedure used in this section does not compare with the radical departure which will be made in section 1.6, it serves to illustrate the flexibility which can be gained by appealing to the parametric nature of mapping.

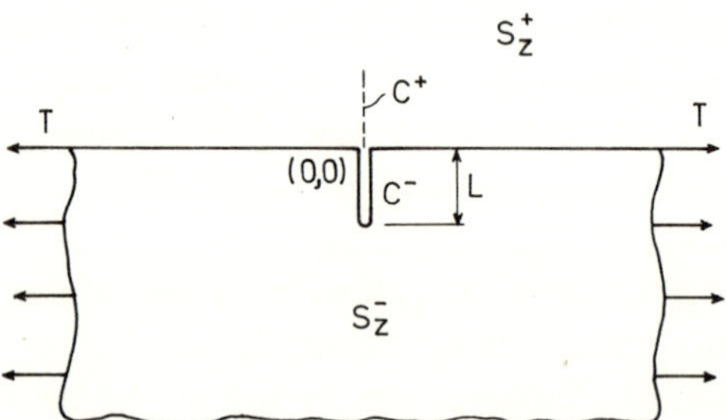


Figure 1.4. Edge crack in a semi-infinite region.

The procedure used in this section was introduced by Bowie [39] and applied to semi-elliptic edge notches in a semi-infinite region under tension. Here, the procedure will be applied to the problem of an edge crack in a semi-infinite sheet under tension. This solution was originally carried out by Bowie and McLaughlin [40] but the results were not reported in the open literature. An accurate solution to this problem was obtained originally by Koiter [30].

Consider the material with an edge crack of length $L$ as occupying the lower half plane $S_z^-$ in Figure 1.4. The boundary of the crack will be denoted by $C^-$. Now extend $\phi(z)$ into $S_z^+$ by the definition (1.13). $\phi(z)$ is now defined in the region exterior to $C = C^+ + C^-$ where $C^+$ is the reflection of $C^-$ across the real axis. If $\phi(z)$ is now considered as analytic exterior to $C$ (except at infinity), then, from the previous continuation arguments, the portion of the boundary lying along the real axis can be considered as traction-free. Apart from the conditions at infinity, the remaining boundary conditions correspond to traction free conditions on $C^-$. Thus,

$$\phi'(z)+\overline{\phi'(z)}-e^{2i\alpha}[(\bar{z}-z)\phi''(z)-\phi'(z)-\bar{\phi}'(z)]=0 \qquad z\in C^- \tag{1.99}$$

where $\alpha$ denotes the angle between the $x$-axis and the outward normal to $C^-$.

It is at this point we introduce an auxiliary $\zeta$-plane and the mapping function

$$z=\omega(\zeta)=(L/2)(\zeta-\zeta^{-1}). \tag{1.100}$$

The unit circle, $\zeta=\sigma=e^{i\theta}$, and its exterior are mapped into $C$ and its exterior, respectively, by equation (1.100). Introduce the new notation $\phi'(z)=\Phi(\zeta)$, $\phi''(z)=\Phi'(\zeta)/\omega'(\zeta)$, etc. The extended stress function $\Phi(\zeta)$ can be written as

$$\Phi(\zeta)=T/4+\Phi_1(\zeta) \tag{1.101}$$

where $T$ is the applied tension at infinity and $\Phi(\zeta)$ is analytic for $|\zeta|>1$ and $O(\zeta^{-2})$ for large $\zeta$. It can easily be verified that equation (1.101) is compatible with the conditions at infinity. Analyticity of $\Phi(\zeta)$ assures traction-free conditions along the real axis of the physical plane. The conditions corresponding to equation (1.99) can be written on the unit circle as

$$G(\sigma)=\Phi_1(\sigma)+\Phi_1(\sigma)+2\Phi_1(\sigma)-2(\sigma-\sigma^{-1})\Phi_1'(\sigma)/(1+\sigma^2)+T=0, \qquad \pi\leqslant\theta\leqslant 2\pi. \tag{1.102}$$

Before an effective representation of $\Phi_1(\zeta)$ can be assumed it is necessary to study the possible singularities existing on the unit circle. It is evident from equation (1.102) that singularities in $\Phi_1(\zeta)$ probably exist at $\sigma=\pm i$. (Note that although $G(\sigma)$ is not defined for $0<\theta<\pi$, singularities at $\sigma=i$ must be considered in the calculation of $G(-i)$ due to the term $\Phi_1(\bar{\sigma})$). Since experience dictates that $\Phi_1(\zeta)$ has a simple pole at the crack tip, $\sigma=-i$, then from equation (1.102) compatibility of singularities implies that $\Phi_1(\zeta)$ has a third order pole at $\sigma=+i$.

It is convenient, therefore, to introduce the function $F(\zeta)$ where

$$F(\zeta)=(\zeta^2+1)(\zeta-i)^2\Phi_1(\zeta). \tag{1.103}$$

The function $F(\zeta)$ will have no singularities of a severe nature on the unit circle and will be analytic exterior to the unit circle except for a pole of order two at infinity. Thus, $F(\zeta)$ may be represented as

$$F(\zeta) = TR_{-2}\zeta^2 + TR_{-1}\zeta + T\sum_{n=0}^{\infty} R_n \zeta^{-n}, \qquad |\zeta| \geqslant 1. \tag{1.104}$$

From obvious stress symmetries, it follows that

$$\begin{aligned} R_{2K} &= S_{2K}, \qquad & K &= -1, 0, 1, 2, \ldots \\ R_{2K-1} &= \mathrm{i}\,T_{2K-1}, \qquad & K &= 0, 1, \ldots \end{aligned} \tag{1.105}$$

where $S_{2K}$ and $T_{2K-1}$ are real.

The boundary conditions (1.102) can now be expressed in series form by utilizing equations (1.103)–(1.105). Satisfaction of these conditions is assured by imposing the conditions

$$\frac{\partial}{\partial R_j}\left\{\int_{\pi}^{2\pi} [G(\sigma)]^2 \,\mathrm{d}\theta\right\} = 0, \qquad j = -2, -1, 0, 1, \ldots . \tag{1.106}$$

The conditions (1.106) provide a linear system of simultaneous equations for the determination of the unknowns $R_j$.

The first fifty six coefficients are tabulated in the Appendix, Table II. The accuracy of the results can be shown by the calculation of the stress intensity factor. It is easily shown that

$$k_1 = T(\tfrac{1}{4}L^{\frac{1}{2}}) \sum_{K=-1}^{\infty} (S_{2K} - T_{2K+1})(-1)^K . \tag{1.107}$$

Using the values in Table II, we find $k_1 = 1.1214\, L^{\frac{1}{2}}$. This can be compared with Koiter's result, $k_1 = 1.1215\, L^{\frac{1}{2}}$.

The philosophy of this approach represents a departure from conventional mapping. A mapping of a standard type of parameter region would involve branch points in the mapping function corresponding to the junction of the crack and the real axis. It is interesting that this aspect of the problem has been formally circumvented in the solution above.

There is a useful class of edge crack problems which can be solved by this approach. As examples, the mapping function (1.74) can be used to analyze a semi-circular edge notch with radial cracks. The problem of an oblique edge crack could be solved by using the mapping function described in Andersson's work. In general, if the mapping function is known for a region exterior to a closed contour possessing symmetry with respect to an axis, a corresponding edge notch problem can be solved by the above approach. When such mapping functions involve branch points, it is assumed that the method of polynomial approximation can be incorporated as part of the procedure.

## 1.6 The modified mapping-collocation method

Recently, an extremely effective technique which combines modified versions of conformal mapping and boundary collocation arguments has been developed for two-dimensional crack problems. This method will be referred to as the MMC (modified mapping-collocation) technique. The MMC technique was originally presented by Bowie and Neal [6] as a procedure for the analysis of an internal crack in a finite geometry for isotropic materials. Extension of the method to orthotropic (or rectilinearly anisotropic) materials was carried out shortly thereafter by Bowie and Freese [7]. Since that time the MMC method has been developed and extended to cover a wide range of problems involving cracks, notches, cutouts, etc. The remainder of this chapter will be devoted to a discussion of this technique along with several numerical illustrations.

*The MMC method for isotropic problems.* The MMC technique depends basically on combining the most attractive features of conformal mapping and boundary collocation arguments. With the flexibility gained by combining these methods, it is possible to arrive at a versatile and effective plan for solving many two-dimensional problems of elasticity.

In the MMC technique a major modification of the conventional mapping procedure is made in the choice of parameter regions. It is no longer necessary to seek the mapping function which maps a rigidly prescribed parameter region into the entire physical region. Instead, only a portion of the parameter region is prescribed and simple mapping forms are chosen to describe key portions of the physical boundary at the discretion of the analyst. Algebraically simple forms of mapping functions can therefore be utilized which, in turn, can be inverted to locate the entire geometry of the parameter region. The carry-over of the analytic function theory into the parameter plane is still retained, and, many of the useful features of the continuation arguments can be preserved in local regions of the problem.

The difficulty of finding the mapping functions for complicated geometries is minimized by the plan above. This compromise must entail some sacrifice, in particular in the direct analytical arguments previously described for the solution of the boundary value problem. This matter is compensated for by the modified boundary collocation scheme which is described later.

As an illustration of the modification of the mapping arguments, consider the original solution by Bowie and Neal [6] for a circular disk with an internal crack, Figure 1.5. Such a problem illustrates the limitations of the conventional mapping technique. Since the region is doubly connected, a natural choice of the parameter region is a concentric circular ring. Even though the exact mapping function happens to be known in this case (Nehari [41]), the difficulties in finding suitable polynomial mapping approximation necessary to the stress analysis are formidable.

For this problem, the simple approach of the MMC method consists of choosing the mapping

$$z = \omega(\zeta) = (L/2)(\zeta + \zeta^{-1}) \,. \tag{1.108}$$

The unit circle and its exterior in the $\zeta$-plane are mapped by equation (1.108)

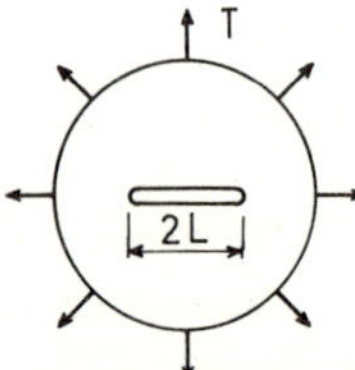


Figure 1.5. Circular disk with internal crack loaded by uniform external tension, $T$.

into the crack and its exterior in the $z$-plane. The points in the physical plane can be simply related to the parameter plane by

$$\zeta = (z/L) + [(z/L)^2 - 1]^{\frac{1}{2}} \,. \tag{1.109}$$

In particular, the outer circular boundary will correspond to a closed curve $\tau$ exterior to the unit circle in the $\zeta$-plane (Figure 1.6).

The parameter region corresponding to equation (1.108) is therefore the annulus bounded by $|\zeta| = 1$ and $\tau$.

Traction-free conditions on the crack can again be handled by applying the continuation arguments of section 1.2. In particular, if the extension described by equations (1.21)–(1.24) is adopted, traction-free conditions on the crack are ensured if the extended stress function $\phi(\zeta)$ is analytic in the annulus bounded by $\tau$ and $\tau'$, where $\tau'$ is the inverted image of $\tau$ with respect to the unit circle.

The determination of $\phi(\zeta)$ requires a form of representation and the satisfaction of conditions on $\tau$ corresponding to the tractions specified on the circular boundary in the physical plane. For this problem, it will be assumed that $\phi(\zeta)$ can be represented in the form of a Laurent series. This appears to be a reasonable assumption although the boundaries $\tau$ and $\tau'$ are not circular. There is no a priori reason in the present problem to suspect that the region of convergence of such a series could not extend over the necessary parameter region. Thus, taking into account obvious stress symmetries,

$$\phi(\zeta) = T \sum_{-\infty}^{\infty} \alpha_n \zeta^{2n+1} \tag{1.110}$$

where the $\alpha_n$'s are real and must be determined from the boundary conditions on $\tau$.

The choice of mapping function (1.108) is by no means unique although it does provide a basis for generating power series expansions with traction-

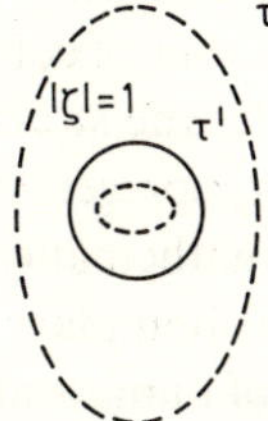


Figure 1.6. Region defined in the parameter plane.

free conditions on the crack. In general, the analyst has the option of the choice of mapping function. The choice will generally depend on the plan of expansion of the stress function and clearly could vary with the particular problem as will be illustrated in section 1.7. The practical restriction on $\omega(\zeta)$ is that its inverse be readily calculable. It is clear that unspecified portions of parametric boundary, e.g., $\tau$, must be found in order to consider the boundary conditions. Since boundary collocation arguments will be used, only a pointwise knowledge of the boundary of the parameter regions is required.

The method of boundary collocation of a stress function for solving crack problems was developed by Gross et al. [34] and Kobayashi [42]. Gross solved the problem of a single edge crack in rectangular tensile sheet by boundary collocation of a stress function derived by Williams [22]. Kobayashi solved the same problem by boundary collocation of a series of complex stress functions derived by generalizing a stress function due to Westergaard. In recent years several crack problems have been studied effectively by boundary collocation arguments.

As a method, boundary collocation has the advantage of simplicity. If the analyst can find a representation of the biharmonic stress function suitable for a particular problem it is an easy matter to program on a digital computer the boundary conditions at specified boundary stations. The corresponding linear system of simultaneous equations can be directly solved for the stress function coefficients for a given truncation of the expansion. Since the conditions are considered at discrete boundary points, irregularities in the boundary geometry create little disturbance in the computational set-up. In this sense, the adoption of this plan to compliment the modified mapping approach, e.g., to complete the analysis on $\tau$ to determine the $\alpha_n$'s in equation (1.110), is very attractive.

There are several, somewhat subtle, pitfalls which must be considered in the application of boundary collocation arguments. Perhaps the most important consideration is the matter of completeness of the representation. Now, as is commonly the case, let the stress function be represented by an infinite number of terms. Such a representation could be complete, incomplete, or even overly complete, mathematically, for the actual solution. On the other hand, the conventional boundary collocation process by its very nature will generally yield a set of numerical values for the coefficients for a truncated form of the representation. Furthermore, large errors in the "off-point" intervals of the boundary can occur in such a manner that "apparent" convergence can be found to an incorrect result. For example, simply take a known function analytic in the region of a circle and compute its boundary values on the circle. Now assume the obviously incorrect form of a Laurent series and apply the conventional-boundary collocation argument. It will be found that the coefficients of the incorrect terms of the expansion will not necessarily approach zero with refinement of mesh spacing on the boundary. Frequently, the coefficients will appear to converge at the expense of an increase in the off-point boundary error. Clearly, the conventional boundary collocation argument does not have the finesse to evaluate over-specification or recognize underspecification of functional representation.

Since the representation of stress functions for complicated geometries must be somewhat heuristic, e.g., equation (1.110), the analyst is faced with the problem of assessment of the accuracy of solution when boundary collocation is used. Furthermore, a rigorous proof of completeness of representation for most problems would be impossible. In fact, the requirement of completeness is not always desirable. Many effective solutions are asymptotic in nature in that the desired information can be accurately determined although the solution as a whole is incorrectly represented. On the other hand, a modification of the conventional boundary collocation is clearly necessary to avoid "apparent" convergence to incorrect results.

In the MMC method, a modification of conventional boundary collocation is introduced by heuristically providing a rational assessment of the effects of residual errors in the boundary conditions on the accuracy of the desired information. The plan is simple, as it should be, and is motivated by Saint Venant's principle. It will be assumed that the desired information is the crack-tip stress intensity factor. Now the effects of residual boundary errors are grouped into two types, namely, remote and local. Errors remote to the crack tip will be assessed on the basis of resultant force and resultant moment in keeping with Saint Venant's argument. Errors local to the crack tip imply that the crack tip is sufficiently close to boundary to be influenced by local errors in the boundary conditions.

The effects of residual boundary errors of the remote type can be studied conveniently by considering the resultant force $f_1 + \mathrm{i}f_2$, e.g., equation (1.6), and the resultant moment $M_0$, e.g., equation (1.7), calculated as a function of the arc length on the boundary. If $f_1 + \mathrm{i}f_2$ is satisfied at two successive boundary stations, then the residual error in stress boundary conditions in the intermediate boundary interval must correspond to a self-equilibrating distribution of force. A measure of the moment effect along a boundary arc $L$ is provided by

$$\begin{aligned} M_0 &= \int_L (xY_n - yX_n)\mathrm{d}s = -\int_L (x\mathrm{d}f_1 + y\mathrm{d}f_2) \\ &= -(xf_1 + yf_2)_L + \int_L (f_1\mathrm{d}x + f_2\mathrm{d}y)\,. \end{aligned} \tag{1.111}$$

It is clear from equation (1.111) that off-point errors in $f_1$ and $f_2$ can build up into an accumulated moment effect due to the last integral. On the other hand, it is evident that the accuracy of $f_1 + \mathrm{i}f_2$ is the dominant factor in general for minimizing remote types of error effect. (A strict adherence to the Saint Venant argument would require a resolution of the moment into components with respect to orthogonal axes, e.g., $M_x$ and $M_y$. Such a resolution is awkward to formulate compared with equation (1.7) and the small gain made in the argument does not seem to justify the considerable additional effort.)

In the MMC method, the following modification of the conventional boundary collocation plan is suggested by the preceding considerations. At each boundary station, the five conditions corresponding to $M_0$, $f_1 + \mathrm{i}f_2$, and $N_1+iT_1$ are imposed where $N_1$ and $T_1$ correspond to the normal and tangential components of the applied stress. Then the coefficients of the stress function can be determined on the basis of satisfying these conditions in a least square sense. With this modification, the effects of remote errors are minimized by $M_0$, $f_1$, and $f_2$ and the effects of local errors are controlled by $N_1$ and $T_1$. The analyst is therefore provided with a rational basis for error assessment.

There is some flexibility in the plan which can be deduced from its own arguments. The role of $M_0$ is generally secondary compared with $f_1$ and $f_2$ as was shown by equation (1.111), thus, $M_0$ can frequently be omitted from the plan. When the completeness of the stress function representation is obvious, accurate results can be expected by using only $f_1$ and $f_2$ provided the crack tip is remote to the physical boundary. In regions where the crack tip is near the boundary, $N_1$ and $T_1$ must be considered in the plan at least over an interval of the boundary near the crack tip. Finally, when there is reason to suspect that the representation is over- or underspecified, as frequently occurs, the fully modified plan should be adopted for the complete range of computation.

To complete the solution of the problem shown in Figure 1.5, truncations of equation (1.110) are considered.

$$\phi(\zeta)=T\sum_{-M}^{N}\alpha_n\zeta^{2n+1}. \tag{1.112}$$

The stress boundary conditions on the circular boundary are

$$\sigma_r=T,\ \tau_{r\theta}=0. \tag{1.113}$$

Thus,

$$N_1-iT_1=\Phi(\zeta)+\overline{\Phi(\zeta)}$$
$$-\frac{e^{2i\theta}}{\omega'(\zeta)}\{\overline{\omega(\zeta)}+\zeta^{-2}\bar{\omega}'(1/\zeta)[\bar{\Phi}(1/\zeta)+\Phi(\zeta)-\bar{\omega}(1/\zeta)\Phi'(\zeta)]\}=T,\ \zeta\in\tau \tag{1.114}$$

where $\Phi(\zeta)=\phi'(\zeta)/\omega'(\zeta)$, etc., and $\omega(\zeta)$ is the mapping function (1.108). The boundary condition in terms of the force resultant becomes

$$\phi(\zeta)-\phi(1/\bar{\zeta})+[\omega(\zeta)-\omega(1/\bar{\zeta})]\overline{\Phi(\zeta)}=f_1+if_2=T\omega(\zeta),\qquad \zeta\in\tau. \tag{1.115}$$

Similarly, a condition corresponding to $M_0$ for $\zeta\in\tau$ can be derived from equation (1.7).

Truncated series forms (equation (1.112)) when substituted into (1.114),

(1.115) and the moment condition, lead to five real conditions at each boundary station for the determination of the $\alpha_n$'s. For given set of boundary stations, the $\alpha_n$'s can be determined on the basis of minimizing the total of each error squared and summed over the boundary stations. The numerical results of this calculation are available in the paper by Bowie and Neal [6] and will not be repeated here.

The numerical results obtained for the previous problem were remarkably consistent with the heuristic arguments which formed the basis of the MMC method. For short and intermediate crack lengths, it was found that reliable crack-tip stress intensities can be obtained by the use of $f_1 + \mathrm{i}f_2$ alone. For the deeper cracks a build-up of the stress errors on the boundary was observed which, in turn, affected the reliability of the computed stress intensity factors. No systematic process involving the choices of the truncation indices, $M$ and $N$, and the point spacing on the boundary could be found to eliminate this difficulty. On the other hand, with the use of both $f_1 + \mathrm{i}f_2$ and $N_1 + \mathrm{i}\,T_1$, reliable results were obtained for very deep cracks. The effect of the moment condition, $M_0$, in general appeared to be secondary although for very deep cracks a slight improvement in accuracy was obtained. The trends indicated here (based on subsequent experience of the MMC method) are typical of the behavior of a well-specified representation of the stress function.

The versatility of the MMC method is quite evident. The outer boundary and loading in Figure 1.5 could have been chosen as any closed contour with arbitrary loading provided geometric irregularities or singular loads are not introduced which would invalidate the form of expansion (1.110). In the latter instance, appropriate changes in the form of expansion would be required. Several isotropic solutions have been carried out since the original method was devised. Bowie and Neal [43] found an accurate solution for the well-known problem of a central crack in a uniformly stressed strip. Bowie and Freese [44] found the solution for a radial crack in a circular ring. Several additional examples are described in section 1.7.

It was brought to the author's attention that Newman [45] independently studied the use of $f_1 + \mathrm{i}f_2$ in the least square sense. He has recently, Newman [46], carried out several applications of his "modified boundary-collocation method" to crack problems. The accuracy of his results is consistent with our own observations. His method, which is an abbreviated version of the MMC method, should be used with caution in situations where both force and stress conditions in the collocation plan are required.

*The MMC method for anisotropy.* The MMC method was applied successfully to the two-dimensional problems of rectilinear anisotropy by Bowie and Freese [7] when they considered the solution for a central crack in an orthotropic rectangular panel under tension (Figure 1.7). The approach is very similar to that of the isotropic case; thus, only a brief outline of the analysis will be summarized here.

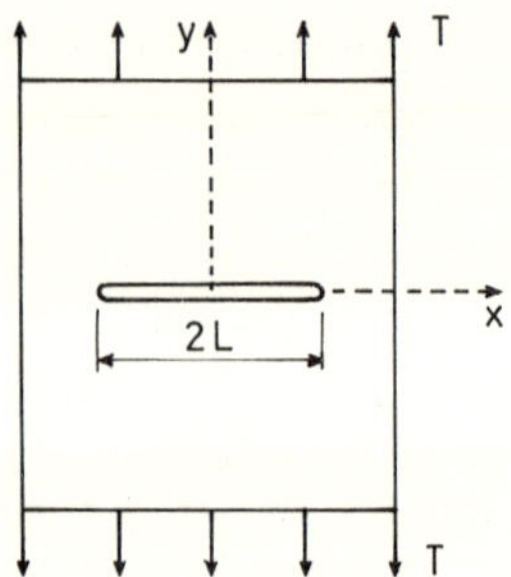


Figure 1.7. Central crack in rectangular orthotropic panel under tension, $\sigma_y = T$.

The mapping plan consists of choosing the simple mappings

$$\begin{aligned} z &= \omega(\zeta) = (L/2)(\zeta + \zeta^{-1}) \\ z_1 &= \omega_1(\zeta_1) = \omega(\zeta_1) = (L/2)(\zeta_1 + \zeta_1^{-1}) \\ z_2 &= \omega_2(\zeta_2) = \omega(\zeta_2) = (L/2)(\zeta_2 + \zeta_2^{-1}) \,. \end{aligned} \tag{1.116}$$

On the unit circle, $\zeta$, $\zeta_1$ and $\zeta_2$ coincide, that is, the three parameter regions defined to map into $z$, $z_1$, $z_2$ intersect on the crack and its images (section 1.3). Otherwise, $\zeta_1$ and $\zeta_2$ are distinct and are found from

$$\begin{aligned} \zeta_1 &= (z_1/L) + [(z_1/L)^2 - 1]^{\frac{1}{2}} \\ \zeta_2 &= (z_2/L) + [(z_2/L)^2 - 1]^{\frac{1}{2}} \,. \end{aligned} \tag{1.117}$$

Traction-free conditions on the crack can be ensured by applying the continuation arguments indicated in equation (1.48) and illustrated in (1.61)–(1.65). In the present problem, the parameter region for which $F(\zeta)$ is defined is a finite annulus. Although the annulus is not circular we again make the reasonable assumption of a Laurent expansion for $F(\zeta)$, thus,

$$F(\zeta) = T \sum_{-\infty}^{\infty} (a_n + \mathrm{i} b_n) \zeta^n \,. \tag{1.118}$$

The coefficients $a_n + \mathrm{i} b_n$ can be determined by applying the modified boundary collocation arguments to the remaining boundary conditions corresponding to the loads acting on the perimeter of the rectangle in Figure 1.7. The force conditions become

$$(1+\mathrm{i}s_1)F(\zeta_1)+(1+\mathrm{i}s_2)[B\bar{F}(1/\zeta_2)+CF(\zeta_2)]+(1+\mathrm{i}\bar{s}_1)\overline{F(\zeta_1)}$$
$$+(1+\mathrm{i}\bar{s}_2)[\bar{B}F(1/\bar{\zeta}_2)+\bar{C}\overline{F(\zeta_2)}]=f_1+\mathrm{i}f_2 \qquad (1.119)$$

and the stress boundary conditions can be easily found from

$$\sigma_x = 2\,\mathrm{Re}\,\{s_1^2F'(\zeta_1)/\omega'(\zeta_1)+s_2^2[-B\bar{F}'(1/\zeta_2)/\zeta_2^2\omega'(\zeta_2)+CF'(\zeta_2)/\omega'(\zeta_2)]\}$$
$$\text{etc.} \qquad (1.120)$$

In equations (1.119) and (1.120) the notation is that defined in section 1.2.

The modified collocation argument can be applied in essentially the same manner as we have described previously for the isotropic case. The choice of boundary stations can be made on the rectangular boundary in the $z$-plane. These stations along with the material properties determine $z_1$ and $z_2$ and, hence from equations (1.117), the corresponding $\zeta_1$ and $\zeta_2$ locations. The numerical trends previously discussed in the isotropic case carry over in a similar manner to the problems we have solved in orthotropy. The moment resultant $M_0$ again plays a secondary role and can be generally omitted in the MMC plan with little loss of accuracy.

In the earliest version of this solution, for the sake of concreteness in the numerical calculations, it was assumed that the principal directions of elastic symmetry coincide with the $x$- and $y$-axes in Figure 1.7. In this case, from obvious stress symmetries,

$$F(\zeta)=T[s_2/(s_2-s_1)]\sum_{-\infty}^{\infty}c_n\zeta^{2n+1} \qquad (1.121)$$

where the $c_n$'s are real. This case falls into Sih and Liebowitz's [47] category of "plane symmetric loading" and equation (1.73) reduces to

$$k_1=(2/L^{\frac{1}{2}})[(s_2-s_1)/s_2]F'(1)$$
$$=(2T/L^{\frac{1}{2}})\sum_{-\infty}^{\infty}(2n+1)c_n\,. \qquad (1.122)$$

It should be noted that geometry and loading symmetries are not sufficient to justify the stress symmetries implied by equation (1.121). The compatibility of the material properties must also be considered. E.g., in the present example, if the principal directions of the material properties are not assumed to coincide with the $x$- and $y$-axes, the reduction of equation (1.118) to

(1.121) would no longer be valid. In this latter situation, equation (1.122) would no longer hold, and, over a range of the parameters, there would be both a $k_1$ and $k_2$ contribution in the calculation of the stress intensity factors.

An interesting example of the orthotropic case is provided by $s_1 = i\beta_1$ and $s_2 = i\beta_2$. In this case,

$$\begin{aligned} \beta_1\beta_2 &= (E_1/E_2)^{\frac{1}{2}} \\ \beta_1+\beta_2 &= 2^{\frac{1}{2}}\{(E_1/E_2)^{\frac{1}{2}} - \nu_{12} + E_1/2\mu_{12}\} \end{aligned} \tag{1.123}$$

where, in equations (1.26)

$$\begin{aligned} &a_{11} = 1/E_1\,, \quad a_{22} = 1/E_2\,, \quad a_{66} = 1/\mu_{12} \\ &a_{16} = a_{26} = 0\,, \quad a_{12} = -\nu_{12}/E_1 = -\nu_{21}/E_2\,. \end{aligned} \tag{1.124}$$

In the above, $E_1$ and $E_2$ are the Young's moduli, $\nu_{12}$ and $\nu_{21}$, the Poisson ratios, and $\mu_{12}$, the shear modulus.

By fixing $\beta_1 = 1$ and considering $\beta_2 = 1 \pm \varepsilon$ in the solution of the problem corresponding to Figure 1.7, two interesting results were obtained. First, it was found that the isotropic solution, $\varepsilon = 0$, can be accurately bracketed by solving for small values of $\pm\varepsilon$. For many practical purposes, therefore, there is no need to consider two separate analyses. The second and perhaps the most significant result was the clear indication of the dependence of the orthotropic stress intensity factors on the elastic constants. In contrast, for boundary value problems of this type, the isotropic stress intensity factors are independent of the material constants. Thus, the equivalence of the anisotropic stress intensity factors for boundary value problems involving self-equilibrating load systems on the crack (which was commonly accepted before this work) is not generally valid. (It can be shown, however, that there are large ranges of the geometry where such an approximation is valid.)

The MMC method has been applied successfully to several additional problems in orthotropic elasticity. Gandhi [48] extended the original solution above to the case of several angular orientations of the crack for various orientations of the material properties. Freese [49] has computerized the procedure for a useful class of notches and cutouts in orthotropic panels.

## 1.7 Several applications of the MMC method

In this section, consider several representative numerical solutions recently obtained by the MMC method. The problems are selected to illustrate useful variations within the technique. Rectangular outer boundaries are chosen only for the sake of concreteness. Only isotropic solutions are specifically considered due to the wide range of material properties in orthotropic analysis. Yet, the carry-over of ideas to orthotropic analysis will be evident. The numerical data are included in the Appendix and generally are not previously available in the open literature.

*Normal edge crack in a rectangular panel.* In Figure 1.8a, we consider a rectangular panel with an edge crack of length $L$ loaded by a uniform tension acting over an interval of length $a$. The total force acting on the interval will be denoted by $P$ and we shall consider $P$ as constant throughout the calculation. With $P$ fixed and allowing $a \to 0$, the limiting case for the familiar problem of the "crackline loaded single-edge-cracked specimen" is obtained.

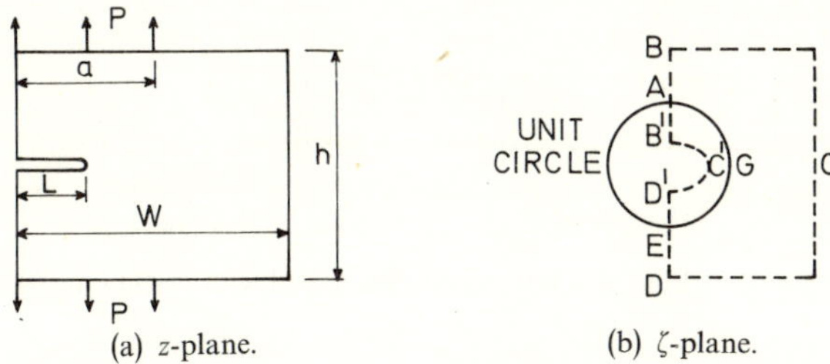


(a) $z$-plane. (b) $\zeta$-plane.

Figure 1.8. Edge crack in rectangular panel with orientation I.

Srawley and Gross [50] have considered the latter problem by a boundary collocation technique.

With the geometry orientation shown in Figure 1.8a, it is natural to choose the mapping function (1.108) to describe the crack, i.e.

$$z = \omega(\zeta) = (L/2)(\zeta + \zeta^{-1}).$$

Then, the interval $\widehat{EGA}$ in the $\zeta$-plane maps into the crack and $\widehat{ABCDE}$ maps into the rectangular boundary. Traction-free conditions on the crack are ensured by adopting the extension argument of equations (1.21)–(1.24). The extended parameter region now becomes the interior of C′B′ABCDED′C′. The extended stress function $\phi(\zeta)$ must now be considered analytic in this simply-connected region.

A power series representation of $\phi(\zeta)$ is at best awkward in the present circumstance. The point $\zeta=0$ is not even in the region. Alternative points of expansion within the region, e.g., $\zeta=1$, are not promising due to the shape of the region. Such expansions, however, could be expected to provide reasonable asymptotic results with the MMC arguments. This was verified numerically. It was found, however, that such expansions had difficulty in matching boundary conditions along $\overline{AB}$ ($\overline{ED}$).

An alternative plan which proved to be very effective is indicated in Figure 1.9. The argument is essentially that used in section 5. First, $\phi(z)$ is

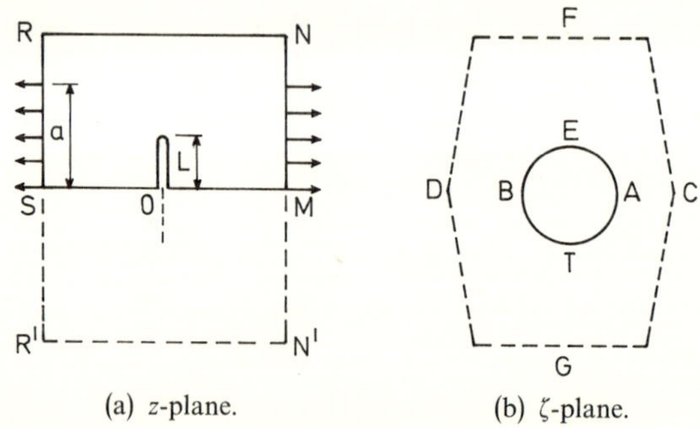


(a) $z$-plane. (b) $\zeta$-plane.

Figure 1.9. Edge crack in rectangular panel with orientation II.

extended across the real axis by equation (1.13). Then, if $\phi(z)$ is analytic within the rectangle RNN′R′ except for the crack and its reflection, traction-free conditions are guaranteed on the segments $\overline{OS}$ and $\overline{OM}$. For parametric representation of this region, the mapping function (1.100) is again introduced, i.e.,

$$z=\omega(\zeta)=(L/2)(\zeta-\zeta^{-1}).$$

The problem then becomes one of finding $\phi(\zeta)$ defined in the annular region illustrated in Figure 1.9b. Boundary conditions on $\widehat{AEB}$ (corresponding to the crack) and on CFD (corresponding to the specified traction on MNRS) can be imposed by the modified collocation process.

The arguments in section 1.5 relative to singularities on the unit circle are again pertinent and it is again appropriate to introduce a function $F(\zeta)$ defined a manner comparable to equation (1.103). It is now reasonable to assume a Laurent expansion for $F(\zeta)$, i.e.,

$$F(\zeta) = \sum_{-\infty}^{\infty} a_n \zeta^n . \tag{1.125}$$

Since boundary collocation arguments are used to determine the $a_n$'s, it was found advisable to impose the force conditions at the crack tip as an explicit relation between the coefficients for a truncated form of equation (1.125). Accuracy at the crack tip is thus directly weighted which is highly desirable, considering that boundary conditions along the crack are enforced by collocation in this formulation.

The above plan proved to be very effective. Freese [49] carried out the numerical solution shown in Table III. The quantity $H$ where

$$H = Wk_1/P(L^{\frac{1}{2}}) \tag{1.126}$$

was computed for various $a/W$ and $L/W$ ratios for a square panel $W/h = 1.0$. For $a/W = 1$, the results compare within one percent with Bowie and Neal [33]. As $a/w \to 0$, the values of $H$ can be obtained by extrapolation to within an estimated accuracy of one percent. The agreement with the corresponding data of Srawley and Gross [50] is to within two percent. The apparent growth in discrepancy for shorter and deeper crack lengths is consistent with their stated limitations of their stress function representation.

*Oblique edge crack in a rectangular panel.* Consider now the case of an oblique edge crack in a rectangular panel as illustrated in Figure 1.10.

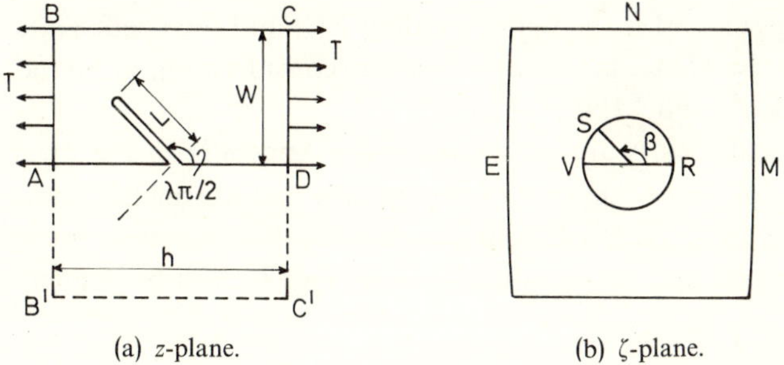


(a) $z$-plane. (b) $\zeta$-plane.

Figure 1.10. Oblique edge crack in a rectangular panel under uniform tension, $T$.

Again, $\phi(z)$ will be considered extended across the real axis by equation (1.13). Then, if $\phi(z)$ is analytic within the rectangle B′BCC′ except for the crack and its reflection, traction-free conditions are guaranteed on the segments $\overline{\text{OA}}$ and $\overline{\text{OD}}$.

This problem was chosen to illustrate the flexibility in the MMC method of the choice of mapping. It will be recalled that a criterion for the choice was the relative simplicity by which the inverse can be found. In the present case, it is desired to find a mapping of the unit circle and its exterior into a bent crack (the inclined crack and its reflection) and its exterior. Such a mapping function is a special case of Andersson's class of mappings [27] and can be written as

$$z = \omega(\zeta) = A\zeta^{-1}(\zeta - 1)^{\lambda}(\zeta + 1)^{2-\lambda}. \tag{1.127}$$

The interval $\widehat{\mathrm{RSV}}$ of the unit circle maps into the physical crack inclined by the angle $\lambda\pi/2$ to the $x$-axis. The point S, corresponding to $\zeta = \mathrm{e}^{\mathrm{i}\beta}$, corresponds to the crack tip and can be calculated from

$$\beta = \mathrm{Arctan}\left\{\frac{(2\lambda - \lambda^2)^{\frac{1}{2}}}{1-\lambda}\right\}. \tag{1.128}$$

The constant $A$ in equation (1.127) is chosen so that

$$|\omega(\mathrm{e}^{\mathrm{i}\beta})| = L. \tag{1.129}$$

The mapping function (1.127) can be inverted by any of several iterative schemes to find the corresponding boundary in the parameter region.

The solution of this problem was carried out by Freese [49]. Again the anticipation of singularities on the unit circle must be considered. A Laurent series is assumed for the corresponding $F(\zeta)$ and the coefficients are determined by the MMC arguments from the consideration of boundary conditions on $\widehat{\mathrm{RSV}}$ and $\widehat{\mathrm{MNE}}$.

Numerical results are presented in the Appendix in the form of curves (Figure 1.16). Since the solution is generally non-symmetric about the crack tip, there will be both a $k_1$ and $k_2$ component of the stress intensity factor. In an obvious manner the quantities $H_1$ and $H_2$ will be defined as

$$\begin{aligned} H_1 &= k_1/T(L^{\frac{1}{2}}) \\ H_2 &= k_2/T(L^{\frac{1}{2}}). \end{aligned} \tag{1.130}$$

Curves of $H_1$ and $H_2$ are shown as a function of $\lambda$ for various crack depths $L/W$ for a rectangle with $h/W = 2.0$. When $\lambda = 1$, obviously $H_2 = 0$ and the value of $H_1$ correspond to the normal edge crack solution previously referenced.

*Mixed boundary value problem for a central crack in a rectangular panel.* As an illustration of the MMC method for mixed boundary value problems, consider the problem of a central crack in a rectangular panel with constrained ends (Figure 1.11). This problem was solved recently by Neal [51] to determine the effects of rigidity of the end constraints on the crack tip stress intensity factors.

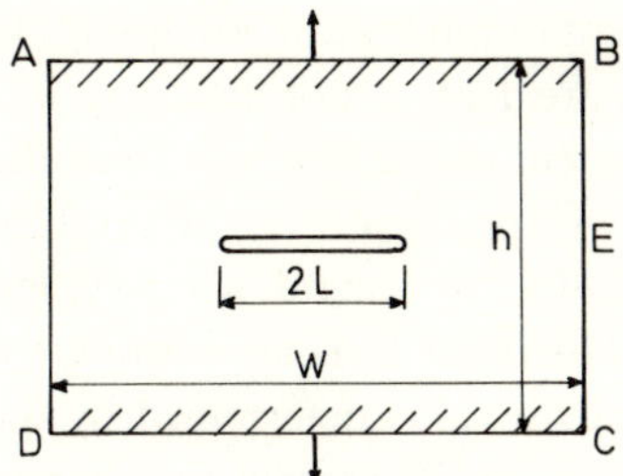


Figure 1.11. Central crack in rectangular panel with constrained ends.

Two extreme types of boundary conditions were considered which will be denoted by case 1 and case 2.

Case 1.

$$\begin{aligned} &u = 0\,, \quad v = v_0 \quad \text{(constant) on } \overline{\text{AB}} \\ &u = 0\,, \quad v = -v_0 \ \text{(constant) on } \overline{\text{CD}} \end{aligned} \tag{1.131}$$

Case 2.

$$\begin{aligned} &\tau_{xy} = 0\,, \quad v = v_0 \quad \text{(constant) on } \overline{\text{AB}} \\ &\tau_{xy} = 0\,, \quad v = -v_0 \ \text{(constant) on } \overline{\text{CD}} \end{aligned} \tag{1.132}$$

where $u$ and $v$ are the horizontal and vertical displacements, respectively. In both cases the crack and segments AD and BC are considered traction-free.

The mapping and extension phase of the analysis was handled in the same manner as described in section 1.6. Along $\overline{\text{BC}}$ and $\overline{\text{AD}}$, the force and stress conditions in the MMC plan are prescribed. For consistency on the ends $\overline{\text{AB}}$ and $\overline{\text{CD}}$, to the conditions (1.131) and (1.132) are added the additional (redundant) conditions.

$$\text{Case 1. } \partial u/\partial x = \partial v/\partial x = 0 \text{ on } \overline{\text{AB}}, \overline{\text{CD}} \tag{1.133}$$

$$\text{Case 2. } f_1 = 0\,, \quad \partial v/\partial x = 0 \text{ on } \overline{\text{AB}}, \overline{\text{CD}}\,. \tag{1.134}$$

In multiply connected regions there is an interesting difficulty associated with constants of integration in $f_1 + \mathrm{i}f_2$. In the present problem, for example, when the extension plan of section 1.6 is chosen for a traction-free condition on the crack, the constant of integration for $f_1 + \mathrm{i}f_2$ has been chosen as zero. Thus, $f_1 + \mathrm{i}f_2$ at E (Figure 1.11) is not arbitrary, in fact, it is the resultant force acting on the segment of the real axis from the crack tip $z = L$ to the point E. In the case of symmetry, if $F$ is the total force acting on AB, then obviously $f_1 + \mathrm{i}f_2 = F/2$ at E. In general, when symmetry is not present, it is necessary to consider $f_1 + \mathrm{i}f_2$ at E as an unknown constant of integration and treat it as an unknown value determined by the solution.

It was found convenient in the present problem to introduce an average tensile stress, $T_A$, acting on $\overline{\text{AB}}$, where

$$T_A = \frac{1}{W} \int_{\text{A}}^{\text{B}} \sigma_y \mathrm{d}x\,. \tag{1.135}$$

Then, obviously

$$f_1 + \mathrm{i}f_2 = WT_A/2 \text{ at E}\,. \tag{1.136}$$

Instead of choosing $v_0$ and determining $T_A$, it is convenient to fix $T_A$ and treat $v_0$ as an unknown constant in the calculations.

Numerical results for a square specimen, $h/W = 1.0$ are presented in the Appendix. In the calculations, $\kappa$ was chosen as $\kappa = 2.2$ (see Equation (1.4)). It is clear from a comparison of $H^{(1)}$ and $H^{(2)}$ that the effect of the manner of end constraint is relatively insignificant. It is interesting to compare the results with $H^{(3)}$ and $H^{(4)}$, the tension solutions for rectangular panels with central cracks with $h/W \geqslant 3.0$ and $h/W = 1.0$, respectively. Evidently the end constraints have altered the tension solution for the short specimen in such a manner that the solution is equivalent to the tension of a long strip with a central crack.

*Radial cracks emanating from a circular cut-out in a rectangular panel.* Now consider the solution for radial cracks emanating from a circular cut-out in a rectangular panel (Figure 1.12). Conceptually this problem is similar to the radial crack in a circular ring carried out by Bowie and Freese [44]. The present problem will be approached from a somewhat different point of view by considering the orthotropic solution for Figure 1.12 and then demonstrating that the isotropic solution can be found numerically by approaching the appropriate limiting values of the orthotropic constants.

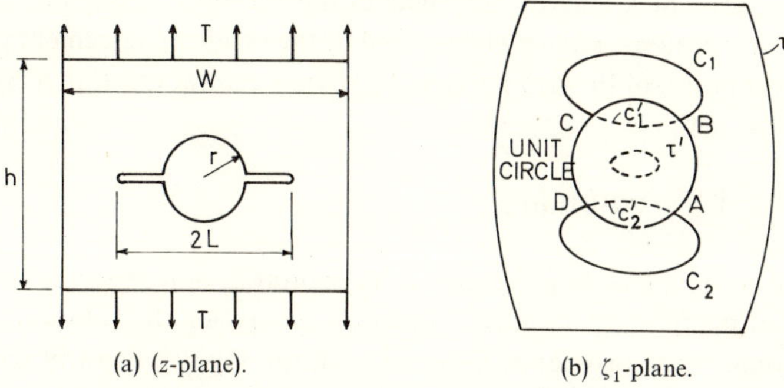


(a) ($z$-plane). (b) $\zeta_1$-plane.

Figure 1.12. Radial cracks emanating from a circular cut-out in a rectangular panel.

The mapping function (1.116) will be adopted, i.e.,

$$z_1 = \omega(\zeta_1) = (L/2)(\zeta_1 + \zeta_1^{-1})$$
$$z_2 = \omega(\zeta_2) = (L/2)(\zeta_2 + \zeta_2^{-1}) .$$

Thus, in the $\zeta_1$-plane the arcs $\widehat{AB}$ and $\widehat{CD}$ of the unit circle map into the two intervals in the $z_1$-plane corresponding to the crack segments (Figure 1.12b). Similarly, the contours $C_1$, $C_2$ and $\tau$ map into the contours in the $z_1$-plane corresponding to the upper and lower halves of the circular cut-out and the rectangular boundary, respectively. These contours can be calculated by using equations (1.117). A similar consideration holds for the second auxiliary plane, the $\zeta_2$-plane.

Traction-free conditions on the crack segments can again be ensured by applying the continuation arguments in equations (1.48) and (1.61)–(1.65). The reflected boundaries are indicated by primes in Figure 1.12b. It is clear that both parameter regions will have a connectivity of four. Thus, $F(\zeta)$ was chosen as

$$F(\zeta) = \sum_{-\infty}^{\infty} a_n \zeta^{2n+1} + \sum_{n=1}^{\infty} b_n \zeta (\zeta^2 + 1)^{-n} . \tag{1.137}$$

Freese [49] carried out this solution for $h/W=2.0$ and $2r/W=0.25$ for several values of $2L/W$ and the results are shown in the Appendix. In these calculations, the isotropic solution was found by setting the orthotropic constants $s_1=1.0i$, $s_2=i$ and $s_1=0.99i$ and $s_2=i$. The numerical results coincide to three significant figures in both cases.

By coincidence, Newman [46] solved this problem by least squares and the force conditions. The agreement is very good as can be seen from Table V. For very deep crack lengths, the effect of the cut-out is negligible with our choice of dimensions. This can be verified by the excellent agreement with the previous solution of Bowie and Neal [43] for a central crack in a rectangle with $h/W=2.00$.

## 1.8 Summary

Outlined in the preceding sections is an evolution of techniques involving conformal mapping and analytic function theory for the solution of two-dimensional crack problems. Utilization of the rapid growth in computer technology is reflected in the nature of the recent approaches, yet, a considerable emphasis on basic analysis is retained. Such a compromise enjoys several redeeming features as compared with strictly numerical procedures. The sterility of routine arithmetic approaches is avoided and much of the interesting beauty of the continuum theory is preserved for the analyst. Furthermore, the analyst is encouraged to develop improvement in analytical methods. In turn, the analyst is rewarded by a deeper insight into the evaluation of solution accuracy, particularly for problems which are locally characterized by mathematical singularities. At the same time, with a compromise on the requirement of strict mathematical rigor, flexible procedures can be adopted which utilize current computer capability, thereby allowing for an educated numerical solution for most problems.

In the author's opinion, the MMC method at the present time appears to be a very promising approach to the numerical solution of most two-dimensional problems of elasticity. An interesting balance of analysis and computer applications is basic to the approach. In addition to the crack problems discussed in this chapter, other notch and cut-out configurations in finite geometries have been considered. A variety of interesting analytical difficulties challenge the analyst, yet, when resolved lead to numerically accurate solutions in both isotropic and orthotropic studies. Several of the latter solutions are currently being used in conjunction with parallel experimental investigations to determine the extent of correlation of homogeneous orthotropic theory with the behavior of composite materials.

Several improvements are needed immediately in stress function representation. For example, a power series representation is woefully inadequate to handle certain geometrical regions. Certainly, also, the mathematical formalization of the conditions for an adequate asymptotic representation of the desired solution implied by the MMC arguments would be of considerable practical value. The extension of the modified collocation arguments to three-dimensional problems would also appear to be a challenging and promising direction for future work. A combination of modified collocation and finite element procedures offers still another potentially effective direction. In general, there is an apparent need for a new set of formalized mathematical theorems with an eye set on the auxiliary application of computer capability of the present and the future.

## 1.9 Appendix

(1) *Solution for periodic edge cracks in a semi-infinite sheet under tension.*

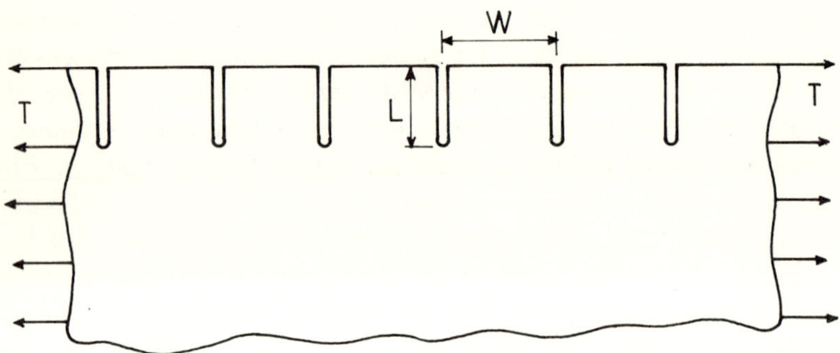


Figure 1.13. Periodic edge cracks in semi-infinite sheet.

TABLE I

*Stress intensity factors, $k_1$, for periodic edge cracks in a semi-infinite tensile strip*

| $L$ | $L/W$ | $k_1/T(L)^{\frac{1}{2}}$ | $L$ | $L/W$ | $k_1/T(L)^{\frac{1}{2}}$ |
|---|---|---|---|---|---|
| 0.00 | 0.000 | 1.12 | 4.36 | 0.693 | 0.48 |
| 0.32 | 0.051 | 1.09 | 4.88 | 0.776 | 0.45 |
| 0.65 | 0.104 | 1.03 | 5.58 | 0.887 | 0.42 |
| 0.96 | 0.153 | 0.95 | 5.99 | 0.952 | 0.41 |
| 1.23 | 0.195 | 0.88 | 6.68 | 1.063 | 0.39 |
| 1.49 | 0.237 | 0.81 | 8.99 | 1.430 | 0.33 |
| 1.76 | 0.281 | 0.75 | 11.29 | 1.797 | 0.30 |
| 2.06 | 0.328 | 0.69 | 13.59 | 2.163 | 0.27 |
| 2.42 | 0.385 | 0.64 | 15.89 | 2.530 | 0.25 |
| 2.89 | 0.459 | 0.58 | 18.20 | 2.896 | 0.23 |
| 3.20 | 0.509 | 0.55 | 20.50 | 3.263 | 0.22 |
| 3.64 | 0.578 | 0.52 | | | |

(2) *Unconventional application of mapping to an edge crack in a semi-infinite region.*

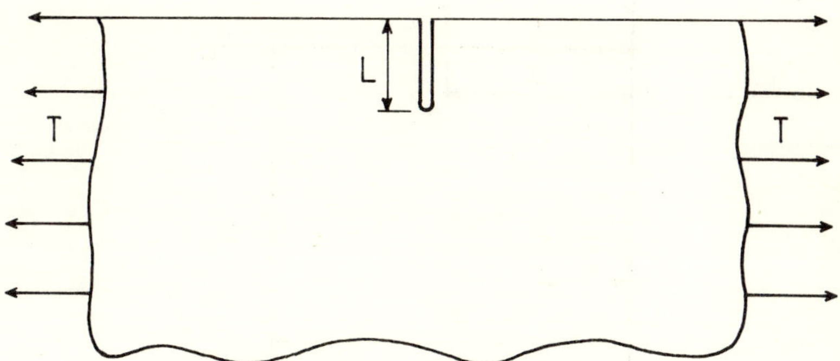


Figure 1.14. Edge crack in semi-infinite region.

TABLE II

*Stress function coefficients for single-edge crack in semi-infinite tensile sheet*

| $K$ | $S_{2K}$ | $T_{2K+1}$ | $K$ | $S_{2K}$ | $T_{2K+1}$ |
|---|---|---|---|---|---|
| −1 | −2.51562 | 0.94674 | 13 | −0.00176 | −0.00031 |
| 0 | 0.94401 | −0.12669 | 14 | −0.00142 | −0.00031 |
| 1 | 0.10043 | 0.08003 | 15 | −0.00114 | −0.00036 |
| 2 | −0.00379 | 0.04932 | 16 | −0.00093 | −0.00032 |
| 3 | −0.01877 | 0.02674 | 17 | −0.00076 | −0.00035 |
| 4 | −0.01763 | 0.01452 | 18 | −0.00062 | −0.00029 |
| 5 | −0.01395 | 0.00798 | 19 | −0.00051 | −0.00032 |
| 6 | −0.01064 | 0.00442 | 20 | −0.00042 | −0.00026 |
| 7 | −0.00801 | 0.00239 | 21 | −0.00036 | −0.00029 |
| 8 | −0.00611 | 0.00125 | 22 | −0.00029 | −0.00022 |
| 9 | −0.00465 | 0.00055 | 23 | −0.00025 | −0.00025 |
| 10 | −0.00362 | 0.00017 | 24 | −0.00020 | −0.00018 |
| 11 | −0.00281 | −0.00009 | 25 | −0.00019 | −0.00023 |
| 12 | −0.00223 | −0.00020 | 26 | −0.00014 | −0.00015 |

(3) *Normal edge crack in a rectangular panel.*

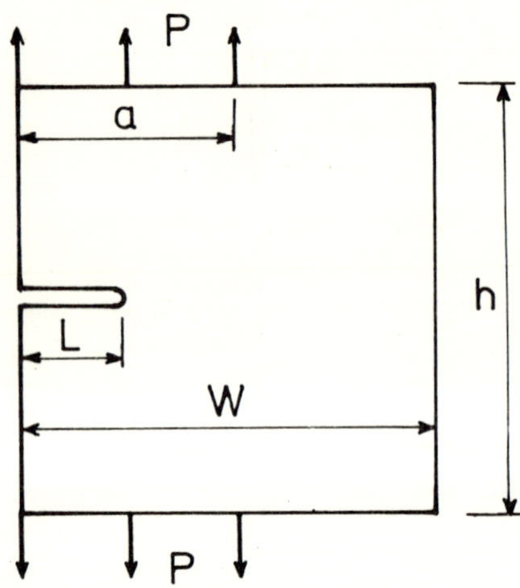


Figure 1.15. Loading of rectangular panel with edge crack.

TABLE III

*Tabular values of* $H = Wk_1/P\,(L)^{\frac{1}{2}}$ *for* $W/h = 1.0$

| $a/W$ \ $L/W$ | 0.1 | 0.2 | 0.3 | 0.4 | 0.5 | 0.6 | 0.7 | 0.8 |
|---|---|---|---|---|---|---|---|---|
| 1.0 | 1.23 | 1.49 | 1.85 | 2.32 | 3.01 | 4.15 | 6.40 | 12.0 |
| 0.9 | 1.43 | 1.72 | 2.13 | 2.67 | 3.45 | 4.73 | 7.22 | 13.4 |
| 0.8 | 1.67 | 1.99 | 2.45 | 3.05 | 3.91 | 5.31 | 8.05 | 14.8 |
| 0.7 | 1.95 | 2.31 | 2.82 | 3.48 | 4.40 | 5.92 | 8.88 | 16.2 |
| 0.6 | 2.31 | 2.70 | 3.25 | 3.95 | 4.93 | 6.54 | 9.71 | 17.6 |
| 0.5 | 2.78 | 3.19 | 3.76 | 4.48 | 5.48 | 7.17 | 10.50 | 19.0 |
| 0.4 | 3.38 | 3.76 | 4.32 | 5.03 | 6.03 | 7.78 | 11.40 | 20.4 |
| 0.3 | 4.09 | 4.43 | 4.92 | 5.57 | 6.57 | 8.39 | 12.20 | 21.8 |
| 0.2 | 4.88 | 5.16 | 5.54 | 6.13 | 7.12 | 9.02 | 13.00 | 23.2 |
| 0.1 | 5.64 | 5.88 | 6.16 | 6.68 | 7.67 | 9.65 | 13.90 | 24.7 |
| 0.0[(1)] | 6.40 | 6.60 | 6.78 | 7.22 | 8.22 | 10.30 | 14.80 | 26.1 |
| 0.0[(2)] | – | 6.70 | 6.68 | 7.14 | 8.14 | 10.10 | 14.60 | – |

[(1)] Obtained by extrapolation.
[(2)] Srawley and Gross [50].

(4) *Oblique edge crack in rectangular panel.*

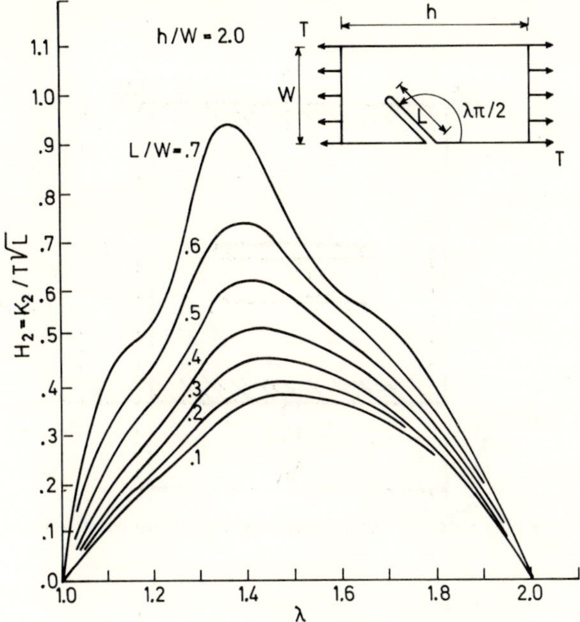


Figure 1.16a. $H_1$ as a function of $\lambda$ for $h/W=2.0$.

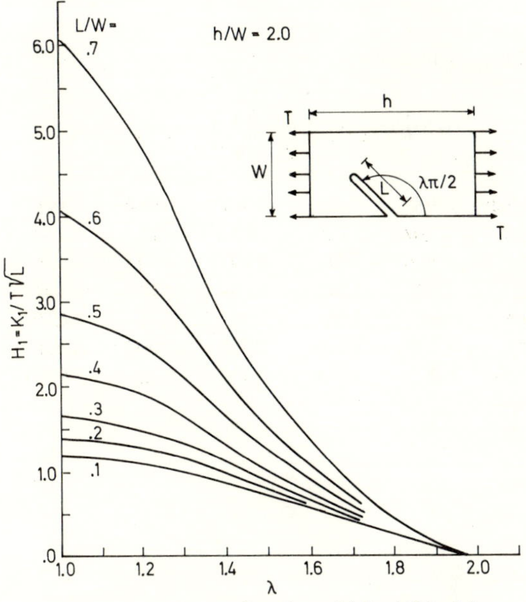


Figure 1.16b. $H_2$ as a function of $\lambda$ for $h/W=2.0$.

(5) *Mixed boundary value problem for a central crack in a rectangular sheet.*

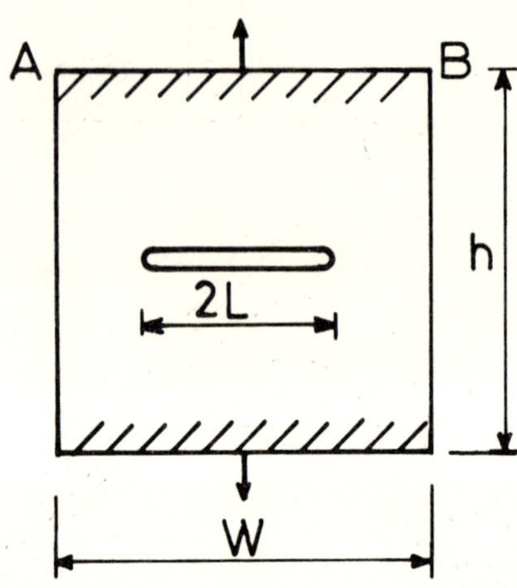


Figure 1.17. Central crack in rectangular panel with constrained ends.

TABLE IV

*Tabular values of H for h/W = 1.0*

$$H = k_1/T_A(L^{\frac{1}{2}}), \quad T_A = \frac{1}{W}\int_A^B \sigma_y \mathrm{d}x, \quad \kappa = 2.2$$

| $2L/W$ | 0.040 | 0.100 | 0.200 | 0.400 | 0.500 | 0.667 | 0.741 | 0.800 |
|---|---|---|---|---|---|---|---|---|
| $H^{(1)}$ | 1.03 | 1.04 | 1.05 | 1.10 | 1.17 | 1.39 | 1.57 | 1.78 |
| $H^{(2)}$ | 1.00 | 1.00 | 1.02 | 1.10 | 1.17 | 1.40 | 1.59 | 1.80 |
| $H^{(3)}$ | 1.00 | 1.01 | 1.02 | 1.11 | 1.19 | 1.41 | 1.59 | 1.81 |
| $H^{(4)}$ | 1.00 | 1.01 | 1.06 | 1.22 | 1.33 | 1.60 | 1.78 | 2.00 |
| $\mu V_0^{(5)}/T_A$ | 0.198 | 0.200 | 0.210 | 0.253 | 0.290 | 0.389 | 0.460 | 0.540 |
| $\mu V_0^{(6)}/T_A$ | 0.200 | 0.203 | 0.213 | 0.255 | 0.291 | 0.391 | 0.464 | 0.545 |

(1) Case 1
(2) Case 2
(3) [43], $h/W \geqslant 3.00$
(4) [43], $h/W = 1.00$
(5) Case 1
(6) Case 2

(6) *Radial cracks emanating from a circular cut-out in a rectangular panel.*

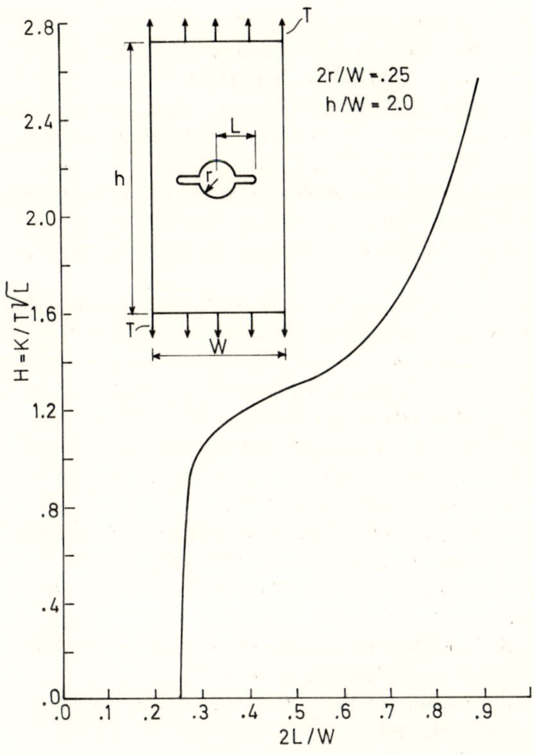


Figure 1.18. The function $H = K/T(L^{\frac{1}{2}})$ for $h/W = 2.00$, $2r/W = 0.25$.

TABLE V

*Tabular values for H for h/W=2.00 and 2r/W=0.25*

| $2L/W$ | 0.25 | 0.27 | 0.30 | 0.35 | 0.40 | 0.50 | 0.60 | 0.70 | 0.80 | 0.90 |
|---|---|---|---|---|---|---|---|---|---|---|
| MMC | 0.0 | 0.85 | 1.08 | 1.18 | 1.22 | 1.28 | 1.40 | 1.57 | 1.89 | 2.58 |
| Newman | 0.0 | 0.85 | 1.08 | 1.18 | 1.22 | 1.28 | 1.40 | 1.57 | 1.90 | 2.62 |

## References


[1] Inglis, C. E., *Trans. Roy. Inst. Naval Architecks*, 60, p. 219 (1913).
[2] Kolosoff, G., Doctoral dissertation, Dorpat (1909).
[3] Irwin, G. R., *J. Appl. Mech.*, 24, p. 361 (1957).
[4] Griffith, A. A., Phil. Trans. Roy. Soc. London, Series A221, p. 163 (1921).
[5] Bowie, O. L., *J. Math. and Phys.*, 35, p. 60 (1956).
[6] Bowie, O. L. and D. M. Neal, *Intl. J. Fracture Mech.*, 6, p. 199 (1970a).
[7] Bowie, O. L. and C. E. Freese, Army Symp. on Solid Mech., AMMRC, Watertown, Mass. (To appear in *Intl. J. Frac. Mech.*) (1970a).
[8] Muskhelishvili, N. I., *Some Basic Problems of Mathematical Theory of Elasticity*, Noordhoff, Groningen, Holland (1953).
[9] Leknitskii, S. G., *Theory of Elasticity of an Anisotropic Elastic Body*, Holden-Day, San Francisco (1963).
[10] Timoshenko, S. and J. N. Goodier, *Theory of Elasticity*, McGraw-Hill, New York (1951).
[11] Savin, G. N., *Stress Concentration Around Holes*, Pergamon, New York (1961).
[12] Timoshenko, S., *Theory of Elasticity*, McGraw-Hill, New York (1934).
[13] Kartzivadze, I. N., *Comp. rend. de l'acad. sc. de l'U.R.S.S.*, 20, p. 95 (1943).
[14] Kolosoff, G., *Z. Math. Physik*, 62 (1914).
[15] Muskhelishvili, N. I., Izv. A. N. SSSR, p. 663 (1919).
[16] Griffith, A. A., *Tech. Rpt. Aeronaut. Res. Comm.* (Great Britain), 2, p. 668 (1927).
[17] Neuber, H., *Ingenieur-Archiv*, 5, p. 242 (1934).
[18] Sneddon, I. N. and H. A. Elliott, *Quart. Appl. Math.*, 4, p. 262 (1946).
[19] Wilmore, T. J., *Quart. J. Mech. and Appl. Math.*, 2, p. 53 (1949).
[20] Sneddon, I. N., *Crack Problems in the Mathematical Theory of Elasticity*. North Carolina State College, Raleigh, N.C. (1961).
[21] Sneddon, I. N., Proc. Roy. Soc. (London) Series A187, p. 229 (1946).
[22] Williams, M. L., *J. Appl. Mech.*, 24, p. 109 (1957).
[23] Irwin, G. R., in: *Structure Mechanics*; *Proc. 1st. Symposium on Naval Structure Mechanics*, Pergamon, New York (1960).
[24] Sih, G. C., P. C. Paris and G. R. Irwin, *Int. J. Fracture Mech.*, 1, p. 189 (1965).
[25] Sih, G. C., P. C. Paris and F. Erdogan, *J. Appl. Mech.*, 29, p. 306 (1962).
[26] Bowie, O. L., *J. Appl. Mech.*, 31, p. 208 (1964a).
[27] Andersson, H., *J. Mech. Phys. Solids*, 17, p. 405 (1969).
[28] Morkovin, V., *Quart. Appl. Math.*, 2, p. 350 (1945).
[29] Bowie, O. L. and C. E. Freese, Unpublished data at AMMRC, (1970b).
[30] Koiter, W. T., *J. Appl. Mech.*, 32, p. 237 (1965).
[31] Bowie, O. L., *J. Appl. Mech.*, 31, p. 726 (1964b).
[32] Bloom, J. M., *Int. J. Fract. Mech.*, Vol. 2, p. 597.
[33] Bowie, O. L. and D. M. Neal, *J. Appl. Mech.*, 32, p. 708, (1965).
[34] Gross, B., J. E. Srawley and W. F. Brown, NASA TND-2395, Lewis Research Center, Cleveland, Ohio (1964).
[35] Akao, H. T., A. S. Kobayashi, *J. of Basic Engineering*, 89 (1967).
[36] Brown, R. C. and A. S. Kobayashi, *The Trend*. University of Washington, 19, p. 8 (1967).
[37] Rich, T. P. and R. Roberts, *J. of Appl. Mech.*, 34, p. 777 (1967).

[38] Wieselmann, P. A., Doctoral Dissertation, Mass. Inst. of Tech., Cambridge, Mass. (1969).
[39] Bowie, O. L., *J. Math. and Phys.*, 45, p. 356 (1966).
[40] Bowie, O. L. and J. J. McLaughlin, AMRA TR66-16, Army Materials Research Agency, Watertown, Mass. (1966).
[41] Nehari, F., *Conformal Mapping*, McGraw-Hill, New York (1952).
[42] Kobayashi, A. S., Document No. D2-23551 Boeing Co., Seattle, Washington (1964).
[43] Bowie, O. L. and D. M. Neal, *Eng. Fracture Mech.*, 2, p. 181 (1970b).
[44] Bowie, O. L. and C. E. Freese, Fourth National Symposium on Fracture Mechanics, Carnegie-Mellon, Pittsburgh, Pa., (To appear in J. Eng. Fract. Mech.) 1970c.
[45] Newman, J. C., Jr., Master's Thesis, Virginia Poly. Inst., Blacksburg, Va. (1969).
[46] Newman, J. C., Jr., NASA TN. D-6376, Langley Research Center, Hampton, Va. (1971).
[47] Sih, G. C. and H. Liebowitz, In *Fracture* (ed. H. Liebowitz) Academic Press, New York (1968).
[48] Gandhi, K., *J. of Strain Analysis*, (To be published (1970)).
[49] Freese, C. E., Private communication (1971).
[50] Srawley, J. E. and B. Gross, NASA TN-D3820 Lewis Research Center, Cleveland, Ohio (1967).

*M. Isida*

# 2 Method of Laurent series expansion for internal crack problems

## 2.1 Introduction

This paper presents a general method of analysis of internal cracks in isotropic homogeneous elastic media. The proposed method is based on the Laurent series expansions of the complex potentials which are consistent with the singlevaluedness of displacements as well as stresses and strains. It applies to the problems of longitudinal shear, plane stress or plane strain and classical plate bending in which the stress state is completely characterized by one or two complex potentials.

The common procedure of the analysis is as follows:

(1) Assume the Laurent series expansions of the complex potentials, and determine the relations among their coefficients from traction-free conditions of the edge of an elliptical hole or a crack. Each of the coefficients of negative powers are expressed as linear combinations of positive powers. These relations can be applied to many problems involving a variety of shape and loading parameters.

(2) Additional relations among the coefficients of the assumed Laurent series are developed for other boundary conditions. In contrast to the above free hole relations, the coefficients of positive powers are expressed in terms of negative powers.

(3) The above simultaneous relations are then treated by a perturbation procedure, where all the unknowns are assumed as power series of a relative crack length $\lambda$ and their coefficients are determined successively by alternative use of the above relations.

(4) The stresses, stress intensity factors and other physical quantities are given by double infinite series, corresponding to the assumed Laurent expansions and the power series introduced in the perturbation procedure,

but the former series can be summed up in closed forms and the results are obtained as power series of $\lambda$ whose coefficients are evaluated exactly.

The method is applied to longitudinal shear, plane extension and classical plate bending problems. The analyses are developed for general elliptical holes which include the crack configuration as a special case.

Numerical results of the stress intensity factors for many crack problems are given in the Appendices.

## 2.2 Longitudinal shear

*Basic formulae.* In the state of longitudinal shear, components of displacement and stress are given in terms of the only complex potential $\phi(z)$, as follows:

$$w = \frac{\tau d}{G} \operatorname{Re}\{\phi(z)\}$$
$$\tau_{xt} - \mathrm{i}\,\tau_{yt} = \tau\phi'(z) \tag{2.1}$$

where $\tau$ and $d$ are some reference stress and length, and $z$ is the dimensionless complex variable defined by

$$z = \frac{X + \mathrm{i}\,Y}{d}. \tag{2.2}$$

Stress components $\tau_{nt}$ and $\tau_{st}$ with respect to new coordinate axes $n$, $s$

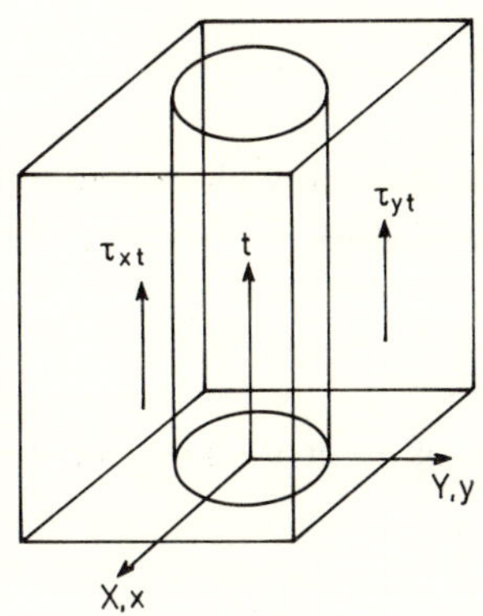


Figure 2.1. Longitudinal shear problem.

generated by rotating the axes $X$, $Y$ are related with $\tau_{xt}$ and $\tau_{yt}$ by the formula

$$\tau_{nt} - \mathrm{i}\,\tau_{st} = \mathrm{e}^{\mathrm{i}\alpha}(\tau_{xt} - \mathrm{i}\,\tau_{yt}) = \tau\,\mathrm{e}^{\mathrm{i}\alpha}\phi'(z) \tag{2.3}$$

where $\alpha$ is the rotation angle.

Now consider a longitudinal hole of an arbitrary profile and derive the necessary conditions to be satisfied by $\phi(z)$ in order that the hole boundary should be free from stress. For convenience, introduce the mapping function

$$z = \Omega(\zeta) \tag{2.4}$$

which conformally transforms the hole boundary and its external region into a unit circle and its external region as shown in Figure 2.2. Taking the outward

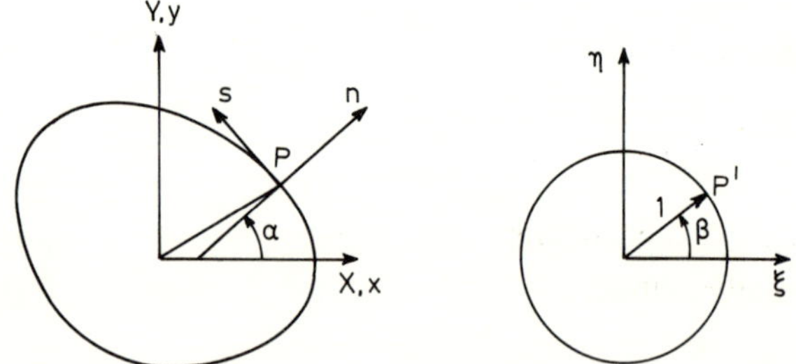


Figure 2.2. Conformal mapping of hole boundary into unit circle.

normal and tangent at a point P on the hole edge as axes $n$ and $s$, the following relation is obtained:

$$\mathrm{e}^{\mathrm{i}\alpha} = \left[\frac{\Omega'(\zeta)\zeta}{|\Omega'(\zeta)\zeta|}\right]_{\zeta = \mathrm{e}^{\mathrm{i}\beta}}. \tag{2.5}$$

The stress components at point P with respect to those new axes are given from equations (2.3) and (2.5) as

$$\tau_{nt} - \mathrm{i}\,\tau_{st} = \tau\left[\frac{\zeta\Phi'(\zeta)}{|\Omega'(\zeta)\zeta|}\right]_{\zeta = \mathrm{e}^{\mathrm{i}\beta}} \tag{2.6}$$

where $\beta$ is the argument of point P′, image of P on $\zeta$-plane, and

$$\phi(z) = \phi[\Omega(\zeta)] = \Phi(\zeta). \tag{2.7}$$

*General form of complex potential for free elliptical hole.* For a free hole $\tau_{nt}$ must vanish on the hole edge and hence from equation (2.6)

$$[\zeta \Phi'(\zeta)+\bar{\zeta}\bar{\Phi}'(\bar{\zeta})]_{\zeta=e^{i\beta}}=0\,. \tag{2.8}$$

In the present problem stresses and strains have no singularities within the domain. Therefore $\Phi'(\zeta)$ must be an analytic function from equation (2.6), and by means of the well known theorem it can be expressed as the following Laurent series of $\zeta$:

$$\Phi'(\zeta)=\alpha_0+\sum_{m=1}^{\infty}(\alpha_m\zeta^m+\beta_m\zeta^{-m})\,. \tag{2.9}$$

Substitute equation (2.9) into (2.8), replace $\bar{\zeta}$ by $\zeta^{-1}$, rearrange in ascending

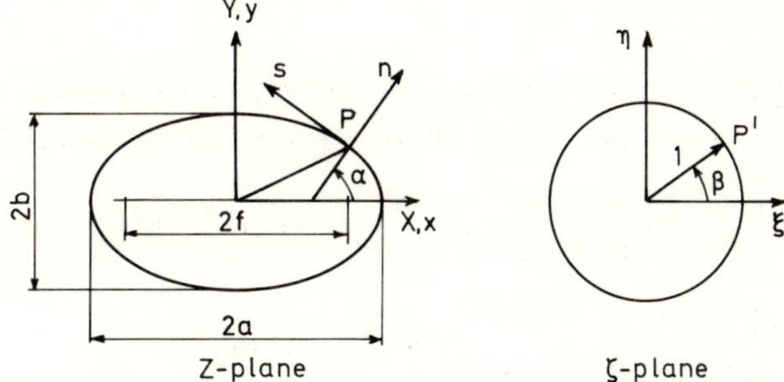

Figure 2.3. Conformal mapping of elliptical hole into unit circle.

power series of $\zeta$ and put the coefficients of all the powers to zero. It is found that

$$\mathrm{Re}[\beta_1]=0\,,\quad \beta_m=-\bar{\alpha}_{m-2}\qquad (m\geqslant 2)\,. \tag{2.10}$$

Integrating equation (2.9), with relations (2.10) and dropping the term in $\log \zeta$ which gives multivalued displacement, the following general form of the complex potential is found:

$$\Phi(\zeta)=\sum_{m=0}^{\infty}(\gamma_m\zeta^{m+1}+\bar{\gamma}_m\zeta^{-(m+1)}) \tag{2.11}$$

where new coefficients $\gamma_m$ are used in place of $\alpha_m/(m+1)$.

Now consider the special case of an elliptical hole having its center at the origin and principal diameters $2a$ and $2b$ along axes $X$ and $Y$. Define the following dimensionless parameters

$$\lambda = \frac{a}{d}, \quad c = \frac{f}{d} = \frac{(a^2-b^2)^{\frac{1}{2}}}{d}$$
$$\varepsilon = \frac{b}{a}, \quad R = \left(\frac{a+b}{a-b}\right)^{\frac{1}{2}} = \left(\frac{1+\varepsilon}{1-\varepsilon}\right)^{\frac{1}{2}} \tag{2.12}$$

and the mapping function (2.4) becomes

$$z = \Omega(\zeta) = \frac{c}{2}\left(R\zeta + \frac{1}{R\zeta}\right). \tag{2.13}$$

On inverting equation (2.13) the result is

$$R\zeta = \frac{z}{c}\left\{1 + \left[1 - \left(\frac{c}{z}\right)^2\right]^{\frac{1}{2}}\right\}$$
$$(R\zeta)^{-1} = \frac{z}{c}\left\{1 - \left[1 - \left(\frac{c}{z}\right)^2\right]^{\frac{1}{2}}\right\}. \tag{2.14}$$

Equation (2.11) is the general expression of $\Phi(\zeta)$ for arbitrary free holes, and can be written in terms of $z$. It can be shown from equation (2.13) that

$$z^{2p} = \left(\frac{c}{2}\right)^{2p}\left[\binom{2p}{p} + \sum_{m=1}^{p}\binom{2p}{p-m}\right]\{(R\zeta)^{2m} + (R\zeta)^{-2m}\}$$
$$z^{2p+1} = \left(\frac{c}{2}\right)^{2p+1}\sum_{m=0}^{p}\binom{2p+1}{p-m}\{(R\zeta)^{2m+1} + (R\zeta)^{-(2m+1)}\}. \tag{2.15}$$

Moreover, the coefficients $\gamma_m$ in equation (2.11) may be expressed in terms of the new parameters $M_p$:

$$\gamma_{2m} = R^{2m+1}\sum_{p=m}^{\infty}\left(\frac{c}{2}\right)^{2p+1}\binom{2p+1}{p-m}M_{2p}$$
$$\gamma_{2m+1} = R^{2m+2}\sum_{p=m}^{\infty}\left(\frac{c}{2}\right)^{2p+2}\binom{2p+2}{p-m}M_{2p+1}. \tag{2.16}$$

Putting (2.16) into equation (2.11) and dropping trivial constant terms lead to

$$\phi(z) = \Phi(\zeta) = \sum_{n=0}^{\infty} M_n z^{n+1} + \sum_{m=0}^{\infty}\gamma_m\zeta^{-(m+1)}$$
$$- \sum_{p=0}^{\infty}\left[M_{2p}\left(\frac{c}{2}\right)^{2p+1}\sum_{m=0}^{p}\binom{2p+1}{p-m}(R\zeta)^{-(2m+1)}\right.$$
$$\left. + M_{2p+1}\left(\frac{c}{2}\right)^{2p+2}\sum_{m=0}^{p}\binom{2p+2}{p-m}(R\zeta)^{-(2m+2)}\right]. \tag{2.17}$$

The negative power terms of $\zeta$ in the righthand-side can be replaced by those of $z$ by means of the following expansions:

$$(R\zeta)^{-2m} = \sum_{n=m}^{\infty} A_{n-m,\,2m}\left(\frac{c}{z}\right)^{2n}$$

$$(R\zeta)^{-(2m+1)} = \sum_{n=m}^{\infty} A_{n-m,\,2m+1}\left(\frac{c}{z}\right)^{2n+1} \tag{2.18}$$

which are derived from the second relation of equation (2.14), and the coefficients in the righthand-sides of equation (2.18) are given explicitly by [1, 2]

$$A_{n-m,\,2m} = \frac{m}{2^{2n}n}\binom{2n}{n-m}$$

$$A_{n-m,\,2m+1} = \frac{2m+1}{2^{2n+1}(2n+1)}\binom{2n+1}{n-m} \tag{2.19}$$

Substituting equation (2.18) into (2.17) and exchanging the order of summations, $\phi(z)$ for free elliptical holes takes the form

$$\phi(z) = \sum_{n=0}^{\infty} \{(F^{\cdot}_{n}+\mathrm{i}F'_{n})\,z^{-(n+1)}+(M^{\cdot}_{n}+\mathrm{i}M'_{n})\,z^{n+1}\} \tag{2.20}$$

dots and primes denoting real and imaginary parts respectively and the coefficients of negative powers of equation (2.20) can be expressed in terms of those of positive powers as follows:

$$F^{\cdot}_{2n} = -\sum_{p=0}^{\infty} \lambda^{2n+2p+2} P^{2n}_{2p} M^{\cdot}_{2p}, \qquad F^{\cdot}_{2n+1} = -\sum_{p=0}^{\infty} \lambda^{2n+2p+4} P^{2n+1}_{2p+1} M^{\cdot}_{2p+1}$$

$$F'_{2n} = -\sum_{p=0}^{\infty} \lambda^{2n+2p+2} Q^{2n}_{2p} M'_{2p}, \qquad F'_{2n+1} = -\sum_{p=0}^{\infty} \lambda^{2n+2p+4} Q^{2n+1}_{2p+1} M'_{2p+1} \tag{2.21}$$

where

$$\begin{Bmatrix} P^{2n}_{2p} \\ Q^{2n}_{2p} \end{Bmatrix} = \frac{(1-\varepsilon^2)^{n+p+1}}{2^{2p+1}} \sum_{m=0}^{n,p} \binom{2p+1}{p-m}(1\mp R^{4m+2})A_{n-m,\,2m+1}$$

$$\begin{Bmatrix} P^{2n+1}_{2p+1} \\ Q^{2n+1}_{2p+1} \end{Bmatrix} = \frac{(1-\varepsilon^2)^{n+p+2}}{2^{2p+2}} \sum_{m=0}^{n,p} \binom{2p+2}{p-m}(1\mp R^{4m+4})A_{n-m,\,2m+2} \tag{2.22}$$

and for the upper bounds of summations smaller values should be taken.

In the special case of a crack ($\varepsilon=0$, $R=1$), equations (2.22) become

$$\begin{aligned} P_{2p}^{2n} &= P_{2p+1}^{2n+1} = 0 \\ Q_{2p}^{2n} &= \frac{1}{2^{2p}} \sum_{m=0}^{n,p} \binom{2p+1}{p-m} A_{n-m,\,2m+1} \\ Q_{2p+1}^{2n+1} &= \frac{1}{2^{2p+1}} \sum_{m=0}^{n,p} \binom{2p+2}{p-m} A_{n-m,\,2m+1}. \end{aligned} \tag{2.23}$$

The values of $Q_{2p}^{2n}$, $Q_{2p+1}^{2n+1}$ ($p, n \leqslant 8$) are given in Tables I and II of Appendix I. For a circular hole the relations (2.21) simplify to

$$F_n^{\bullet} = \lambda^{2n+2} M_n^{\bullet}, \quad F_n' = -\lambda^{2n+2} M_n'. \tag{2.24}$$

In order to illustrate the procedure of getting closed form solutions from the Laurent expansions, consider a simple example of an infinite body with an elliptical hole subjected to $\tau_{yt}^{\infty}=\tau$ at infinity. From the symmetry of the problem, assume

$$\phi(z) = \mathrm{i} \sum_{n=0}^{\infty} (F_{2n}' z^{-(2n+1)} + M_{2n}' z^{2n+1}). \tag{2.25}$$

The boundary conditions at infinity yield at once

$$M_0' = -1, \quad M_{2n}' = 0 \qquad (n \geqslant 1) \tag{2.26}$$

and the free crack relations (2.21) become

$$F_{2n}' = -c^{2n+2} \frac{1+R^2}{2} A_{n,1} \cdot M_0'. \tag{2.27}$$

Thus the potential function is written as

$$\phi(z) = \mathrm{i} M_0' \left\{ z - \tfrac{1}{2}(1+R^2)\, c \sum_{n=0}^{\infty} A_{n,1} \left(\frac{c}{z}\right)^{2n+1} \right\} \tag{2.28}$$

and is summed up in a closed form

$$\phi(z) = \mathrm{i} M_0' [z - \tfrac{1}{2}(1+R^2)\{z - (z^2-c^2)^{\frac{1}{2}}\}] \tag{2.29}$$

by using the equations (2.18) and (2.14). Note that equation (2.29) can be also transformed into the expression in the elliptical coordinates $(\xi, \eta)$ using a mapping function

$$z = \frac{c}{2}(\mathrm{e}^{\xi+\mathrm{i}\eta} + \mathrm{e}^{-(\xi+\mathrm{i}\eta)}), \quad x = c\cosh\xi\cos\eta, \quad y = c\sinh\xi\sin\eta. \tag{2.30}$$

*Random elliptical holes and cracks in infinite body.* Consider the longitudinal shear of an infinite body containing random distribution of $N$ longitudinal elliptical holes, some of which may be circular holes or cracks as special cases. Let axes $X$, $Y$ be chosen as shown in Figure 2.4. Denote the polar

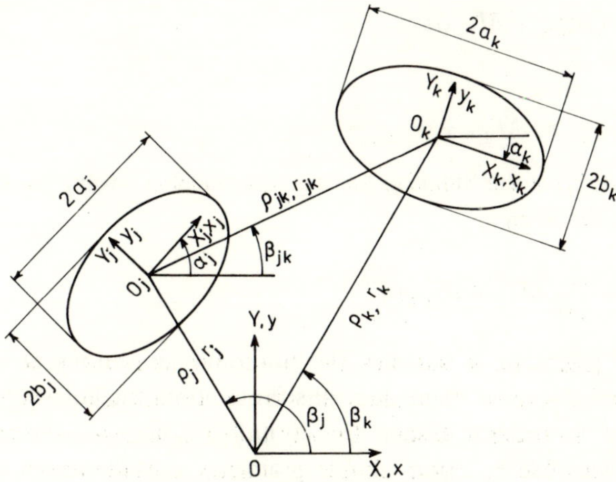

Figure 2.4. Random elliptical holes.

coordinates $(\rho_j, \beta_j)$ of the center $O_j$ of the $j$-th hole, its major and minor diameters $2a_j$, $2b_j$ and inclination angle $\alpha_j$ as in Figure 2.4. Take also $N$ coordinate systems $(X_j, Y_j)$ $(j=1, 2, \ldots, N)$ with their origins at $O_j$ and with $X_j$-axes along the major diameters of the corresponding holes.

Define dimensionless coordinates, complex variables and parameters by

$$x = \frac{X}{d}, \quad y = \frac{Y}{d}, \quad z = x + \mathrm{i}y$$

$$x_j = \frac{X_j}{d}, \quad y_j = \frac{Y_j}{d}, \quad z_j = x_j + \mathrm{i}y_j \tag{2.31}$$

$$\lambda_j = \frac{a_j}{d}, \quad r_j = \frac{\rho_j}{d}, \quad r_{jk} = \frac{\rho_{jk}}{d} \qquad (j, k = 1, 2, \ldots, N)$$

with a reference length $d$ which is arbitrarily specified, then the dimensionless radius vector between $O_j$ and $O_k$ is given by

$$r_{jk}\,\mathrm{e}^{\mathrm{i}\beta_{jk}} = r_k\,\mathrm{e}^{\mathrm{i}\beta_k} - r_j\,\mathrm{e}^{\mathrm{i}\beta_j}\,. \tag{2.32}$$

Now the total complex potential $\phi$ of the problem defined by equations (2.1) is assumed as the following sum of $(N+1)$ functions:

$$\phi = \phi_0(z) + \sum_{k=1}^{N} \phi_k(z_k) \tag{2.33}$$

where $\phi_0(z)$ corresponds to the applied stresses $\tau_{xt}^{\infty}$ and $\tau_{yt}^{\infty}$ at infinity and is given by

$$\phi_0(z) = (M^{\bullet}_{0,0} + \mathrm{i} M'_{0,0}) z$$

with (2.34)

$$M^{\bullet}_{0,0} = \tau_{xt}^{\infty}/\tau \, , \; M'_{0,0} = -\tau_{yt}^{\infty}/\tau \, .$$

Further, $\phi_k(z_k)$ is the function having singularities within the $k$-th hole and written in the form

$$\phi_k(z_k) = \sum_{n=0}^{\infty} (F^{\bullet}_{n,k} + \mathrm{i} F'_{n,k}) z_k^{-(n+1)} \, . \tag{2.35}$$

The total potential $\phi$ satisfies the boundary conditions at infinity since $\phi_k(z_k)$ give no stresses there, and thus those along the hole edges have to be considered. In order to discuss the $j$-th hole it is necessary to express $\phi$ with the only variable $z_j$. From simple geometry considerations of Figure 2.4 the relations

$$\begin{aligned} z &= z_j \mathrm{e}^{\mathrm{i}\alpha_j} + r_j \mathrm{e}^{\mathrm{i}\beta_j} \\ z_k &= z_j \mathrm{e}^{\mathrm{i}(\alpha_j - \alpha_k)} - r_{jk} \, \mathrm{e}^{\mathrm{i}(\beta_{jk} - \alpha_k)} \end{aligned} \tag{2.36}$$

are obtained. Substituting equations (2.36) into (2.34) and (2.35), the total potential is written in terms of $z_j$ as follows:

$$\phi = \sum_{n=0}^{\infty} \{ (F^{\bullet}_{n,j} + \mathrm{i} F'_{n,j}) z_j^{-(n+1)} + (M^{\bullet}_{n,j} + \mathrm{i} M'_{n,j}) z_j^{n+1} \} \tag{2.37}$$

where

$$\begin{aligned} M^{\bullet}_{n,j} &= (M^{\bullet}_{0,0} \cos \alpha_j - M'_{0,0} \sin \alpha_j) \Delta_n^0 + \sum_{p=0}^{\infty} \sum_{k \neq j}^{N} (e_{n,j}^{p,k} F^{\bullet}_{p,k} - f_{n,j}^{p,k} F'_{p,k}) \\ M'_{n,j} &= (M^{\bullet}_{0,0} \sin \alpha_j + M'_{0,0} \cos \alpha_j) \Delta_n^0 + \sum_{p=0}^{\infty} \sum_{k \neq j}^{N} (f_{n,j}^{p,k} F^{\bullet}_{p,k} + e_{n,j}^{p,k} F'_{p,k}) \end{aligned} \tag{2.38}$$

$\Delta_n^0$ being the Kronecker delta, and

$$\begin{Bmatrix} e_{n,j}^{p,k} \\ f_{n,j}^{p,k} \end{Bmatrix} = \frac{(-1)^{p+1}}{(r_{jk})^{n+p+2}} \binom{n+p+1}{p} \begin{pmatrix} \cos \\ \sin \end{pmatrix} \{ (n+1)(\alpha_j - \beta_{jk}) + (p+1)(\alpha_k - \beta_{jk}) \} \tag{2.39}$$

The complex potential is thus reduced to the same form as equations (2.20) of the preceding section, and the free hole conditions (2.21) and (2.22) are rewritten as follows:

$$\begin{aligned}
F^{\bullet}_{2n,j} &= -\sum_{p=0}^{\infty} \lambda_j^{2n+2p+2} (P^{2n}_{2p})_j M^{\bullet}_{2p,j} \\
F^{\bullet}_{2n+1,j} &= -\sum_{p=0}^{\infty} \lambda_j^{2n+2p+4} (P^{2n+1}_{2p+1})_j M^{\bullet}_{2p+1,j} \\
F'_{2n,j} &= -\sum_{p=0}^{\infty} \lambda_j^{2n+2p+2} (Q^{2n}_{2p})_j M'_{2p,j} \\
F'_{2n+1,j} &= -\sum_{p=0}^{\infty} \lambda_j^{2n+2p+4} (Q^{2n+1}_{2p+1})_j M'_{2p+1,j}
\end{aligned} \tag{2.40}$$

where

$$\left\{\begin{matrix}(P^{2n}_{2p})_j \\ (Q^{2n}_{2p})_j\end{matrix}\right\} = \frac{(1-\varepsilon_j^2)^{n+p+1}}{2^{2p+1}} \sum_{m=0}^{n,p} \binom{2p+1}{p-m} (1 \mp R_j^{4m+2}) A_{n-m,\,2m+1}$$

$$\left\{\begin{matrix}(P^{2n+1}_{2p+1})_j \\ (Q^{2n+1}_{2p+1})_j\end{matrix}\right\} = \frac{(1-\varepsilon_j^2)^{n+p+2}}{2^{2p+2}} \sum_{m=0}^{n,p} \binom{2p+2}{p-m} (1 \mp R_j^{4m+4}) A_{n-m,\,2m+2} \tag{2.41}$$

$$\varepsilon_j = \frac{b_j}{a_j}, \quad R_j^2 = \frac{1+\varepsilon_j}{1-\varepsilon_j}.$$

The next step in the analysis is to determine the unknown coefficients $F^{\bullet}_{n,j}$, $F'_{n,j}$, $M^{\bullet}_{n,j}$, $M'_{n,j}$ $(n=0, 1, 2, \ldots, ;\ j=1, 2, \ldots, N)$ from the relations (2.38) and (2.40). Use is made of the perturbation technique. For convenience let

$$\lambda_j = s_j \lambda \tag{2.42}$$

with a new parameter $\lambda$ and constants $s_j$ which represent the ratio of hole lengths, and all the unknowns are assumed as the following power series of the only parameter $\lambda$:

$$\begin{aligned}
F_{2n,j} &= \sum_{p=n+1}^{\infty} F^{(2p)}_{2n,j} \lambda^{2p}, \quad F_{2n+1,j} = \sum_{p=n+2}^{\infty} F^{(2p)}_{2n+1,j} \lambda^{2p} \\
M_{n,j} &= M^{(0)}_{n,j} \Delta^0_n + \sum_{p=1}^{\infty} M^{(2p)}_{n,j} \lambda^{2p}.
\end{aligned} \tag{2.43}$$

Substitute equations (2.43) into (2.38) and (2.40), equate the terms of the same powers in the both sides, there results

$$M^{\bullet(0)}_{0,j} = M^{\bullet}_{0,0}\cos\alpha_j - M'_{0,0}\sin\alpha_j\,, \quad M'^{(0)}_{0,j} = M^{\bullet}_{0,0}\sin\alpha_j + M'_{0,0}\cos\alpha_j$$

$$M^{\bullet(0)}_{n,j} = M'^{(0)}_{n,j} = 0 \qquad (n \geqslant 1)$$

$$F^{\bullet(2n+2)}_{2n,j} = -s_j^{2n+2}(P_0^{2n})_j M^{\bullet(0)}_{0,j}\,, \quad F'^{(2n+2)}_{2n,j} = -s_j^{2n+2}(Q_0^{2n})_j M'^{(0)}_{0,j}$$

$$M^{\bullet(2)}_{n,j} = \sum_{k\neq j}^{N} (e^{0,k}_{n,j} F^{\bullet(2)}_{0,k} - f^{0,k}_{n,j} F'^{(2)}_{0,k})$$

$$M'^{(2)}_{n,j} = \sum_{k\neq j}^{N} (f^{0,k}_{n,j} F^{\bullet(2)}_{0,k} + e^{0,k}_{n,j} F'^{(2)}_{0,k})$$

$$F^{\bullet(2n+2q)}_{2n,j} = -\sum_{p=0}^{q-1} s_j^{2n+2p+2}(P^{2n}_{2p})_j M^{\bullet(2q-2p-2)}_{2p,j}$$

$$F'^{(2n+2q)}_{2n,j} = -\sum_{p=0}^{q-1} s_j^{2n+2p+2}(Q^{2n}_{2p})_j M'^{(2q-2p-2)}_{2p,j}$$

$$F^{\bullet(2n+2q)}_{2n+1,j} = -\sum_{p=0}^{q-2} s_j^{2n+2p+4}(P^{2n+1}_{2p+1})_j M^{\bullet(2q-2p-4)}_{2p+1,j}$$

$$F'^{(2n+2q)}_{2n+1,j} = -\sum_{p=0}^{q-2} s_j^{2n+2p+4}(Q^{2n+1}_{2p+1})_j M'^{(2q-2p-4)}_{2p+1,j}$$

$$M^{\bullet(2q)}_{n,j} = \sum_{k\neq j}^{N} \sum_{p=0}^{2q-2} (e^{p,k}_{n,j} F^{\bullet(2q)}_{p,k} - f^{p,k}_{n,j} F'^{(2q)}_{p,k})$$

$$M'^{(2q)}_{n,j} = \sum_{k\neq j}^{N} \sum_{p=0}^{2q-2} (f^{p,k}_{n,j} F^{\bullet(2q)}_{p,k} + e^{p,k}_{n,j} F'^{(2q)}_{p,k})$$

$$(j = 1, 2, \ldots, N\ ;\ q = 2, 3, 4, \ldots\ .) \qquad (2.44)$$

By using the above relations in this order, taking $j=1, 2, \ldots, N$ for each relation, the expansion coefficients of equations (2.43) can be calculated successively.

Now from equations (2.6) and (2.13), the stresses along the $j$-th hole are calculated by the formula

$$(\tau_{nt} - \mathrm{i}\,\tau_{st})_j = \tau\left[\frac{(z_j^2 - c_j^2)^{\frac{1}{2}}}{|(z_j^2 - c_j^2)^{\frac{1}{2}}|}\,\phi'(z_j)\right]. \qquad (2.45)$$

In case when this hole is a crack, the stress intensity factor at crack tip $x_j = \lambda_j$ is given by [3]

$$k_{3,j} = \mathrm{i}\tau(2d)^{\frac{1}{2}} \lim_{z_j\to\lambda_j} [(z_j-\lambda_j)^{\frac{1}{2}}\phi'(z_j)]\,. \tag{2.46}$$

Substitute equation (2.43) into (2.37) and use the calculated results of $F^{\bullet(2p)}_{2n,j}$ etc., then $\phi'(z_j)$ in equations (2.45) or (2.46) can be written as the following power series of $\lambda$:

$$\begin{aligned}\phi'(z_j) = \sum_{q=0}^{\infty} \lambda^{2q} \sum_{p=0}^{q} s_j^{2p} \Bigg[ M^{\bullet(2q-2p)}_{2p,j} \Big\{ Z_j^{2p} + \sum_{n=0}^{\infty} (2n+1)(P^{2n}_{2p})_j Z_j^{-(2n+2)} \Big\} \\ + \mathrm{i} M'^{(2q-2p)}_{2p,j} \Big\{ Z_j^{2p} + \sum_{n=0}^{\infty} (2n+1)(Q^{2n}_{2p})_j Z_j^{-(2n+2)} \Big\} \Bigg] \\ + \sum_{q=0}^{\infty} \lambda^{2q+1} \sum_{p=0}^{q} s_j^{2p-1} \Bigg[ M^{\bullet(2q-2p)}_{2p+1,j} \Big\{ Z_j^{2p+1} \\ + \sum_{n=0}^{\infty} (2n+2)(P^{2n+1}_{2p+1})_j Z_j^{-(2n+3)} \Big\} \\ + \mathrm{i} M'^{(2q-2p)}_{2p+1,j} \Big\{ Z_j^{2p+1} + \sum_{n=0}^{\infty} (2n+2)(Q^{2n+1}_{2p+1})_j Z_j^{-(2n+3)} \Big\} \Bigg]\end{aligned} \tag{2.47}$$

where

$$Z_j = \frac{z_j}{\lambda_j}\,. \tag{2.48}$$

The infinite series including $(P^{2n}_{2p})_j$ etc. can be summed up in closed forms by means of the definitions (2.41) and the relations

$$\begin{aligned}\sum_{n=m}^{\infty} (2n+1) A_{n-m,\,2m+1} Z^{-(2n+2)} &= \frac{2m+1}{(Z^2-1)^{\frac{1}{2}}} \{Z-(Z^2-1)^{\frac{1}{2}}\}^{2m+1} \\ \sum_{n=m}^{\infty} (2n+2) A_{n-m,\,2m+2} Z^{-(2n+3)} &= \frac{2m+2}{(Z^2-1)^{\frac{1}{2}}} \{Z-(Z^2-1)^{\frac{1}{2}}\}^{2m+2}\end{aligned} \tag{2.49}$$

which are easily derived from (2.18) and (2.14). In this way the stresses along the hole edges or the stress intensity factors at crack tips are given as power series of $\lambda$ whose coefficients are evaluated exactly.

It follows that equation (2.46) for a crack is reduced to the simple expression

$$\begin{aligned}k_{3,j} = -\tau(a_j)^{\frac{1}{2}} \Bigg[ \sum_{n=0}^{\infty} \lambda^{2n} \sum_{p=0}^{n} \frac{(p+1)}{2^{2p}} \binom{2p+1}{p} s_j^{2p} M'^{(2n-2p)}_{2p,j} \\ + \sum_{n=0}^{\infty} \lambda^{2n+1} \sum_{p=0}^{n} \frac{(p+1)}{2^{2p}} \binom{2p+1}{p} s_j^{2p+1} M'^{(2n-2p)}_{2p+1,j} \Bigg].\end{aligned} \tag{2.50}$$

A computer program is then prepared according to the above theory. It automatically computes the results for the circumferential stresses or the stress intensity factors of all the holes or cracks with various geometric and loading parameters.

The stress intensity factor for the crack is thus given by a power series of the form

$$k_{3,j} = \tau(a_j)^{\frac{1}{2}} F_{3,j}, \quad F_{3,j} = \sum_{n=0}^{M} C_{n,j} \lambda^n \tag{2.51}$$

in which the coefficients $C_{n,j}$ are given exactly by closed form expressions as mentioned above.

*Infinite row of periodic cracks.* The analysis is simplified in the special case of periodic cracks. As an example consider an infinite row of equal parallel

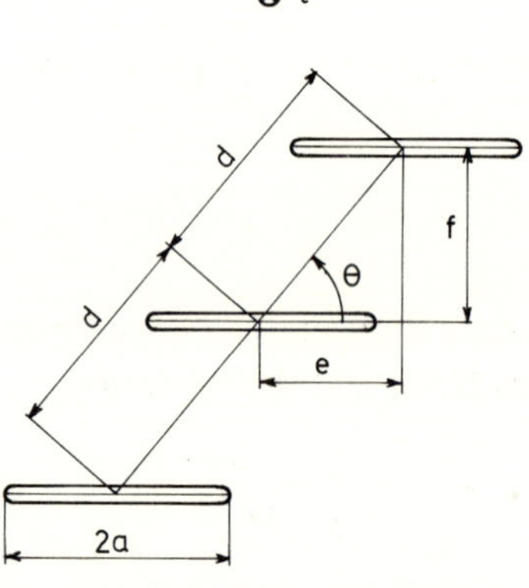


Figure 2.5. Infinite row of equal parallel cracks.

cracks as shown in Figure 2.5. The complex potential must have the same type of singularities within each crack, and can be written by the sum of the principal parts as follows:

$$\begin{aligned} &\tau_{xt} - i\tau_{yt} = \tau\phi' \\ &\phi = -iz_0 + \sum_{k=-\infty}^{\infty} \sum_{n=0}^{\infty} F_n z_k^{-(n+1)} \end{aligned} \tag{2.52}$$

where $\tau$ is the shearing stress at infinity and $z_k$ is the complex variable with the origin at crack center $O_k$.

Substituting $z_k = z_0 - k\,\mathrm{e}^{\mathrm{i}\theta}$ and expanding, $\phi$ is written as the Laurent series of $z_0$

$$\phi \quad = \sum_{n=0}^{\infty} (F_n z_0^{-(n+1)} + M_n z_0^{n+1}) \tag{2.53}$$

$$\left.\begin{aligned} M_{2n} &= -\mathrm{i}\Delta_n^0 - \sum_{p=0}^{\infty} \binom{2n+2p+1}{2p} S_{2n+2p+2}\,\mathrm{e}^{-\mathrm{i}(2n+2p+2)\theta} F_{2p} \\ M_{2n+1} &= \sum_{p=0}^{\infty} \binom{2n+2p+3}{2p+1} S_{2n+2p+4}\,\mathrm{e}^{-\mathrm{i}(2n+2p+4)\theta} F_{2p+1} \end{aligned}\right\} \tag{2.54}$$

$$S_{2q} \quad = 2\sum_{n=1}^{q} \frac{1}{n^{2q}} = \frac{(-1)^{q+1} B_{2q}(2\pi)^{2q}}{(2q)!} \tag{2.55}$$

where $B_{2q}$ are the Bernoulli's numbers with the following first values and recurrent formula

$$\begin{aligned} &B_2 = \tfrac{1}{6}, \quad B_4 = -\tfrac{1}{30}, \ldots \\ &B_{2q} = \frac{1}{2q+1}\left\{q - \tfrac{1}{2} - \sum_{m=1}^{q-1} \binom{2q+1}{2m} B_{2m}\right\}. \end{aligned} \tag{2.56}$$

The free crack conditions for the potential $\phi$ are given from equations (2.21)

$$\begin{aligned} F_n^{\bullet} \quad &= 0 \\ F'_{2n} \quad &= \lambda^{2n+2} Q_0^{2n} - \sum_{p=0}^{\infty} \lambda^{2n+2p+2} Q_{2p}^{2n} M'_{2p} \\ F'_{2n+1} &= -\sum_{p=0}^{\infty} \lambda^{2n+2p+4} Q_{2p+1}^{2n+1} M'_{2p+1}\,. \end{aligned} \tag{2.57}$$

The simultaneous relations (2.54) and (2.57) give at once

$$F_{2n+1} = M_{2n+1} = 0 \qquad (n = 0, 1, 2, \ldots) \tag{2.58}$$

and the potential function for the problem is simply given as follows:

$$\phi \quad = \sum_{n=0}^{\infty} \{\mathrm{i}F'_{2n} z_0^{-(2n+1)} + (M_{2n}^{\bullet} + \mathrm{i}M'_{2n}) z_0^{2n+1}\} \tag{2.59}$$

$$\begin{aligned} &M_{2n}^{\bullet} = -\sum_{p=0}^{\infty} f_{2p}^{2n} F'_{2p}\,, \quad M'_{2n} = -\Delta_n^0 - \sum_{p=0}^{\infty} e_{2p}^{2n} F'_{2p} \\ &F'_{2n} \quad = -\sum_{p=0}^{\infty} \lambda^{2n+2p+2} Q_{2p}^{2n} M'_{2p} \end{aligned} \tag{2.60}$$

where

$$\left\{\begin{matrix} e_{2p}^{2n} \\ f_{2p}^{2n} \end{matrix}\right\} = \binom{2n+2p+1}{2p} S_{2n+2p+2} \cdot \left\{\begin{matrix} \cos \\ \sin \end{matrix}\right\} (2n+2p+2)\theta \tag{2.61}$$

and $Q_{2p}^{2n}$ are given by equation (2.23) and Table I of Appendix I. The unknowns $F'_{2n}$ and $M'_{2n}$ are determined by a perturbation procedure.

Now assume the following power series

$$F'_{2n} = \sum_{q=n+1}^{\infty} F'^{(2q)}_{2n} \lambda^{2q}, \quad M'_{2n} = \sum_{q=0}^{\infty} M'^{(2q)}_{2n} \lambda^{2q} \tag{2.62}$$

and as many coefficients of equations (2.62) as required are easily calculated successively. Hence, the stress intensity factor is calculated from

$$k_3 = \tau(a^{\frac{1}{2}}) \left\{ 1 - \sum_{n=1}^{\infty} \lambda^{2n} \sum_{p=0}^{n} \frac{p+1}{2^{2p}} \binom{2p+1}{p} M'^{(2n-2p)}_{2p} \right\} = \tau(a^{\frac{1}{2}}) F_3$$

with

$$F_3 = 1 + \sum_{n=2}^{M} C_n \lambda^{2n} . \tag{2.63}$$

Numerical results based on the analyses of this section are given in Figures 2.14 and 2.15 of Appendix IV.

## 2.3 Plane problems

*Stress combinations in transformed variable.* In plane problems the Airy's stress function is given in terms of two complex potentials $\phi(z)$ and $\psi(z)$ as follows:

$$\chi = \sigma d^2 \operatorname{Re} \{\bar{z}\phi(z) + \psi(z)\} . \tag{2.64}$$

And components of the stress, rotation, displacement and resultant force are expressed as follows:

$$\begin{aligned} \sigma_x + \sigma_y + \frac{8iG}{1+\kappa} \omega_{xy} &= 4\sigma\phi'(z) \\ \sigma_x - \sigma_y - 2i\tau_{xy} &= -2\sigma\{\bar{z}\phi''(z) + \psi''(z)\} \end{aligned} \tag{2.65}$$

$$2G(u - iv) = \sigma d\{\kappa\bar{\phi}(\bar{z}) - \bar{z}\phi'(z) - \psi'(z)\} \tag{2.66}$$

$$P_y + iP_x = -\sigma d\{\bar{\phi}(\bar{z}) + \bar{z}\phi'(z) + \psi'(z)\} \tag{2.67}$$

where

$$\kappa = \frac{3-\nu}{1+\nu} \text{ (plane stress)}, \quad \kappa = 3-4\nu \text{ (plane strain)} \tag{2.68}$$

and $P_y$, $P_x$ are components of the resultant of the stresses acted upon the lefthand-side by the righthand-side of the domain along any path leading to point $(x, y)$ from a definite point.

In what follows the general expressions of $\phi(z)$ and $\psi(z)$ appropriate for the free elliptical hole as shown in Figure 2.3 will be derived. Since the stresses and rotation are single-valued within the plate, $\phi'(z)$ and $\psi''(z)$ must be analytic functions and can be expanded in some Laurent series. To start with, the following expressions are expressed in terms of $\zeta$

$$\begin{aligned} &\phi'\{\Omega(\zeta)\} = \alpha_0 + \sum_{m=1}^{\infty} (\alpha_m \zeta^m + \beta_m \zeta^{-m}) \\ &\frac{\mathrm{d}\psi'\{\Omega(\zeta)\}}{\mathrm{d}\zeta} = \frac{c}{2}\left\{\gamma_0 + \sum_{m=1}^{\infty} (\gamma_m \zeta^m + \delta_m \zeta^{-m})\right\}. \end{aligned} \tag{2.69}$$

Moreover, since these complex potentials must give single-valued displacement, equations (2.69) and their integrated expressions, together with equation (2.66), give the relation

$$\delta = \kappa(R\bar{\beta}_1 - R^{-1}\bar{\alpha}_1) = 0\,. \tag{2.70}$$

The stress components with respect to axes $n$, $s$ shown in Figure 2.3 are related to those associated with axes $X$, $Y$ by the formulae

$$\begin{aligned} &\sigma_n + \sigma_s = \sigma_x + \sigma_y \\ &\sigma_n - \sigma_s - 2\mathrm{i}\,\tau_{ns} = \mathrm{e}^{2\mathrm{i}\alpha}(\sigma_x - \sigma_y - 2\mathrm{i}\,\tau_{xy})\,. \end{aligned} \tag{2.71}$$

Using equations (2.65) and the relation

$$\mathrm{e}^{2\mathrm{i}\alpha} = \frac{\zeta\Omega'(\zeta)}{\bar{\zeta}\bar{\Omega}'(\bar{\zeta})} \tag{2.72}$$

in equation (2.71), stress components along the hole edge are given by

$$\begin{aligned} &\sigma_n + \sigma_s = 2\sigma\{\phi'(z) + \bar{\phi}'(\bar{z})\}_{\zeta=\mathrm{e}^{\mathrm{i}\beta}} \\ &\sigma_n - \sigma_s - 2\mathrm{i}\,\tau_{ns} = -2\sigma\left[\frac{\zeta}{\bar{\zeta}\bar{\Omega}'(\bar{\zeta})}\left\{\bar{\Omega}(\bar{\zeta})\frac{\mathrm{d}\phi'}{\mathrm{d}\zeta} + \frac{\mathrm{d}\psi'}{\mathrm{d}\zeta}\right\}\right]_{\zeta=\mathrm{e}^{\mathrm{i}\beta}}. \end{aligned} \tag{2.73}$$

*General form of stress function for a free elliptical hole.* For a free boundary the sum of the both expressions of equations (2.73) must vanish. Equating the coefficients of all the terms to zero and after some algebra, the following relations are found:

$$\beta_{2m+3} = R^{-(2m+1)}\alpha_1 + \sum_{p=0}^{m} R^{2p-2m}\{(2p+2)(\bar{\alpha}_{2p+3} + R^{-2}\bar{\alpha}_{2p+1}) + R^{-1}\bar{\gamma}_{2p+1}\}$$

$$\beta_{2m+2} = 2R^{-(2m+2)}\alpha_0^{\cdot} + (2m+1)\bar{\alpha}_{2m+2} + \sum_{p=0}^{m} R^{2p-2m-2}(4p\bar{\alpha}_{2p} + R\bar{\gamma}_{2p})$$

$$\delta_1 = 0$$

$$\begin{aligned}\delta_{2m+3} = {} & (2m+2)(R^{-(2m+1)} + R^{-(2m+5)})\alpha_1 + \{R + (2m+2)^2 R^{-3}\}\bar{\alpha}_{2m+1} \\ & + \{(2m+2)^2 - 1\}R^{-1}\bar{\alpha}_{2m+3} + (2m+2)R^{-2}\bar{\gamma}_{2m+1} \\ & + (R^3 + R^{-1})\sum_{p=0}^{m-1} R^{2p-2m}\{(2p+2)(\bar{\alpha}_{2p+3} + R^{-2}\bar{\alpha}_{2p+1}) + R^{-1}\bar{\gamma}_{2p+1}\}\end{aligned}$$

$$\begin{aligned}\delta_{2m+2} = {} & 2(2m+1)(R^{-(2m-1)} + R^{-(2m+3)})\alpha_0^{\cdot} + 2m\{2mR + (4m+2)R^{-3}\}\bar{\alpha}_{2m} \\ & + 2m(2m+2)R^{-1}\bar{\alpha}_{2m+2} + (2m+1)R^{-2}\bar{\gamma}_{2m} \\ & + (2m+1)(R^3 + R^{-1})\sum_{p=0}^{m-1} R^{2p-2m-2}(4p\bar{\alpha}_{2p} + R\bar{\gamma}_{2p}) \quad (m=0,1,2\ldots).\end{aligned} \tag{2.74}$$

Making use of equations (2.69) together with relations (2.70) and (2.74) the general expressions of the complex potentials in terms of $\zeta$ are obtained. In order to write them in terms of $z$, $\alpha_n$ and $\gamma_n$ will be written in terms of the new parameters $M_p$ and $K_p$.

$$\begin{aligned}\alpha_{2m} &= R^{2m}\sum_{p=0}^{\infty}(2p+1)\left(\frac{c}{2}\right)^{2p}\binom{2p}{p-m}M_{2p} \\ \alpha_{2m+1} &= R^{2m+1}\sum_{p=0}^{\infty}(2p+2)\left(\frac{c}{2}\right)^{2p+1}\binom{2p+1}{p-m}M_{2p+1} \\ \gamma_{2m} &= (2m+1)R^{2m+1}\sum_{p=0}^{\infty}(2p+2)\left(\frac{c}{2}\right)^{2p}\binom{2p+1}{p-m}K_{2p} \\ \gamma_{2m+1} &= (2m+2)R^{2m+2}\sum_{p=0}^{\infty}(2p+3)\left(\frac{c}{2}\right)^{2p+1}\binom{2p+2}{p-m}K_{2p+1}\,.\end{aligned} \tag{2.75}$$

Substituting equations (2.75) into (2.69), derivatives of complex potentials are written by terms of positive powers of $z$ and negative powers of $\zeta$. Using expansions (2.18) and after integration, the complex potentials appropriate for free elliptical holes become

$$\phi(z) = \sum_{n=0}^{\infty} \{(F_n^{\bullet} + \mathrm{i} F_n')z^{-(n+1)} + (M_n^{\bullet} + \mathrm{i} M_n')z^{n+1}\}$$
$$\psi(z) = -D_0^{\bullet} \log z + \sum_{n=1}^{\infty} (D_n^{\bullet} + \mathrm{i} D_n')z^{-n} + \sum_{n=0}^{\infty} (K_n^{\bullet} + \mathrm{i} K_n')z^{n+2} \tag{2.76}$$

with the relations among the coefficients

$$D_{2n}^{\bullet} = \sum_{p=0}^{\infty} \lambda^{2n+2p+2}(P_{2p}^{2n} K_{2p}^{\bullet} + R_{2p}^{2n} M_{2p}^{\bullet})$$

$$F_{2n}^{\bullet} = -\sum_{p=0}^{\infty} \lambda^{2n+2p+2}(Q_{2p}^{2n} K_{2p}^{\bullet} + S_{2p}^{2n} M_{2p}^{\bullet})$$

$$D_{2n+1}^{\bullet} = \sum_{p=0}^{\infty} \lambda^{2n+2p+4}(P_{2p+1}^{2n+1} K_{2p+1}^{\bullet} + R_{2p+1}^{2n+1} M_{2p+1}^{\bullet})$$

$$F_{2n+1}^{\bullet} = -\sum_{p=0}^{\infty} \lambda^{2n+2p+4}(Q_{2p+1}^{2n+1} K_{2p+1}^{\bullet} + S_{2p+1}^{2n+1} M_{2p+1}^{\bullet})$$

$$D_{2n}' = -\sum_{p=0}^{\infty} \lambda^{2n+2p+2}(T_{2p}^{2n} K_{2p}' + V_{2p}^{2n} M_{2p}') \tag{2.77}$$

$$F_{2n}' = \sum_{p=0}^{\infty} \lambda^{2n+2p+2}(Q_{2p}^{2n} K_{2p}' + W_{2p}^{2n} M_{2p}')$$

$$D_{2n+1}' = -\sum_{p=0}^{\infty} \lambda^{2n+2p+4}(T_{2p+1}^{2n+1} K_{2p+1}' + V_{2p+1}^{2n+1} M_{2p+1}')$$

$$F_{2n+1}' = \sum_{p=0}^{\infty} \lambda^{2n+2p+4}(Q_{2p+1}^{2n+1} K_{2p+1}' + W_{2p+1}^{2n+1} M_{2p+1}')$$

where

$$P_{2p}^{0} = (1-\varepsilon^2)^{p+1} \frac{(p+1)}{2^{2p}} \binom{2p+1}{p}, \quad R_{2p}^{0} = P_{2p}^{0} \frac{(R^2 + R^{-2})}{2}$$

$$\begin{Bmatrix} P_{2p}^{2n} \\ T_{2p}^{2n} \end{Bmatrix} = \frac{(1-\varepsilon^2)^{n+p+1}}{2^{2p+2}} \left[ \begin{Bmatrix} 4(p+1)\binom{2p+1}{p}\left(1+\frac{1}{n}\right) A_{n,1} \\ 0 \end{Bmatrix} + \sum_{m=1}^{p+1,n} \binom{2p+2}{p-m+1} \{(2p+1)R^{4m} \mp 1\} A_{n-m,\,2m} \right]$$

$$\begin{Bmatrix} R^{2n}_{2p} \\ V^{2n}_{2p} \end{Bmatrix} = \frac{(1-\varepsilon^2)^{n+p+1}}{2^{2p+1}} \left[ \begin{Bmatrix} (p+1)\binom{2p+1}{p}(R^2+R^{-2})\left(1+\frac{1}{n}\right)A_{n,1} \\ 0 \end{Bmatrix} + 2p \sum_{m=0}^{n,p} \binom{2p+1}{p-m} R^{4m+2} A_{n-m,\,2m+1} \right] \qquad (n \geqslant 1)$$

$$Q^{2n}_{2p} = (1-\varepsilon^2)^{n+p+1} \frac{(p+1)}{2^{2p}} \sum_{m=0}^{n,p} \binom{2p+1}{p-m} R^{4m+2} A_{n-m,\,2m+1}$$

$$\begin{Bmatrix} S^{2n}_{2p} \\ W^{2n}_{2p} \end{Bmatrix} = \frac{(1-\varepsilon^2)^{n+p+1}}{2^{2p+1}} \left[ (p+1)\binom{2p+1}{p} A_{n,1} \pm \sum_{m=0}^{n,p} \binom{2p+1}{p-m} A_{n-m,\,2m+1} + 2(2p+1) \sum_{m=1}^{n+1,p} \binom{2p}{p-m} R^{4m} A_{n-m+1,\,2m} \right]$$

$$\begin{Bmatrix} P^{2n+1}_{2p+1} \\ T^{2n+1}_{2p+1} \end{Bmatrix} = \frac{(1-\varepsilon^2)^{n+p+2}}{2^{2p+3}} \left[ \pm(p+2)\binom{2p+3}{p+1} A_{n,1} + \sum_{m=0}^{n,p+1} \binom{2p+3}{p-m+1} \{(2p+2)R^{4m+2} \mp 1\} A_{n-m,\,2m+1} \right]$$

$$\begin{Bmatrix} R^{2n+1}_{2p+1} \\ V^{2n+1}_{2p+1} \end{Bmatrix} = \frac{(1-\varepsilon^2)^{n+p+2}}{2^{2p+2}} \left[ \pm(p+1)\binom{2p+1}{p}(R \pm R^{-1})^2 A_{n,1} + (2p+1) \sum_{m=1}^{n+1,p+1} \binom{2p+2}{p-m+1} R^{4m} A_{n-m+1,\,2m} \right]$$

$$Q^{2n+1}_{2p+1} = (1-\varepsilon^2)^{n+p+2} \frac{(2p+3)}{2^{2p+2}} \sum_{m=1}^{n+1,p+1} \binom{2p+2}{p-m+1} R^{4m} A_{n-m+1,\,2m}$$

$$\begin{Bmatrix} S^{2n+1}_{2p+1} \\ W^{2n+1}_{2p+1} \end{Bmatrix} = \frac{(1-\varepsilon^2)^{n+p+2}}{2^{2p+2}} \left[ \pm \sum_{m=1}^{n+1,p+1} \binom{2p+2}{p-m+1} A_{n-m+1,\,2m} + 2(2p+2) \sum_{m=0}^{n+1,p} \binom{2p+1}{p-m} R^{4m+2} A_{n-m+1,\,2m+1} \right] \qquad (n \geqslant 0)$$

(2.78)

In the special case of a crack, ε and $R$ in equations (2.78) are put to zero and one respectively and

$$P_p^n = R_p^n\,, \quad Q_p^n = S_p^n = T_p^n\,, \quad V_p^n = W_p^n \tag{2.79}$$

with exceptions of trivial coefficients $T_{2p}^0$ and $V_{2p}^0$. Values of $P_{2p}^{2n}$ etc. for the case of a crack for $p, n \leqslant 8$ are given in Tables III to VIII of Appendix I. For a circular hole equations (2.77) are reduced to the simple relations

$$\begin{aligned} D_0^{\cdot} &= 2M_0^{\cdot}\lambda^2\,, \quad D_n = (n-1)\overline{K}_{n-2}\lambda^{2n} + n\overline{M}_n\lambda^{2n+2} \qquad (n \geqslant 1) \\ F_n &= -(n+2)\overline{K}_n\lambda^{2n+2} - (n+3)\overline{M}_{n+2}\lambda^{2n+4}\,. \end{aligned} \tag{2.80}$$

Now let us show by a simple example how the Laurent series expansions lead to the closed form solutions. Assume a wide plate with an elliptical hole under the stresses at infinity

$$\sigma_x^\infty = \alpha\sigma\,, \quad \sigma_y^\infty = \beta\sigma\,, \quad \tau_{xy}^\infty = \gamma\sigma\,. \tag{2.81}$$

These stress conditions give

$$K_0 = \frac{\beta-\alpha+2\mathrm{i}\gamma}{4}\,, \quad M_0^{\cdot} = \frac{\beta+\alpha}{4}\,, \quad K_n = M_n = 0 \qquad (n \geqslant 1) \tag{2.82}$$

and hence from equations (2.77) and (2.78)

$$F_{2n}^{\cdot} = -c^{2n+2}(R^2K_0^{\cdot} + M_0^{\cdot})A_{n,1}\,, \quad F_{2n}' = c^{2n+2}R^2K_0'A_{n,1}\,, \quad F_{2n+1} = 0$$
$$(n = 0, 1, 2\ \ldots)\,. \tag{2.83}$$

The circumferential stress $\sigma_s$ is given by 4 Re $\{\phi'(z)\}$ and the corresponding infinite series are summed up in a closed form solution as follows:

$$\begin{aligned} \sigma_s &= 4M_0^{\cdot} + 4\,\mathrm{Re}\left\{(R^2\overline{K}_0 + M_0^{\cdot}) \sum_{n=0}^{\infty} (2n+1)A_{n,1}\left(\frac{c}{z}\right)^{2n+2}\right\} \\ &= \beta+\alpha+\mathrm{Re}\left[\{(R^2+1)\beta-(R^2-1)\alpha-2\mathrm{i}\gamma R^2\}\left\{\frac{z}{(z^2-c^2)^{\frac{1}{2}}} - 1\right\}\right]. \end{aligned} \tag{2.84}$$

Another form of equation (2.84) may be obtained by means of the mapping function (2.30). This expresses $\sigma_s$ in the elliptical coordinates:

$$\begin{aligned} \sigma_s = \sigma\Bigg[&\left\{\frac{(\mathrm{e}^{2\xi_1}+1)\sinh 2\xi_1}{\cosh 2\xi_1 - \cos 2\eta} - \mathrm{e}^{2\xi_1}\right\}\beta - \left\{\frac{(\mathrm{e}^{2\xi_1}-1)\sinh 2\xi_1}{\cosh 2\xi_1 - \cos 2\eta} - \mathrm{e}^{2\xi_1}\right\}\alpha \\ &- \frac{2\mathrm{e}^{2\xi_1}\sin 2\eta}{\cosh 2\xi_1 - \cos 2\eta}\gamma\Bigg]. \end{aligned} \tag{2.85}$$

*Tension of an infinite strip with an eccentric crack.* Consider an eccentrically cracked long strip subjected to uniform tensile stress $\sigma$ at infinity. Take

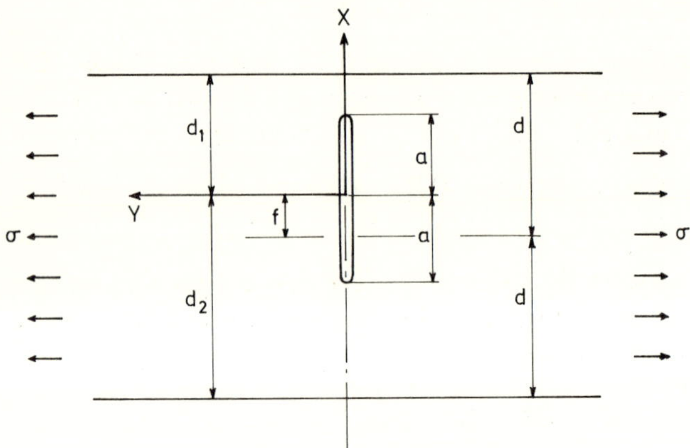


Figure 2.6. Tension of long strip with eccentric crack.

axes $X$, $Y$ and define geometric parameters as shown in Figure 2.6, and the following dimensionless coordinates and parameters:

$$\delta = \frac{f}{d}, \quad \lambda = \frac{a}{d_1} \qquad \qquad (2.86)$$
$$x = \frac{X}{d}, \quad y = \frac{Y}{d}, \quad z = x + iy .$$

First the stress function is assumed as

$$\chi = \chi_0 + \chi_1$$
$$\chi_0 = \sigma d^2 \operatorname{Re}\{\bar{z}\phi_0(z) + \psi_0(z)\} \qquad \qquad (2.87)$$
$$\chi_1 = \sigma d^2 \operatorname{Re}\{\bar{z}\phi(z) + \psi(z)\}$$

$\chi_0$ corresponds to the uniform tensile stress at infinity, and

$$\phi_0(z) = M_{0,0}z , \quad \psi_0(z) = K_{0,0}z^2 , \quad M_{0,0} = K_{0,0} = \tfrac{1}{4} . \qquad (2.88)$$

The complex potentials $\phi(z)$ and $\psi(z)$ are written as Laurent series of the forms (2.76). From stress symmetry with respect to the $X$-axis, the coefficients must be real, and writing simply $F_n^{\bullet} = F_n$ etc. this yields

$$\phi(z)=\sum_{n=0}^{\infty}(F_n z^{-(n+1)}+M_n z^{n+1})$$
$$\psi(z)=-D_0\log z+\sum_{n=1}^{\infty}D_n z^{-n}+\sum_{n=0}^{\infty}K_n z^{n+2}\,. \tag{2.89}$$

The free crack relations (2.77) become

$$D_n=\lambda^{n+2}P_0^n(K_{0,0}+M_{0,0})+\sum_{p=0}^{\infty}\lambda^{n+p+2}P_p^n(K_p+M_p)$$
$$F_n=-\lambda^{n+2}Q_0^n(K_{0,0}+M_{0,0})-\sum_{p=0}^{\infty}\lambda^{n+p+2}Q_p^n(K_p+M_p) \tag{2.90}$$

where terms for odd values of $(n+p)$ are missing, and values of $P_p^n$ and $Q_p^n$ are calculated from equations (2.78), putting $R=1$ and $\varepsilon=0$. Numerical values for $n,p\leqslant 8$ are also given in Tables III to VI of Appendix I.

Now there exist some relations among the coefficients of equations (2.89) since the straight edges are free from tractions. The necessary relations were first derived by C. B. Ling [4] for the circular hole problem, and are written in the present notations as follows:

$$K_n=\sum_{p=0}^{\infty}(a_p^n D_p+b_p^n F_p)\,,\quad M_n=\sum_{p=0}^{\infty}(c_p^n D_p+d_p^n F_p) \tag{2.91}$$

where

$$a_{2p}^n=\frac{1}{(n+2)!\,(2p-1)!}\Big\{P_{n+2p+2}+nT_{n+2p+1}+\frac{2}{1-\delta}(-V_{n+2p+2}+\delta T_{n+2p+2})\Big\}$$

$$a_{2p+1}^n=\frac{1}{(n+2)!\,(2p)!}\Big\{-P_{n+2p+3}+(n+2)U_{n+2p+2}+\frac{2}{1-\delta}(-W_{n+2p+3}+\delta U_{n+2p+3})\Big\}$$

$$b_{2p}^n=\frac{1}{(n+2)!\,(2p)!}\Big[-(n+2p+2)P_{n+2p+2}-2npT_{n+2p+1}$$
$$+\frac{2}{1-\delta}\{(n+2p+2)V_{n+2p+2}-\delta(n+2)U_{n+2p+2}-2\delta pT_{n+2p+2}\}$$
$$+\Big(\frac{2}{1-\delta}\Big)^2\{-T_{n+2p+3}+\delta(V_{n+2p+3}+W_{n+2p+3})-\delta^2U_{n+2p+3}\}\Big]$$

$$b_{2p+1}^n = \frac{1}{(n+2)!(2p+1)!}\left[(n+2p+3)P_{n+2p+3} - (n+2)(2p+3)U_{n+2p+2}\right.$$
$$+ \frac{2}{1-\delta}\{(n+2p+3)W_{n+2p+3} - \delta n T_{n+2p+3} - \delta^2(2p+3)U_{n+2p+3}\}$$
$$\left. + \left(\frac{2}{1-\delta}\right)^2\{-U_{n+2p+4} + \delta(V_{n+2p+4} + W_{n+2p+4}) - \delta^2 T_{n+2p+4}\}\right]$$

$$c_{2p}^n = \frac{T_{n+2p+1}}{(n+1)!(2p-1)!}, \quad c_{2p+1}^n = \frac{U_{n+2p+2}}{(n+1)!(2p)!}$$

$$d_{2p}^n = \frac{1}{(n+1)!(2p)!}\left\{-P_{n+2p+2} - 2pT_{n+2p+1}\right.$$
$$\left. + \frac{2}{1-\delta}(V_{n+2p+2} - \delta U_{n+2p+2})\right\}$$

$$d_{2p+1}^n = \frac{1}{(n+1)!(2p+1)!}\left\{P_{n+2p+3} - (2p+3)U_{n+2p+2}\right.$$
$$\left. + \frac{2}{1-\delta}(W_{n+2p+3} - \delta T_{n+2p+3})\right\} \tag{2.92}$$

with definition $(-1)!=0!=1$. And $P_k$, $T_k$, $U_k$, $V_k$, $W_k$ are given by the following integrals:

$$P_{2k} = \left(\frac{1-\delta}{2}\right)^{2k}\int_0^\infty \frac{m^{2k-1}(\mathrm{e}^{-m} + \cosh \delta m)}{\sinh m}\,\mathrm{d}m$$

$$P_{2k+1} = \left(\frac{1-\delta}{2}\right)^{2k+1}\int_0^\infty \frac{m^{2k}\sinh \delta m}{\sinh m}\,\mathrm{d}m$$

$$T_{2k} = U_{2k} = \frac{(1-\delta)^{2k+1}}{2}\int_0^\infty m^{2k}\sinh 2\delta m\left(\frac{1}{\Lambda_1} + \frac{1}{\Lambda_2}\right)$$

$$T_{2k+1} = (1-\delta)^{2k+2}\int_0^\infty m^{2k+1}\left(\frac{\cosh^2 \delta m}{\Lambda_1} + \frac{\sinh^2 \delta m}{\Lambda_2}\right)\mathrm{d}m$$

$$U_{2k+1} = (1-\delta)^{2k+2}\int_0^\infty m^{2k+1}\left(\frac{\sinh^2 \delta m}{\Lambda_1} + \frac{\cosh^2 \delta m}{\Lambda_2}\right)\mathrm{d}m$$

$$V_{2k} = (1-\delta)^{2k+1}\int_0^\infty m^{2k}\left(\frac{\coth m \cdot \cosh^2 \delta m}{\Lambda_1} + \frac{\tanh m \cdot \sinh^2 \delta m}{\Lambda_2}\right)\mathrm{d}m$$

$$W_{2k} = (1-\delta)^{2k+1} \int_0^\infty m^{2k} \left( \frac{\tanh m \cdot \sinh^2 \delta m}{\Lambda_1} + \frac{\coth m \cdot \cosh^2 \delta m}{\Lambda_2} \right) dm$$

$$V_{2k+1} = \frac{(1-\delta)^{2k+2}}{2} \int_0^\infty m^{2k+1} \sinh 2\delta m \left( \frac{\coth m}{\Lambda_1} + \frac{\tanh m}{\Lambda_2} \right) dm$$

$$W_{2k+1} = \frac{(1-\delta)^{2k+2}}{2} \int_0^\infty m^{2k+1} \sinh 2\delta m \left( \frac{\tanh m}{\Lambda_1} + \frac{\coth m}{\Lambda_2} \right) dm$$

$$\Lambda_1 = \sinh 2m + 2m\,, \quad \Lambda_2 = \sinh 2m - 2m\,. \tag{2.93}$$

These integrals can be evaluated by numerical integration with sufficient accuracy using the Laguerre–Gauss formulae [5].

For the case of a central crack ($\delta = 0$) the above relations can be shown to agree with the results obtained by R. C. J. Howland [6]. In another special case when $d_2$ approaches infinity, which corresponds to a cracked semiinfinite plate, the following simple relations hold:

$$\lim_{\delta \to 1} (P_{k+1}, T_k, U_k, V_k, W_k) = \frac{k!}{2^{k+1}} \tag{2.94}$$

and the resulting forms of $a_p^n$, $b_p^n$, $c_p^n$ and $d_p^n$ again agree with the author's previous results for this case [7].

Equations (2.90) and (2.91) give simultaneous equations to determine the unknowns. For convenience let

$$\begin{aligned} K_{0,0} + M_{0,0} &= H_{0,0}\,, \quad & K_n + M_n &= H_n \\ a_p^n + c_p^n &= e_p^n\,, & b_p^n + d_p^n &= f_p^n \end{aligned} \tag{2.95}$$

and hence

$$\begin{aligned} D_n &= H_{0,0} P_0^n \lambda^{n+2} + \sum_{p=0}^{\infty} H_p P_p^n \lambda^{n+p+2} \\ F_n &= -H_{0,0} Q_0^n \lambda^{n+2} - \sum_{p=0}^{\infty} H_p Q_p^n \lambda^{n+p+2} \\ H_n &= \sum_{p=0}^{\infty} (e_p^n D_p + f_p^n F_p)\,. \end{aligned} \tag{2.96}$$

The unknown coefficients $D_n$, $F_n$ and $H_n$ in equations (2.96) may be solved with the aid of a perturbation technique. Assume that all the unknowns can be expressed as the following power series:

$$H_n=\sum_{p=1}^{\infty} H_n^{(2p)}\lambda^{2p},\quad D_{2n}=\sum_{p=n+1}^{\infty} D_{2n}^{(2p)}\lambda^{2p},\quad F_{2n}=\sum_{p=n+1}^{\infty} F_{2n}^{(2p)}\lambda^{2p}$$

$$D_{2n+1}=\sum_{p=n+3}^{\infty} D_{2n+1}^{(2p)}\lambda^{2p},\quad F_{2n+1}=\sum_{p=n+3}^{\infty} F_{2n+1}^{(2p)}\lambda^{2p}. \tag{2.97}$$

Put these into equations (2.96), rearrange the righthand-side in ascending power series of $\lambda$ and equate the coefficients of the same powers in the both sides. The results are

$$D_{2n}^{(2n+2)}=P_0^{2n}H_{0,0},\quad F_{2n}^{(2n+2)}=-Q_0^{2n}H_{0,0},\quad H_n^{(2)}=e_0^nD_0^{(2)}+f_0'^nF_0^{(2)}$$

$$D_{2n}^{(2n+2q+2)}=\sum_{p=0}^{q-1} P_{2p}^{2n}H_{2p}^{(2q-2p)},\qquad F_{2n}^{(2n+2q+2)}=-\sum_{p=0}^{q-1} Q_{2p}^{2n}H_{2p}^{(2q-2p)}$$

$$D_{2n+1}^{(2n+2q+2)}=\sum_{p=0}^{q-2} P_{2p+1}^{2n+1}H_{2p+1}^{(2q-2p-2)},\quad F_{2n+1}^{(2n+2q+2)}=-\sum_{p=0}^{q-2} Q_{2p+1}^{2n+1}H_{2p+1}^{(2q-2p-2)}$$

$$H_n^{(2q+2)}=\sum_{p=0}^{2q}\left(e_p^nD_p^{(2q+2)}+f_p^nF_p^{(2q+2)}\right)\qquad (q=1,2,\dots) \tag{2.98}$$

Successive calculation of the expansion coefficients in equations (2.97) may now be carried by taking as many terms as required.

The stress intensity factor at the crack tip on the minimum section is given by the formula [8]

$$k_1=2\sigma(2d)^{\frac{1}{2}}\lim_{z\to\lambda}\{(z-\lambda)^{\frac{1}{2}}\phi'(z)\}. \tag{2.99}$$

Combining equations (2.89) and (2.97), rearranging the result in power series of $\lambda$ and dropping trivial terms, $\phi'(z)$ is expressed as follows:

$$\phi'(z)=\sum_{q=0}^{\infty}\lambda^{2q}\sum_{p=0}^{q}H_{2p}^{(2q-2p)}\sum_{n=0}^{\infty}(2n+1)Q_{2p}^{2n}Z^{-(2n+2)}$$
$$+\lambda^{2q+1}\sum_{p=0}^{q}H_{2p+1}^{(2q-2p)}\sum_{n=0}^{\infty}(2n+2)Q_{2p+1}^{2n+1}Z^{-(2n+3)} \tag{2.100}$$

where

$$Z=\frac{z}{a}. \tag{2.101}$$

The infinite series including $Q_{2p}^{2n}$, $Q_{2p+1}^{2n+1}$ are summed up in closed forms by the definitions (2.78) and the previous relations (2.49).

Finally the stress intensity factors are given as the following power series whose coefficients are evaluated exactly:

$$k_{1,A} = \sigma(a^{\frac{1}{2}})F_{1,A}(\delta, \lambda), \qquad k_{1,B} = \sigma(a^{\frac{1}{2}})F_{1,B}(\delta, \lambda)$$
$$F_{1,A}(\delta, \lambda) = 1 + \sum_{n=2}^{M} C_n\lambda^n, \quad F_{1,B}(\delta, \lambda) = 1 + \sum_{n=2}^{M} C_n(-\lambda)^n. \tag{2.102}$$

Numerical calculations are performed for various values of $\delta$ and the series are obtained up to the term in $\lambda^{19}$. For the two special cases, centrally cracked strip ($\delta=0$) and cracked half plane ($\delta\to 1$), special computing programs have been prepared and the series are computed up to the term in $\lambda^{70}$. Numerical results of this section are summarized in Appendix III.

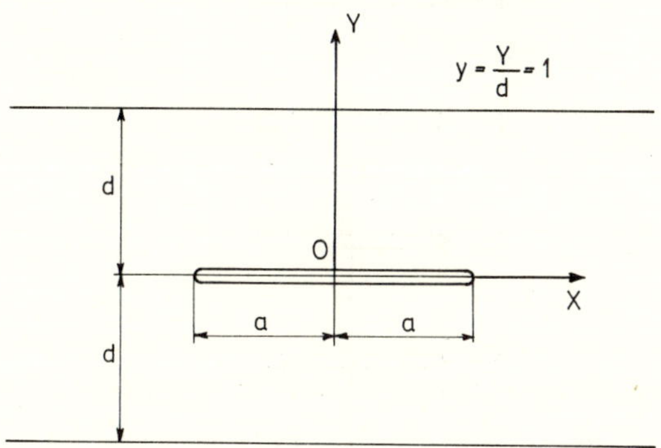


Figure 2.7. Infinite strip with longitudinal crack.

*Other problems of finite plates* [9]. An infinite strip with a longitudinal crack as shown in Figure 2.7 is treated under the following six boundary conditions along the straight edges $Y = \pm d$:

(1) Tension

(1.a) Uniform stress $\quad (\sigma_y = \sigma_0,\ \tau_{xy} = 0)$

(1.b) Uniform displacement without shear $\quad (v = \pm v_0,\ \tau_{xy} = 0)$ (2.103a)

(1.c) Uniform displacements $\quad (v = \pm v_0,\ u = 0)$

(2) In-plane bending

(2.a) Linear stress $\quad \left(\sigma_y = \sigma_0 \dfrac{X}{d},\ \tau_{xy} = 0\right)$

(2.b) Rotation without shear $\quad \left(v = \pm v_0 \dfrac{X}{d},\ \tau_{xy} = 0\right)$ (2.103b)

(2.c) Rotation by clamped ends $\quad \left(v = \pm v_0 \dfrac{X}{d},\ u = 0\right)$.

It is noteworthy that cases (1.b) and (2.b) correspond exactly to the tension and in-plane bending of infinite plates containing periodic parallel cracks as shown in Figure 2.8.

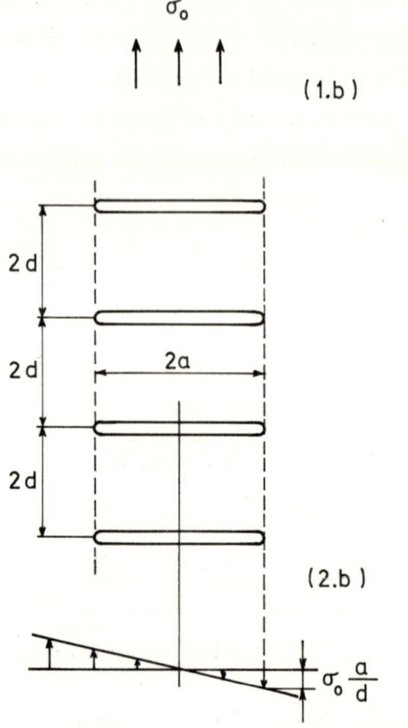


Figure 2.8. Equal parallel cracks as special cases.

For tension cases equations (2.86) to (2.90) also apply except that $d_1 = d$ in (2.86) and the following definitions

$$\begin{aligned} &(1.a)\ K_{0,0} = M_{0,0} = \tfrac{1}{4}, \quad \sigma = \sigma_0 \\ &(1.b)\ K_{0,0} = M_{0,0} = \tfrac{1}{4}, \quad \sigma = \frac{Ev_0}{d} \\ &(1.c)\ K_{0,0} = \frac{1-\nu}{4}, \quad M_{0,0} = \frac{1+\nu}{4}, \quad \sigma = \frac{Ev_0}{(1-\nu^2)d} \end{aligned} \tag{2.104}$$

are used. For in-plane bending equations (2.88) are replaced by

$$\phi_0(z) = M_{1,0} z^2, \quad \psi_0(z) = K_{1,0} z^3 \tag{2.105}$$

with $K_{1,0}$, $M_{1,0}$ being defined for each case as

$$\begin{aligned}
&(2.a)\ K_{1,0}=\tfrac{1}{24}, \qquad M_{1,0}=\tfrac{1}{8}, \qquad \sigma=\sigma_0\\
&(2.b)\ K_{1,0}=\tfrac{1}{24}, \qquad M_{1,0}=\tfrac{1}{8}, \qquad \sigma=\frac{Ev_0}{d}\\
&(2.c)\ K_{1,0}=\frac{1-3v}{24}, \quad M_{1,0}=\frac{1+v}{8}, \quad \sigma=\frac{Ev_0}{(1-v^2)d}
\end{aligned} \tag{2.106}$$

and equations (2.90) are replaced by the relations

$$\begin{aligned}
D_n &= \lambda^{n+3} P_0^n(K_{1,0}+M_{1,0}) + \sum_{p=0}^{\infty} \lambda^{n+p+2} P_p^n(K_p+M_p)\\
F_n &= -\lambda^{n+3} Q_0^n(K_{1,0}+M_{1,0}) - \sum_{p=0}^{\infty} \lambda^{n+p+2} Q_p^n(K_p+M_p).
\end{aligned} \tag{2.107}$$

The stress function $\chi_0$ is chosen so as to give the assigned boundary stresses or displacements (2.103). Therefore the second function $\chi_1$ should satisfy the following conditions along $Y=\pm d$.

$$\begin{aligned}
&\text{Cases (1.a) and (2.a)}: \sigma_y=0, \quad \tau_{xy}=0\\
&\text{Cases (1.b) and (2.b)}: v=0, \quad \tau_{xy}=0\\
&\text{Cases (1.c) and (2.c)}: v=0, \quad u=0
\end{aligned} \tag{2.108}$$

and they require that some relations exist among the coefficients of equations (2.89). Following the procedure outlined in a paper by R. C. J. Howland [6], the results are given for the foregoing six cases:

$$H_n = K_n + M_n = \sum_{p=0}^{\infty} (e_p^n D_p + f_p^n F_p) \tag{2.109}$$

in which for Cases (1.a) and (2.a), $e_p^n$ and $f_p^n$ are given by

$$\begin{aligned}
e_p^n &= \frac{\varepsilon_{n,p}}{(n+2)!(p-1)!}\int_0^{\infty} \frac{m^{n+p+1}(se^{-m}+m)}{sc+m}\,dm\\
f_p^n &= \frac{2\varepsilon_{n,p}}{(n+2)!\,p!}\int_0^{\infty} \frac{m^{n+p+1}\{-m^2+p(se^{-m}+m)\}}{sc+m}\,dm
\end{aligned} \tag{2.110a}$$

For Cases (1.b) and (2.b), the results are

$$\begin{aligned}
e_p^n &= -\frac{\varepsilon_{n,p}}{(n+2)!(p-1)!}\int_0^{\infty} \frac{e^{-m} m^{n+p+1}}{s}\,dm\\
f_p^n &= \frac{2\varepsilon_{n,p}}{(n+2)!\,p!}\int_0^{\infty} \frac{m^{n+p+1}(m-ps\,e^{-m})}{s^2}\,dm
\end{aligned} \tag{2.110b}$$

Finally, the following results

$$e_p^n = \frac{\varepsilon_{n,p}}{(n+2)!\,(p-1)!}\int_0^\infty \frac{e^{-m}m^{n+p+1}[\{1-\nu-(1+\nu)m\}s-\{2+(1+\nu)m\}c]}{(3-\nu)cs-(1+\nu)m}\,dm$$

$$f_p^n = \frac{2\varepsilon_{n,p}}{(n+2)!\,p!}\int_0^\infty \frac{e^{-m}m^{n+p+1}dm}{(3-\nu)cs-(1+\nu)m}\left[\{ms(1+\nu)+2c\}\left(\frac{1-\nu}{1+\nu}+m-p\right)\right.$$
$$\left. + \{mc(1+\nu)-(1-\nu)s\}\left(-\frac{2}{1+\nu}+m-p\right)\right] \qquad (2.110c)$$

pertain to Cases (1.c) and (2.c). Adopted in the above equations are the notations

$$s=\sinh m,\ c=\cosh m,\ 0!=(-1)!=1,\ \varepsilon_{n,p}=\begin{cases}(-1)^{\frac{1}{2}(n+p)} & \text{(tension)}\\ (-1)^{\frac{1}{2}(n+p)-1} & \text{(bending)}\end{cases} \qquad (2.111)$$

The simultaneous relations (2.90) or (2.107) and (2.109) corresponding to the given boundary conditions are solved by a perturbation procedure similar to that in the preceding section. The stress intensity factors are then calculated from the power series formulae.

Thus far only the problems of the cracked strip have been treated. In the cases of rectangular plates having finite length and width, it is difficult to satisfy the boundary conditions of the outer edge analytically. However, a number of approximate methods are available for obtaining reasonably accurate solutions. Analyses have been done for the three boundary conditions by determining the coefficients of the Laurent series with a boundary collocation procedure based on the stress resultants and mean displacements.

The numerical results of the stress intensity factors in this section are displayed in Figures 2.16 to 2.18 of Appendix IV. Additional results may be found elsewhere [9].

*Stiffened panels.* Consider the panels made by joining two different plates and with stringers along their connecting boundaries. Figure 2.9a shows the symmetric case where two kinds of strips of the same width $2d$ are bonded alternatively, and Figure 2.9b is the unsymmetric case where two half planes are bonded together. In the following they are simply termed as “Case (a)” and “Case (b)” respectively.

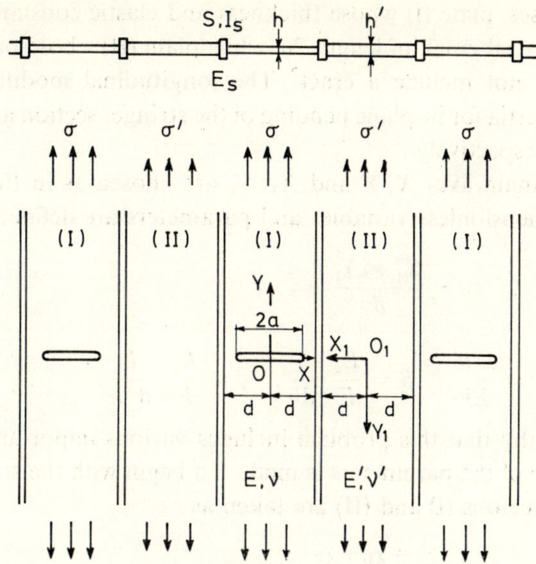


Figure 2.9a. Joined strips stiffened by stringers along boundaries.

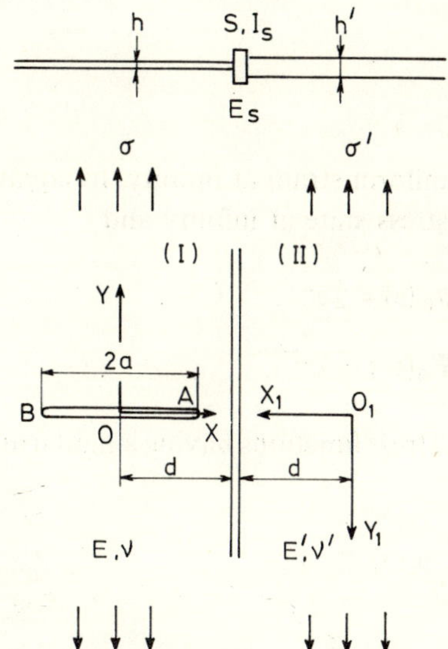


Figure 2.9b. Joined half planes stiffened by stringer along boundary.

In both cases, plate (I) whose thickness and elastic constants are $h$, $E$, $\nu$ contains a central crack of length $2a$, while plate (II) whose parameters are $h'$, $E'$, $\nu'$ does not include a crack. The longitudinal modulus, area and moment of inertia for in-plane bending of the stringer section are denoted by $E_s$, $S$ and $I_s$ respectively.

The coordinate axes $X$, $Y$ and $X_1$, $Y_1$ are chosen as in the figure. The following dimensionless variables and parameters are defined:

$$z = \frac{X+\mathrm{i}Y}{d}, \quad z_1 = \frac{X_1+\mathrm{i}Y_1}{d}$$

$$\lambda = \frac{a}{d}, \quad \alpha = \frac{E'h'}{Eh}, \quad \beta = \frac{E_s}{E}\cdot\frac{S}{dh}, \quad \gamma = \frac{E_s}{E}\cdot\frac{I_s}{d^3 h}. \tag{2.112}$$

It is noteworthy that this problem includes various important cases when proper choice of the parameters is made. To begin with the stress functions $\chi$ and $\chi'$ for regions (I) and (II) are taken as

$$\begin{aligned} &\chi = \chi_0+\chi_1+\chi_2, \quad \chi' = \chi'_0+\chi'_2 \\ &\chi_k = \sigma d^2 \operatorname{Re}\{\bar{z}\phi_k(z)+\psi_k(z)\} \qquad (k=0, 1, 2) \\ &\chi'_k = \sigma' d^2 \operatorname{Re}\{\bar{z}_1 \Phi_k(z)+\Psi_k(z_1)\} \qquad (k=0, 2) \end{aligned} \tag{2.113}$$

where

$$\frac{\sigma}{E} = \frac{\sigma'}{E'} = \varepsilon_\infty \tag{2.114}$$

with $\varepsilon_\infty$ being the uniform strain at infinity. In equations (2.113), $\chi_0$ and $\chi'_0$ correspond to the stress state at infinity and

$$\begin{aligned} &\phi_0(z) = \tfrac{1}{4}z, \qquad \psi_0(z) = \tfrac{1}{4}z^2 \\ &\Phi_0(z_1) = \tfrac{1}{4}z_1, \quad \Psi_0(z_1) = \tfrac{1}{4}z_1^2. \end{aligned} \tag{2.115}$$

Moreover, $\chi_1$ is the stress functions having singularities within the crack and may be written as

$$\begin{aligned} &\phi_1(z) = \sum_{n=0}^{\infty} F_n z^{-(n+1)} \\ &\psi_1(z) = -D_0 \log z + \sum_{n=1}^{\infty} D_n z^{-n} \end{aligned} \tag{2.116}$$

where $D_n$ and $F_n$ are real coefficients. Here and in the following, terms in $D_n$ and $F_n$ with odd subscripts are missing in Case (a).

The third type of the stress functions $\chi_2$ and $\chi_2'$ are introduced in order to satisfy the boundary conditions, together with $\chi_0$, $\chi_0'$, $\chi_1$, and are assumed to take the integral forms

$$\phi_2(z) = \tfrac{1}{2}\int_0^\infty B(m)S(mz)\mathrm{d}m$$

$$\psi_2(z) = \int_0^\infty \left\{A(m)C(mz) + \frac{B(m)}{2} zS(mz)\right\}\mathrm{d}m$$

$$\Phi_2(z_1) = \tfrac{1}{2}\int_0^\infty H(m)S(mz_1)\mathrm{d}m$$

$$\Psi_2(z_1) = \int_0^\infty \left\{G(m)C(mz_1) + \frac{H(m)}{2} z_1 S(mz_1)\right\}\mathrm{d}m \tag{2.117}$$

where $S(mz)$, $C(mz)$ etc. are known functions defined by

$$\text{Case (a)}: S(X) = \sinh X\,,\quad C(X) = \cosh X$$
$$\text{Case (b)}: S(X) = C(X) = \mathrm{e}^X \tag{2.118}$$

The unknown coefficients $D_n$, $F_n$ in equations (2.116) and functions $A(m)$, $B(m)$, $G(m)$, $H(m)$ in equations (2.117) will be determined from the boundary conditions along the crack edge and plate boundaries.

If the stringer thickness is assumed to be smaller in comparison with $d$, then the following boundary conditions prevail along plate edges:

$$(u)_{x=1,\,y=y} = -(u')_{x_1=1,\,y_1=-y}$$
$$(v)_{x=1,\,y=y} = (v')_{x_1=1,\,y_1=-y} \tag{2.119a}$$

which describe continuity of displacement.

Equilibrium of stringer element (Figure 2.10) requires

$$h(\tau_{xy})_{x=1,\,y=y} + h'(\tau_{x_1y_1})_{x_1=1,\,y_1=-y} = S\frac{\partial\sigma_{ys}}{\partial Y}$$

$$-h(\sigma_x)_{x=1,\,y=y} + h'(\sigma_{x_1})_{x_1=1,\,y_1=-y} = E_s I_s \frac{\partial^4 u_s}{\partial Y^4} \tag{2.119b}$$

In equations (2.119b) $u_s$ is replaced by $(u)_{x=1,\,y=y}$, and

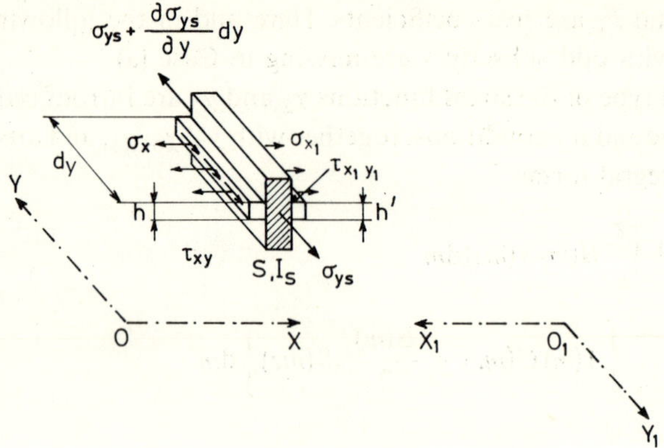


Fig. 2.10. Equilibrium of stringer element.

$$\sigma_{ys}=E_s\varepsilon_s=E_s(\varepsilon_y)_{x=1,\,y=y}=\frac{E_s}{E}(\sigma_y-\nu\sigma_x)_{x=1,\,y=y}$$

where the subscript "$s$" denotes the corresponding quantities of the stringer. The boundary conditions (2.119) are expressed by stresses and displacements of plate (I) at $x=1$ and plate (II) at $x_1=1$. Using equations (2.65), (2.66), and (2.114) to (2.118), boundary conditions (2.119) are reduced to the following integrals:

$$\int_0^\infty [(1+\nu)msA+\{(\nu-1)s+(1+\nu)mc\}B+(1+\nu')msG$$
$$+\{(\nu'-1)s+(1+\nu')mc\}H]\cos my\,\mathrm{d}m=\frac{Eu_1(y)}{\sigma d}$$

$$\int_0^\infty [(1+\nu)mcA+\{2c+(1+\nu)ms\}B-(1+\nu')mcG$$
$$-\{2c+(1+\nu')ms\}H]\sin my\,\mathrm{d}m=-\frac{Ev_1(y)}{\sigma d}$$

$$\int_0^\infty [m\{s+(1+\nu)\beta mc\}A+\{s+mc+\beta m(2c+(1+\nu)ms)\}B$$
$$+\alpha\{msG+(mc+s)H\}]$$
$$\times\, m\sin my\,\mathrm{d}m=\frac{1}{\sigma}\left[-\tau_1(y)+\beta\frac{\mathrm{d}}{\mathrm{d}y}\{\sigma_2(y)-\nu\sigma_1(y)\}\right]$$

$$\int_0^\infty [\{c+(1+\nu)\gamma m^3 s\}A+\{s+\gamma m^2((1+\nu)mc-(1-\nu)s)\}B$$
$$-\alpha(cG+sH)]m^2\cos my\,\mathrm{d}m=\frac{1}{\sigma}\left[\sigma_1(y)+\frac{\gamma}{\mathrm{d}}\frac{d^4\{Eu_1(y)\}}{\mathrm{d}y^4}\right]. \quad (2.120)$$

where

$$s=S(m)\,,\quad c=C(m) \tag{2.121}$$

and $A(m)$, $B(m)$, $G(m)$, $H(m)$ are simply written as $A, B, G, H$. With the stress components $\sigma_1(y), \sigma_2(y), \tau_1(y), u_1(y), v_1(y)$ denoting the values of $\sigma_x, \sigma_y, \tau_{xy}, u, v$ at $x=1$, the stresses and displacements corresponding to the function $\chi_1$ are

$$\sigma_1(y)=(\sigma_x)_{x=1} = -\sigma\left[\frac{D_0+2F_0}{1+y^2}\cos 2\theta+\sum_{n=1}^{\infty}\left\{\frac{(n+1)(nD_n+2F_n)}{(1+y^2)^{\frac{1}{2}n+1}}\right.\right.$$
$$\left.\left.+\frac{n(n-1)F_{n-2}}{(1+y^2)^{\frac{1}{2}n}}\right\}\cos(n+2)\theta\right]$$

$$\sigma_2(y)=(\sigma_y)_{x=1} = \sigma\left[\frac{D_0-2F_0}{1+y^2}\cos 2\theta+\sum_{n=1}^{\infty}\left\{\frac{(n+1)(nD_n-2F_n)}{(1+y^2)^{\frac{1}{2}n+1}}\right.\right.$$
$$\left.\left.+\frac{n(n-1)F_{n-2}}{(1+y^2)^{\frac{1}{2}n}}\right\}\cos(n+2)\theta\right]$$

$$\tau_1(y)=(\tau_{xy})_{x=1} = -\sigma\left[\frac{D_0}{1+y^2}\sin 2\theta+\sum_{n=1}^{\infty}\left\{\frac{n(n+1)D_n}{(1+y^2)^{\frac{1}{2}n+1}}\right.\right.$$
$$\left.\left.+\frac{n(n-1)F_{n-2}}{(1+y^2)^{\frac{1}{2}n}}\right\}\sin(n+2)\theta\right]$$

$$u_1(y)=(u)_{x=1} = \frac{\sigma d}{E}\left[\frac{(1+\nu)D_0+(3-\nu)F_0}{(1+y^2)^{\frac{1}{2}}}\cos\theta+\sum_{n=1}^{\infty}\left\{\frac{(1+\nu)nD_n+(3-\nu)F_n}{(1+y^2)^{\frac{1}{2}(n+1)}}\right.\right.$$
$$\left.\left.+\frac{(1+\nu)(n-1)F_{n-2}}{(1+y^2)^{\frac{1}{2}(n-1)}}\right\}\cos(n+1)\theta\right]$$

$$v_1(y)=(v)_{x=1} = \frac{\sigma d}{E}\left[\frac{(1+\nu)D_0-(3-\nu)F_0}{(1+y^2)^{\frac{1}{2}}}\sin\theta+\sum_{n=1}^{\infty}\left\{\frac{(1+\nu)nD_n-(3-\nu)F_n}{(1+y^2)^{\frac{1}{2}(n+1)}}\right.\right.$$
$$\left.\left.+\frac{(1+\nu)(n-1)F_{n-2}}{(1+y^2)^{\frac{1}{2}(n-1)}}\right\}\sin(n+1)\theta\right] \tag{2.122}$$

in which $\theta=\tan^{-1}y$. In order to give $A(m)$ etc. in explicit forms, apply Fourier transformation to equations (2.120). For instance the first relation becomes

$$(1+\nu)msA+\{(\nu-1)s+(1+\nu)mc\}B+(1+\nu')msG$$
$$+\{(\nu'-1)s+(1+\nu')mc\}H = \frac{2E}{\pi\sigma d}\int_0^\infty u_1(y)\cos my\,dy\,.$$

Use is also made of the integral formulae [6]

$$\int_0^\infty \frac{\cos n\theta \cos my\,\mathrm{d}y}{(1+y^2)^{\frac{1}{2}n}} = \int_0^\infty \frac{\sin n\theta \sin my\,\mathrm{d}y}{(1+y^2)^{\frac{1}{2}n}} = \frac{\pi \mathrm{e}^{-m} m^{n-1}}{2(n-1)!}$$
$$\int_0^\infty \frac{\cos(n+2)\theta \cos my\,\mathrm{d}y}{(1+y^2)^{\frac{1}{2}n}} = \int_0^\infty \frac{\sin(n+2)\theta \sin my\,\mathrm{d}y}{(1+y^2)^{\frac{1}{2}n}}$$
$$= \frac{\pi \mathrm{e}^{-m} m^{n-1}(2m-n)}{2n!} \qquad (\theta = \tan^{-1} y) \qquad (2.123)$$

All the relations of (2.120) can be treated in the same way and four linear equations are obtained, from which the unknown functions $A(m)$ etc. can be determined as linear combinations of unknown coefficients $D_n$ and $F_n$. The resulting expressions can be substituted into equations (2.117) and the latter are expanded into power series of $z$ or $z_1$ by using the relations

$$\text{Case (a)} \quad S(X) = \sum_{n=0}^{\infty} \frac{X^{2n+1}}{(2n+1)!}, \quad C(X) = \sum_{n=0}^{\infty} \frac{X^{2n}}{(2n)!}$$
$$(2.124)$$
$$\text{Case (b)} \quad S(X) = C(X) = \sum_{n=0}^{\infty} \frac{X^n}{n!}$$

Combining the resulting expressions with (2.115) and (2.116), the usual forms of the stress functions are obtained:

$$\chi \ = \sigma d^2 \,\mathrm{Re}\,\{\bar{z}\,\phi(z)+\psi(z)\} \qquad (2.125)$$

$$\phi(z) = \tfrac{1}{4}z + \sum_{n=0}^{\infty} (F_n z^{-(n+1)} + M_n z^{n+1})$$
$$\psi(z) = \tfrac{1}{4}z^2 - D_0 \log z + \sum_{n=1}^{\infty} D_n z^{-n} + \sum_{n=0}^{\infty} K_n z^{n+2} \qquad (2.126)$$

$$K_n \ = \sum_{p=0}^{\infty} (a_p^n D_p + b_p^n F_p), \quad M_n = \sum_{p=0}^{\infty} (c_p^n D_p + d_p^n F_p)\,. \qquad (2.127)$$

Similar expression for $\chi'$ can be found. Further analyses can be proceeded in the same way as previously for the strip, equations (2.95) to (2.102).

Note that the present problems include the following important cases according to special choice of the parameters:

(1) Strip and half plane reinforced by stringers along edges ($\alpha=0$)
(2) Strip and half plane with free edges ($\alpha=\beta=\gamma=0$)
(3) Periodic colinear cracks ($\alpha=1$, $\beta=\gamma=0$)
(4) Stiffened panels with a crack in each region (I) and (II) ($\alpha=1$, $\gamma=\infty$)
(5) Unstiffened plates with periodic colinear cracks or a pair of cracks ($\alpha=1$, $\beta=0$, $\gamma=\infty$).

Figures 2.19 to 2.24 of Appendix IV show some numerical results of joined plates and stiffened strips.

*Random elliptical holes and cracks in wide plate.* Consider a wide plate containing an arbitrary distribution of stress free elliptical holes, some of which may be circular holes or cracks. The analysis is rather complicated but similar to that of an infinite body. The same notations are used unless otherwise specified.

The plate is assumed to be subjected to the following stresses at infinity.

$$\sigma_x^\infty = \sigma\left(\alpha + \frac{\delta Y}{d}\right), \quad \sigma_y^\infty = \sigma\left(\beta + \frac{\mu X}{d}\right), \quad \tau^\infty = \sigma\gamma \tag{2.128}$$

where $\sigma$ is a reference stress and $\alpha$, $\beta$, $\gamma$, $\delta$, $\mu$ are constants as shown in Figure 2.11. First the stress function is written as the following sum

$$\chi = \chi_0 + \sum_{k=1}^{N} \chi_k \tag{2.129}$$

in which $\chi_0$ corresponds to the stress at infinity

$$\chi_0 = \sigma d^2 \operatorname{Re}\{\bar{z}\phi_0(z) + \psi_0(z)\}$$

and

$$\phi_0(z) = \tfrac{1}{4}(\beta+\alpha)z + \tfrac{1}{8}(\mu - \mathrm{i}\delta)z^2 \tag{2.130}$$

$$\psi_0(z) = \tfrac{1}{4}(\beta-\alpha+2\mathrm{i}\gamma)z^2 + \tfrac{1}{24}(\mu+\mathrm{i}\delta)z^3 .$$

The stress functions $\chi_k$ ($k=1, 2, \ldots, N$) possess singularities within the holes and can be expanded in the following Laurent series:

$$\chi_k = \sigma d^2 \operatorname{Re}\{\bar{z}_k \phi_k(z_k) + \psi_k(z_k)\}$$

and

$$\phi_k(z_k) = \sum_{n=0}^{\infty} (F^{\bullet}_{n,k} + \mathrm{i}F'_{n,k}) z_k^{-(n+1)} \tag{2.131}$$

$$\psi_k(z_k) = -D^{\bullet}_{0,k} \log z_k + \sum_{n=1}^{\infty} (D^{\bullet}_{n,k} + \mathrm{i}D'_{n,k}) z_k^{-n} .$$

The function (2.129) is chosen so as to satisfy the boundary conditions at infinity, and thus only those along the hole edges will have to be considered. To discuss the $j$-th crack the total stress function is expressed with the

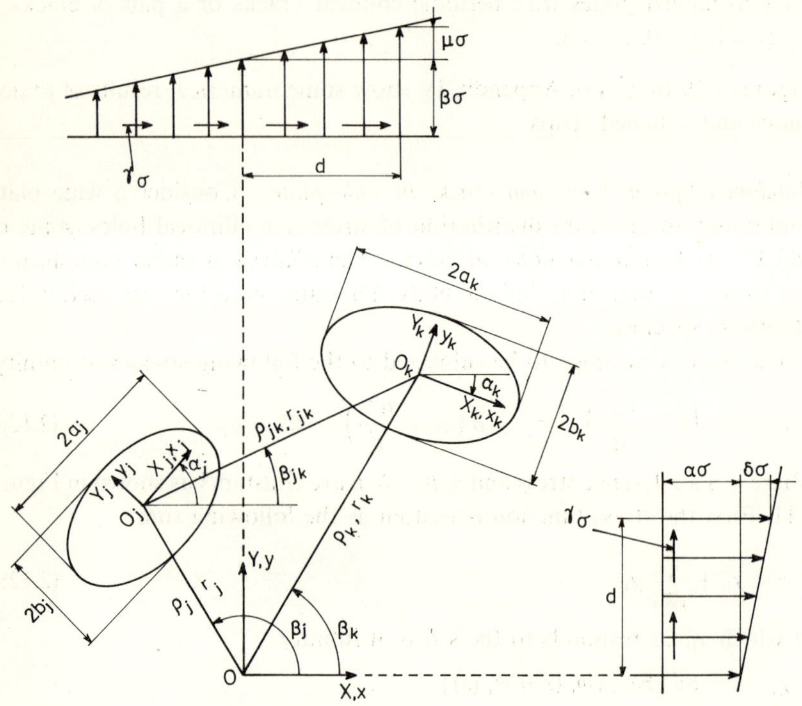


Figure 2.11. Random elliptical holes in wide plate.

only variable $z_j$, by using relations (2.36), together with (2.129) to (2.131). The results are written as follows:

$$\chi = \sigma d^2 \operatorname{Re}\{\bar{z}_j \Phi_j(z_j) + \Psi_j(z_j)\}$$

$$\Phi_j(z_j) = \sum_{n=0}^{\infty} \{(F^{\cdot}_{n,j} + \mathrm{i} F'_{n,j}) z_j^{-(n+1)} + (M^{\cdot}_{n,j} + \mathrm{i} M'_{n,j}) z_j^{n+1}\} \tag{2.132}$$

$$\Psi_j(z_j) = -D^{\cdot}_{0,j} \log z_j + \sum_{n=1}^{\infty} (D^{\cdot}_{n,j} + \mathrm{i} D'_{n,j}) z_j^{-n} + \sum_{n=0}^{\infty} (K^{\cdot}_{n,j} + \mathrm{i} K'_{n,j}) z_j^{n+2} .$$

The following relations among the coefficients have been adopted:

$$M^{\bullet}_{n,j} = \frac{\Delta^n_0}{4}\{\beta+\alpha+r_j(\mu\cos\beta_j+\delta\sin\beta_j)\} + \frac{\Delta^n_1}{8}(\mu\cos\alpha_j+\delta\sin\alpha_j)$$
$$+\sum_{p=0}^{\infty}\sum_{k\neq j}^{N}(e^{p,k}_{n,j}F^{\bullet}_{p,k}+f^{p,k}_{n,j}F'_{p,k})$$

$$M'_{n,j} = \frac{\Delta^n_0}{4}r_j(\mu\sin\beta_j-\delta\cos\beta_j) + \frac{\Delta^n_1}{8}(\mu\sin\alpha_j-\delta\cos\alpha_j)$$
$$+\sum_{p=0}^{\infty}\sum_{k\neq j}^{N}(-f^{p,k}_{n,j}F^{\bullet}_{p,k}+e^{p,k}_{n,j}F'_{p,k})$$

$$K^{\bullet}_{n,j} = \frac{\Delta^n_0}{4}\{(\beta-\alpha)\cos 2\alpha_j-2\gamma\sin 2\alpha_j+r_j\cos 2\alpha_j(\mu\cos\beta_j-\delta\sin\beta_j)\}$$
$$+\frac{\Delta^n_1}{24}(\mu\cos 3\alpha_j-\delta\sin 3\alpha_j)$$
$$+\sum_{p=0}^{\infty}\sum_{k\neq j}^{N}(a^{p,k}_{n,j}D^{\bullet}_{p,k}+b^{p,k}_{n,j}D'_{p,k}+c^{p,k}_{n,j}F^{\bullet}_{p,k}+d^{p,k}_{n,j}F'_{p,k})$$

$$K'_{n,j} = \frac{\Delta^n_0}{4}\{(\beta-\alpha)\sin 2\alpha_j+2\gamma\cos 2\alpha_j+r_j\sin 2\alpha_j(\mu\cos\beta_j-\delta\sin\beta_j)\}$$
$$+\frac{\Delta^n_1}{24}(\mu\sin 3\alpha_j+\delta\cos 3\alpha_j)$$
$$+\sum_{p=0}^{\infty}\sum_{k\neq j}^{N}(-b^{p,k}_{n,j}D^{\bullet}_{p,k}+a^{p,k}_{n,j}D'_{p,k}-d^{p,k}_{n,j}F^{\bullet}_{p,k}+c^{p,k}_{n,j}F'_{p,k}) \qquad (2.133)$$

where

$$a^{0,k}_{n,j} = \frac{\cos\{(n+2)(\beta_{jk}-\alpha_j)\}}{(n+2)(r_{jk})^{n+2}}, \quad b^{0,k}_{n,j} = \frac{\sin\{(n+2)(\beta_{jk}-\alpha_j)\}}{(n+2)(r_{jk})^{n+2}}$$

$$\begin{Bmatrix} a^{p,k}_{n,j} \\ b^{p,k}_{n,j} \end{Bmatrix} = (-1)^p\binom{n+p+1}{n+2}\frac{\begin{Bmatrix}\cos\\ \sin\end{Bmatrix}\{p(\alpha_j-\alpha_k)+(n+p+2)(\beta_{jk}-\alpha_j)\}}{(r_{jk})^{n+p+2}}$$

$$\begin{Bmatrix} c^{p,k}_{n,j} \\ d^{p,k}_{n,j} \end{Bmatrix} = (-1)^p\binom{n+p+2}{n+2}\frac{\begin{Bmatrix}\cos\\ \sin\end{Bmatrix}\{(p+2)(\alpha_j-\alpha_k)+(n+p+4)(\beta_{jk}-\alpha_j)\}}{(r_{jk})^{n+p+2}}$$

$$\begin{Bmatrix} e^{p,k}_{n,j} \\ f^{p,k}_{n,j} \end{Bmatrix} = (-1)^{p+1}\binom{n+p+1}{n+1}\frac{\begin{Bmatrix}\cos\\ \sin\end{Bmatrix}\{(p+2)(\alpha_j-\alpha_k)+(n+p+2)(\beta_{jk}-\alpha_j)\}}{(r_{jk})^{n+p+2}}. \qquad (2.134)$$

The stress function is thus reduced to the same forms as equations (2.76) and the free hole relations can be written down, just by adding the subscripts "$j$" to the necessary quantities in equations (2.77) and (2.78). The relations from (2.77) and (2.133) for $j=1, 2, \ldots, N$ are then treated with a perturbation procedure, and further analysis is performed in the same way as for the strip. The stress intensity factors of the crack tip $X=a_j$ are given by the following power series

$$(k_1 - \mathrm{i}k_2)_{A_j} = 2(2d)^{\frac{1}{2}} \lim_{z_j \to \lambda_j} [(z_j - \lambda_j)^{\frac{1}{2}} \phi_j'(z_j)] = \sigma(a_j)^{\frac{1}{2}} (F_{1,j} - \mathrm{i}F_{2,j})$$

with

$$F_{1,j} = \sum_{n=0}^{M} C_{n,j}^{(1)} \lambda_j^n , \quad F_{2j} = \sum_{n=0}^{M} C_{n,j}^{(2)} \lambda_j^n . \tag{2.135}$$

The corresponding values for the opposite crack tip $x_j = -\lambda_j$ may be obtained upon replacing $\phi_j(z_j)$ by $\phi_j(-z_j)$ and $\lambda_j$ by $(-\lambda_j)$ in equations (2.135). The coefficients $C_{n,j}$ are evaluated exactly by closed form expressions.

Typical results of stress intensity factors based on the above analysis are given in Figures 2.25 to 2.33 of Appendix IV. Other results may be found in Ref. [10].

*Infinite row of periodic cracks.* The analysis is simplified in the special case of periodic cracks. Consider a wide plate containing an infinite row of equal parallel cracks subjected to stresses at infinity

$$\sigma_y^\infty = \beta\sigma , \quad \tau^\infty = \gamma\sigma . \tag{2.136}$$

The complex potentials are then expressed in the forms which include singularities within all the cracks in accordance with the periodicity of the problem. Reexpanding them as in the case of longitudinal shear, the following results are obtained:

$$\chi = \sigma d^2 \,\mathrm{Re}\, \{\bar{z}_0 \phi(z_0) + \psi(z_0)\}$$

$$\phi(z_0) = \sum_{n=0}^{\infty} \{F_{2n} z_0^{-(2n+1)} + M_{2n} z_0^{2n+1}\}$$

$$\psi(z_0) = -D_0 \log z_0 + \sum_{n=1}^{\infty} D_{2n} z_0^{-2n} + \sum_{n=0}^{\infty} K_{2n} z_0^{2n+2} \tag{2.137}$$

with

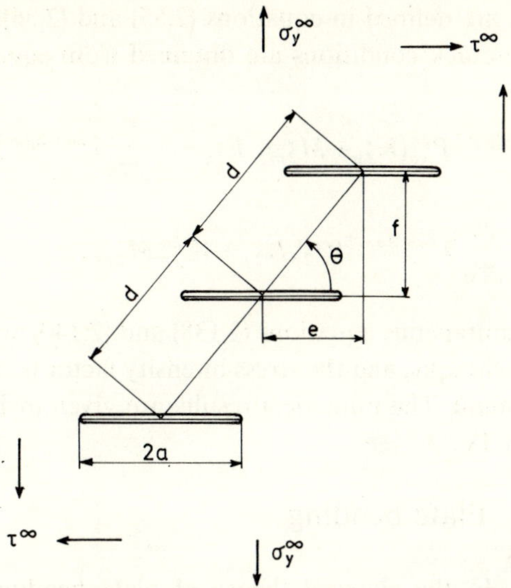


Figure 2.12. Infinite row of equal parallel cracks.

$$K_{2n}^{\cdot} = \tfrac{1}{4}\beta\Delta_n^0 + \sum_{p=0}^{\infty} (a_{2p}^{2n} D_{2p}^{\cdot} + b_{2p}^{2n} D_{2p}' + c_{2p}^{2n} F_{2p}^{\cdot} + d_{2p}^{2n} F_{2p}')$$

$$K_{2n}' = \tfrac{1}{2}\gamma\Delta_n^0 + \sum_{p=0}^{\infty} (-b_{2p}^{2n} D_{2p}^{\cdot} + a_{2p}^{2n} D_{2p}' - d_{2p}^{2n} F_{2p}^{\cdot} + c_{2p}^{2n} F_{2p}')$$

$$M_{2n}^{\cdot} = \tfrac{1}{4}\beta\Delta_n^0 - \sum_{p=0}^{\infty} (e_{2p}^{2n} F_{2p}^{\cdot} + f_{2p}^{2n} F_{2p}')$$

$$M_{2n}' = \sum_{p=0}^{\infty} (f_{2p}^{2n} F_{2p}^{\cdot} - e_{2p}^{2n} F_{2p}') \tag{2.138}$$

and

$$\begin{Bmatrix} a_0^{2n} \\ b_0^{2n} \end{Bmatrix} = \frac{S_{2n+2}}{(2n+2)} \begin{Bmatrix} \cos \\ \sin \end{Bmatrix} (2n+2)\theta$$

$$\begin{Bmatrix} a_{2p}^{2n} \\ b_{2p}^{2n} \end{Bmatrix} = \binom{2n+2p+1}{2p-1} S_{2n+2p+2} \begin{Bmatrix} \cos \\ \sin \end{Bmatrix} (2n+2p+2)\theta \qquad (p \geqslant 1)$$

$$\begin{Bmatrix} c_{2p}^{2n} \\ d_{2p}^{2n} \end{Bmatrix} = \binom{2n+2p+2}{2p} S_{2n+2p+2} \begin{Bmatrix} \cos \\ \sin \end{Bmatrix} (2n+2p+4)\theta$$

$$\begin{Bmatrix} e_{2p}^{2n} \\ f_{2p}^{2n} \end{Bmatrix} = \binom{2n+2p+1}{2p} S_{2n+2p+2} \begin{Bmatrix} \cos \\ \sin \end{Bmatrix} (2n+2p+2)\theta \tag{2.139}$$

Other quantities are defined in equations (2.55) and (2.56).

The stress free crack conditions are obtained from equations (2.77) and (2.79) as

$$D_{2n}^{;}=\sum_{p=0}^{\infty}\lambda^{2n+2p+2}P_{2p}^{2n}(K_{2p}^{;}+M_{2p}^{;}),\ F_{2n}^{;}=-\sum_{p=0}^{\infty}\lambda^{2n+2p+2}Q_{2p}^{2n}(K_{2p}^{;}+M_{2p}^{;})$$

$$F_{2n}'=-D_{2n}'=\sum_{p=0}^{\infty}\lambda^{2n+2p+2}(Q_{2p}^{2n}K_{2p}'+W_{2p}^{2n}M_{2p}'). \tag{2.140}$$

As before the simultaneous equations (2.138) and (2.140) are then solved by a perturbation technique, and the stress intensity factor is calculated from a power series formula. The numerical results are given in Figures 2.34 and 2.35 of Appendix IV.

## 2.4 Plate bending

*Basic formulae.* In the classical theory of plate bending, displacement, components of moment and shearing force are given in terms of two complex potentials as follows [11]:

$$\begin{aligned}
&Dw=T_0d^2\,\mathrm{Re}\{\bar{z}\phi(z)+\psi(z)\},\quad D=\frac{Eh^3}{12(1-\nu^2)}\\
&T_x+T_y=-4(1+\nu)T_0\,\mathrm{Re}\{\phi'(z)\}\\
&T_x-T_y-2\mathrm{i}\,T_{xy}=-2(1-\nu)T_0\{\bar{z}\phi''(z)+\psi''(z)\}\\
&Q_x-\mathrm{i}Q_y=-\frac{4T_0}{d}\phi''(z)
\end{aligned} \tag{2.141}$$

where

$$\begin{aligned}
&T_x=\int_{-\frac{1}{2}h}^{\frac{1}{2}h}\sigma_x t\,\mathrm{d}t,\quad T_y=\int_{-\frac{1}{2}h}^{\frac{1}{2}h}\sigma_y t\,\mathrm{d}t,\quad T_{xy}=\int_{-\frac{1}{2}h}^{\frac{1}{2}h}\tau_{xy}t\,\mathrm{d}t\\
&Q_x=\int_{-\frac{1}{2}h}^{\frac{1}{2}h}\tau_{xt}\,\mathrm{d}t,\quad Q_y=\int_{-\frac{1}{2}h}^{\frac{1}{2}h}\tau_{yt}\,\mathrm{d}t
\end{aligned} \tag{2.142}$$

and $T_0$ is a reference moment.

Corresponding quantities with respect to the new axes which are taken along the normal and tangent to the hole edge as shown in Figure 2.2 of section 2.2, are related to the above Cartesian components by the formulae

$$T_n + T_s = T_x + T_y$$
$$T_n - T_s - 2\mathrm{i}\, T_{ns} = \mathrm{e}^{2\mathrm{i}\alpha}(T_x - T_y - 2\mathrm{i}\, T_{xy}) \tag{2.143}$$
$$Q_n - \mathrm{i}\, Q_s = \mathrm{e}^{\mathrm{i}\alpha}(Q_x - \mathrm{i}\, Q_y)\,.$$

in which $\alpha$ is the angle between the normal $n$ and the $x$-axis. For free hole problems, the moment $T_n$ and effective shearing force $Q_n + (\partial T_{ns}/\partial s)$ should vanish on the hole edge. This requires the following relation

$$T_n - \mathrm{i}\left(T_{ns} + \int_s Q_{ns}\,\mathrm{d}s\right)$$
$$= -[(\nu-1)\phi'(z) + (3+\nu)\bar{\phi}'(\bar{z}) + (1-\nu)\mathrm{e}^{2\mathrm{i}\alpha}\{\bar{z}\phi''(z) + \psi''(z)\}] = 0 \tag{2.144}$$

*Free crack in symmetric problems.* The general forms of the complex potentials for free elliptical holes can be derived in the same way as in section 2.3 for plane problems. Laurent expansions (2.69) are assumed first. Substituting them in equations (2.144) together with (2.13) and (2.72), and replacing $\bar{\zeta}$ by $\zeta^{-1}$, the results are in terms of a power series in $\zeta$. The potentials can then be transformed into the $z$-plane by using equations (2.75) and (2.15).

The final expressions for the case of a free crack in symmetric problems are given by

$$\phi(z) = \sum_{n=0}^{\infty} (F_n z^{-(n+1)} + M_n z^{n+1})$$
$$\psi(z) = -D_0 \log z + \sum_{n=1}^{\infty} D_n z^{-n} + \sum_{n=0}^{\infty} K_n z^{n+2}\,. \tag{2.145}$$

Here, all the coefficients are real and related to each other by the equations:

$$D_{2n} = \sum_{p=0}^{\infty} \lambda^{2n+2p+2}(P_{2p}^{2n} K_{2p} + R_{2p}^{2n} M_{2p})$$

$$F_{2n} = \sum_{p=0}^{\infty} \lambda^{2n+2p+2}(Q_{2p}^{2n} K_{2p} + S_{2p}^{2n} M_{2p})$$

$$D_{2n+1} = \sum_{p=0}^{\infty} \lambda^{2n+2p+4}(P_{2p+1}^{2n+1} K_{2p+1} + R_{2p+1}^{2n+1} M_{2p+1})$$

$$F_{2n+1} = \sum_{p=0}^{\infty} \lambda^{2n+2p+4}(Q_{2p+1}^{2n+1} K_{2p+1} + S_{2p+1}^{2n+1} M_{2p+1}) \tag{2.146}$$

where

$$P^0_{2p} = \left(1 - \frac{1}{\mu}\right)\frac{2p+2}{2^{2p+2}}\binom{2p+1}{p}$$

$$R^0_{2p} = \frac{2p+1}{2^2p+2}\left[\left(\mu - \frac{1}{\mu}\right)\binom{2p}{p-1} - \frac{(\mu-1)^2}{\mu}\binom{2p}{p}\right]$$

$$P^{2n}_{2p} = \frac{2p+2}{2^{2p+1}}\left\{\frac{1}{2n}\sum_{m=0}^{n,p}\left(1 - \frac{2m+1}{\mu}\right)\binom{2p+1}{p-m}A_{n-m,\,2m+1} - \frac{1}{\mu}\sum_{m=0}^{n-1,p}\frac{2m+1}{2m+2}\binom{2p+1}{p-m}A_{n-m-1,\,2m+2}\right\}$$

$$R^{2n}_{2p} = \frac{2p+1}{2^{2p+2}}\left[\left(1 - \frac{1}{\mu}\right)2\binom{2p}{p}A_{n,1} - \frac{4}{\mu}\sum_{m=0}^{n-1,p}\frac{2m}{2m+2}\binom{2p}{p-m}A_{n-m-1,\,2m+2}\right.$$
$$+ \frac{1}{n}\left\{-\frac{(1-\mu)^2}{\mu}\binom{2p}{p}A_{n,1}\right.$$
$$+ \sum_{m=0}^{n,p-1}\left(\frac{\mu}{2m+1} - \frac{2m+1}{\mu}\right)\binom{2p}{p-m-1}A_{n-m,\,2m+1}$$
$$\left.\left.- \sum_{m=1}^{n,p}\left(\frac{\mu}{2m+1} + \frac{6m-1}{\mu}\right)\binom{2p}{p-m}A_{n-m,\,2m+1}\right\}\right] \quad (n \geqslant 1)$$

$$Q^{2n}_{2p} = \frac{2p+2}{\mu 2^{2p+1}}\sum_{m=0}^{n,p}\binom{2p+1}{p-m}A_{n-m,\,2m+1}$$

$$S^{2n}_{2p} = \frac{2p+1}{2^{2p+1}}\left\{\left(\frac{1}{\mu} - 1\right)\binom{2p}{p}A_{n,1} + \frac{1}{\mu}\sum_{m=0}^{n,p}\frac{4m}{4m+1}\binom{2p}{p-m}A_{n-m,\,2m+1} + \frac{2}{2n+1}\sum_{m=1}^{n+1,p}\left(1 + \frac{2m-1}{\mu}\right)\binom{2p}{p-m}A_{n-m+1,\,2m}\right\}$$

$$P^{2n+1}_{2p+1} = \frac{2p+3}{2^{2p+2}}\left\{-\frac{1}{\mu}\sum_{m=1}^{n,p+1}\frac{2m}{2m+1}\binom{2p+2}{p-m+1}A_{n-m,\,2m+1} + \frac{1}{2n+1}\sum_{m=0}^{n,p}\left(1 - \frac{2m+2}{\mu}\right)\binom{2p+2}{p-m}A_{n-m,\,2m+2}\right\}$$

$$R_{2p+1}^{2n+1} = \frac{2p+2}{2^{2p+2}}\left[\binom{2p+1}{p}A_{n,1} - \frac{2}{\mu}\sum_{m=1}^{n,p+1}\frac{2m-1}{2m+1}\binom{2p+1}{p-m+1}A_{n-m,\,2m+1}\right.$$
$$-\frac{1}{2n+1}\sum_{m=0}^{n,p}\frac{1}{p+m+2}\binom{2p+1}{p-m}\left\{\mu + \frac{1}{\mu}((8m+4)p\right.$$
$$\left.\left.+4(m^2+3m+1))\right\}A_{n-m,\,2m+2}\right]$$

$$Q_{2p+1}^{2n+1} = \frac{2p+3}{\mu\cdot 2^{2p+2}}\sum_{m=0}^{n,p}\binom{2p+2}{p-m}A_{n-m,\,2m+2}$$

$$S_{2p+1}^{2n+1} = \frac{2p+2}{2^{2p+2}}\left[-\binom{2p+1}{p}\left(1+\frac{1}{n+1}\right)A_{n+1,1}\right.$$
$$+\frac{2}{2n+2}\sum_{m=0}^{n+1,p}\left(1+\frac{2m}{\mu}\right)\binom{2p}{p-m}A_{n-m+1,\,2m+1}$$
$$\left.+\frac{1}{\mu}\sum_{m=0}^{n,p}\frac{4m+2}{2m+2}\binom{2p+1}{p-m}A_{n-m,\,2m+2}\right]. \tag{2.147}$$

The notation

$$\mu = \frac{3+\nu}{1-\nu} \tag{2.148}$$

has been used.

Various symmetric crack problems can be treated by using the Laurent expansions (2.145) in a manner analogous to the plane problems, since the basic structure of the equations (2.64) to (2.67) for plane extension is similar to that of equations (2.141) to (2.143) applied to bending. Finally, the stress intensity factor $k_B$ is calculated from the limiting expression

$$k_B = \frac{-12(2d)^{\frac{1}{2}}(3+\nu)T_0}{h^2}\lim_{z\to\lambda}\left[(z-\lambda)^{\frac{1}{2}}\phi'(z)\right]. \tag{2.149}$$

The dashed line in Figure 2.34 of Appendix IV shows the dimensionless factor $F_B$ $(=k_B\cdot h^2/(6T_y^\infty a^{\frac{1}{2}}))$ for infinite parallel cracks in a wide plate subjected to $T_y^\infty$ at infinity.

## 2.5 Applicability of the method

In the foregoing analyses the stress intensity factors or their dimensionless factors $F(\lambda)$ have been given by power series of a relative crack length $\lambda$, and their coefficients have been evaluated from closed form expressions. Therefore they must agree with the Maclaurin's expansions which should exist uniquely. Actually the results of $F(\lambda)$ agree with the expansions of the closed form solutions for several special cases such as problems (a) and (b) of Appendix II.

Thus the only source of the numerical errors is the truncation of the infinite series, but no practical way is available for estimating the upper bounds of these errors at the present time. However, according to the author's experience in various problems treated in this paper as well as previous works based on the perturbation technique, it appears reasonable to estimate upper bounds of the numerical errors by assuming some geometric series for the uncalculated terms of higher orders.

On this basis the proposed method seems to give practically exact values as long as the smallest circles enclosing each hole or crack are not too close to each other and to the other boundaries. More precisely the method seems to be valid if the relative crack length $\lambda$ is less than 0.9 with few exceptions, as suggested by several examples in Appendix II. This condition is sufficient for some cases treated in section 2.3. In the cases of parallel cracks, $\lambda$ may take any large values and hence the present analysis will no longer apply. The situation may be improved if the perturbation procedure is replaced by a direct solution of simultaneous equations, but it will be inevitably accompanied by the loss of wide applicability and other difficulties such as the error estimation, increase of necessary computer capacity and so on.

A few comments with regard to the extension of the proposed method will be made. Effect of body forces which has been neglected here can be easily taken into consideration. Expressions of the complex potentials for free elliptical holes in the presence of in-plane gravitational forces can be derived. This was done for the problem of an elliptic-sectioned tunnel under a horizontal surface [12]. Further applications could be made for problems such as the cracked rotating disks.

The present method of solution may also be applied to solve crack problems involving surface tractions such as the Dugdale crack model. With certain modifications, problems with multi-valued displacement fields may be treated. An example of this is the case of cracks in initially stressed fields.

## 2.6 Appendix I

*Coefficients of free crack relations* (2.21), (2.77) *and* (2.146).

TABLE I — $Q_{2p}^{2n}$ (longitudinal shear)

| $n \backslash p$ | 0 | 1 | 2 | 3 | 4 | 5 | 6 | 7 | 8 |
|---|---|---|---|---|---|---|---|---|---|
| 0 | 0.500000 | 0.375000 | 0.312500 | 0.273438 | 0.246094 | 0.225586 | 0.209473 | 0.196381 | 0.185471 |
| 1 | 0.125000 | 0.125000 | 0.117188 | 0.109375 | 0.102539 | 0.096680 | 0.091644 | 0.087280 | |
| 2 | 0.062500 | 0.070313 | 0.070313 | 0.068359 | 0.065918 | 0.063446 | 0.061096 | | |
| 3 | 0.039062 | 0.046875 | 0.048828 | 0.048828 | 0.048065 | 0.046997 | | | |
| 4 | 0.027344 | 0.034180 | 0.036621 | 0.037384 | 0.037384 | | | | |
| 5 | 0.020508 | 0.026367 | 0.028839 | 0.029907 | | | | | |
| 6 | 0.016113 | 0.021149 | 0.023499 | | | | | | |
| 7 | 0.013092 | 0.017456 | | | | | | | |
| 8 | 0.010910 | | | | | | | | |

TABLE II — $Q_{2p+1}^{2n+1}$ (longitudinal shear)

| $n \backslash p$ | 0 | 1 | 2 | 3 | 4 | 5 | 6 | 7 | 8 |
|---|---|---|---|---|---|---|---|---|---|
| 0 | 0.125000 | 0.125000 | 0.117188 | 0.109375 | 0.102539 | 0.096680 | 0.091644 | 0.087280 | 0.083462 |
| 1 | 0.062500 | 0.070313 | 0.070313 | 0.068359 | 0.065918 | 0.063446 | 0.061096 | 0.058914 | |
| 2 | 0.039063 | 0.046875 | 0.048828 | 0.048828 | 0.048065 | 0.046997 | 0.045822 | | |
| 3 | 0.027344 | 0.034180 | 0.036621 | 0.037384 | 0.037384 | 0.037010 | | | |
| 4 | 0.020508 | 0.026367 | 0.028839 | 0.029907 | 0.030281 | | | | |
| 5 | 0.016113 | 0.021149 | 0.023499 | 0.024673 | | | | | |
| 6 | 0.013092 | 0.017456 | 0.019638 | | | | | | |
| 7 | 0.010910 | 0.014729 | | | | | | | |
| 8 | 0.009274 | | | | | | | | |

TABLE III — $P_{2p}^{2n}$ (plane problem)

| n \ p | 0 | 1 | 2 | 3 | 4 | 5 | 6 | 7 | 8 |
|---|---|---|---|---|---|---|---|---|---|
| 0 | 1.000000 | 1.500000 | 1.875000 | 2.187500 | 2.460938 | 2.707032 | 2.932618 | 3.142090 | 3.338470 |
| 1 | 0.250000 | 0.500000 | 0.703125 | 0.875000 | 1.025391 | 1.160156 | 1.283020 | 1.396485 | |
| 2 | 0.093750 | 0.210938 | 0.316406 | 0.410156 | 0.494385 | 0.571015 | 0.641510 | | |
| 3 | 0.052083 | 0.125000 | 0.195313 | 0.260417 | 0.320435 | 0.375977 | | | |
| 4 | 0.034180 | 0.085449 | 0.137329 | 0.186920 | 0.233650 | | | | |
| 5 | 0.024609 | 0.063281 | 0.103821 | 0.143555 | | | | | |
| 6 | 0.018799 | 0.049347 | 0.082245 | | | | | | |
| 7 | 0.014962 | 0.039900 | | | | | | | |
| 8 | 0.012274 | | | | | | | | |

TABLE IV — $P_{2p+1}^{2n+1}$ (plane problem)

| n \ p | 0 | 1 | 2 | 3 | 4 | 5 | 6 | 7 | 8 |
|---|---|---|---|---|---|---|---|---|---|
| 0 | 0.562500 | 0.937500 | 1.230469 | 1.476563 | 1.691895 | 1.885254 | 2.061997 | 2.225646 | 0.000000 |
| 1 | 0.156250 | 0.292969 | 0.410156 | 0.512696 | 0.604248 | 0.687332 | 0.763703 | 0.834618 | |
| 2 | 0.082031 | 0.164063 | 0.239258 | 0.307617 | 0.370102 | 0.427673 | 0.481133 | | |
| 3 | 0.052734 | 0.109863 | 0.164795 | 0.216293 | 0.264359 | 0.309300 | | | |
| 4 | 0.037598 | 0.080566 | 0.123367 | 0.164490 | 0.203556 | | | | |
| 5 | 0.028564 | 0.062485 | 0.097199 | 0.131218 | | | | | |
| 6 | 0.022659 | 0.050354 | 0.079308 | | | | | | |
| 7 | 0.018547 | 0.041731 | | | | | | | |
| 8 | 0.015547 | | | | | | | | |

TABLE V — $Q_{2p}^{2n}$ (plane problem)

| n \ p | 0 | 1 | 2 | 3 | 4 | 5 | 6 | 7 | 8 |
|---|---|---|---|---|---|---|---|---|---|
| 0 | 0.500000 | 0.750000 | 0.937500 | 1.093750 | 1.230469 | 1.353516 | 1.466309 | 1.571045 | 1.669235 |
| 1 | 0.125000 | 0.250000 | 0.351563 | 0.437500 | 0.512696 | 0.580078 | 0.641510 | 0.698243 | |
| 2 | 0.062500 | 0.140625 | 0.210938 | 0.273438 | 0.329590 | 0.380676 | 0.427673 | | |
| 3 | 0.039063 | 0.093750 | 0.146484 | 0.195313 | 0.240326 | 0.281983 | | | |
| 4 | 0.027344 | 0.068359 | 0.109863 | 0.149536 | 0.186920 | | | | |
| 5 | 0.020508 | 0.052734 | 0.086517 | 0.119629 | | | | | |
| 6 | 0.016113 | 0.042297 | 0.070496 | | | | | | |
| 7 | 0.013092 | 0.034912 | | | | | | | |
| 8 | 0.010910 | | | | | | | | |

TABLE VI — $Q_{2p+1}^{2n+1}$ (plane problem)

| n \ p | 0 | 1 | 2 | 3 | 4 | 5 | 6 | 7 | 8 |
|---|---|---|---|---|---|---|---|---|---|
| 0 | 0.187500 | 0.312500 | 0.410156 | 0.492188 | 0.563965 | 0.628418 | 0.687332 | 0.741882 | 0.792887 |
| 1 | 0.093750 | 0.175781 | 0.246094 | 0.307617 | 0.362549 | 0.412399 | 0.458222 | 0.500771 | |
| 2 | 0.058594 | 0.117188 | 0.170898 | 0.219727 | 0.264359 | 0.305481 | 0.343666 | | |
| 3 | 0.041016 | 0.085449 | 0.128174 | 0.168228 | 0.205612 | 0.240566 | | | |
| 4 | 0.030762 | 0.065918 | 0.100937 | 0.134583 | 0.166546 | | | | |
| 5 | 0.024170 | 0.052872 | 0.082245 | 0.111031 | | | | | |
| 6 | 0.019638 | 0.043640 | 0.068733 | | | | | | |
| 7 | 0.016365 | 0.036821 | | | | | | | |
| 8 | 0.013910 | | | | | | | | |

TABLE VII — $W_{2p}^{2n}$ (plane problem)

| n \ p | 0 | 1 | 2 | 3 | 4 | 5 | 6 | 7 | 8 |
|---|---|---|---|---|---|---|---|---|---|
| 0 | 0.000000 | 0.375000 | 0.625000 | 0.820313 | 0.984375 | 1.127930 | 1.256836 | 1.374665 | 1.483765 |
| 1 | 0.000000 | 0.125000 | 0.234375 | 0.328125 | 0.410156 | 0.483399 | 0.549866 | 0.610962 | |
| 2 | 0.000000 | 0.070313 | 0.140625 | 0.205078 | 0.263672 | 0.317230 | 0.366577 | | |
| 3 | 0.000000 | 0.046875 | 0.097656 | 0.146484 | 0.192261 | 0.234985 | | | |
| 4 | 0.000000 | 0.034180 | 0.073242 | 0.112152 | 0.149536 | | | | |
| 5 | 0.000000 | 0.026367 | 0.057678 | 0.089722 | | | | | |
| 6 | 0.000000 | 0.021149 | 0.046997 | | | | | | |
| 7 | 0.000000 | 0.017456 | | | | | | | |
| 8 | 0.000000 | | | | | | | | |

TABLE VIII — $W_{2p+1}^{2n+1}$ (plane problem)

| n \ p | 0 | 1 | 2 | 3 | 4 | 5 | 6 | 7 | 8 |
|---|---|---|---|---|---|---|---|---|---|
| 0 | 0.062500 | 0.187500 | 0.292969 | 0.382813 | 0.461426 | 0.531738 | 0.595688 | 0.654602 | 0.709425 |
| 1 | 0.031250 | 0.105469 | 0.175781 | 0.239258 | 0.296631 | 0.348953 | 0.397125 | 0.441857 | |
| 2 | 0.019531 | 0.070313 | 0.122070 | 0.170898 | 0.216293 | 0.258484 | 0.297844 | | |
| 3 | 0.013672 | 0.051270 | 0.091553 | 0.130844 | 0.168228 | 0.203556 | | | |
| 4 | 0.010254 | 0.039551 | 0.072098 | 0.104675 | 0.136265 | | | | |
| 5 | 0.008057 | 0.031723 | 0.058746 | 0.086357 | | | | | |
| 6 | 0.006546 | 0.026184 | 0.049095 | | | | | | |
| 7 | 0.005455 | 0.022093 | | | | | | | |
| 8 | 0.004637 | | | | | | | | |

TABLE IX — $P_{2p}^{2n}$ (plate bending, $\nu=0.3$)

| $n$ \ $p$ | 0 | 1 | 2 | 3 | 4 | 5 | 6 | 7 | 8 |
|---|---|---|---|---|---|---|---|---|---|
| 0 | 0.393939 | 0.590909 | 0.738636 | 0.861743 | 0.969460 | 1.066406 | 1.155274 | 1.237793 | 1.315155 |
| 1 | 0.022727 | 0.045455 | 0.063920 | 0.079545 | 0.093217 | 0.105469 | 0.116638 | 0.126953 | |
| 2 | −0.000947 | −0.002131 | −0.003196 | −0.004143 | −0.004994 | −0.005768 | −0.006480 | | |
| 3 | −0.003157 | −0.007576 | −0.011837 | −0.015783 | −0.019420 | −0.022786 | | | |
| 4 | −0.003107 | −0.007768 | −0.012484 | −0.016993 | −0.021241 | | | | |
| 5 | −0.002734 | −0.007031 | −0.011536 | −0.015951 | | | | | |
| 6 | −0.002360 | −0.006195 | −0.010325 | | | | | | |
| 7 | −0.002040 | −0.005441 | | | | | | | |
| 8 | −0.001777 | | | | | | | | |

TABLE X — $P_{2p+1}^{2n+1}$ (plate bending, $\nu=0.3$)

| $n$ \ $p$ | 0 | 1 | 2 | 3 | 4 | 5 | 6 | 7 | 8 |
|---|---|---|---|---|---|---|---|---|---|
| 0 | 0.107955 | 0.179924 | 0.236151 | 0.283381 | 0.324707 | 0.361817 | 0.395737 | 0.427144 | 0.456510 |
| 1 | 0.004735 | 0.008878 | 0.012429 | 0.015536 | 0.018311 | 0.020828 | 0.023143 | 0.025291 | |
| 2 | −0.003196 | −0.006392 | −0.009322 | −0.011985 | −0.014420 | −0.016663 | −0.018745 | | |
| 3 | −0.004084 | −0.008508 | −0.012762 | −0.016750 | −0.020472 | −0.023952 | | | |
| 4 | −0.003832 | −0.008212 | −0.012575 | −0.016766 | −0.020748 | | | | |
| 5 | −0.003396 | −0.007428 | −0.011555 | −0.015599 | | | | | |
| 6 | −0.002975 | −0.006612 | −0.010414 | | | | | | |
| 7 | −0.002612 | −0.005877 | | | | | | | |
| 8 | −0.002306 | | | | | | | | |

TABLE XI — $R_{2p}^{2n}$ (plate bending, $\nu = 0.3$)

| n \ p | 0 | 1 | 2 | 3 | 4 | 5 | 6 | 7 | 8 |
|---|---|---|---|---|---|---|---|---|---|
| 0 | −0.731602 | −0.253247 | 0.035173 | 0.246212 | 0.415483 | 0.558594 | 0.683733 | 0.795724 | 0.897645 |
| 1 | −0.042208 | −0.019481 | 0.003044 | 0.022727 | 0.039950 | 0.055246 | 0.069031 | 0.081613 | |
| 2 | 0.001759 | 0.000913 | −0.000152 | −0.001184 | −0.002140 | −0.003021 | −0.003835 | | |
| 3 | 0.005862 | 0.003247 | −0.000564 | −0.004509 | −0.008323 | −0.011936 | | | |
| 4 | 0.005771 | 0.003329 | −0.000594 | −0.004855 | −0.009103 | | | | |
| 5 | 0.005078 | 0.003013 | −0.000549 | −0.004557 | | | | | |
| 6 | 0.004383 | 0.002655 | −0.000492 | | | | | | |
| 7 | 0.003789 | 0.002332 | | | | | | | |
| 8 | 0.003300 | | | | | | | | |

TABLE XII – $R_{2p+1}^{2n+1}$ (plate bending, $\nu = 0.3$)

| n \ p | 0 | 1 | 2 | 3 | 4 | 5 | 6 | 7 | 8 |
|---|---|---|---|---|---|---|---|---|---|
| 0 | −0.316423 | −0.385078 | −0.415610 | −0.436583 | −0.454080 | −0.469953 | −0.484901 | −0.499228 | −0.513085 |
| 1 | −0.051159 | −0.087206 | −0.110022 | −0.127629 | −0.142447 | −0.155502 | −0.167326 | −0.178227 | |
| 2 | −0.018593 | −0.041079 | −0.057254 | −0.070599 | −0.082231 | −0.092674 | −0.102226 | | |
| 3 | −0.009000 | −0.024663 | −0.036829 | −0.047344 | −0.056781 | −0.065414 | | | |
| 4 | −0.005078 | −0.016771 | −0.026342 | −0.034907 | −0.042774 | | | | |
| 5 | −0.003153 | −0.012306 | −0.020090 | −0.027244 | | | | | |
| 6 | −0.002092 | −0.009504 | −0.015996 | | | | | | |
| 7 | −0.001455 | −0.007616 | | | | | | | |
| 8 | −0.001050 | | | | | | | | |

TABLE XIII — $Q_{2p}^{2n}$ (plate bending, $\nu=0.3$)

| n \ p | 0 | 1 | 2 | 3 | 4 | 5 | 6 | 7 | 8 |
|---|---|---|---|---|---|---|---|---|---|
| 0 | 0.106061 | 0.159091 | 0.198864 | 0.232008 | 0.261009 | 0.287109 | 0.311035 | 0.333252 | 0.354080 |
| 1 | 0.026515 | 0.053030 | 0.074574 | 0.092803 | 0.108754 | 0.123047 | 0.136078 | 0.148112 | |
| 2 | 0.013258 | 0.029830 | 0.044744 | 0.058002 | 0.069913 | 0.080750 | 0.090719 | | |
| 3 | 0.008286 | 0.019886 | 0.031072 | 0.041430 | 0.050978 | 0.059814 | | | |
| 4 | 0.005800 | 0.014500 | 0.023304 | 0.031720 | 0.039650 | | | | |
| 5 | 0.004350 | 0.011186 | 0.018352 | 0.025376 | | | | | |
| 6 | 0.003418 | 0.008972 | 0.014954 | | | | | | |
| 7 | 0.002777 | 0.007406 | | | | | | | |
| 8 | 0.002314 | | | | | | | | |

TABLE XIV — $Q_{2p+1}^{2n+1}$ (plate bending, $\nu=0.3$)

| n \ p | 0 | 1 | 2 | 3 | 4 | 5 | 6 | 7 | 8 |
|---|---|---|---|---|---|---|---|---|---|
| 0 | 0.039773 | 0.066288 | 0.087003 | 0.104403 | 0.119629 | 0.133301 | 0.145798 | 0.157369 | 0.168188 |
| 1 | 0.019886 | 0.037287 | 0.052202 | 0.065252 | 0.076904 | 0.087479 | 0.097199 | 0.106224 | |
| 2 | 0.012429 | 0.024858 | 0.036251 | 0.046609 | 0.056076 | 0.064799 | 0.072899 | | |
| 3 | 0.008700 | 0.018126 | 0.027188 | 0.035685 | 0.043615 | 0.051029 | | | |
| 4 | 0.006525 | 0.013983 | 0.021411 | 0.028548 | 0.035328 | | | | |
| 5 | 0.005127 | 0.011215 | 0.017446 | 0.023552 | | | | | |
| 6 | 0.004166 | 0.009257 | 0.014580 | | | | | | |
| 7 | 0.003471 | 0.007811 | | | | | | | |
| 8 | 0.002951 | | | | | | | | |

TABLE XV — $S_{2p}^{2n}$ (plate bending, $\nu=0.3$)

| $n$ \ $p$ | 0 | 1 | 2 | 3 | 4 | 5 | 6 | 7 | 8 |
|---|---|---|---|---|---|---|---|---|---|
| 0 | −0.196970 | −0.068182 | 0.009470 | 0.066288 | 0.111861 | 0.150391 | 0.184082 | 0.214234 | 0.241674 |
| 1 | −0.049242 | −0.022727 | 0.003551 | 0.026515 | 0.046609 | 0.064453 | 0.080536 | 0.095215 | |
| 2 | −0.024621 | −0.012784 | 0.002131 | 0.016572 | 0.029963 | 0.042297 | 0.053691 | | |
| 3 | −0.015388 | −0.008523 | 0.001480 | 0.011837 | 0.021848 | 0.031331 | | | |
| 4 | −0.010772 | −0.006214 | 0.001110 | 0.009063 | 0.016993 | | | | |
| 5 | −0.008079 | −0.004794 | 0.000874 | 0.007250 | | | | | |
| 6 | −0.006348 | −0.003845 | 0.000712 | | | | | | |
| 7 | −0.005157 | −0.003174 | | | | | | | |
| 8 | −0.004298 | | | | | | | | |

TABLE XVI — $S_{2p+1}^{2n+1}$ (plate bending, $\nu=0.3$)

| $n$ \ $p$ | 0 | 1 | 2 | 3 | 4 | 5 | 6 | 7 | 8 |
|---|---|---|---|---|---|---|---|---|---|
| 0 | −0.035985 | −0.009470 | 0.015980 | 0.038116 | 0.057484 | 0.074707 | 0.090256 | 0.104472 | −0.372801 |
| 1 | −0.017992 | −0.005327 | 0.009588 | 0.023822 | 0.036954 | 0.049026 | 0.060170 | −0.129234 | |
| 2 | −0.011245 | −0.003551 | 0.006658 | 0.017016 | 0.026946 | 0.036316 | −0.066013 | | |
| 3 | −0.007872 | −0.002589 | 0.004994 | 0.013028 | 0.020958 | −0.039349 | | | |
| 4 | −0.005904 | −0.001998 | 0.003933 | 0.010422 | −0.025252 | | | | |
| 5 | −0.004639 | −0.001602 | 0.003204 | −0.016693 | | | | | |
| 6 | −0.003769 | −0.001322 | −0.010934 | | | | | | |
| 7 | −0.003141 | −0.009205 | | | | | | | |
| 8 | −0.005152 | | | | | | | | |

## 2.7 Appendix II

*Comparison of the present results with closed form solutions.* To know the general trend of the obtained series $F(\lambda)$ consider the two special cases whose exact solutions are given by the following expressions:

(a) A pair of equal colinear cracks in infinite bodies.

The dimensionless factors $F(\lambda)$ are the same for tension, in-plane shear, longitudinal shear and plate bending, and the common results are [13]

$$F_A = \frac{(1+\lambda)^2 \dfrac{E(m)}{K(m)} - (1-\lambda)^2}{2\lambda(1-\lambda)^{\frac{1}{2}}} \qquad \text{(inner crack tip)}$$

$$F_B = \frac{(1+\lambda)^{\frac{3}{2}} \left\{ 1 - \dfrac{E(m)}{K(m)} \right\}}{2\lambda} \qquad \text{(outer crack tip)} \tag{2.150}$$

$$m = \frac{2(\lambda^{\frac{1}{2}})}{1+\lambda}, \qquad \lambda = \frac{\text{crack length}}{\text{crack spacing}}$$

where $K(m)$ and $E(m)$ are complete elliptical integrals of the first and second kinds.

(b) Infinite row of equal colinear cracks [14]

$$F(\lambda) = \left( \frac{2}{\pi\lambda} \tan \frac{\pi\lambda}{2} \right)^{\frac{1}{2}}. \tag{2.151}$$

Series of $F(\lambda)$ for those cases obtained by the above analyses have agreed with the expansions of equations (2.150) and (2.151). The magnitudes of the series coefficients are generally found to decrease with higher powers of $\lambda$, and the numerical results by the present analyses are assumed to give practically exact values unless $\lambda$ is less than and not so close to one in most cases.

For rough estimation of the numerical errors it is also convenient to examine the partial sums of the series. Table XVII shows the partial sums for the above two cases together with those for other three problems below.

*Wide plate containing infinite row of equal parallel cracks.*

(c) Tension
(d) In-plane shear
(e) Plate bending.

TABLE XVII — Partial sums of $F(\lambda)$

| Case | $\lambda$ \ $M$ | 10 | 20 | 30 | 40 | 50 | 60 | 70 | exact solution |
|---|---|---|---|---|---|---|---|---|---|
| (a) | 0.7 | 1.1308 | 1.1332 | 1.1333 | 1.1333 | 1.1333 | 1.1333 | 1.1333 | 1.1333 |
| | 0.8 | 1.2133 | 1.2277 | 1.2288 | 1.2289 | 1.2289 | 1.2289 | 1.2289 | 1.2289 |
| | 0.9 | 1.3522 | 1.4269 | 1.4460 | 1.4515 | 1.4531 | 1.4536 | 1.4538 | 1.4539 |
| | 0.95 | 1.4555 | 1.6199 | 1.6927 | 1.7285 | 1.7470 | 1.7569 | 1.7622 | 1.7689 |
| (b) | 0.7 | 1.3247 | 1.3358 | 1.3360 | 1.3360 | 1.3360 | 1.3360 | 1.3360 | 1.3360 |
| | 0.8 | 1.5069 | 1.5603 | 1.5646 | 1.5649 | 1.5650 | 1.5650 | 1.5650 | 1.5650 |
| | 0.9 | 1.7930 | 2.0258 | 2.0873 | 2.1053 | 2.1108 | 2.1125 | 2.1130 | 2.1133 |
| | 0.95 | 1.9938 | 2.4669 | 2.6827 | 2.7912 | 2.8483 | 2.8792 | 2.8962 | 2.9180 |
| (c) | 0.5 | 0.7985 | 0.7895 | 0.7896 | 0.7896 | 0.7896 | 0.7896 | 0.7896 | |
| | 0.6 | 0.7828 | 0.7294 | 0.7349 | 0.7343 | 0.7344 | 0.7344 | 0.7344 | |
| | 0.65 | 0.8097 | 0.6855 | 0.7140 | 0.7075 | 0.7090 | 0.7086 | 0.7087 | |
| | 0.7 | 0.8828 | 0.5889 | 0.7300 | 0.6629 | 0.6948 | 0.6796 | 0.6868 | |
| (d) | 0.7 | 1.1494 | 1.1532 | 1.1532 | 1.1532 | 1.1532 | 1.1532 | 1.1532 | |
| | 0.8 | 1.1743 | 1.1880 | 1.1877 | 1.1877 | 1.1877 | 1.1877 | 1.1877 | |
| | 0.9 | 1.1827 | 1.2255 | 1.2236 | 1.2220 | 1.2227 | 1.2225 | 1.2226 | |
| | 0.95 | 1.1746 | 1.2485 | 1.2449 | 1.2354 | 1.2425 | 1.2388 | 1.2405 | |
| (e) | 0.7 | 0.9003 | 0.9019 | 0.9019 | 0.9019 | 0.9019 | 0.9019 | 0.9019 | |
| | 0.8 | 0.8728 | 0.8788 | 0.8778 | 0.8779 | 0.8779 | 0.8779 | 0.8779 | |
| | 0.9 | 0.8388 | 0.8622 | 0.8509 | 0.8542 | 0.8534 | 0.8535 | 0.8535 | |
| | 0.95 | 0.8175 | 0.8657 | 0.8288 | 0.8468 | 0.8390 | 0.8421 | 0.8410 | |

The table suggests that the $F$-series calculated up to $\lambda^{70}$ give practically exact values if $\lambda \leqslant 0.9 \sim 0.95$ with an exception of case (c).

## 2.8 Appendix III

*Stress intensity factor for tension of infinite strip with eccentric internal crack.*

Typical numerical results obtained from equation (2.102) truncating at $M=19$ are shown in Figure 2.13. It is noteworthy that $F_{1,A}(\delta, \lambda)$ is almost independent of the eccentricity for $\delta \leqslant 0.4$. In other words, the stress intensity factor is close to that of a centrally cracked strip whose width is $2d_1$. Their numerical values and further discussions appear in [15].

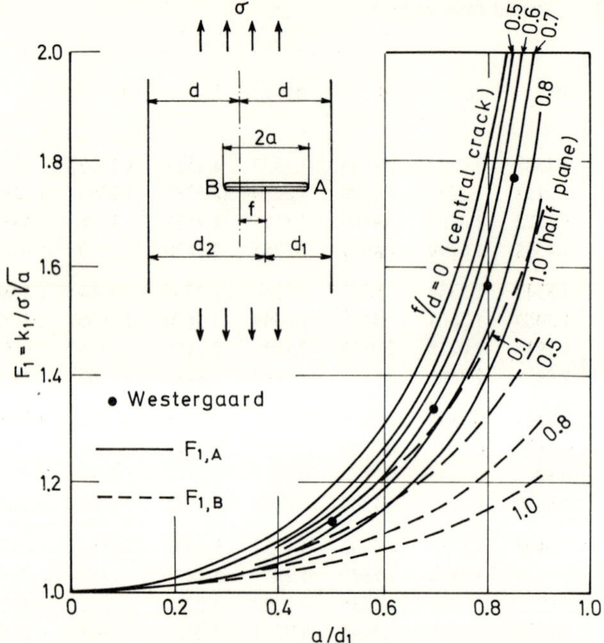


Figure 2.13. Tension of long strip with eccentric crack.

For the important case of a central crack ($\delta = 0$), the following empirical formulae are proposed on the basis of the above results [16]:

(1) Polynomial $\quad F(\lambda) = 1 + 0.128\,\lambda - 0.288\,\lambda^2 + 1.525\,\lambda^3 \qquad (2.152)$

(2) Secant formula $\quad F(\lambda) = (\sec \tfrac{1}{2}\pi\lambda)^{\frac{1}{2}}\,. \qquad (2.153)$

Equation (2.152) has been derived by least square curve fitting, while equation (2.153) has been assumed by simple analogy of the tangent formula for periodic colinear cracks given by equations (2.151) of Appendix II. The remarkable accuracy of equation (2.153) suggests the possibility of a similar formula for general eccentric cracks.

The following expressions are recommended for their simplicity and reasonable accuracy:

$$(1)\ F_A(\delta, \lambda) = \left(\sec \frac{\pi\lambda}{2+\delta}\right)^{\frac{1}{2}} \qquad \begin{pmatrix} \lambda \leqslant 0.8 - 2\delta & \text{for } \delta \leqslant 0.2 \\ \lambda \leqslant 0.4 & \text{for } \delta > 0.2 \end{pmatrix}$$

$$(2)\ F_A(\delta, \lambda) = \left(\sec \frac{\pi\lambda}{2} \cdot \frac{\sin 2\lambda\delta}{2\lambda\delta}\right)^{\frac{1}{2}} \quad (\lambda \leqslant 0.7 \sim 0.8 \text{ for all } \delta)\,. \qquad (2.154)$$

TABLE XVIII — Values of $F(\lambda)$ for centrally cracked strip

| $\lambda$ | 0.1 | 0.2 | 0.3 | 0.4 | 0.5 | 0.6 | 0.7 | 0.8 | 0.9 |
|---|---|---|---|---|---|---|---|---|---|
| Author (up to $\lambda^{70}$) | 1.0060 | 1.0246 | 1.0577 | 1.1094 | 1.1867 | 1.3033 | 1.4882 | 1.8160 | 2.58 |
| Author (up to $\lambda^{19}$) [15] | 1.0060 | 1.0246 | 1.0577 | 1.1094 | 1.1867 | 1.3033 | 1.4881 | 1.811 | 2.47 |
| Eq. (2.153) | 1.0062 | 1.0254 | 1.0594 | 1.1118 | 1.1892 | 1.3043 | 1.4841 | 1.799 | 2.53 |
| Eq. (2.152) | 1.006 | 1.026 | 1.054 | 1.109 | 1.183 | 1.303 | 1.472 | 1.70 | 2.0 |

Errors of these formulae will be less than one per cent within the range shown in parentheses.

Further calculations have been made for the two special cases, centrally cracked strip ($\delta=0$) and cracked half plane ($\delta\to 1$), and the series (2.102) were obtained up to the term in $\lambda^{70}$. Table XVIII shows the comparison of $F$-values from various sources.

## 2.9 Appendix IV

*Other numerical results of $F(\lambda)$.*

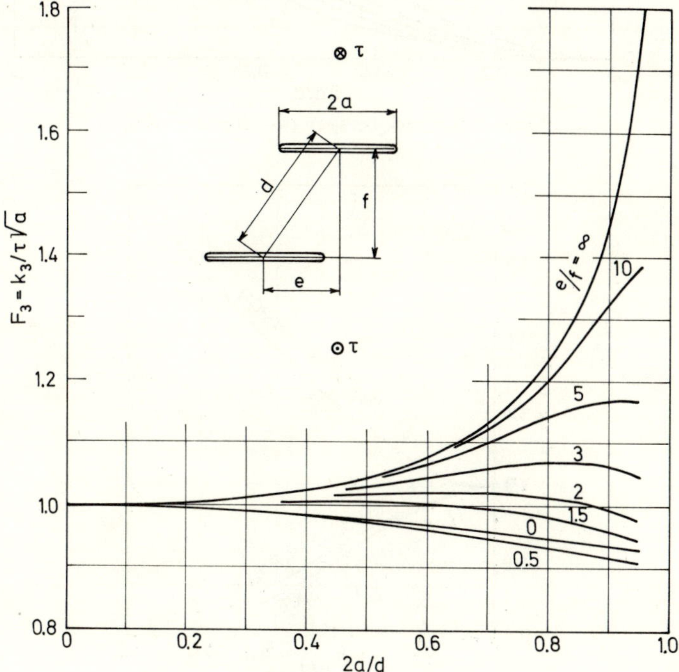


Figure 2.14. A pair of equal parallel cracks.

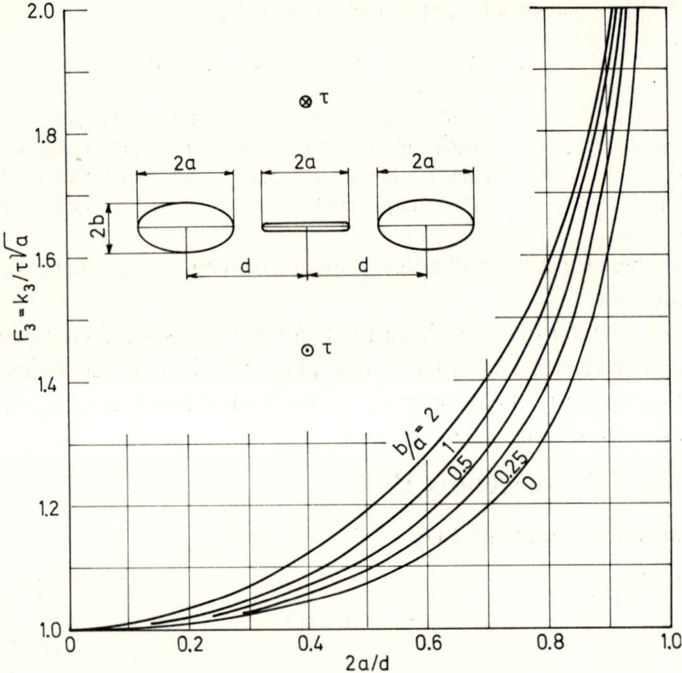


Figure 2.15. Crack between two elliptical holes.

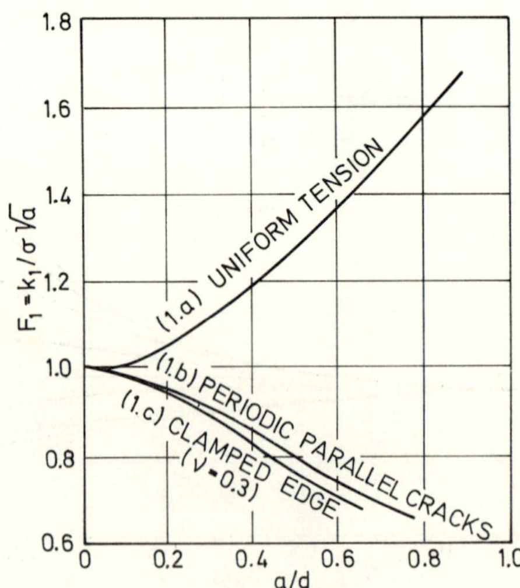


Figure 2.16. Strip with longitudinal crack subjected to tension along straight edges.

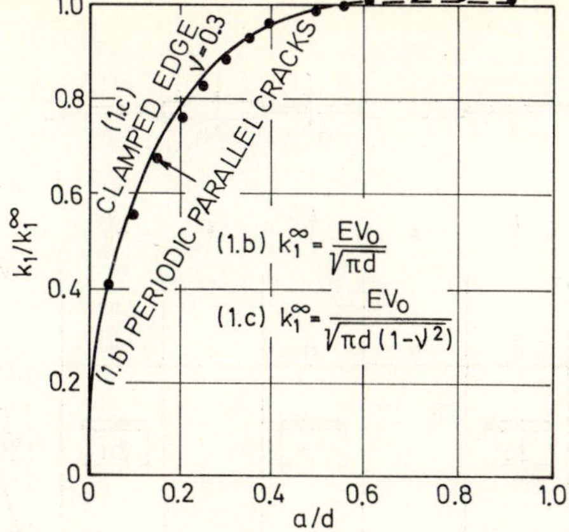


Figure 2.17. Fractions of stress intensity factors on the basis of those for semi-infinite crack.

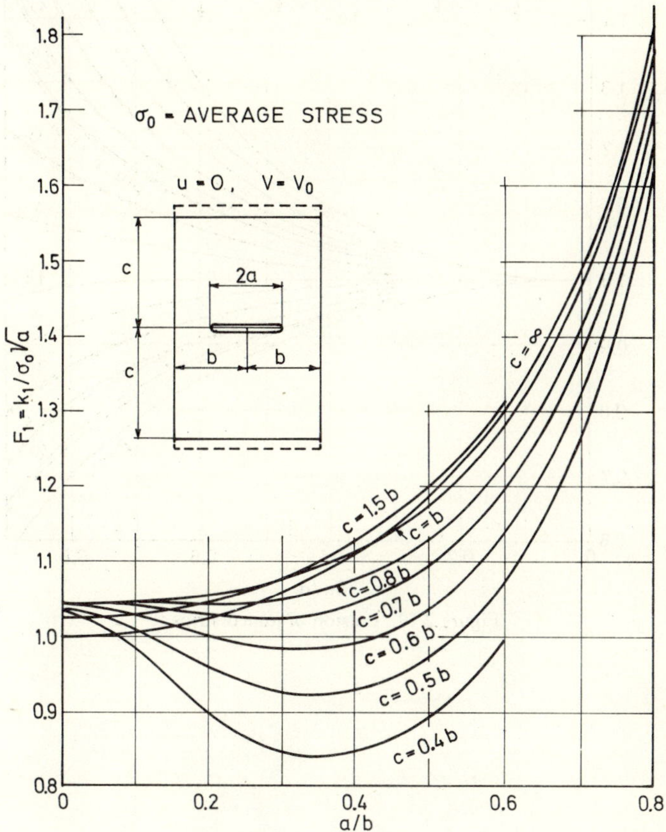


Figure 2.18. Rectangular plate subjected to tension by clamped ends.

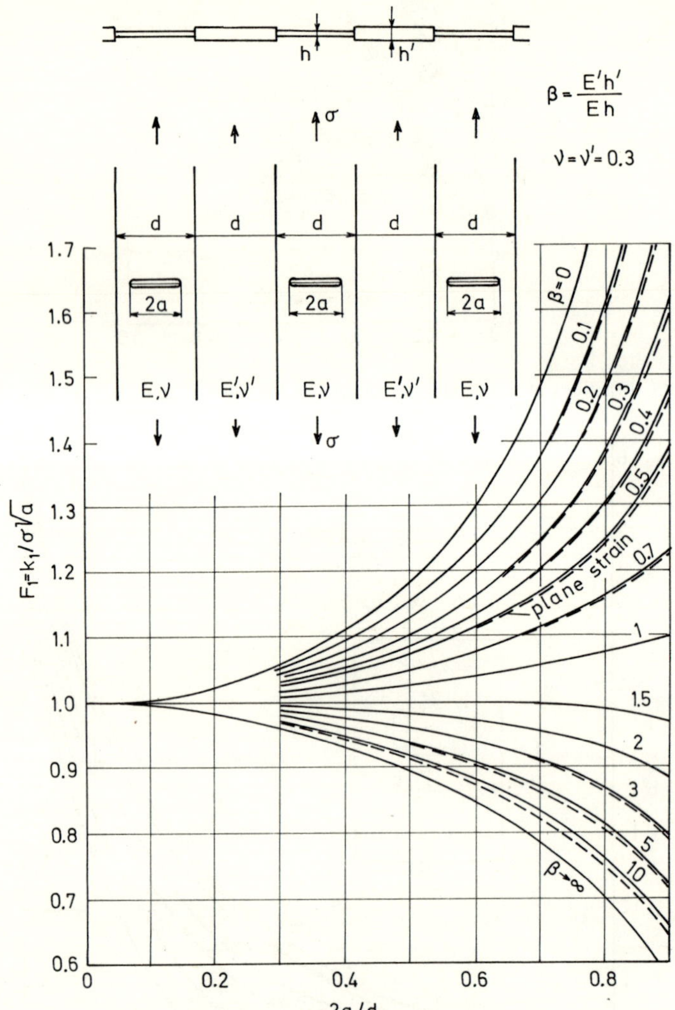


Figure 2.19. Tension of joined strips.

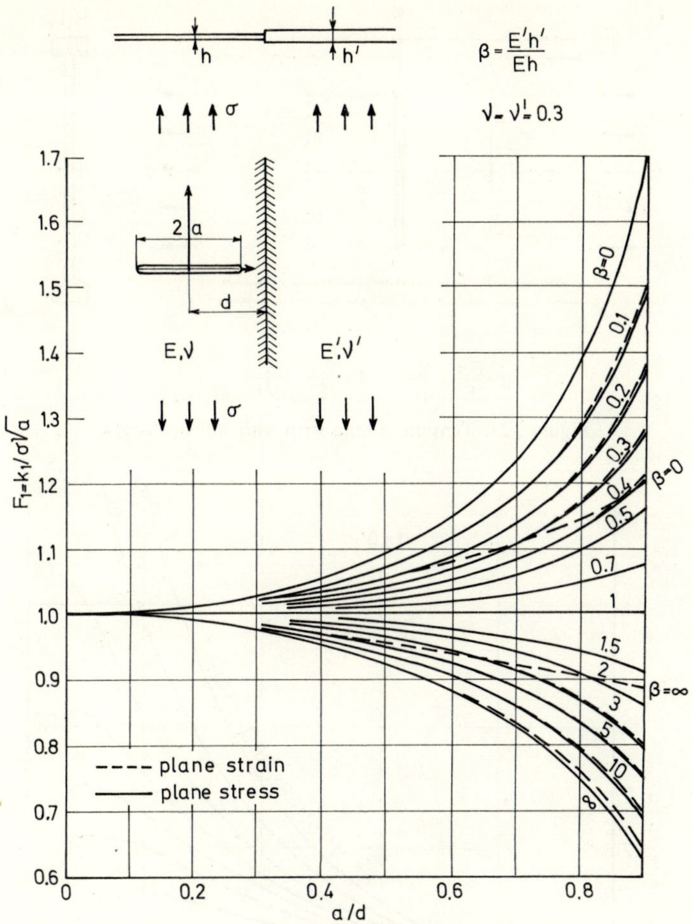


Figure 2.20. Tension of joined half planes.

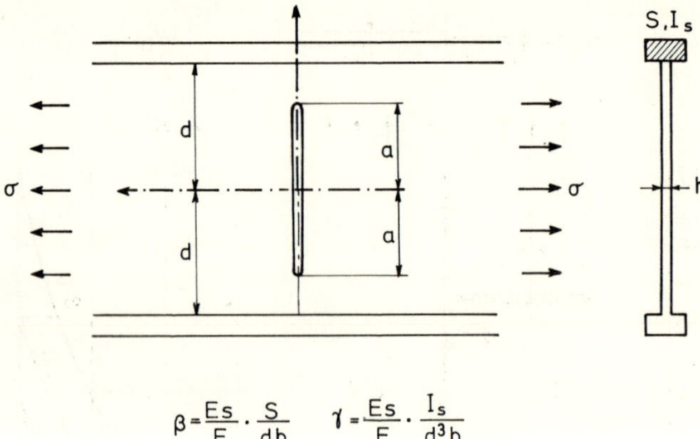


Figure 2.21. Tension of long strip with stiffened edges.

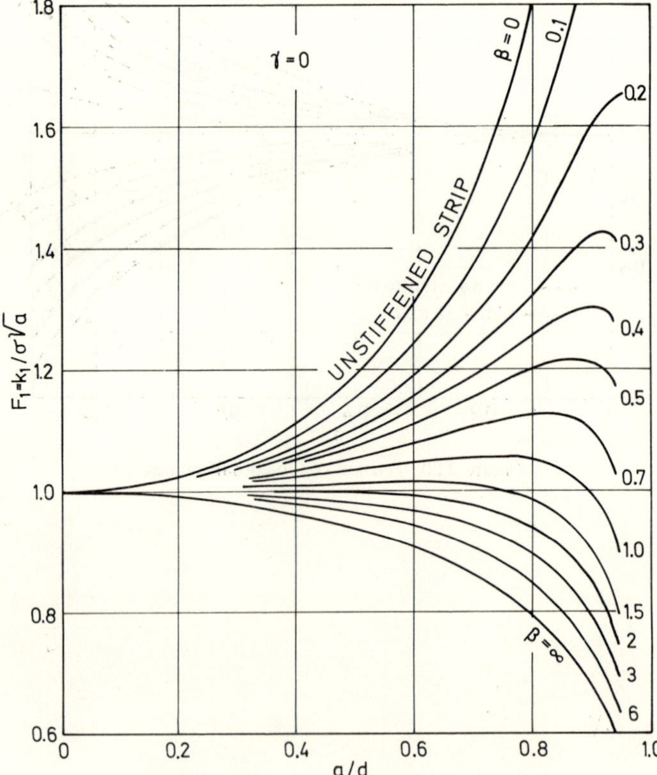


Figure 2.22. Results for $\gamma = 0$, $\nu = 0.3$.

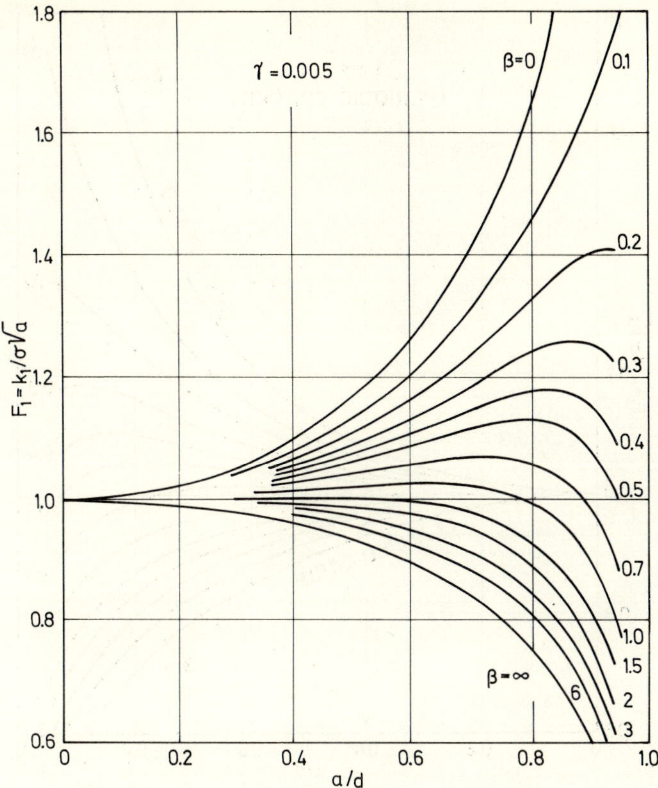


Figure 2.23. Results for $\gamma = 0.005$, $\nu = 0.3$.

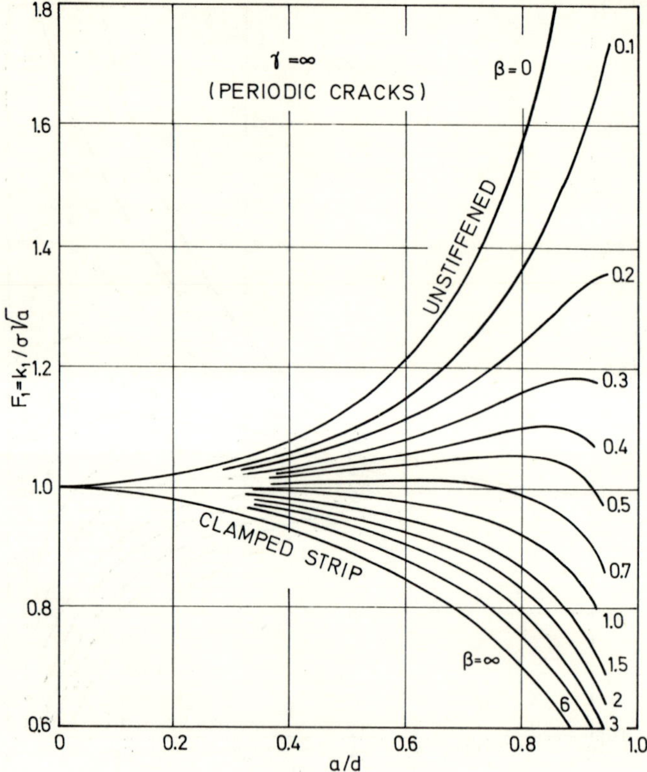


Figure 2.24. Results for $\gamma=\infty$, $\nu=0.3$.

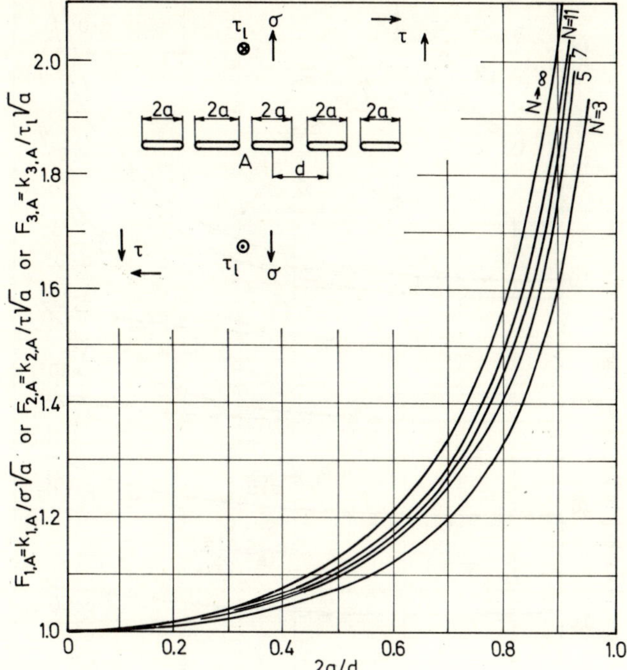


Figure 2.25. Equal colinear cracks.

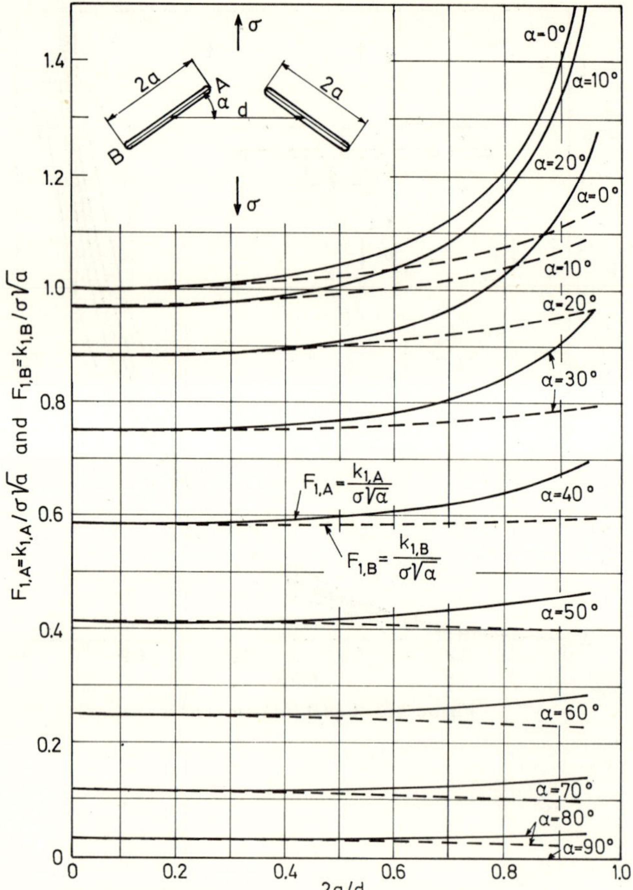


Figure 2.26. A pair of equal cracks inclined to each other.

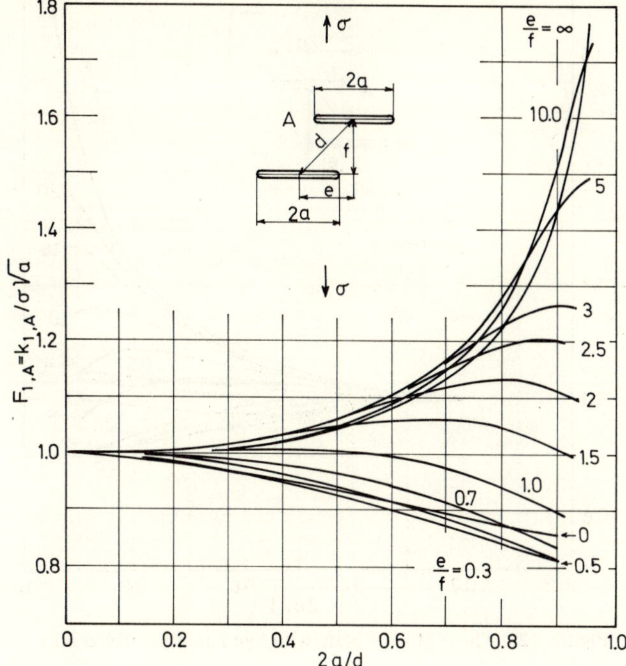


Figure 2.27. Tension of wide plate with two equal parallel cracks.

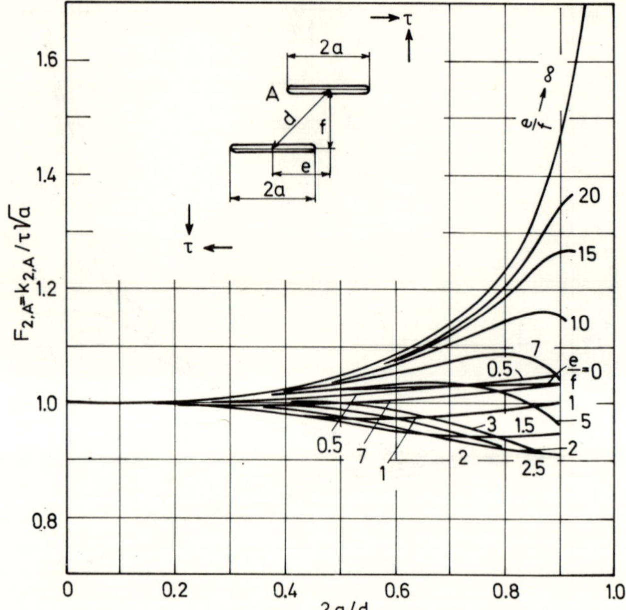


Figure 2.28. Shear of wide plate with two equal parallel cracks.

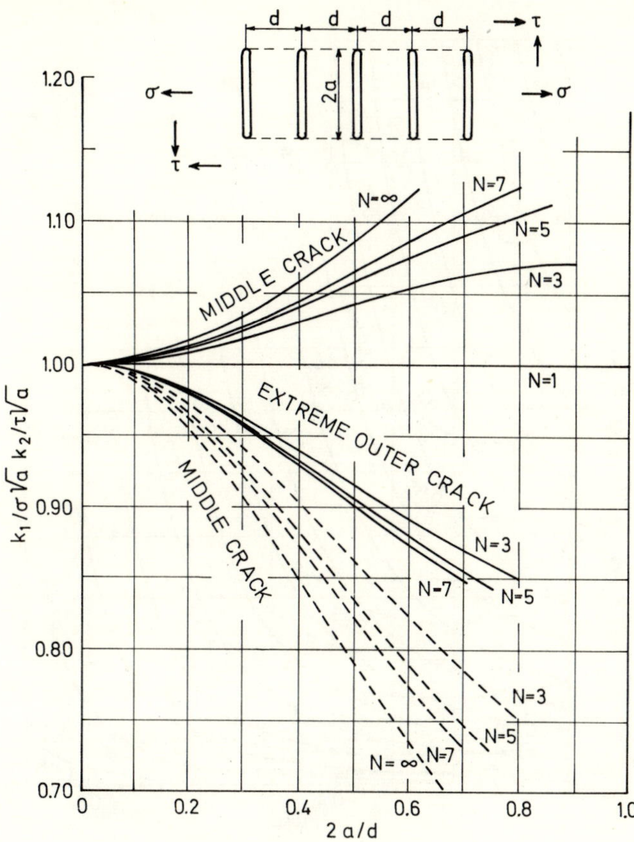


Figure 2.29. Equal parallel cracks.

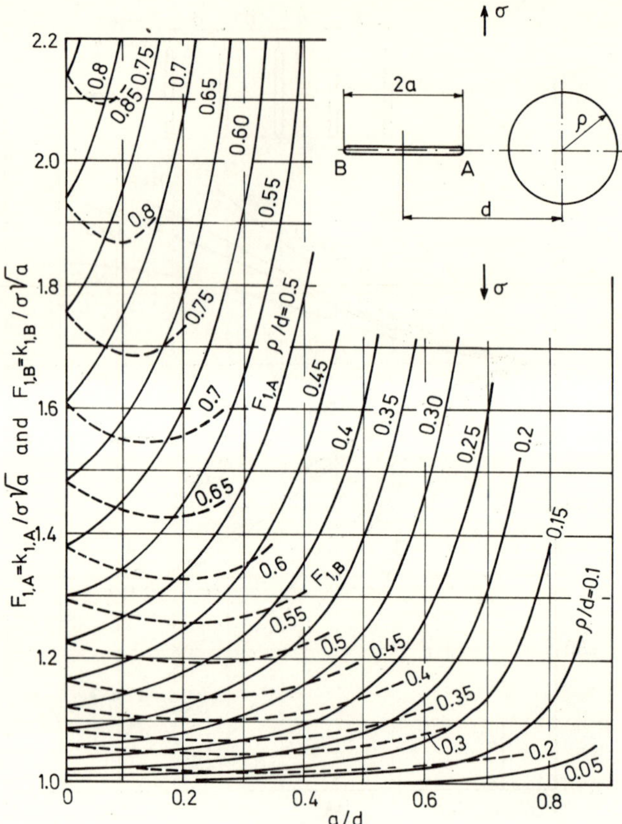


Figure 2.30. Crack approaching circular hole.

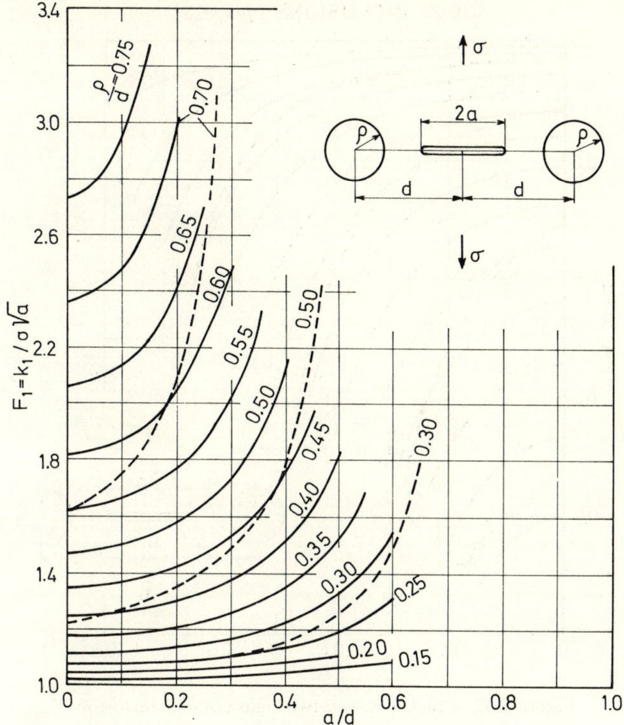


Figure 2.31. Crack between two circular holes.

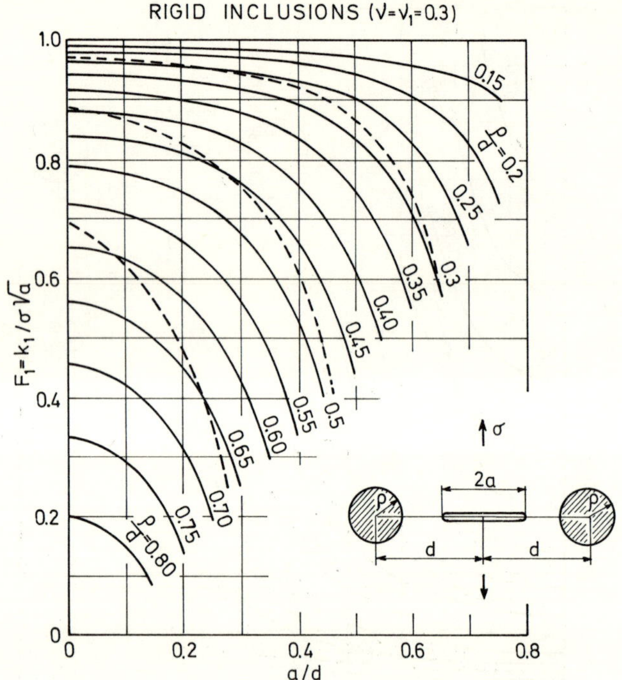


Figure 2.32. Crack between two rigid circular inclusions.

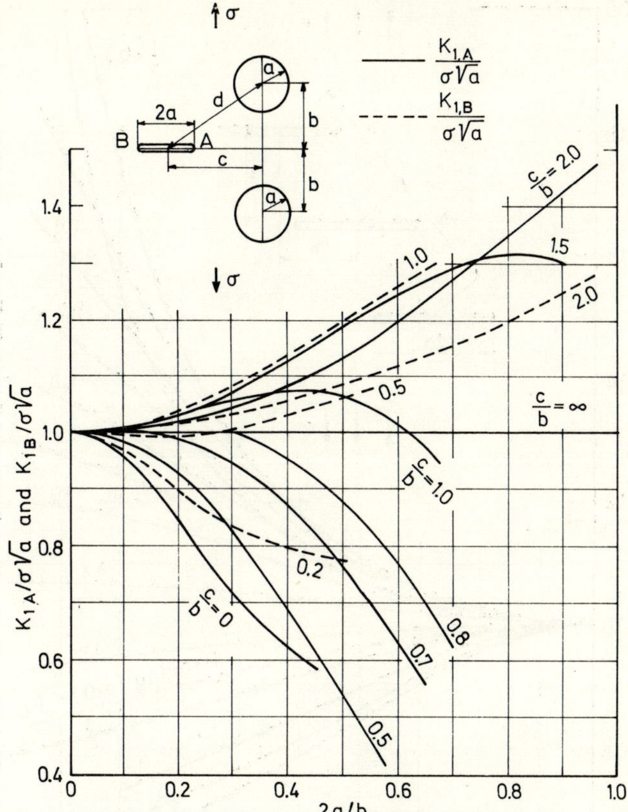


Figure 2.33. Crack approaching two circular holes.

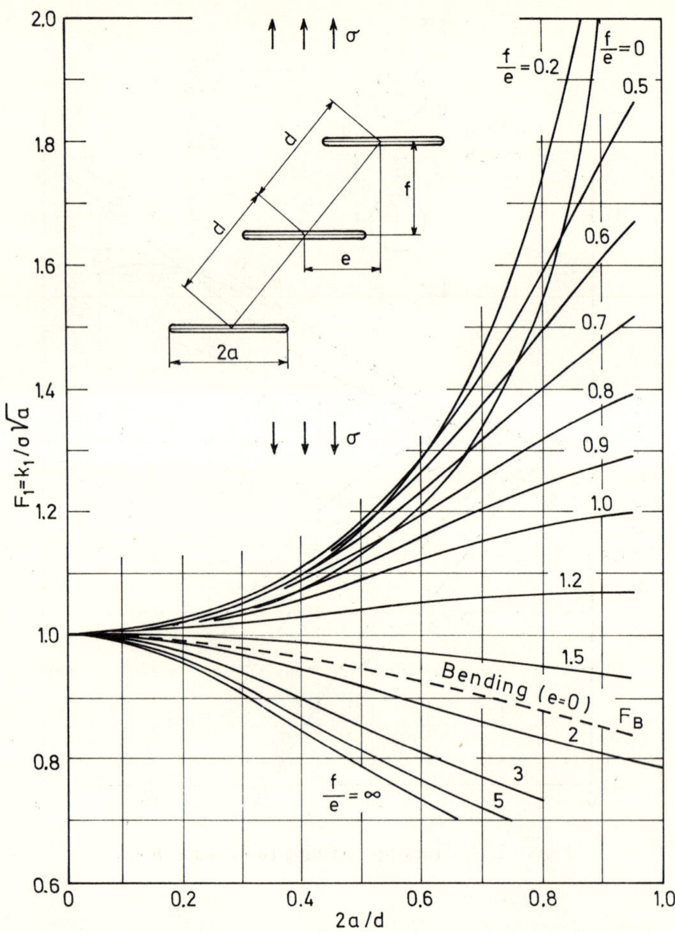


Figure 2.34. Tension of plate with infinite row of equal parallel cracks.

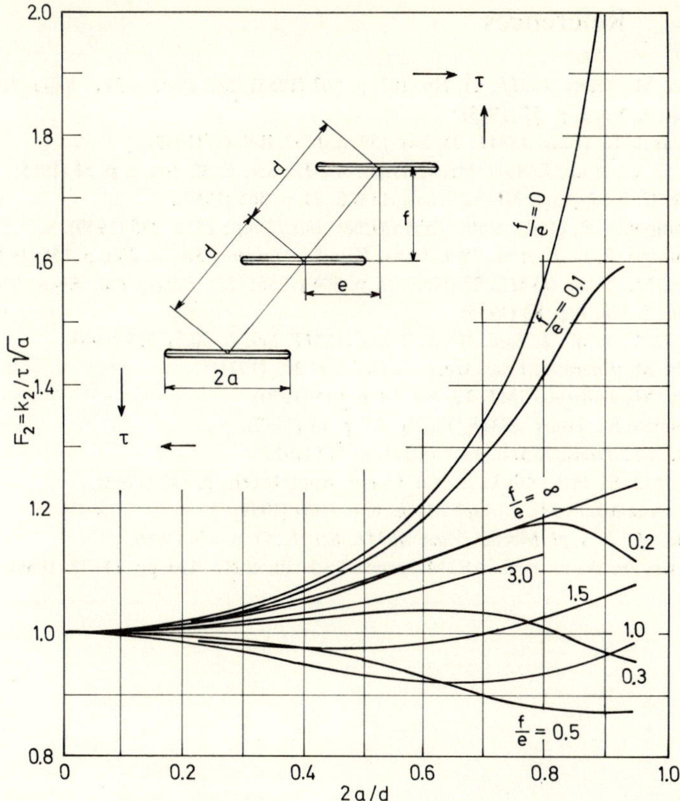


Figure 2.35. In-plane shear of plate with infinite row of equal parallel cracks.

## References

[1] Isida, M., *Trans. JSME*, 21, No. 107, p. 502 (1955); *Sci. Papers, Fac. Engg. Tokushima Univ.*, 4, No. 1, p. 35 (1953).
[2] Hayashi, T., *Trans. JSME*, 25, No. 159, p. 1133, Ref. (8) (1959).
[3] Sih, G. C., et al., *J. Appl. Mech., Trans. ASME, Ser. E*, 32, No. 1, p. 51 (1965).
[4] Ling, C. B., *J. Appl. Mech., Trans. ASME*, 24, p. 365 (1957).
[5] Rabinowitz, P., et al., *Math. Tables Other Aids Comp.*, 13, p. 285 (1959).
[6] Howland, R. C. J., et al., *Phil. Trans. Roy. Soc. London, Ser. A*, 232, p. 155 (1933).
[7] Isida, M., *Trans. JSME*, 22, No. 123, p. 804 (1956); *Sci. Papers Fac. Engg., Tokushima Univ.*, 5, No. 1, p. 83 (1955).
[8] Sih, G. C., et al., *J. Appl. Mech., Trans. ASME, Ser. E*, 30, p. 528 (1963).
[9] Isida, M., *Intern. J. Frac. Mech.*, 7, No. 3, p. 301 (1971).
[10] Isida, M., *Bulletin JSME*, 13, No. 59, p. 635 (1970).
[11] Washizu, K., *Trans. JSME*, 18, No. 68, p. 41 (1952).
[12] Isida, M., *Trans. JSME*, 23, No. 131, p. 474 (1957).
[13] Erdogan, F., Proc. 4th U. S. Nat. Congr. Appl. Mech., p. 547 (1962).
[14] Westergaard, F. M., *J. Appl. Mech.*, 6, p. A-49 (1939).
[15] Isida, M., *J. Appl. Mech., Trans. ASME, Ser. E*, 33, p. 674 (1966).
[16] Brown, Jr., W. F., et al., ASTM Special Tech. Phb., No. 410, pp. 11, 77 (1966).

*J. P. Benthem and W. T. Koiter*

# 3 *Asymptotic approximations to crack problems*

## 3.1 Introduction

Asymptotic methods play an important role in all branches of applied mathematics in the evaluation of the solutions of problems depending on a parameter with a certain range. Usually the solution is obtained first in a more or less explicit form, and is then simplified by the appropriate asymptotic expansions when the parameter tends to one or both limiting values of its interval. On the other hand, in elasticity one often encounters problems for which no explicit solution is available for arbitrary values of the parameter. Numerical techniques may enable us to obtain accurate numerical solutions for specific values of the parameter. Such numerical methods, however, are often less convenient to discuss the asymptotic behaviour of the solution when the parameter tends to a limiting value of its range.

In some problems it has proved advantageous to consider from the outset the cases of the limiting values of the parameter. This advantage occurs only, of course, if the problem takes a simpler form for both limiting values of the parameter, although these forms are naturally quite different. The more difficult problem of a complete solution is evaded, and it is hoped that the solutions for both ends of the range of the parameter are each valid in a sufficiently wide neighbourhood, thus enabling a satisfactory interpolation for intermediate values of the parameter. This approach has been advocated strongly in [33] and supported with evidence of its application in four widely different problems, but apart from [6] it does not seem to have been applied to crack problems.

In some cases accurate (numerical) solutions of the problem for intermediate values of the parameter will remain highly desirable, or even indispensable, but even then the asymptotic solutions at the ends of the range are extremely useful to complete the picture, the more so because the numerical techniques often become awkward at these end points. The present article has as its main purpose the encouragement of the fullest possible use of asymptotic methods in the analysis of cracks.

The procedure of interpolation between the asymptotic solutions at the ends of the interval is often facilitated, if the variable parameter is chosen in such a way that it varies from zero to unity in its range, and if the results are presented in such a form that they are of order unity throughout the range. The particular choice of interpolation procedure is to a large extent a matter of taste. Engineers are likely to favour curve fitting by hand after the asymptotic solutions have been drawn [33]. Numerical interpolation can often be achieved by polynomials, provided the available data at the ends are sufficiently well-behaved. In some cases, where steep slopes occur at one end of the range, an interpolation by a conic section fitting of the available data has been preferred (Section 3.5) and in one case the result of the polynomial interpolation is compared with the asymptotic solution near the ends in a diagram with a transformed parameter in an attempt to obtain a better picture of the accuracy achieved (Section 3.6). Wherever possible the results have been checked against available numerical data in the literature, and in one case an exact solution serves as a check on the interpolation procedure (Section 3.6).

The interpolation procedure is based on the simplified form of the problem in question for the end values of the parameter, and a summary of available information on these simplified basic problems is presented in Sections 3.2–3.4. This summary is concise where the information is easily accessible in the literature, but more explanation is offered where this appears necessary or helpful to the reader. All results are presented in the form of interpolation formulae in Sections 3.5–3.7 and in Figures 3.4–3.12. Each figure is self-explanatory, and for practical applications it is recommended to use these graphs.

In order to keep the Chapter within reasonable bounds attention is concentrated on the *stress intensity factor* which is defined, for the various loading cases somewhat unconventionally, in order to keep it of order unity. Information on the energy release due to a crack is not included. This preference of the stress intensity factor is based on the argument that this local property is less amenable to approximate analysis than the global energy release. Satisfactory results for the more sensitive property ensure *a fortiori* that the energy release problem will yield to the same treatment.

Moreover, quite a few energy problems have already been discussed in the literature [4, 7, 12, 24, 30], some of them by asymptotic approximations. Finally, reference can be made to Irwin's well-known general formulae for the energy release rate associated with a crack growth rate. The total energy release due to the cracks may thus be obtained by integration of Irwin's formulae with respect to the crack length.

## 3.2 Basic plane problems

*Single crack in infinite plane.* The stresses $\sigma_x$, $\sigma_y$, $\tau_{xy}=\tau$ and the displacement components $u$, $v$ in the $x$–$y$ plane are represented by the complex stress functions $\phi(z)$ and $\psi(z)$ of Kolosov and Muskhelishvili [3, §32], where $z=x+\mathrm{i}y$. Denoting the shear modulus by $G$, Poisson's ratio by $\nu$, and introducing the number $\kappa=3-4\nu$ in the case of plane strain, and $\kappa=(3-\nu)/(1+\nu)$ in the case of generalized plane stress,

$$\begin{aligned} \sigma_x+\sigma_y &= 2[\phi'(z)+\overline{\phi'(z)}]\,, \\ \sigma_y-\sigma_x+2\mathrm{i}\tau &= 2[\bar{z}\phi''(z)+\psi'(z)]\,, \\ 2G(u+\mathrm{i}v) &= \kappa\phi(z)-z\overline{\phi'(z)}-\overline{\psi(z)}\,, \end{aligned} \tag{3.1}$$

where primes and bars denote derivatives and conjugate values respectively.

Let the crack lie along the real axis from $x=-c$ to $x=c$. The complex plane is cut along this segment and all square roots will represent their principal branches. Consider the disturbance of the state of stress due to the removal from the edges of the crack of the tractions occurring in the solid infinite plane.

In the case of *uniform tension* (Mode I) parallel to the $y$-axis at infinity ($\sigma_y \to p$ for $|z|\to\infty$) the disturbed state of stress is described by

$$\begin{aligned} \phi(z) &= \tfrac{1}{2}p(z^2-c^2)^{\frac{1}{2}}-\tfrac{1}{2}pz\,, \\ \psi(z) &= \phi(z)-z\phi'(z)\,. \end{aligned} \tag{3.2}$$

Along the real axis

$$\begin{aligned} |x|<c:&\ \sigma_y = -p\,, \\ |x|>c:&\ \sigma_y = p\,\frac{|x|}{(x^2-c^2)^{\frac{1}{2}}} - p\,, \end{aligned} \tag{3.3}$$

and the *dimensionless stress intensity factor K*, defined by

$$\sigma_y \to Kp \frac{c^{\frac{1}{2}}}{(2r)^{\frac{1}{2}}}, \qquad (k_1 = Kpc^{\frac{1}{2}}), \tag{3.4}$$

has the value $K=1$.

For large values of $|z|$ all stresses tend to zero as $pc^2|z|^{-2}$. On the real axis in particular

$$\sigma_y \to \tfrac{1}{2}p \frac{c^2}{x^2} \quad \text{for} \quad |x| \to \infty, \tag{3.5}$$

and on the imaginary axis

$$\sigma_y \to -\tfrac{3}{2}p \frac{c^2}{y^2} \quad \text{for} \quad |y| \to \infty. \tag{3.6}$$

In the case of *bending* the undisturbed state of stress in the infinite plane is assumed in the form $\sigma_y = px/c$. The disturbed state of stress in this case is described by

$$\begin{aligned} \phi(z) &= \frac{p}{4c}\left[z(z^2-c^2)^{\frac{1}{2}}-z^2\right], \\ \psi(z) &= \phi(z)-z\phi'(z). \end{aligned} \tag{3.7}$$

Along the real axis

$$\begin{aligned} |x|<c: \quad \sigma_y &= -p\frac{x}{c} \\ |x|>c: \quad \sigma_y &= \frac{p}{2c}\left[\frac{x|x|}{(x^2-c^2)^{\frac{1}{2}}} + \frac{|x|}{x}(x^2-c^2)^{\frac{1}{2}}\right] - p\frac{x}{c}, \end{aligned} \tag{3.8}$$

and the *dimensionless stress intensity factor K*, defined by

$$\sigma_y \to Kp \frac{c^{\frac{1}{2}}}{(2r)^{\frac{1}{2}}}, \qquad (k_1 = Kpc^{\frac{1}{2}}), \tag{3.9}$$

has the value $K=\frac{1}{2}$. For large values of $|z|$ all stresses tend to zero as $pc^3|z|^{-3}$.

Strictly speaking, the bending case can only occur in combination with tension (or in cracks with a small radius of curvature at the tips) because a mutual penetration of the crack edges in the range $-c<x<0$ is not permitted.

In the case of *uniform shear* (Mode II) at infinity ($\tau \to t$ for $|z| \to \infty$) the disturbed state of stress is described by

$$\begin{aligned}\phi(z) &= -\tfrac{1}{2}it(z^2-c^2)^{\frac{1}{2}}+\tfrac{1}{2}itz\,,\\ \psi(z) &= -\phi(z)-z\phi'(z)\,.\end{aligned} \qquad (3.10)$$

Along the real axis

$$\begin{aligned}|x|<c: \quad &\tau = -t\,,\\ |x|>c: \quad &\tau = t\,\frac{|x|}{(x^2-c^2)^{\frac{1}{2}}} - t\,.\end{aligned} \qquad (3.11)$$

The *dimensionless stress intensity factor* $K$, defined by

$$\tau \to Kt\,\frac{c^{\frac{1}{2}}}{(2r)^{\frac{1}{2}}}\,, \qquad (k_2 = Ktc^{\frac{1}{2}})\,, \qquad (3.12)$$

has the value $K=1$.

For large values of $|z|$ all stresses tend to zero as $tc^2|z|^{-2}$. On the real axis in particular

$$\tau \to \tfrac{1}{2}t\,\frac{c^2}{x^2} \quad \text{for} \quad |x| \to \infty\,, \qquad (3.13)$$

and on the imaginary axis

$$\tau \to \tfrac{1}{2}t\,\frac{c^2}{y^2} \quad \text{for} \quad |y| \to \infty\,. \qquad (3.14)$$

The reader's attention is drawn to the fact that the alternative solution in [1, pages 35–37] does not apply to the present shear problem. The reason is the imposition of the boundary condition $v(x, 0)=0$ for $y=0$, $|x|>c$, instead of the correct condition $\sigma_y=0$. Analogous statements for the penny-shaped crack in the case of uniform shear [1, pages 156, 157] can be made.

*Dam between two half-planes.* A dam of length $2a$ between two half-planes is obtained by cutting the infinite plane along the real axis from $x=-\infty$ to $x=-a$ and from $x=a$ to $x=\infty$. The stress distribution is again described by equations (3.1).

The stress distribution in the case of a *central normal force N* per unit thickness coincides with Sadowsky's solution for a smooth rigid stamp on a half-plane [2] (*cf.* also [3, §116]). The stress distribution is given by the principal branch of the functions

$$\phi(z) = \frac{N}{2\pi} \arc \sin \frac{z}{a},$$
$$\psi(z) = \phi(z) - z\phi'(z). \tag{3.15}$$

The stress distribution in the dam is given by

$$\sigma_y = \frac{N}{\pi(a^2-x^2)^{\frac{1}{2}}}, \tag{3.16}$$

and the *dimensionless stress intensity factor* $K$, defined by

$$\sigma_y \to K \frac{N}{2a^{\frac{1}{2}}} \frac{1}{(2r)^{\frac{1}{2}}}, \qquad \left(k_1 = \frac{KN}{2a^{\frac{1}{2}}}\right), \tag{3.17}$$

has the value $K=2/\pi$. For large values of $|z|$ all stresses tend to zero as $N|z|^{-1}$.

In the case of *bending* [3, §116] let $M$ denote the bending moment transmitted by the dam per unit thickness. The stress distribution is again given by equations (3.1) where $\phi(z)$ and $\psi(z)$ are the principal branches of

$$\phi(z) = \frac{M}{\pi a^2} [-(a^2-z^2)^{\frac{1}{2}} + iz],$$
$$\psi(z) = \phi(z) - z\phi'(z). \tag{3.18}$$

The stress distribution in the dam is given by

$$\sigma_y = \frac{2M}{\pi a^2} \frac{x}{(a^2-x^2)^{\frac{1}{2}}}, \tag{3.19}$$

and the *dimensionless stress intensity factor* $K$, defined by

$$\sigma_y \to K \frac{3M}{2a(a^{\frac{1}{2}})} \frac{1}{(2r)^{\frac{1}{2}}}, \qquad \left(k_1 = \frac{3KM}{2a(a^{\frac{1}{2}})}\right), \tag{3.20}$$

has the value $K=4/3\pi$. For large values of $|z|$ all stresses tend to zero as $M|z|^{-2}$.

In the case of *shear* let $S$ denote the shear force transmitted by the dam per unit thickness. The complex stress functions in equations (3.1) are now specified by the principal branches of

$$\phi(z) = -\frac{iS}{2\pi} \arc \sin \frac{z}{a},$$
$$\psi(z) = -\phi(z) - z\phi'(z). \tag{3.21}$$

The stress distribution in the dam is given by

$$\tau = \frac{S}{\pi(a^2 - x^2)^{\frac{1}{2}}}, \tag{3.22}$$

and the *dimensionless stress intensity factor K*, defined by

$$\tau \to K\frac{S}{2a^{\frac{1}{2}}}\frac{1}{(2r)^{\frac{1}{2}}}, \qquad \left(k_2 = \frac{KS}{2a^{\frac{1}{2}}}\right), \tag{3.23}$$

has the value $K = 2/\pi$. For large values of $|z|$ all stresses tend to zero as $S|z|^{-1}$.

*Edge crack in half-plane.* [4, 5] Consider a half-plane $x > 0$ with an edge crack $0 < x < c$ along the $x$-axis and perpendicular to the edge (Figure 3.1).

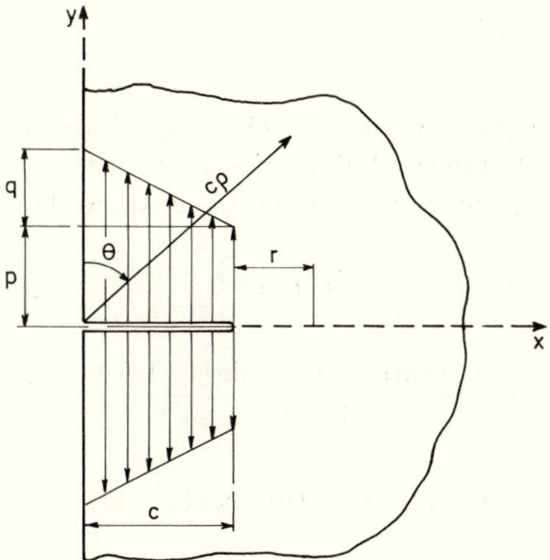


Figure 3.1. Edge-crack in a half-plane with linear load distribution $\sigma_y = p + q(c - x)/c$.

There is a linear distribution of tensile stresses $\sigma_y = p + q(c - x)/c$ in the solid half-plane, and the tractions along the crack edges are removed by a linear load distribution of opposite sign. The disturbance due to the crack is now governed by a discontinuous boundary value problem of the quarter-plane $x > 0$, $y > 0$

$$x=0:\quad \sigma_x=\tau_{xy}=0\;;$$

$$y=0:\begin{cases}0<x<c: & \sigma_y=-p-q(c-x)/c\,,\quad \tau_{xy}=0\,,\\ x>c: & v=0\,,\quad \tau_{xy}=0\,.\end{cases} \tag{3.24}$$

Introduce polar coordinates $c\rho$, $\theta$ (Figure 3.1), and express the stresses in terms of an Airy stress function $f(\rho,\theta)$

$$c^2\sigma_\rho=\frac{1}{\rho^2}\frac{\partial^2 f}{\partial\theta^2}+\frac{1}{\rho}\frac{\partial f}{\partial\rho},\quad c^2\sigma_\theta=\frac{\partial^2 f}{\partial\rho^2},\quad c^2\tau_{\rho\theta}=\frac{1}{\rho^2}\frac{\partial f}{\partial\theta}-\frac{1}{\rho}\frac{\partial^2 f}{\partial\rho\,\partial\theta}. \tag{3.25}$$

The Mellin transforms of the stress function

$$F(s,\theta)=\int_0^\infty f(\rho,\theta)\rho^{s-2}\mathrm{d}\rho\,, \tag{3.26}$$

of the (unknown) stresses $\sigma_y(x>c,\ y=0)=\sigma_\theta(\rho>1,\ \theta=\pi/2)$

$$P_-(s)=\int_1^\infty \sigma_\theta(\rho,\pi/2)\rho^s\mathrm{d}\rho\,, \tag{3.27}$$

regular in a half-plane $\mathrm{Re}(s)<\mu_1$ (where $\mu_1>1$) are introduced, and a solution of the transformed (biharmonic) boundary value problem is obtained in a form which still contains the unknown function $P_-(s)$

$$\begin{aligned}F(s,\theta)=c^2\left[-\frac{p+q}{s+1}+\frac{q}{s+2}+P_-(s)\right]\frac{s+1}{\cos\pi s+2s^2-1}\cdot\\ \cdot\left\{-\cos\frac{\pi}{2}s\left[\frac{\sin(s-1)\theta}{s-1}-\frac{\sin(s+1)\theta}{s+1}\right]+\right.\\ \left.+\frac{1}{s}\sin\frac{\pi}{2}s\,[\cos(s-1)\theta-\cos(s+1)\theta]\right\}.\end{aligned} \tag{3.28}$$

The Mellin transform of the displacement component $-v(x,0)=v_\theta(\rho,\pi/2)$

$$\int_0^\infty \rho^{s-1}v_\theta(\rho,\pi/2)\mathrm{d}\rho=\int_0^1 \rho^{s-1}v_\theta(\rho,\pi/2)\mathrm{d}\rho=V_+(s)\,, \tag{3.29}$$

regular in a half-plane $\mathrm{Re}(s)>0$, may now be evaluated in terms of the transformed stress function, and there results the Wiener–Hopf equation [4]

$$EV_+(s)=2c\,\frac{\sin\pi s}{s(\cos\pi s+2s^2-1)}\left[-\frac{p+q}{s+1}+\frac{q}{s+2}+P_-(s)\right], \tag{3.30}$$

holding in a strip $0<\mathrm{Re}(s)<\mu_1$ of the complex $s$-plane, where $E$ is Young's modulus in the case of generalized plane stress.

In order to solve equation (3.30), with a number $B>1$, conveniently selected at $B=2\pi/(\pi^2-4)=1.0705$, write

$$H(s)=\frac{-s\sin\pi s}{(B^2-s^2)^{\frac{1}{2}}(\cos\pi s+2s^2-1)}. \tag{3.31}$$

This function $H(s)$ is regular and has no zeros in a strip $-\mu^*<\mathrm{Re}(s)<\mu^*$, where $\mu^*=\mathrm{Min}(\mu_1, B)>1$, and it tends to unity for $|s|\to\infty$ in this strip. Let $H(s)=H_+(s)/H_-(s)$ where

$$\begin{matrix}\log H_+(s)\\ \\ \log H_-(s)\end{matrix}\;=\;-\frac{1}{2\pi\mathrm{i}}\int_{\mu-\mathrm{i}\infty}^{\mu+\mathrm{i}\infty}\frac{\log H(z)}{z-s}\,\mathrm{d}z\,,\qquad \begin{matrix}-\mu^*<\mu<\mathrm{Re}(s)\\ \\ \mu^*>\mu>\mathrm{Re}(s)\end{matrix} \tag{3.32}$$

and the solution of the Wiener–Hopf equation, regular for $\mathrm{Re}(s)<\mu^*$, is given by

$$P_-(s)=\frac{p+q}{s+1}\left[1+\frac{(B+1)^{\frac{1}{2}}}{H_-(-1)}\frac{sH_-(s)}{(B-s)^{\frac{1}{2}}}\right]-\frac{q}{s+2}\left[1+\frac{(B+2)^{\frac{1}{2}}}{2H_-(-2)}\frac{sH_-(s)}{(B-s)^{\frac{1}{2}}}\right]. \tag{3.33}$$

The stresses may now be evaluated by means of the inverse Mellin transforms associated with equations (3.25)

$$\begin{aligned}c^2\sigma_\rho &= \frac{1}{2\pi\mathrm{i}}\int_{\mu-\mathrm{i}\infty}^{\mu+\mathrm{i}\infty}\left[\frac{\mathrm{d}^2}{\mathrm{d}\theta^2}F(s,\theta)-(s-1)F(s,\theta)\right]\rho^{-s-1}\mathrm{d}s\,,\\ c^2\sigma_\theta &= \frac{1}{2\pi\mathrm{i}}\int_{\mu-\mathrm{i}\infty}^{\mu+\mathrm{i}\infty}s(s-1)F(s,\theta)\rho^{-s-1}\mathrm{d}s\,,\\ c^2\tau_{\rho\theta} &= \frac{1}{2\pi\mathrm{i}}\int_{\mu-\mathrm{i}\infty}^{\mu+\mathrm{i}\infty}s\frac{\mathrm{d}}{\mathrm{d}\theta}F(s,\theta)\rho^{-s-1}\mathrm{d}s\,,\end{aligned} \tag{3.34}$$

where $-\mu^*<\mu<\mu^*$.

The stress component $\sigma_y$ $(x>c,\ y=0)$ is obtained by the inverse transformation of equation (3.27)

$$\sigma_y(x>c,\ y=0)=\frac{1}{2\pi\mathrm{i}}\int_{\mu-\mathrm{i}\infty}^{\mu+\mathrm{i}\infty}P_-(s)\rho^{-s-1}\mathrm{d}s\,, \tag{3.35}$$

where $\rho>1$ and $\mu<\mu^*$. Equation (3.33) can be rewritten in the form

$$P_-(s) = (p+q)\frac{(B+1)^{\frac{1}{2}}}{H_-(-1)}\frac{1}{(B-s)^{\frac{1}{2}}} - q\frac{(B+2)^{\frac{1}{2}}}{2H_-(-2)}\frac{1}{(B-s)^{\frac{1}{2}}} +$$

$$+\frac{p+q}{s+1}\left[1+\frac{(B+1)^{\frac{1}{2}}}{H_-(-1)}\{sH_-(s)-(s+1)\}\frac{1}{(B-s)^{\frac{1}{2}}}\right]-$$

$$-\frac{q}{s+2}\left[1+\frac{(B+2)^{\frac{1}{2}}}{2H_-(-2)}\{sH_-(s)-(s+2)\}\frac{1}{(B-s)^{\frac{1}{2}}}\right], \tag{3.36}$$

where only the first line contributes to the singular behaviour near $\rho=1$. This singular part is easily evaluated in closed form by the formula

$$\frac{1}{2\pi i}\int_{\mu-i\infty}^{\mu+i\infty}\frac{1}{(B-s)^{\frac{1}{2}}}\rho^{-s-1}ds = \pi^{-\frac{1}{2}}\rho^{-B-1}(\log\rho)^{-\frac{1}{2}}, \qquad \rho>1\,.$$

With the numerical values $B=1.0705$, $\log H_-(-1)=0.0234$, $\log H_-(-2)=0.0235$ [4], the singular behaviour of $\sigma_y(x, 0)$ for $x=c\rho=c+r$, $r\to 0$, is described by

$$\sigma_y \to (1.122p+0.439q)\frac{c^{\frac{1}{2}}}{(2r)^{\frac{1}{2}}}\,. \tag{3.37}$$

In the case of pure tension ($q=0$) the *dimensionless stress intensity factor K*, defined by

$$\sigma_y \to Kp\frac{c^{\frac{1}{2}}}{(2r)^{\frac{1}{2}}}, \qquad (k_1 = Kpc^{\frac{1}{2}})\,, \tag{3.38}$$

has the value $K=[2(B+1)/\pi]^{\frac{1}{2}}/H_-(-1)=1.122$.

Since $P_-(s)$ is regular for $\mathrm{Re}(s)<\mu^*$, where $\mu^*>1$, the stresses in the far field $\rho\to\infty$ are governed by the simple pole of equation (3.28) at $s=1$, and these stresses therefore tend to zero proportional to $\rho^{-2}$. The stress component $\sigma_y(x=0,\ y\to\infty)=\sigma_\rho(\rho\to\infty,\ \theta=0)$ in the case of pure tension will be needed. From equations (3.28), (3.33), and (3.34)

$$\sigma_y(0, y) \to [p-2P_-(1)]\frac{c^2}{y^2} = -p\frac{(B+1)^{\frac{1}{2}}}{H_-(-1)}\frac{H_-(1)}{(B-1)^{\frac{1}{2}}}\frac{c^2}{y^2} =$$

$$= -2.516p\frac{c^2}{y^2} \quad \text{for} \quad |y|\to\infty\,. \tag{3.39}$$

*Edge dam between two quarter-planes.* [6] Consider a half-plane $x>0$ with a semi-infinite crack $x>a$ perpendicular to the edge (Figure 3.2), in which the dam transmits per unit thickness a normal force $N$ and a moment $M$

$$N = \int_0^a \sigma_y(x, 0)\,\mathrm{d}x\,, \quad M = \int_0^a x\sigma_y(x, 0)\,\mathrm{d}x\,. \tag{3.40}$$

In view of the symmetry of the problem with respect to the $x$-axis the shear stress $\tau_{xy}$ vanishes for $y=0$, and the dam $0<x<a$ between the upper and lower quarter-planes will remain straight. The relative rotation of the dam with respect to the point at infinity of the upper quarter-plane (in a clockwise sense) is denoted by $\alpha$★.

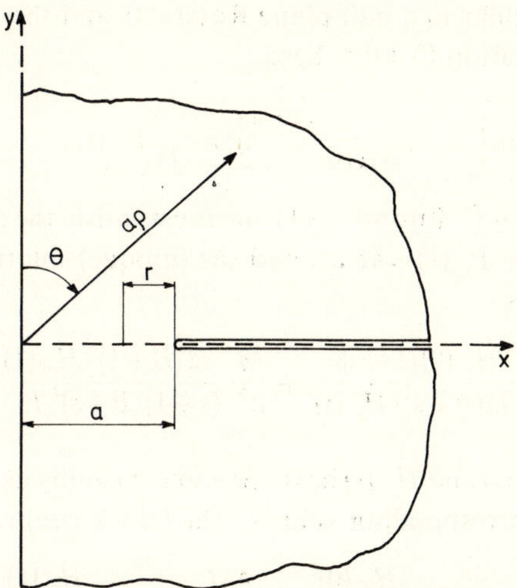


Figure 3.2. Edge-dam between two quarter-planes.

Introducing polar coordinates $a\rho$, $\theta$ (Figure 3.2), the stresses are given by equations (3.25), where $c$ is now replaced by $a$. The Mellin transform of the (unknown) stresses $\sigma_y(0<x<a,\ y=0)=\sigma_\theta(\rho<1,\ \theta=\pi/2)$

$$P_+(s) = \int_0^1 \sigma_\theta(\rho, \pi/2)\,\rho^s\,\mathrm{d}\rho \tag{3.41}$$

is now regular in a half-plane $\mathrm{Re}\,(s) > -1$, and the transformed stress function is now given by the counterpart of equation (3.28)

$$F(s, \theta) = a^2 P_+(s)\,\frac{s+1}{\cos \pi s + 2s^2 - 1}\left[-\cos\frac{\pi}{2}s\left\{\frac{\sin(s-1)\theta}{s-1} - \frac{\sin(s+1)\theta}{s+1}\right\} + \right.$$
$$\left. + \frac{1}{s}\sin\frac{\pi}{2}s\left\{\cos(s-1)\theta - \cos(s+1)\theta\right\}\right]. \tag{3.42}$$

★ Only the case $\alpha=0$ has been discussed previously [6].

The Mellin transform of the displacement component $-v(x, 0)=v_\theta(\rho, \pi/2)$ is in this case

$$\int_0^\infty \rho^{s-1} v_\theta(\rho, \pi/2)\mathrm{d}\rho = \int_0^1 \alpha a \rho^s \mathrm{d}\rho + \int_1^\infty \rho^{s-1} v_\theta(\rho, \pi/2)\mathrm{d}\rho =$$

$$= \frac{\alpha a}{s+1} + V_-(s)\,, \tag{3.43}$$

where $V_-(s)$ is regular in a half-plane $\mathrm{Re}(s) < 0$, and the counterpart of the Wiener–Hopf equation (3.30) is here

$$E\left[\frac{\alpha a}{s+1} + V_-(s)\right] = 2a\,\frac{\sin \pi s}{s(\cos \pi s + 2s^2 - 1)}\,P_+(s)\,. \tag{3.44}$$

In view of equations (3.40) and (3.41) one must satisfy the additional requirements $P_+(0)=N/a$, $P_+(1)=M/a^2$, and the (unique) solution of (3.44) under these conditions is

$$P_+(s) = \frac{N}{a}\,\frac{B^{\frac{1}{2}} H_+(0)(1-s)}{(s+1)(B+s)^{\frac{1}{2}} H_+(s)} + \frac{M}{a^2}\,\frac{2(B+1)^{\frac{1}{2}} H_+(1)s}{(s+1)(B+s)^{\frac{1}{2}} H_+(s)}\,, \tag{3.45}$$

where $B$, $H(s)$, $H_+(s)$ and $H_-(s)$ have the same meaning as in equations (3.31) and (3.32). The corresponding value of the (clockwise) rotation is

$$\alpha = -\,\frac{4N}{Ea}\,[B(B+1)]^{\frac{1}{2}}\,\frac{H_+(0)}{H_-(-1)} + \frac{4M}{Ea^2}\,(B+1)\,\frac{H_+(1)}{H_-(-1)}\,. \tag{3.46}$$

*No rotation** occurs, if the line of action of the resulting force $N$ is

$$x = a\left(\frac{B}{B+1}\right)^{\frac{1}{2}} \frac{H_+(0)}{H_+(1)} = 0.736\,a\,. \tag{3.47}$$

In this case equation (3.45) is reduced to

$$P_+(s) = \frac{N}{a}\,\frac{B^{\frac{1}{2}} H_+(0)}{(B+s)^{\frac{1}{2}} H_+(s)}\,, \tag{3.48}$$

and the singular behaviour of $\sigma_y(x, 0)$ for $x=a-r$, $r\to 0$ is described by

$$\sigma_y \to \pi^{-\frac{1}{2}} B^{\frac{1}{2}} H_+(0)\,\frac{N}{a}\left(\frac{a}{r}\right)^{\frac{1}{2}} = \frac{2}{(\pi^2-4)^{\frac{1}{2}}}\,\frac{N}{a}\,\frac{a^{\frac{1}{2}}}{(2r)^{\frac{1}{2}}}\,. \tag{3.49}$$

The *nondimensional stress intensity factor* $K$, defined by

* The solution of [25] applies only to this case.

$$\sigma_y \to K \frac{N}{a^{\frac{1}{2}}} \frac{1}{(2r)^{\frac{1}{2}}}, \qquad \left(k_1 = K \frac{N}{a^{\frac{1}{2}}}\right), \tag{3.50}$$

has the value $K = 2/(\pi^2 - 4)^{\frac{1}{2}} = 0.826$.

In the case of *pure bending* ($N=0$) the singular behaviour of $\sigma_y(x, 0)$ is described by

$$\sigma_y \to \frac{2}{\pi^{\frac{1}{2}}} \frac{M}{a^2} (B+1)^{\frac{1}{2}} H_+(1) \left(\frac{a}{r}\right)^{\frac{1}{2}}, \tag{3.51}$$

and the *nondimensional stress intensity factor* K, defined by

$$\sigma_y \to K \frac{6M}{a^2} \frac{a^{\frac{1}{2}}}{(2r)^{\frac{1}{2}}}, \qquad \left(k_1 = \frac{6M}{a(a^{\frac{1}{2}})}\right), \tag{3.52}$$

has the value $K = \frac{1}{3}[2(B+1)/\pi]^{\frac{1}{2}} H_+(1) = 0.374$, *exactly* one third of the value in equation (3.38).

In the far field $\rho \to \infty$ the stresses are governed by the pole in the half-plane $\mathrm{Re}(s) > -1$ nearest to this boundary. In the case of a normal force $N$ the stress component $\sigma_\rho$ has the slowest decay, proportional to $N/a\rho$, and in the case $N=0$ all stresses decay proportional to $M/a^2\rho^2$. These statements are easily verified by inspection of equations (3.34), where $c$ has been replaced by $a$, (3.42) and (3.45).

*An infinite row of long parallel cracks.* The *long* cracks $y = \pm(2j+1)b$, $0 < x < 2c$, $j = 1, 2, \ldots$, and $c/b \gg 1$, are parallel to the $x$-axis (Figure 3.3). The cases of uniform tension $\sigma_y = p$ and uniform shear $\tau_{xy} = t$ in the solid plane will be considered. The disturbance may be obtained by considering a single strip $-b < y < b$ under uniform loads ($\sigma_y = -p$ and $\tau_{xy} = -t$ respectively) along the segments of the edges $0 < x < 2c$. Since the mutual interference between the crack tips at $x=0$ and those at $x=2c$ is small of *exponential* order, i.e. of order $\exp(-\mu c/b)$, where the positive number $\mu$ is of order of magnitude unity, we may consider the edges $0 < x < 2c$ to be *semi-infinite* $(0 < x < \infty)$.

The stresses are expressed in terms of an Airy stress function $f(x, y)$. The boundary conditions in the case of *tension* are

$$y = \pm b, \quad \begin{array}{l} x < 0: \quad \tau_{xy} = 0, \; v = 0; \\ x > 0: \quad \tau_{xy} = 0, \; \sigma_y = -p. \end{array} \tag{3.53}$$

In the case of shear the appropriate boundary conditions are

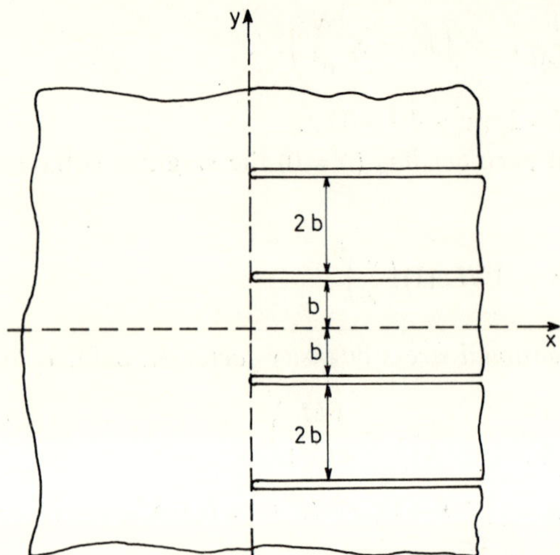


Figure 3.3. An infinite row of semi-infinite parallel cracks in an infinite plane.

$$y = \pm b\,, \quad \begin{array}{l} x<0: \quad \sigma_y = 0\,, \;\; u = 0\;; \\ x>0: \quad \sigma_y = 0\,, \;\; \tau_{xy} = -t\,. \end{array} \tag{3.54}$$

Since the bending moment in the strip must vanish at $x=c$, there is an additional requirement in this case:

$$\int_{-\infty}^{0} \tau_{xy}(x, b)\mathrm{d}x = tc\,. \tag{3.55}$$

Fourier transforms of the stress function, and of the unknown stresses $\sigma_y(x<0, b)$ in the case of tension, $\tau_{xy}(x<0, b)$ in the case of shear are introduced [4, 7],

$$F(\lambda, y) = (2\pi)^{-\frac{1}{2}} \int_{-\infty}^{\infty} f(x, y)\mathrm{e}^{\mathrm{i}\lambda x}\mathrm{d}x\,, \tag{3.56}$$

$$P_-(\lambda) \;= (2\pi)^{-\frac{1}{2}} \int_{-\infty}^{0} \sigma_y(x, b)\mathrm{e}^{\mathrm{i}\lambda x}\mathrm{d}x\,, \tag{3.57}$$

$$T_-(\lambda) \;= (2\pi)^{-\frac{1}{2}} \int_{\infty}^{0} \tau_{xy}(x, b)\mathrm{e}^{\mathrm{i}\lambda x}\mathrm{d}x\,. \tag{3.58}$$

The transform (3.56) is regular in a strip $0<\mathrm{Im}\,(\lambda)<\beta$, the transforms (3.57) and (3.58) are regular in a lower half-plane $\mathrm{Im}\,(\lambda)<\beta$. Likewise, introduce the

Fourier transforms of the unknown displacement components $v\,(x>0,\ y=b)$ in the case of tension, and $u\,(x>0,\ y=b)$ in the case of shear

$$V_+(\lambda) = (2\pi)^{-\frac{1}{2}} \int_0^\infty v(x, b)\,\mathrm{e}^{\mathrm{i}\lambda x}\,\mathrm{d}x\,, \tag{3.59}$$

$$U_+(\lambda) = (2\pi)^{-\frac{1}{2}} \int_0^\infty u(x, b)\,\mathrm{e}^{\mathrm{i}\lambda x}\,\mathrm{d}x\,. \tag{3.60}$$

These transforms are regular in a half-plane $\mathrm{Im}\,(\lambda)>0$.

The boundary value problem in generalized plane stress in the case of *tension* is now reduced to the Wiener–Hopf equation

$$EV_+(\lambda) = \left[P_-(\lambda) - \frac{\mathrm{i}p}{\lambda(2\pi)^{\frac{1}{2}}}\right] \frac{2(\cosh 2\lambda b - 1)}{\lambda(\sinh 2\lambda b + 2\lambda b)}\,, \tag{3.61}$$

holding in a strip $0<\mathrm{Im}\,(\lambda)<\beta$. In the case of *shear* the similar Wiener–Hopf equation reads

$$EU_+(\lambda) = \left[T_-(\lambda) - \frac{\mathrm{i}t}{\lambda(2\pi)^{\frac{1}{2}}}\right] \frac{2(\cosh 2\lambda b - 1)}{\lambda(\sinh 2\lambda b - 2\lambda b)}\,, \tag{3.62}$$

and its solution has to satisfy the additional requirement (3.55), viz. $T_-(0)=tc/(2\pi)^{\frac{1}{2}}$.

In the case of *tension* the function

$$H^*(w) = (w^2+4)^{\frac{1}{2}}\,\frac{\cosh 2w - 1}{w(\sinh 2w + 2w)}\,, \tag{3.63}$$

is regular and non-zero in a strip $-\gamma^*<\mathrm{Im}\,(w)<\gamma^*$ and tends to unity for $|w|\to\infty$ in this strip. Let $H^*(w)=H^*_+(w)/H^*_-(w)$, where

$$\left.\begin{matrix}\log H^*_+(w)\\ \\ \log H^*_-(w)\end{matrix}\right\} = \frac{1}{2\pi\mathrm{i}}\int_{\mathrm{i}\gamma-\infty}^{\mathrm{i}\gamma+\infty} \frac{\log H(z)}{z-w}\,\mathrm{d}z\,, \quad \begin{matrix}-\gamma^*<\gamma<\mathrm{Im}\,(w)\\ \\ \gamma^*>\gamma>\mathrm{Im}\,(w)\,.\end{matrix} \tag{3.64}$$

Likewise, in the case of *shear* the function

$$H^{**}(w) = (w^2+\tfrac{9}{4})^{\frac{1}{2}}\,\frac{\sinh 2w - 2w}{w(\cosh 2w - 1)}\,, \tag{3.65}$$

is regular and non-zero in a strip $-\gamma^{**}<\mathrm{Im}\,(w)<\gamma^{**}$ and tends to unity for $|w|\to\infty$ in this strip. Again let $H^{**}(w)=H^{**}_+(w)/H^{**}_-(w)$ with definitions similar to (3.64).

The unique solution of equation (3.61) in the case of *tension* is now given by

$$P_{-}(\lambda) = \frac{\mathrm{i}p}{\lambda(2\pi)^{\frac{1}{2}}}\left[1 - \frac{(\lambda b - 2\mathrm{i})^{\frac{1}{2}}}{(-2\mathrm{i})^{\frac{1}{2}} H_{-}^{*}(\lambda b)}\right], \tag{3.66}$$

where the square roots are defined by the principal branch. From the inverse transform for $\sigma_y(x, b)$ the stress singularity for $x = -r$, $r \to 0$ is obtained in the form*

$$\sigma_y \to Kp \frac{b^{\frac{1}{2}}}{(2r)^{\frac{1}{2}}}, \qquad (k_1 = Kpb^{\frac{1}{2}}), \tag{3.67}$$

where $K = \pi^{-\frac{1}{2}}$ is the *nondimensional stress intensity factor in the case of tension.*

The solution of equation (3.62) in the case of *shear*, under the subsidiary condition (3.55), is also unique [7]

$$T_{-}(\lambda) = \frac{\mathrm{i}t}{\lambda(2\pi)^{\frac{1}{2}}}\left[1 - \frac{(-\frac{3}{2}\mathrm{i})^{\frac{1}{2}}\{1 + \mathrm{i}\lambda(c + 0.2865b)\}}{(\lambda b - \frac{3}{2}\mathrm{i})^{\frac{1}{2}} H_{-}^{**}(\lambda b)}\right], \tag{3.68}$$

where the square roots again denote their principal values, and the numerical coefficient has been obtained from

$$\frac{1}{3} - \frac{1}{2\pi}\int_{-\infty}^{\infty} z^{-2} \log H^{**}(z)\mathrm{d}z = 0.2865 \tag{3.69}$$

by numerical integration. From the inverse transform for $\tau_{xy}(x, b)$ *the stress singularity in the case of shear* for $x = -r$, $r \to 0$ is obtained in the form

$$\tau_{xy} \to \left(\frac{3}{\pi}\right)^{\frac{1}{2}} t \left(\frac{c}{b} + 0.2865\right) \frac{b^{\frac{1}{2}}}{(2r)^{\frac{1}{2}}}. \tag{3.70}$$

## 3.3 Basic anti-plane problems

The anti-plane problem is characterized by a single displacement component $w(x, y)$ normal to the $x$–$y$ plane, and independent of the coordinate perpendicular to this plane. The non-vanishing stress components are the shear stresses $\tau_x$ and $\tau_y$ in the $x$–$y$ plane. The anti-plane state of stress is described by a single analytic function $\phi(z)$, where $z = x + \mathrm{i}y$

$$\begin{aligned} Gw &= \phi(z) + \overline{\phi(z)}, \\ \tau_x - \mathrm{i}\tau_y &= 2\phi'(z). \end{aligned} \tag{3.71}$$

* The stress singularities have not been discussed previously in [4, 7].

*Crack in infinite space or edge crack in half-space* (*Mode III*). The crack extends from $x=-c$ to $x=c$ in the plane $y=0$ and extends to infinity in both directions perpendicular to the $x$–$y$ plane. In the case of a uniform shear stress $\tau_x=0$, $\tau_y=t$ at infinity, the function $\phi(z)$ in equations (3.71) is given by

$$\phi(z) = -\tfrac{1}{2}\mathrm{i}t\,(z^2-c^2)^{\frac{1}{2}}\,, \tag{3.72}$$

where the square root has its cut along the segment $-c<z<c$. Along the $x$-axis

$$\tau_y = t\,\frac{|x|}{(x^2-c^2)^{\frac{1}{2}}} \quad \text{for} \quad |x|>c\,, \tag{3.73}$$

and the *nondimensional stress intensity factor* $K$, defined by $x=c+r$, $r\to 0$ and

$$\tau_y \to Kt\,\frac{c^{\frac{1}{2}}}{(2r)^{\frac{1}{2}}}\,, \qquad (k_3 = Ktc^{\frac{1}{2}})\,, \tag{3.74}$$

has the value $K=1$. The disturbance of the uniform shear stress field decays for $|z|\to\infty$ proportional to $t|z|^{-2}$.

Since the shear stress $\tau_x$ vanishes for $x=0$, one may cut the space along the plane $x=0$ without disturbing the stress distribution. The solution (3.72)–(3.74) therefore applies also to the case of a half-space $x>0$ with an edge crack $0<x<c$ in the plane $y=0$.

*Dam between two half-spaces or edge dam between two quarter-spaces.* The dam between the two half-spaces $y<0$ and $y>0$ extends in the plane $y=0$ from $x=-a$ to $x=a$, and extends to infinity in both directions perpendicular to the $x$–$y$ plane. The dam transmits a shear force $2S$ per unit length in the direction normal to the plane. The function $\phi(z)$ in equations (3.71) is now given by

$$\phi(z) = -\mathrm{i}\,\frac{S}{\pi}\,\text{arc sin}\,\frac{z}{a}\,, \tag{3.75}$$

where the function has its cuts along $-\infty<z<-a$ and $a<z<\infty$. On the $x$-axis

$$\tau_y = \frac{2S}{\pi}\,\frac{1}{(a^2-x^2)^{\frac{1}{2}}} \quad \text{for} \quad |x|<a\,, \tag{3.76}$$

and the *nondimensional stress intensity factor* $K$, defined by $x=a-r$, $r\to 0$ and

$$\tau_y \to K \frac{S}{a^{\frac{1}{2}}} \frac{1}{(2r)^{\frac{1}{2}}}, \qquad \left(k_3 = \frac{KS}{a^{\frac{1}{2}}}\right),$$

has the value $K = 2/\pi$. The stresses decay for $|z| \to \infty$ proportional to $S|z|^{-1}$.

Here again one may cut the space along the plane $x=0$ without disturbing the stress distribution. The present solution therefore applies also to the case of a half-space $x>0$ with a semi-infinite crack $a<x<\infty$ in the plane $y=0$, transmitting a force $S$ per unit length through the dam $0<x<a$.

## 3.4 Basic space problems

*Penny-shaped crack in infinite space.* Cylindrical coordinates $\rho$, $\phi$, $z$ will be used. The penny-shaped crack is described by the circle $\rho=c$ in the plane $z=0$. The undisturbed state of stress in solid space is described by a single non-vanishing normal stress component in the cases of *tension* and *bending*

$$\sigma_z = p\left(\frac{\rho}{c}\right)^n \cos n\phi, \qquad n=0 \text{ or } 1; \tag{3.77}$$

the case of *torsion* is specified by a single non-vanishing shear stress component in solid space

$$\tau_{z\phi} = t\frac{\rho}{c}. \tag{3.78}$$

The disturbance due to the crack is obtained by applying a load of opposite sign to the crack faces $z=0$, $0<\rho<c$ in the infinite solid with vanishing stresses at infinity*.

The problem is reduced to a problem for a half-space $z>0$ under surface loads in the circle $0<\rho<c$ which are the negative of (3.77) or (3.78). In the cases of *tension and bending* the purely normal loads $\sigma_z$ are unknown for $\rho>c$, but we have here the alternative boundary condition that the axial displacement component $u_z$ must vanish for $z=0$, $\rho>c$. In the case of *torsion* the surface loads are shear stresses $\tau_{z\phi}$, the opposite of (3.78) in the circle $0<\rho<c$, and unknown for $\rho>c$. The alternative boundary condition is here that the tangential displacement component $u_\phi$ must vanish for $z=0$,

* The tension problem has been discussed by many writers [9], but we have found no reference to the bending problem. The torsion case has been dealt with in [10] and [30, part II]. The discussion in [1, p. 158] is incorrect since the equilibrium equations are violated by the undisturbed field.

$\rho > c$. Both problems are reduced to mixed boundary value problems for a single harmonic function.

The stress distribution in a half-space under purely normal surface loads may be described by a single harmonic function $f(\rho, \phi, z)$. The relevant stress and displacement components are given by [8, pp. 92, 93]

$$\sigma_z = z\frac{\partial^3 f}{\partial z^3} - \frac{\partial^2 f}{\partial z^2}, \quad 2Gu_z = z\frac{\partial^2 f}{\partial z^2} - 2(1-\nu)\frac{\partial f}{\partial z}. \tag{3.79}$$

In the loading condition specified by the negative of (3.77) in the circle $0 < \rho < c$ of the plane $z = 0$

$$f(\rho, \phi, z) = f^*(\rho, z)\cos n\phi . \tag{3.80}$$

The Hankel transform

$$F^*(\gamma, z) = \int_0^\infty \rho f^*(\rho, z) J_n(\gamma\rho)\mathrm{d}\rho , \tag{3.81}$$

is introduced and the transformed harmonic equation has the solution $F^*(\gamma, z) = A(\gamma)\exp(-\gamma z)$. Evaluating the stress and displacement components (3.79) by the inverse Hankel transformation, and substituting the result in the boundary conditions at $z = 0$, a pair of dual integral equations for the unknown function $A(\gamma)$ is obtained

$$\begin{aligned} -\frac{\sigma_z(\rho, \phi, 0)}{\cos n\phi} &= \int_0^\infty \gamma^3 A(\gamma) J_n(\gamma\rho)\mathrm{d}\gamma = p\left(\frac{\rho}{c}\right)^n \quad \text{for} \quad 0 < \rho < c , \\ \frac{Gu_z(\rho, \phi, 0)}{(1-\nu)\cos n\phi} &= \int_0^\infty \gamma^2 A(\gamma) J_n(\gamma\rho)\mathrm{d}\gamma = 0 \qquad \text{for} \quad \rho > c . \end{aligned} \tag{3.82}$$

The explicit solution of (3.82) in terms of infinite integrals is given in [28, p. 339] and it may be evaluated by gamma and Bessel function formulae [29, vol. 2, p. 22]

$$\gamma^2 A(\gamma) = pc^{\frac{3}{2}}(2\gamma)^{-\frac{1}{2}}\frac{\Gamma(n+1)}{\Gamma(n+\frac{3}{2})} J_{n+\frac{3}{2}}(\gamma c) . \tag{3.83}$$

The normal stress $\sigma_z$ at the plane $z = 0$ may again be evaluated from equations (3.82) by means of a Bessel function formula [29, vol. 2, p. 48]. For $\rho < c$ it is confirmed that this stress is the negative of (3.77), for $\rho > c$

$$\sigma_z(\rho, \phi, 0) = \frac{p}{2(\pi^{\frac{1}{2}})}\frac{\Gamma(n+1)}{\Gamma(n+\frac{5}{2})} F\left(n+\tfrac{3}{2}, \tfrac{3}{2}; n+\tfrac{5}{2}; \frac{c^2}{\rho^2}\right)\left(\frac{c}{\rho}\right)^{n+3} \cos n\phi , \tag{3.84}$$

where $F(a, b; c; z)$ denotes the hypergeometric function. It is noted in passing that this result holds for all positive integral values of $n$ for loads specified by the negative of (3.77) in the circle $0<\rho<c$ of the plane $z=0$.

In the case of *tension* ($n=0$) the hypergeometric function may be expressed in terms of elementary functions, and after some laborious algebra

$$\sigma_z(\rho, \phi, 0) = \frac{2p}{\pi}\left[\frac{c}{(\rho^2-c^2)^{\frac{1}{2}}} - \arc\sin\frac{c}{\rho}\right]. \tag{3.85}$$

At infinity the stresses decay proportional to $pc^3/R^3$, where $R^2=\rho^2+z^2$ (on the plane $z=0$ even as $pc^4/\rho^4$). The *nondimensional stress intensity factor* $K$, defined by

$$\sigma_z \to Kp\frac{c^{\frac{1}{2}}}{(2r)^{\frac{1}{2}}}, \qquad (k_1 = Kp\,c^{\frac{1}{2}}), \tag{3.86}$$

for $\rho=c+r$, $r\to 0$ has the well-known value $K=2/\pi$.

In the case of *bending* ($n=1$) the simplified form of equation (3.84) is

$$\sigma_z(\rho, \phi, 0) = \frac{2p}{3\pi}\left[\frac{\rho^2+c^2}{\rho(\rho^2-c^2)^{\frac{1}{2}}} + \frac{2}{\rho}(\rho^2-c^2)^{\frac{1}{2}} - \frac{3\rho}{c}\arc\sin\frac{c}{\rho}\right]\cos\phi, \tag{3.87}$$

and the *nondimensional stress intensity factor* $K$, defined by equation (3.86), has the value $K=4/3\pi$. In this case the stresses decay at infinity proportional to $pc^4/R^4$, where $R^2=\rho^2+z^2$ (on the plane $z=0$ as $pc^5/\rho^5$).

In the *torsion* problem the function $u_\phi(\rho, z)\cos\phi$ is harmonic. With the introduction of the Hankel transform

$$U_\phi(\gamma, z) = \int_0^\infty \rho u_\phi(\rho, z) J_1(\gamma\rho)\mathrm{d}\rho, \tag{3.88}$$

the solution of the transformed harmonic equation is $U_\phi(\gamma,z)=B(\gamma)\exp(-\gamma z)$. Evaluating the stress component $\tau_{z\phi}(\rho, z)=G\,\partial u_\phi/\partial z$ and the displacement component $u_\phi(\rho, z)$ by the inverse Hankel transformation, and substituting the result in the boundary conditions at the plane $z=0$, a pair of dual integral equation is again obtained:

$$\begin{aligned} -\tau_{z\phi}(\rho, 0) &= G\int_0^\infty \gamma^2 B(\gamma) J_1(\gamma\rho)\mathrm{d}\gamma = t\frac{\rho}{c} \quad \text{for} \quad 0<\rho<c, \\ u_\phi(\rho, 0) &= \int_0^\infty \gamma B(\gamma) J_1(\gamma\rho)\mathrm{d}\gamma = 0 \quad \text{for} \quad \rho>c. \end{aligned} \tag{3.89}$$

These equations are formally identical to equations (3.82) in the case $n=1$, if $GB(\gamma)$ and $t$ in (3.89) are identified with $\gamma A(\gamma)$ and $p$ in (3.82). It follows that $\tau_{z\phi}(\rho > c, 0)$ in the case of *torsion* is also given by equation (3.87) after replacement of $p$ by $t$, and suppression of the factor $\cos\phi$. The *dimensionless stress intensity factor* $K$ in the *torsion* case, defined by

$$\tau_{z\phi} \to Kt \frac{c^{\frac{1}{2}}}{(2r)^{\frac{1}{2}}}, \qquad (k_3 = Kt\,c^{\frac{1}{2}}), \tag{3.90}$$

for $\rho = c + r$, $r \to 0$, has again the value $K = 4/3\pi$, and the stresses decay at infinity proportional to $tc^4/R^4$, where $R^2 = \rho^2 + z^2$.

*Penny-shaped dam between two half-spaces.* The space is cut in the plane $z=0$ outside the circle $\rho = a$, and the half-spaces $z < 0$ are connected only through the dam at $z=0$, $\rho < a$. The cases where the dam transmits a tensile force $N$, a bending moment $M$ or a torque $T$ will be considered. The solutions to these problems coincide with those of the half-space problems where the load is applied by a rigid flat circular stamp (in the absence of friction in the loading conditions of tension and bending, and with a bonded stamp in the torque loading condition). The solutions to the latter problems are well-known [9, 11, 12, 13, 31], and only the results which are relevant to our purposes will be listed.

In the case of a *tensile load* $N$ the stress distribution in the dam is given by

$$\sigma_z = \frac{N}{2\pi a} \frac{1}{(a^2 - \rho^2)^{\frac{1}{2}}}, \tag{3.91}$$

and the *nondimensional stress intensity factor* $K$, defined by

$$\sigma_z \to K \frac{N}{\pi a^2} \frac{a^{\frac{1}{2}}}{(2r)^{\frac{1}{2}}}, \qquad \left(k_1 = \frac{KN}{\pi a(a^{\frac{1}{2}})}\right) \tag{3.92}$$

for $\rho = a - r$, $r \to 0$, has the value $K = \frac{1}{2}$. At infinity the stresses decay as $NR^{-2}$, where $R^2 = \rho^2 + z^2$.

In the case of a *bending moment* $M$ the stress distribution in the dam is given by

$$\sigma_z = \frac{3M}{2\pi a^3} \frac{\rho}{(a^2 - \rho^2)^{\frac{1}{2}}} \cos\phi\,, \tag{3.93}$$

and the *nondimensional stress intensity factor* $K$ in

$$\sigma_z \to K \frac{4M}{\pi a^3} \frac{a^{\frac{1}{2}}}{(2r)^{\frac{1}{2}}}, \qquad \left(k_1 = \frac{4KM}{\pi a^2 (a^{\frac{1}{2}})}\right), \tag{3.94}$$

for $\rho = a - r$, $r \to 0$, has the value $K = \frac{3}{8}$. In this case the stresses decay at infinity as $MR^{-3}$.

In the case of a *torque* $T$ the stress distribution in the dam is given by

$$\tau_{z\phi} = \frac{3T}{4\pi a^3} \frac{\rho}{(a^2 - \rho^2)^{\frac{1}{2}}}, \tag{3.95}$$

and the *dimensionless stress intensity factor* $K$ in

$$\tau_{z\phi} \to K \frac{2T}{\pi a^3} \frac{a^{\frac{1}{2}}}{(2r)^{\frac{1}{2}}}, \qquad \left(k_3 = \frac{2KT}{\pi a^2 (a^{\frac{1}{2}})}\right), \tag{3.96}$$

for $\rho = a - r$, $r \to 0$, has again the value $K = \frac{3}{8}$. In this case the stresses decay again at infinity as $TR^{-3}$.

## 3.5 Strip problems

*Strip with central crack.* Consider a strip of infinite length and width $2b$ with a central crack of width $2c = 2(b-a)$, in generalized plane stress or plane strain, loaded by a central *tensile force* $N$ or a *bending moment* $M$, each per unit thickness (Figure 3.4).

In the case of *tensile loading*, the leading term of the asymptotic expansion of the stress singularity for $c/b \to 0$ is obtained by substituting $p = N/2b$ in equation (3.4). The longitudinal edges of the strip in the solution (3.2) for the infinite plane under a tensile stress $p = N/2b$ at infinity carry tractions of order $pc^2/b^2$ for $c/b \to 0$. The removal of these edge tractions from the strip (without a central crack) results in an additional (nominal) stress of the same order at the location of the crack. It follows that the second term in the asymptotic expansion of the stress singularity is of order $c^2/b^2$ times the leading term

$$\sigma_y \to \frac{N}{2b} \frac{c^{\frac{1}{2}}}{(2r)^{\frac{1}{2}}} \left[1 + O\left(\frac{c^2}{b^2}\right)\right] \quad \text{for} \quad c/b \to 0\,. \tag{3.97}$$

At the other end of the range of the parameter $c/b$, viz. $c/b \to 1$ or $a/b \to 0$, the solution of Section 3.2 applies. From equation (3.49) for a tensile force $\frac{1}{2}N$ per dam

$$\sigma_y \to \frac{2}{(\pi^2 - 4)^{\frac{1}{2}}} \frac{N}{2a} \frac{a^{\frac{1}{2}}}{(2r)^{\frac{1}{2}}} \quad \text{for} \quad a/b \to 0\,. \tag{3.98}$$

It is now convenient to introduce a *nondimensional stress intensity factor* $K(c/b)$ as a function of the parameter $c/b$ in such a way that it is of order unity in the entire range $0 < c/b < 1$ and tends to the limiting values 1 and

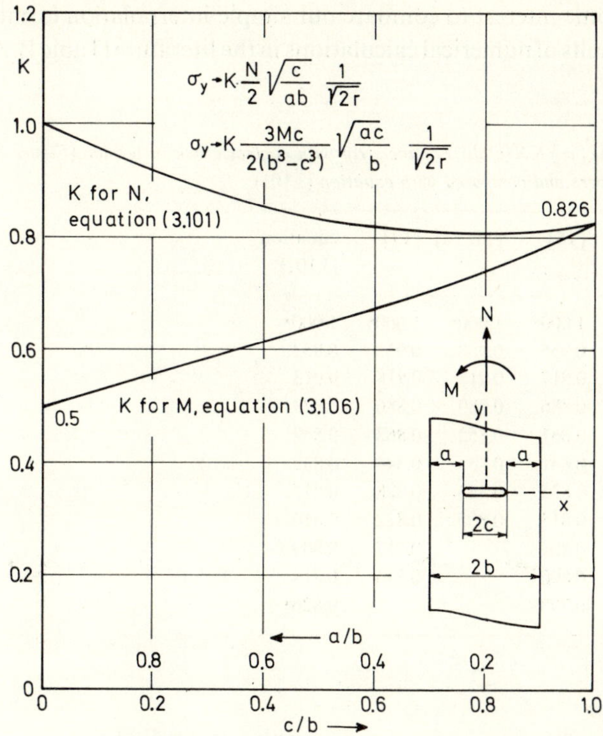


Figure 3.4. $K$-values for strip with central crack.

$2/(\pi^2-4)^{\frac{1}{2}}=0.826$ at the ends $c/b=0$ and $c/b=1$. This purpose is achieved by writing the stress singularity in the form

$$\sigma_y \to K(c/b)\,\frac{N}{2(ab)^{\frac{1}{2}}}\,\frac{c^{\frac{1}{2}}}{(2r)^{\frac{1}{2}}}\,, \qquad \left(k_1 = \frac{KN\,c^{\frac{1}{2}}}{2(ab)^{\frac{1}{2}}}\right) \tag{3.99}$$

where equations (3.97) and (3.98) give

$$K \to 1 - \frac{1}{2}\frac{c}{b} + O\left(\frac{c^2}{b^2}\right) \ \text{ for } \frac{c}{b} \to 0\,, \quad K \to 0.826 \ \text{ for } \frac{c}{b} \to 1\,. \tag{3.100}$$

The simplest possible polynomial interpolation between these asymptotic results in the *tension* case is [6]

$$K = 1 - \frac{1}{2}\frac{c}{b} + 0.326\,\frac{c^2}{b^2}, \tag{3.101}$$

and it is represented by the curve in Figure 3.4.

It is of some interest to compare our simple interpolation formula (3.101) with the results of numerical calculations in the literature (Table I). According

TABLE I

*The K-value ($k_1 = \frac{1}{2}KN(c/ab)^{\frac{1}{2}}$) of the strip with central crack in tension (Figure 3.4), adapted from some papers and compared with equation (3.101).*

| $c/b$ | [15] | [16] | [17, 18] | [19] | equation (3.101) |
|---|---|---|---|---|---|
| 0 | 1.000* | 1.000* | 1.000* | 1.000* | 1.000* |
| 0.1 | 0.954 | 0.955 | 0.958 | 0.955 | 0.953 |
| 0.2 | 0.916 | 0.917 | 0.917 | 0.918 | 0.913 |
| 0.3 | 0.885 | 0.886 | 0.880 | 0.886 | 0.879 |
| 0.4 | 0.859 | 0.861 | 0.853 | 0.863 | 0.852 |
| 0.5 | 0.839 | 0.841 | 0.835 | 0.844 | 0.832 |
| 0.6 | 0.824 | 0.825 | 0.823 | 0.828 | 0.817 |
| 0.7 | 0.815 | 0.813 | 0.805 | 0.822 | 0.810 |
| 0.8 | 0.810 | 0.804 | | 0.817 | 0.809 |
| 0.9 | | 0.800 | | 0.816 | 0.814 |
| 1.0 | | 0.798 | | | 0.826* |

* Exact values.

to Rooke [14] the numerical results of Isida [15] are likely to be the most reliable, but also included are Feddersen's "intelligent guess" [16], which has been conjectured (erroneously) to represent the exact solution [20], the results of Formann and Kobayashi [17, 18], and the results of the (possibly most accurate) recent numerical solution of the basic integral equation by Sneddon and Srivastav [19]. The authors feel entitled to claim that the simple interpolation achieves a one percent accuracy and is adequate for all practical purposes.

In the case of loading by a *bending moment M* equation (3.9) may be applied for $c/b \to 0$ by taking $p = 3Mc/2b^3$. The solution (3.7) for the infinite plane now contains tractions along the longitudinal edges of the strip of order $pc^3/b^3$, and their removal from the strip (without a central crack) now results in an additional (nominal) stress at the location of the crack tip of order $pc^4/b^4$

because it vanishes at the crack center. The second term of the asymptotic expansion of the stress singularity for $c/b \to 0$ is therefore of order $c^4/b^4$ times the leading term. Hence we have in the bending case

$$\sigma_y \to \frac{1}{2} \frac{3Mc}{2b} \frac{c^{\frac{1}{2}}}{(2r)^{\frac{1}{2}}} \left[1 + O\left(\frac{c^4}{b^4}\right)\right] \quad \text{for} \quad c/b \to 0 . \tag{3.102}$$

At the other end of the range $a/b \to 0$ we may again apply the edge dam solution, now for a tensile force $M/2b$ per dam, and

$$\sigma_y \to \frac{2}{(\pi^2-4)^{\frac{1}{2}}} \frac{M}{2ab} \frac{a^{\frac{1}{2}}}{(2r)^{\frac{1}{2}}} \quad \text{for} \quad a/b \to 0 . \tag{3.103}$$

In the *bending case* the *nondimensional stress intensity factor* $K$ is defined by writing the stress singularity in the form

$$\sigma_y \to K \frac{3cM}{2(b^3-c^3)} \left(\frac{ac}{b}\right)^{\frac{1}{2}} \frac{1}{(2r)^{\frac{1}{2}}}, \left(k_1 = K \frac{3cM}{2(b^3-c^3)} \left(\frac{ac}{b}\right)^{\frac{1}{2}}\right). \tag{3.104}$$

The asymptotic formulae for $K$ in this case are

$$\begin{aligned} K &\to \frac{1}{2}\left[1 + \frac{1}{2}\frac{c}{b} + \frac{3}{8}\frac{c^2}{b^2} - \frac{11}{16}\frac{c^3}{b^3} + O\left(\frac{c^4}{b^4}\right)\right] \quad \text{for} \quad c/b \to 0 , \\ K &\to \frac{2}{(\pi^2-4)^{\frac{1}{2}}} = 0.826 \quad \text{for} \quad a/b \to 0 . \end{aligned} \tag{3.105}$$

The simplest polynomial interpolation between these asymptotic results

$$K = \frac{1}{2}\left[1 + \frac{1}{2}\frac{c}{b} + \frac{3}{8}\frac{c^2}{b^2} - \frac{11}{16}\frac{c^3}{b^3} + 0.464\,\frac{c^4}{b^4}\right]. \tag{3.106}$$

is also represented by a curve in Figure 3.4.

Since the interpolation in the bending case is based on more exact data than the similar interpolation for tensile loading, one may expect at least the same accuracy. Even though it has not been possible to compare formula or curve with the results of accurate numerical calculations, the authors feel entitled to claim a one percent accuracy.

*Strip with symmetric edge cracks.* Consider an infinite strip of width $2b$ with symmetric edge cracks of depth $c$, in generalized plane stress or plane strain, loaded by a central *tensile force* $N$ or a *bending moment* $M$, each per unit thickness of the strip (Figure 3.5).

In the case of *tensile loading*, the leading term of the asymptotic expansion of the stress singularity for $c/b \to 0$ is obtained by substituting $p = N/2b$ in equation (3.38). In the half-plane of the edge crack problem the stresses along a line parallel to the edge at a distance $2b$ are at most of order $pc^2/b^2$, and their removal for the solid strip results in an additional (nominal) stress at

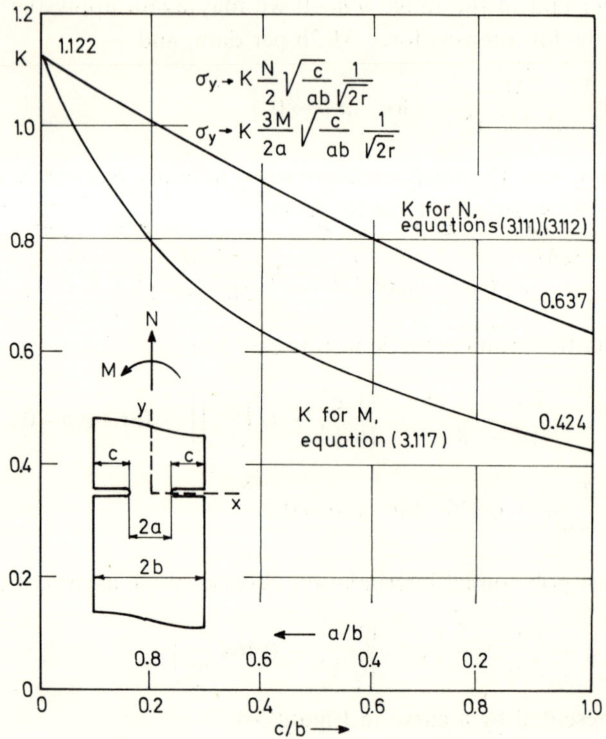


Figure 3.5. *K*-values for strip with symmetric edge-cracks.

the location of the edge crack of the same order of magnitude. Hence for the asymptotic expansion of the stress singularity,

$$\sigma_y \to 1.122 \frac{N}{2b} \frac{c^{\frac{1}{2}}}{(2r)^{\frac{1}{2}}} \left[1 + O\left(\frac{c^2}{b^2}\right)\right] \quad \text{for} \quad c/b \to 0 . \tag{3.107}$$

At the other end of the range of the parameter $c/b \to 1$, or $a/b \to 0$, equation (3.17) may be applied. The longitudinal edges of the strip in the solution

(3.15) for the half-plane carry tractions of order $N/b$. Their removal from the semi-infinite strip $-b<x<b$, $y>0$ introduces a curvature of the edge $y=0$ of order $N/Eb^2$. Since the dam must remain straight, additional stresses are induced by the removal of the curvature. It follows from [3, §116a] that the stress singularity due to a curvature $1/R$ is of order $(Ea/R)(a/r)^{\frac{1}{2}}$. The asymptotic formula for the stress singularity is therefore, from equation (3.17),

$$\sigma_y \to \frac{2}{\pi}\frac{N}{2a}\frac{a^{\frac{1}{2}}}{(2r)^{\frac{1}{2}}}\left[1+O\left(\frac{a^2}{b^2}\right)\right] \quad \text{for} \quad a/b \to 0\,. \tag{3.108}$$

A *nondimensional stress intensity factor K* for *tension* is defined by the formula

$$\sigma_y \to K\frac{N}{2(ab)^{\frac{1}{2}}}\frac{c^{\frac{1}{2}}}{(2r)^{\frac{1}{2}}}, \qquad \left(k_1=\tfrac{1}{2}KN\left(\frac{c}{ab}\right)^{\frac{1}{2}}\right), \tag{3.109}$$

and from equations (3.107) and (3.108) the asymptotic formulae follow

$$\begin{aligned} K &\to 1.122\left(1-\frac{1}{2}\frac{c}{b}\right)+O\left(\frac{c^2}{b^2}\right) \quad \text{for} \quad c/b\to 0\,,\\ K &\to \frac{2}{\pi}\left(1+\frac{1}{2}\frac{a}{b}\right)+O\left(\frac{a^2}{b^2}\right) \quad \text{for} \quad a/b\to 0\,. \end{aligned} \tag{3.110}$$

The simplest possible (cubic) interpolation is therefore [6]

$$K = 1.122\left(1-\frac{1}{2}\frac{c}{b}\right)-0.015\frac{c^2}{b^2}+0.091\frac{c^3}{b^3}\,. \tag{3.111}$$

An alternative (transcendental) interpolation formula, suggested by the exact solution in the case of a row of collinear cracks (cf. formula (3.127))

$$K = \left[1+0.122\sin^2\frac{\pi a}{2b}\right]\left(\frac{2a}{\pi c}\tan\frac{\pi c}{2b}\right)^{\frac{1}{2}} \tag{3.112}$$

differs from equation (3.111) by about 0.3 percent. The result is also given in Figure 3.5.

In view of the fact that the interpolation formula (3.111) is based on *exact* values for the function itself and for its first derivative at both end points of the interval $0<c/b<1$, and that the interpolation does not deviate far from a straight line, the authors feel entitled to claim an accuracy of better than one percent. Some support for this claim is found in the comparison in [6] with results of numerical calculations [21], with a claimed accuracy of one percent. The largest discrepancy found between Bowie's results and equation (3.111) was 0.8 percent, and it may be largely due to errors in his numerical results.

In the case of loading by a *bending moment M* the edge crack solution may again be applied for $c/b \to 0$. The leading term in the stress singularity is given by equation (3.37), where $p = 3M(b-c)/2b^3$ and $q = 3Mc/2b^3$, and the second term is again of order $c^2/b^2$ times the leading term. Therefore

$$\sigma_y \to 1.122\,\frac{3M}{2b^2}\left(1-0.609\,\frac{c}{b}\right)\left[1+O\left(\frac{c^2}{b^2}\right)\right]\frac{c^{\frac{1}{2}}}{(2r)^{\frac{1}{2}}} \quad \text{for} \quad c/b \to 0\,. \tag{3.113}$$

On the other hand, for $a/b \to 0$ one may apply equation (3.20). In the half-plane solution (3.18) the longitudinal edges of the semi-infinite strip carry stresses of order $M/b^2$, and their removal induces a curvature of the edge $y=0$ of order $M/Eb^3$. This curvature vanishes at the center of the dam, however, and it is of order $Ma/Eb^4$ at the ends. The additional terms in the stress singularity due to the annihilation of the curvature along the dam are therefore of order $a^4/b^4$ times the primary singularity, and

$$\sigma_y \to \frac{4}{3\pi}\,\frac{3M}{2a^2}\,\frac{a^{\frac{1}{2}}}{(2r)^{\frac{1}{2}}}\left[1+O\left(\frac{a^4}{b^4}\right)\right] \quad \text{for} \quad a/b \to 0\,. \tag{3.114}$$

Defining the *nondimensional stress intensity factor K* for *bending* by

$$\sigma_y \to K\,\frac{3M}{2a(ab)^{\frac{1}{2}}}\,\frac{c^{\frac{1}{2}}}{(2r)^{\frac{1}{2}}}\,, \qquad \left(k_1 = K\,\frac{3M}{2a}\left(\frac{c}{ab}\right)^{\frac{1}{2}}\right), \tag{3.115}$$

the asymptotic formulae

$$\begin{aligned} K &\to 1.122\left(1-2.109\,\frac{c}{b}\right)+O\left(\frac{c^2}{b^2}\right) \quad \text{for} \quad c/b \to 0\,, \\ K &\to \frac{4}{3\pi}\left(1+\frac{1}{2}\frac{a}{b}+\frac{3}{8}\frac{a^2}{b^2}+\frac{5}{16}\frac{a^3}{b^3}\right)+O\left(\frac{a^4}{b^4}\right) \quad \text{for} \quad a/b \to 0 \end{aligned} \tag{3.116}$$

are obtained. The quintic interpolation formula

$$K = \frac{4}{3\pi}\left(1+\frac{1}{2}\frac{a}{b}+\frac{3}{8}\frac{a^2}{b^2}+\frac{5}{16}\frac{a^3}{b^3}\right) - 0.470\,\frac{a^4}{b^4} + 0.663\,\frac{a^5}{b^5} \tag{3.117}$$

is also given in Figure 3.5. In view of the large number of exact data incorporated in equation (3.117), viz. the function and its first derivative at both end points $a/b=0$ and $a/b=1$, and in addition the second and third derivatives for $a/b=0$, the authors expect again a one percent accuracy, in spite of the stronger curvature of the curve in Figure 3.5.

*Strip with a single edge crack.* Consider an infinite strip of width $b$ with a single crack of depth $c = b - a$ perpendicular to an edge, loaded in plane strain or generalized plane stress by a central *tensile force N* or a *bending moment M*, each per unit thickness of the strip (Figure 3.6).*

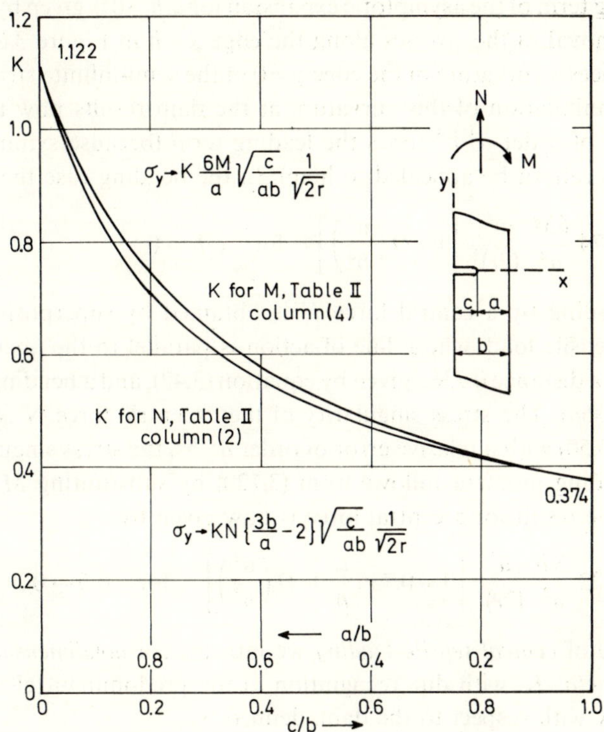


Figure 3.6. *K*-values for strip with edge-crack.

In the limit $c/b \rightarrow 0$ the asymptotic solutions of the previous case remain valid. Replacing $2b$ in equations (3.107) and (3.113) by $b$, there results for tensile loading

$$\sigma_y \rightarrow 1.122\,\frac{N}{b}\,\frac{c^{\frac{1}{2}}}{(2r)^{\frac{1}{2}}}\left[1 + O\left(\frac{c^2}{b^2}\right)\right] \quad \text{for} \quad c/b \rightarrow 0\,, \tag{3.118}$$

and in the case of loading by a bending moment

* We are indebted to one of our students, Mr. G. T. M. Janssen, for an independent check of our analysis of this case.

$$\sigma_y \to 1.122 \frac{6M}{b^2}\left(1-1.217\frac{c}{b}\right)\frac{c^{\frac{1}{2}}}{(2r)^{\frac{1}{2}}}\left[1+O\left(\frac{c^2}{b^2}\right)\right] \quad \text{for} \quad c/b \to 0 . \tag{3.119}$$

At the other end of the range, $c/b \to 1$, $a/b \to 0$, one may again appeal to the solutions for the edge dam. The simplest case is loading by a bending moment. The leading term of the asymptotic expansion for $a/b \to 0$ is given by equation (3.51). Removal of the stresses along the edge $x=b$ in Figure 3.6, of order $M/b^2$, induces a curvature of the edge $y=0$ of the semi-infinite strip of order $M/Eb^3$. Annihilation of this curvature at the dam results now in a stress singularity of order $a^3/b^3$ times the leading term (because symmetry considerations cannot be appealed to here). In the bending case therefore

$$\sigma_y \to 0.374 \frac{6M}{a^2}\frac{a^{\frac{1}{2}}}{(2r)^{\frac{1}{2}}}\left[1+O\left(\frac{a^3}{b^3}\right)\right] \quad \text{for} \quad a/b \to 0 . \tag{3.120}$$

Tensile loading by a central force $N$ is obtained by superposition of an excentric tensile load, whose line of action is parallel to the uncut edge of the strip at a distance $0.736a$ given by equation (3.47), and a bending moment $N(\frac{1}{2}b-0.736a)$. The stress singularity of the excentric force $N$ is given by equation (3.50) with a relative error of order $a^2/b^2$, the stress singularity due to the bending moment follows from (3.120) by substituting $M=N(\frac{1}{2}b-0.736a)$. The result for a central force is now given by

$$\sigma_y \to 1.122 \frac{Nb}{a^2}\frac{a^{\frac{1}{2}}}{(2r)^{\frac{1}{2}}}\left[1-0.736\frac{a}{b}+O\left(\frac{a^3}{b^3}\right)\right] \quad \text{for} \quad a/b \to 0 . \tag{3.121}$$

In the case of *central tensile loading* we introduce a *nondimensional stress intensity factor* $K$, with due recognition of the predominant effects of the excentricity with respect to the dam, defined by

$$\sigma_y \to K \frac{N(3b-2a)}{a(ab)^{\frac{1}{2}}}\frac{c^{\frac{1}{2}}}{(2r)^{\frac{1}{2}}}, \qquad \left(k_1 = KN\frac{(3b-2a)c^{\frac{1}{2}}}{a(ab)^{\frac{1}{2}}}\right). \tag{3.122}$$

The asymptotic expansions for this stress intensity factor are

$$\begin{aligned} &K \to 1.122\left[1-\frac{7}{2}\frac{c}{b}+O\left(\frac{c^2}{b^2}\right)\right] \quad \text{for} \quad c/b \to 0 , \\ &K \to 0.374\left[1+0.431\frac{a}{b}+0.294\frac{a^2}{b^2}+O\left(\frac{a^3}{b^3}\right)\right] \quad \text{for} \quad a/b \to 0 . \end{aligned} \tag{3.123}$$

In view of the large value of the derivative at $c/b=0$ a polynomial interpolation is less suitable in the present case. An interpolation by the conic

TABLE II

*The K-values of the strip with edge-crack (Figure 3.6), compared with those of literature.*

| $c/b$ | Load $N$ | | Load $M$ | |
|---|---|---|---|---|
| | [17, 26] | C.s.i. equations (3.123) | [17, 27] | C.s.i. equations (3.125) |
| | (1) | (2) | (3) | (4) |
| 0 | 1.123 | 1.122* | 1.123 | 1.122* |
| 0.05 | 0.955 | 0.961 | 0.991 | 0.990 |
| 0.10 | 0.844 | 0.849 | 0.892 | 0.890 |
| 0.15 | 0.764 | 0.767 | 0.816 | 0.811 |
| 0.20 | 0.702 | 0.703 | 0.755 | 0.747 |
| 0.25 | 0.651 | 0.653 | 0.703 | 0.693 |
| 0.30 | 0.609 | 0.611 | 0.658 | 0.649 |
| 0.35 | 0.573 | 0.577 | 0.619 | 0.610 |
| 0.40 | 0.544 | 0.548 | 0.585 | 0.577 |
| 0.45 | 0.520 | 0.523 | 0.555 | 0.549 |
| 0.50 | 0.500 | 0.501 | 0.529 | 0.523 |
| 0.55 | 0.482 | 0.482 | 0.506 | 0.501 |
| 0.60 | 0.463 | 0.464 | 0.483 | 0.481 |
| 0.65 | | 0.449 | | 0.463 |
| 0.70 | | 0.435 | | 0.446 |
| 0.75 | | 0.423 | | 0.432 |
| 0.80 | | 0.411 | | 0.418 |
| 0.85 | | 0.401 | | 0.406 |
| 0.90 | | 0.391 | | 0.394 |
| 0.95 | | 0.382 | | 0.384 |
| 1.00 | | 0.374* | | 0.374* |

C.s.i.: Conic section interpolation.
* Exact values.

section has been employed here which has in common with equations (3.123) the values and tangents at both ends of the interval, as well as the curvature at $c/b=1$. The results are presented in Table II and Figure 3.6. Listed also in Table II are the results of an interpolation formula [17, 32], based on numerical calculations [26], and valid only in a limited range. The agreement is excellent, except for very small values of $c/b$, where the formula of [17, 32] is slightly in error, due to an inaccurate value of the derivative at $c/b=0$. The authors feel entitled to claim that their results have an accuracy far better than one percent.

In the case of loading by a *bending moment M* define the *nondimensional stress intensity factor K* by the relation

$$\sigma_y \to K \frac{6M}{a(ab)^{\frac{1}{2}}} \frac{c^{\frac{1}{2}}}{(2r)^{\frac{1}{2}}}, \qquad \left(k_1 = K \frac{6M}{a}\left(\frac{c}{ab}\right)^{\frac{1}{2}}\right), \tag{3.124}$$

and the asymptotic expansions are

$$\begin{aligned} K &\to 1.122\left[1 - 2.717\frac{c}{b} + O\left(\frac{c^2}{b^2}\right)\right] \qquad \text{for} \quad c/b \to 0, \\ K &\to 0.374\left[1 + \frac{1}{2}\frac{a}{b} + \frac{3}{8}\frac{a^2}{b^2} + O\left(\frac{a^3}{b^3}\right)\right] \qquad \text{for} \quad a/b \to 0. \end{aligned} \tag{3.125}$$

Here again a conic section interpolation has been employed, and the results are presented in Table II and Figure 3.6. Also listed in Table II are the results of an interpolation formula [17, 32], based on numerical calculations [27], again valid in a limited range. The discrepancies do not exceed 1.4 percent, but the authors expect their results to be as accurate as those of the (less simple) tensile force case.

## 3.6 Crack configurations in a plane or half-plane

*An infinite row of collinear cracks.* Consider an infinite plane in plane strain or generalized plane stress with cracks of length $2c$ along the $x$-axis with a regular spacing $2b$ (Figure 3.7). The solution in the case of uniform *tension* $\sigma_y = p$ at infinity has been given by Westergaard [23], the solution in the case of uniform *shear* $\tau_{xy} = t$ at infinity has been given in [24]. Both cases are described in terms of the same *dimensionless stress intensity factor K*, defined by

$$\begin{pmatrix}\sigma_y \\ \tau_{xy}\end{pmatrix} \to K \begin{pmatrix}p \\ t\end{pmatrix} \left(\frac{b}{a}\right)^{\frac{1}{2}} \frac{c^{\frac{1}{2}}}{(2r)^{\frac{1}{2}}}, \quad \begin{pmatrix}k_1 \\ k_2\end{pmatrix} = K \begin{pmatrix}p \\ t\end{pmatrix} \left(\frac{bc}{a}\right)^{\frac{1}{2}}, \tag{3.126}$$

with the same numerical value

$$K = \left(\frac{2a}{\pi c} \tan \frac{\pi c}{2b}\right)^{\frac{1}{2}}. \tag{3.127}$$

This *exact* solution may serve as a test case for the asymptotic approximations.

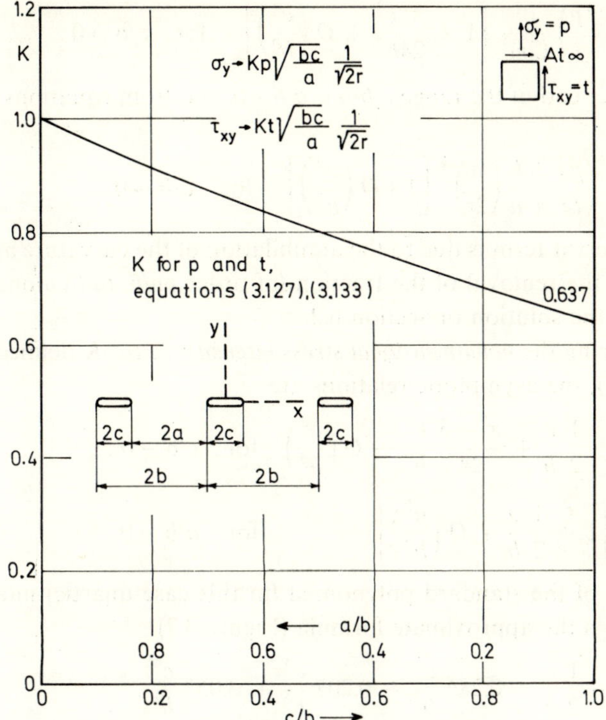


Figure 3.7. *K*-values for collinear cracks.

In the case of a single crack $-c < x < c$ in an infinite plane the disturbance caused at the location of the crack $-c+2nb < x < c+2nb$ $(n = \pm 1, \pm 2, \ldots)$ in the cases of tension and shear is given by equations (3.5) and (3.13)

$$\begin{pmatrix}\sigma_y\\ \tau_{xy}\end{pmatrix} \to \begin{pmatrix}p\\ t\end{pmatrix}\left[\frac{c^2}{8n^2b^2} + O\left(\frac{c^3}{n^3b^3}\right)\right] \quad \text{for} \quad c/b \to 0\,. \tag{3.128}$$

Conversely, the disturbance due to all cracks with $n$ non-zero at the location of the crack $-c < x < c$ is given by

$$\begin{pmatrix}\sigma_y\\ \tau_{xy}\end{pmatrix} \to \begin{pmatrix}p\\ t\end{pmatrix}\frac{c^2}{4b^2}\left[\sum_{n=1}^{\infty}\frac{1}{n^2} + O\left(\frac{c^2}{b^2}\right)\right] = \begin{pmatrix}p\\ t\end{pmatrix}\frac{\pi^2c^2}{24b^2}\left[1 + O\left(\frac{c^2}{b^2}\right)\right]$$

$$\text{for} \quad c/b \to 0\,. \tag{3.129}$$

It follows that the stress singularity for a row of collinear cracks obeys the asymptotic formula

$$\begin{pmatrix}\sigma_y\\ \tau_{xy}\end{pmatrix} \to \begin{pmatrix}p\\ t\end{pmatrix} \frac{c^{\frac{1}{2}}}{(2r)^{\frac{1}{2}}} \left[1 + \frac{\pi^2 c^2}{24b^2} + O\left(\frac{c^4}{b^4}\right)\right] \quad \text{for} \quad c/b \to 0 . \qquad (3.130)$$

At the other end of the range $c/b \to 1$, $a/b \to 0$, and from equations (3.17) and (3.23)

$$\begin{pmatrix}\sigma_y\\ \tau_{xy}\end{pmatrix} \to \begin{pmatrix}p\\ t\end{pmatrix} \frac{2}{\pi} \frac{b}{a} \left(\frac{a}{2r}\right)^{\frac{1}{2}} \left[1 + O\left(\frac{a^2}{b^2}\right)\right] \quad \text{for} \quad a/b \to 0 , \qquad (3.131)$$

where the error term is due to the annihilation of the curvature at the dam, caused by the removal of the traction (of order $pa/b$, $ta/b$) along the lines $x = \pm b$ in the solution of Section 6.2.

Introducing the *nondimensional stress intensity factor* $K$ defined by equation (3.126), the asymptotic relations are

$$\begin{aligned} K &\to 1 - \frac{1}{2}\frac{c}{b} + \frac{\pi^2 - 3}{24}\frac{c^2}{b^2} + O\left(\frac{c^3}{b^3}\right) \quad \text{for} \quad c/b \to 0 , \\ K &\to \frac{2}{\pi}\left[1 + \frac{1}{2}\frac{a}{b} + O\left(\frac{a^2}{b^2}\right)\right] \quad \text{for} \quad a/b \to 0 . \end{aligned} \qquad (3.132)$$

The result of the standard polynomial (in this case quartic) interpolation procedure is the approximate formula (Figure 3.7).

$$K = 1 - \frac{1}{2}\frac{c}{b} + 0.286\frac{c^2}{b^2} - 0.207\frac{c^3}{b^3} + 0.058\frac{c^4}{b^4} . \qquad (3.133)$$

In this test case the approximation is *excellent.* The discrepancy with the exact result (3.127) does not exceed 0.2 percent anywhere.

It may also be worthwhile to note in the present case that one may even ignore the knowledge of the second derivative at $c/b = 0$, and employ the cubic interpolation which has in common with equation (3.133) both values and first derivatives at the end points

$$K = 1 - \frac{1}{2}\frac{c}{b} + 0.228\frac{c^2}{b^2} - 0.091\frac{c^3}{b^3} , \qquad (3.134)$$

with an accuracy of 0.5 percent. It should be borne in mind, however, that these excellent results are related to the fact that the curve in Figure 3.7 does not deviate too far from a straight line.

*An infinite row of parallel cracks.* Consider an infinite plane in plane strain or generalized plane stress with cracks $-c < x < c$ along the lines $y = 2nb$ ($n = 0, \pm 1, \pm 2, \ldots$) with uniform tension $\sigma_y = p$ or uniform shear $\tau_{xy} = t$ at infinity (Figure 3.8).

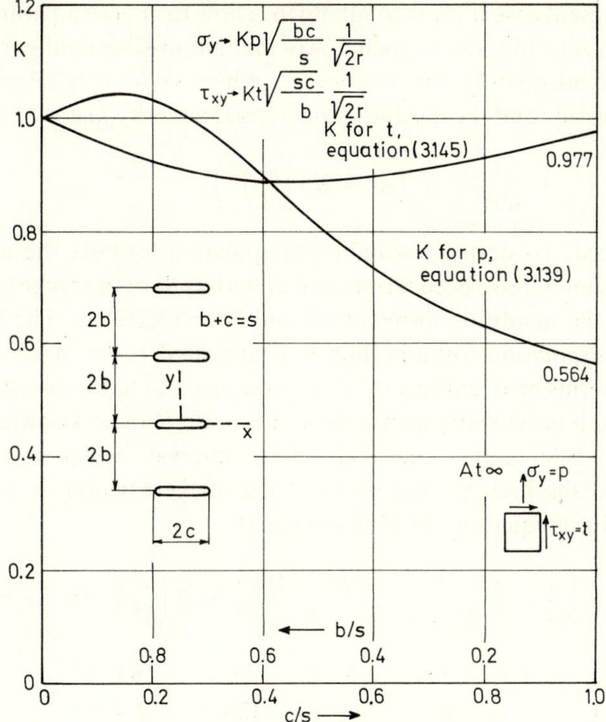


Figure 3.8. *K*-values for parallel cracks.

In the *tension* case the disturbance due to a single crack at $y=0$ at the location of the crack at $y=2nb$ is given by equation (3.6)

$$\sigma_y \to -p\left[\frac{3c^2}{8n^2b^2} + O\left(\frac{c^3}{b^3}\right)\right] \quad \text{for} \quad c/b \to 0\,. \tag{3.135}$$

By the same argument as in the previous case, the stress singularity in the case of a row of parallel cracks is

$$\sigma_y \to p\,\frac{c^{\frac{1}{2}}}{(2r)^{\frac{1}{2}}}\left[1 - \frac{\pi^2c^2}{8b^2} + O\left(\frac{c^4}{b^4}\right)\right] \quad \text{for} \quad c/b \to 0\,. \tag{3.136}$$

At the other end of the range of the parameter $c/b \to \infty$ the asymptotic approximation for the stress singularity is given by equation (3.67), and the relative error of this approximation is, in view of the remarks for long parallel cracks, of exponential order $\exp(-\mu c/b)$.

In the present case it is not at all obvious how to choose a parameter with range from zero to unity in such a way that all possible configurations are represented adequately. The fraction $c/s$, where $s=b+c$, is selected, somewhat arbitrarily, and the *nondimensional stress intensity factor* $K$, defined by

$$\sigma_y \to Kp\left(\frac{bc}{s}\right)^{\frac{1}{2}}\frac{1}{(2r)^{\frac{1}{2}}}, \qquad \left(k_1 = Kp\left(\frac{bc}{s}\right)^{\frac{1}{2}}\right) \tag{3.137}$$

is introduced. To derive a suitable interpolation formula the asymptotic expansions for $K$ have been terminated at both ends of the range ($c/s \to 0$ and $b/s \to 0$) at the quadratic terms in $c/s$ and $b/s$ respectively. The associated quintic interpolation formula, however, appeared to be inaccurate when plotted in terms of an entirely different parameter, to be discussed presently. In this case it is evidently necessary to make use of more knowledge of the asymptotic behaviour at the ends of the interval. Adequate results are achieved in the present case on the basis of the asymptotic expansions, following from equations (3.136) and (3.67),

$$K \to 1 + \frac{1}{2}\frac{c}{s} - \frac{\pi^2-3}{8}\frac{c^2}{s^2} - \frac{5(\pi^2-1)}{16}\frac{c^3}{s^3} + O\left(\frac{c^4}{s^4}\right) \quad \text{for} \quad c/s \to 0,$$

$$K \to \frac{1}{\pi^{\frac{1}{2}}}\left[1 + \frac{1}{2}\frac{b}{s} + \frac{3}{8}\frac{b^2}{s^2} + \frac{5}{16}\frac{b^3}{s^3} + \frac{35}{128}\frac{b^4}{s^4} + \frac{63}{256}\frac{b^5}{s^5} + \right. \tag{3.138}$$

$$\left. + \frac{231}{1024}\frac{b^6}{s^6} + O\left(\frac{b^7}{s^7}\right)\right] \quad \text{for} \quad b/s \to 0.$$

The associated interpolating polynomial of degree ten is (Figure 3.8)

$$K = \frac{1}{\pi^{\frac{1}{2}}}\left[1 + \frac{1}{2}\frac{b}{s} + \frac{3}{8}\frac{b^2}{s^2} + \frac{5}{16}\frac{b^3}{s^3} + \frac{35}{128}\frac{b^4}{s^4} + \frac{63}{256}\frac{b^5}{s^5} + \frac{231}{1024}\frac{b^6}{s^6}\right] +$$

$$+ 15.342\frac{b^7}{s^7} - 47.020\frac{b^8}{s^8} + 45.279\frac{b^9}{s^9} - 14.255\frac{b^{10}}{s^{10}}. \tag{3.139}$$

As a check on the accuracy consider the alternative, entirely different parameter $\alpha$, defined by $\alpha = \arctan(\pi c/b)$, where $0 < \alpha < \pi/2$, and a different *dimensionless stress intensity factor* $K^*$,

$$\sigma_y \to K^*p\left(\frac{bc}{s^*}\right)^{\frac{1}{2}}\frac{1}{(2r)^{\frac{1}{2}}}, \tag{3.140}$$

where $s^{*2}=b^2+\pi^2c^2$. The asymptotic formulae for this modified factor $K^*$ are

$$\begin{aligned} K^* &\to (\cos\alpha)^{-\frac{1}{2}}\left[1-\tfrac{1}{8}\tan^2\alpha+O(\tan^4\alpha)\right] \quad \text{for} \quad \alpha\to 0\,, \\ K^* &\to (\sin\alpha)^{-\frac{1}{2}} \quad \text{for} \quad \alpha\to\pi/2\,. \end{aligned} \tag{3.141}$$

The approximate formula (3.139) may be mapped on a $K^*$-$\alpha$ diagram by the relations

$$c/s=c/(b+c)=(\pi\cot\alpha+1)^{-1}\,, \qquad K^*=(s^*/s)^{\frac{1}{2}} \quad K=\left(\frac{\pi}{\pi\cos\alpha+\sin\alpha}\right)^{\frac{1}{2}}K\,.$$

The result of this mapping is given in Figure 3.9, together with the asymptotic

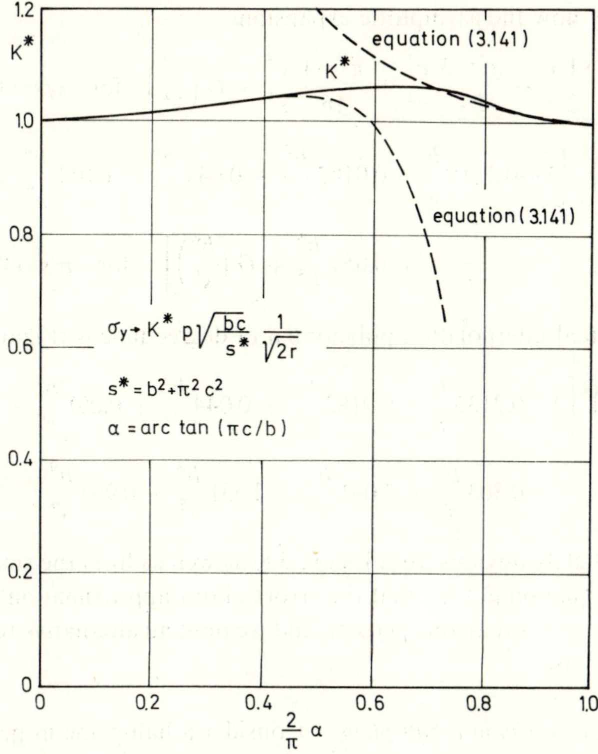


Figure 3.9. Alternative dimensionless stress intensity factor $K$ and parameter $a$ for parallel cracks, tension case. Dotted lines, asymptotic expansions (3.141).

formulae (3.141). It is probable that the actual $K^*$ curve remains at the lower side of the right asymptote and the error in the interpolating curve is estimated to be 1 or 2%.

In the case of a uniform shear $\tau_{xy}=t$ at infinity the counterpart of equation (3.136) is derived in the same way from (3.14)

$$\tau_{xy} \to t \frac{c^{\frac{1}{2}}}{(2r)^{\frac{1}{2}}} \left[1 + \frac{\pi^2 c^2}{24b^2} + O\left(\frac{c^4}{b^4}\right)\right] \quad \text{for} \quad c/b \to 0 . \tag{3.142}$$

At the other end of the range $c/b \to \infty$ the stress singularity has the asymptotic approximation (3.70) with a relative error of exponential order. Define the *nondimensional stress intensity factor* $K$ (note in a way different from (3.137))

$$\tau_{xy} \to Kt \left(\frac{sc}{b}\right)^{\frac{1}{2}} \frac{1}{(2r)^{\frac{1}{2}}}, \qquad \left(k_2 = Kt\left(\frac{sc}{b}\right)^{\frac{1}{2}}\right) \tag{3.143}$$

and employ now the asymptotic expansions

$$K \to 1 - \frac{1}{2}\frac{c}{s} + \frac{\pi^2-3}{24}\frac{c^2}{s^2} + \frac{\pi^2-1}{16}\frac{c^3}{s^3} + O\left(\frac{c^4}{s^4}\right) \quad \text{for} \quad c/s \to 0 ,$$

$$K \to \left(\frac{3}{\pi}\right)^{\frac{1}{2}} \left[1 - 0.2135\frac{b}{s} + 0.0182\frac{b^2}{s^2} + 0.044\frac{b^3}{s^3} + 0.051\frac{b^4}{s^4} + \right. \tag{3.144}$$

$$\left. + 0.051\frac{b^5}{b^5} + O\left(\frac{b^6}{s^6}\right)\right] \quad \text{for} \quad b/s \to 0 .$$

The associated interpolation polynomial of degree nine is (Figure 3.8)

$$K = \left(\frac{3}{\pi}\right)^{\frac{1}{2}} \left[1 - 0.2135\frac{b}{s} + 0.0182\frac{b^2}{s^2} + 0.044\frac{b^3}{s^3} + 0.051\frac{b^4}{s^4} + 0.051\frac{b^5}{s^5}\right]$$

$$-0.303\frac{b^6}{s^6} + 2.041\frac{b^7}{s^7} - 2.631\frac{b^8}{s^8} + 0.963\frac{b^9}{s^9} . \tag{3.145}$$

It is immediately obvious from Figure 3.8, as well as from the smaller coefficients in equation (3.145), that the errors of our approximation are small, most likely far less than one percent, and we omit an alternative representation in this case.

*Parallel edge cracks in a half-plane.* Consider a half-plane in plane strain or generalized plane stress with an infinite row of normal edge cracks of depth $c$ and spacing $2b$, loaded in *tension* by a uniform normal stress $\sigma_y=p$

at infinity. The disturbance at the location of one crack caused by all other cracks may be evaluated in the same way as before, now from equation (3.39)

$$\sigma_y = -p\frac{c^2}{b^2}\left[2.069 + O\left(\frac{c^2}{b^2}\right)\right] \quad \text{for} \quad c/b \to 0\,. \tag{3.146}$$

The stress singularity at the crack tip in the case of widely spaced cracks is therefore by equation (3.37)

$$\sigma_y \to 1.122\, p\frac{c^{\frac{1}{2}}}{(2r)^{\frac{1}{2}}}\left[1 - 2.069\frac{c}{b} + O\left(\frac{c^4}{b^4}\right)\right] \quad \text{for} \quad c/b \to 0\,. \tag{3.147}$$

At the other end of the range $c/b \to \infty$ the stress singularity is again given by (3.67). Introduce again the *nondimensional stress intensity factor* $K$ defined by (3.137), where $s = b + c$, and use the asymptotic expansions

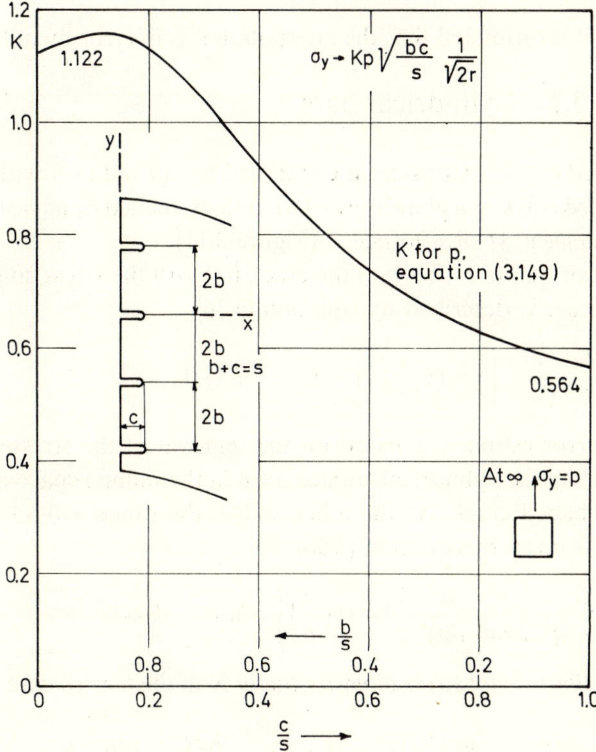


Figure 3.10. $K$-values for parallel edge-cracks.

$$K \to 1.122\left[1+\frac{1}{2}\frac{c}{s}-1.694\frac{c^2}{s^2}-4.860\frac{c^3}{s^3}+O\left(\frac{c^4}{s^4}\right)\right] \quad \text{for} \quad c/s\to 0\,,$$

$$K \to \frac{1}{\pi^{\frac{1}{2}}}\left[1+\frac{1}{2}\frac{b}{s}+\frac{3}{8}\frac{b^2}{s^2}+\frac{5}{16}\frac{b^3}{s^3}+\frac{35}{128}\frac{b^4}{s^4}\right. \tag{3.148}$$

$$\left.+\frac{63}{256}\frac{b^5}{s^5}+\frac{231}{1024}\frac{b^6}{s^6}+O\left(\frac{b^7}{s^7}\right)\right] \quad \text{for} \quad b/s\to 0\,.$$

The associated polynomial interpolation formula of degree ten is given by (Figure 3.10)

$$K = \frac{1}{\pi^{\frac{1}{2}}}\left[1+\frac{1}{2}\frac{b}{s}+\frac{3}{8}\frac{b^2}{s^2}+\frac{5}{16}\frac{b^3}{s^3}+\frac{35}{128}\frac{b^4}{s^4}+\frac{63}{256}\frac{b^5}{s^5}+\frac{231}{1024}\frac{b^6}{s^6}\right]+$$

$$+22.501\frac{b^7}{s^7}-63.502\frac{b^8}{s^8}+58.045\frac{b^9}{s^9}-17.577\frac{b^{10}}{s^{10}}\,. \tag{3.149}$$

Here again it is estimated that the interpolation is in error by only 1 or 2%.

## 3.7 Cylindrical bars

*Penny-shaped crack.* Consider a cylindrical bar of radius $b$ with a central penny-shaped crack in a plane normal to the axis, loaded by a *tensile force* $N$, a *bending moment* $M$ or a *torque* $T$ (Figure 3.11).

For a small radius $c=b-a$ of the crack ($c/b\to 0$) the stress singularity in the *tension* case is described by equation (3.86)

$$\sigma_z \to \frac{2}{\pi}\frac{N}{\pi b^2}\frac{c^{\frac{1}{2}}}{(2r)^{\frac{1}{2}}}\left[1+O\left(\frac{c^3}{b^3}\right)\right] \quad \text{for} \quad c/b\to 0\,, \tag{3.150}$$

where the error estimate is based on the removal of the stresses of order $(N/b^2)c^3/R^3$ on the cylindrical surface $\rho=b$ in the infinite space problem of the penny-shaped crack. At the other end of the range, $c/b\to 1$, the stress singularity is given by equation (3.50)

$$\sigma_z \to \frac{2}{(\pi^2-4)^{\frac{1}{2}}}\frac{N}{2\pi ab}\frac{a^{\frac{1}{2}}}{(2r)^{\frac{1}{2}}}\left[1+O\left(\frac{a}{b}\right)\right] \quad \text{for} \quad a/b\to 0\,. \tag{3.151}$$

Let the *nondimensional stress intensity factor* $K$ in the *tension* case be defined by

$$\sigma_z \to K\frac{N}{\pi(b^2-c^2)}\left(\frac{ac}{b}\right)^{\frac{1}{2}}\frac{1}{(2r)^{\frac{1}{2}}}\,, \quad \left(k_1=\frac{KN}{\pi(b^2-c^2)}\left(\frac{ac}{b}\right)^{\frac{1}{2}}\right), \tag{3.152}$$

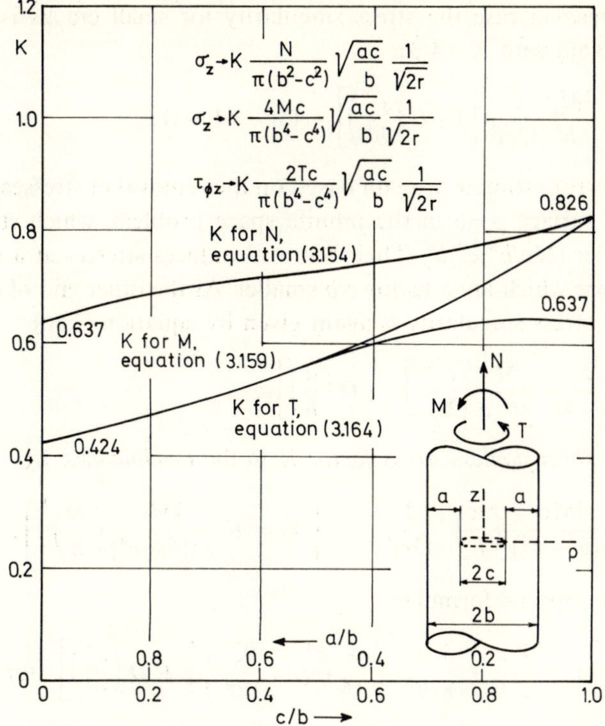


Figure 3.11. *K*-values for cylindrical bar with penny-shaped central crack.

and then the asymptotic formulae are

$$K \to \frac{2}{\pi}\left[1 + \frac{1}{2}\frac{c}{b} - \frac{5}{8}\frac{c^2}{b^2} + O\left(\frac{c^3}{b^3}\right)\right] \quad \text{for} \quad c/b \to 0\,,$$

$$K \to \frac{2}{(\pi^2-4)^{\frac{1}{2}}}\left[1 + O\left(\frac{a}{b}\right)\right] \quad \text{for} \quad a/b \to 0\,. \tag{3.153}$$

The approximate formula, obtained by cubic interpolation, is in this case (Figure 3.11)

$$K = \frac{2}{\pi}\left(1 + \frac{1}{2}\frac{c}{b} - \frac{5}{8}\frac{c^2}{b^2}\right) + 0.268\,\frac{c^3}{b^3}\,. \tag{3.154}$$

Since the curve does not deviate too much from a straight line it may again be claimed to have good accuracy, of the order of one percent.

In the *bending* case the stress singularity for small cracks is given by equation (3.86) with $K=4/3\pi$

$$\sigma_z \to \frac{4}{3\pi}\frac{4Mc}{\pi b^4}\frac{c^{\frac{1}{2}}}{(2r)^{\frac{1}{2}}}\left[1+O\left(\frac{c^5}{b^5}\right)\right] \quad \text{for} \quad c/b \to 0\,, \tag{3.155}$$

where the error estimate is again based on the removal of stresses along the cylindrical surface $\rho=b$ in the infinite space problem, which stresses are now of order $(Mc/b^3)c^4/b^4$. Their removal induces stresses at a distance $c$ from the axis which are a factor $c/b$ smaller. At the other end of the range, $c/b \to 1$, the stress singularity is again given by equation (3.50)

$$\sigma_z \to \frac{2}{(\pi^2-4)^{\frac{1}{2}}}\frac{M}{\pi b^2 a}\frac{a^{\frac{1}{2}}}{(2r)^{\frac{1}{2}}}\left[1+O\left(\frac{a}{b}\right)\right]\,. \tag{3.156}$$

Define the *dimensionless stress factor* $K$ in the *bending* case by

$$\sigma_z \to K\frac{4Mc}{\pi(b^4-c^4)}\left(\frac{ac}{b}\right)^{\frac{1}{2}}\frac{1}{(2r)^{\frac{1}{2}}}\,, \qquad \left[k_1 = K\frac{4Mc}{\pi(b^4-c^4)}\left(\frac{ac}{b}\right)^{\frac{1}{2}}\right], \tag{3.157}$$

with the asymptotic formulae

$$K \to \frac{4}{3\pi}\left[1+\frac{1}{2}\frac{c}{b}+\frac{3}{8}\frac{c^2}{b^2}+\frac{5}{16}\frac{c^3}{b^3}-\frac{93}{128}\frac{c^4}{b^4}+O\left(\frac{c^5}{b^5}\right)\right] \quad \text{for} \quad c/b \to 0\,,$$

$$K \to \frac{2}{(\pi^2-4)^{\frac{1}{2}}}\left[1+O\left(\frac{a}{b}\right)\right] \quad \text{for} \quad a/b \to 0\,, \tag{3.158}$$

and the quintic interpolation formula

$$K = \frac{4}{3\pi}\left[1+\frac{1}{2}\frac{c}{b}+\frac{3}{8}\frac{c^2}{b^2}+\frac{5}{16}\frac{c^3}{b^3}-\frac{93}{128}\frac{c^4}{b^4}+0.483\frac{c^5}{b^5}\right]. \tag{3.159}$$

Since this approximation in the *bending* case is based on more exact data than equation (3.154) in the tension case, an even higher accuracy may be expected.

Finally, in the torsion case the stress singularity for small cracks follows with $K=4/3\pi$ from (3.90)

$$\tau_{z\phi} \to \frac{4}{3\pi}\frac{2Tc}{\pi b^4}\frac{c^{\frac{1}{2}}}{(2r)^{\frac{1}{2}}}\left[1+O\left(\frac{c^5}{b^5}\right)\right] \quad \text{for} \quad c/b \to 0\,, \tag{3.160}$$

where the error estimate is based on the same argument as in the bending case. At the other end of the range, $c/b \to 1$, from equation (3.77)

$$\tau_{z\phi} \to \frac{2}{\pi} \frac{T}{2\pi b^2 a} \frac{a^{\frac{1}{2}}}{(2r)^{\frac{1}{2}}} \left[1 + O\left(\frac{a}{b}\right)\right] \quad \text{for} \quad a/b \to 0 . \tag{3.161}$$

Define the *nondimensional stress intensity factor* $K$ in the *torsion* case by

$$\tau_{z\phi} \to K \frac{2Tc}{\pi(b^4 - c^4)} \left(\frac{ac}{b}\right)^{\frac{1}{2}} \frac{1}{(2r)^{\frac{1}{2}}}, \qquad \left(k_3 = K \frac{2Tc}{\pi(b^4 - c^4)} \left(\frac{ac}{b}\right)^{\frac{1}{2}}\right), \tag{3.162}$$

and then the asymptotic formulae are

$$K \to \frac{4}{3\pi}\left[1 + \frac{1}{2}\frac{c}{b} + \frac{3}{8}\frac{c^2}{b^2} + \frac{5}{16}\frac{c^3}{b^3} - \frac{93}{128}\frac{c^4}{b^4} + O\left(\frac{c^5}{b^5}\right)\right] \quad \text{for} \quad c/b \to 0 ,$$

$$K \to \frac{2}{\pi}\left[1 + O\left(\frac{a}{b}\right)\right] \quad \text{for} \quad a/b \to 0 . \tag{3.163}$$

With the quintic interpolation formula

$$K = \frac{4}{3\pi}\left[1 + \frac{1}{2}\frac{c}{b} + \frac{3}{8}\frac{c^2}{b^2} + \frac{5}{16}\frac{c^3}{b^3} - \frac{93}{128}\frac{c^4}{b^4} + 0.038\frac{c^5}{b^5}\right]. \tag{3.164}$$

Here again an accuracy better than one percent is expected.

*Ring-shaped edge crack or groove.* Consider a cylindrical bar of radius $b$ with a ring-shaped crack $a = b - c < \rho < b$ in a plane normal to the axis, loaded by a *tensile force* $N$, *bending moment* $M$ or a *torque* $T$ (Figure 3.12).

For small values of the crack depth $c$ from equation (3.37) in the *tension* case

$$\sigma_z \to 1.122 \frac{N}{\pi b^2} \frac{c^{\frac{1}{2}}}{(2r)^{\frac{1}{2}}} \left[1 + O\left(\frac{c}{b}\right)\right] \quad \text{for} \quad c/b \to 0 . \tag{3.165}$$

Unlike in the similar plane case of edge cracks the error is here of order $c/b$, due to the application of the basic plane problem to an axisymmetric case. An evaluation of the first order correction in this case requires a considerable additional analytical effort which would hardly be worthwhile for the present purposes.

At the other end of the range of the parameter, $c/b \to 1$, from equation (3.92)

$$\sigma_z \to \frac{1}{2} \frac{N}{\pi a^2} \frac{a^{\frac{1}{2}}}{(2r)^{\frac{1}{2}}} \left[1 + O\left(\frac{a^3}{b^3}\right)\right] \quad \text{for} \quad a/b \to 0 . \tag{3.166}$$

The error estimate is again obtained by considering the removal of the tractions along the cylindrical surface $\rho = b$ in the infinite space problem.

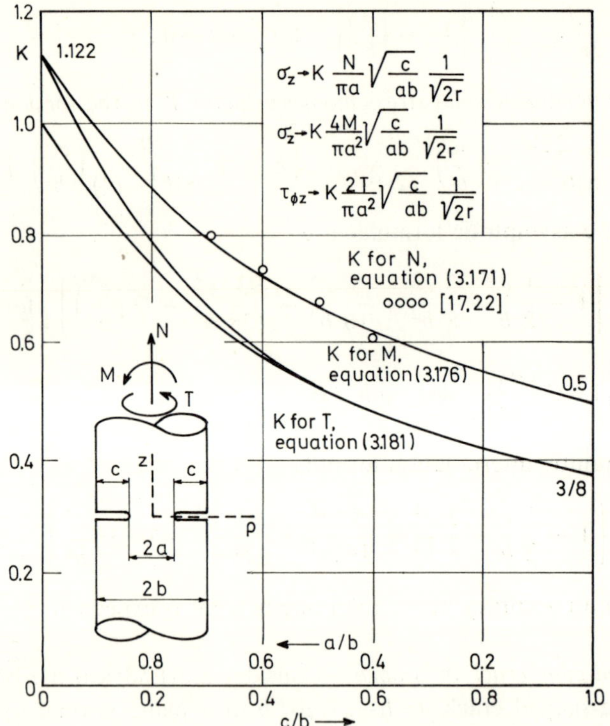


Figure 3.12. *K*-values for cylindrical bar with ring-shaped edge-crack or groove.

These tractions are of order $N/b^2$, and their removal results in a curvature of the plane $z=0$ of order $N/Eb^3$. Annihilation of this curvature in its turn induces additional stresses in the dam of order $a^3/b^3$ times the primary stresses (cf. [3, §116a]).

The *nondimensional stress intensity factor* $K$ in the *tension* case is defined by

$$\sigma_z \to K\,\frac{N}{\pi a^2}\left(\frac{ac}{b}\right)^{\frac{1}{2}}\frac{1}{(2r)^{\frac{1}{2}}}, \qquad \left(k_1 = K\,\frac{N}{\pi a^2}\left(\frac{ac}{b}\right)^{\frac{1}{2}}\right), \tag{3.167}$$

and the asymptotic formulae are

$$\begin{aligned} K &\to 1.122\left[1+O\left(\frac{c}{b}\right)\right] \quad \text{for} \quad c/b\to 0\,, \\ K &\to \frac{1}{2}\left[1+\frac{1}{2}\frac{a}{b}+\frac{3}{8}\frac{a^2}{b^2}+O\left(\frac{a^3}{b^3}\right)\right] \quad \text{for} \quad a/b\to 0\,. \end{aligned} \tag{3.168}$$

The result of a cubic interpolation is the approximation

$$K = \frac{1}{2}\left[1 + \frac{1}{2}\frac{a}{b} + \frac{3}{8}\frac{a^2}{b^2} + 0.368\,\frac{a^3}{b^3}\right]. \tag{3.169}$$

The values of equation (3.169) were compared with an interpolation formula from [17], based on numerical computations [22], and with a limited range of validity, whose accuracy is claimed to be within one percent. The approximation (3.169) is some 3 to 4 percent larger in the range $0.3 < c/b < 0.5$, and this is most likely due to a too small slope of the function at the end point $c/b = 0$, where the exact value of this slope is not available. It shall be seen that the result for bending will be more accurate, and the *difference* in slopes between the bending and tension cases is known *exactly* from equations (3.37) and (3.113), and from definition (3.174) for the stress intensity factor in bending

$$\left[\frac{\mathrm{d}K}{\mathrm{d}(c/b)}\right]_{\text{bending}} - \left[\frac{\mathrm{d}K}{\mathrm{d}(c/b)}\right]_{\text{tension}} = -0.683 \tag{3.170}$$

at the point $c/b = 0$. If it is assumed, for reasons which will be come clear presently, that the derivative in the bending case is accurate, the value $-1.542$ is obtained from equation (3.176) for the derivative in the tension case. The approximate formula (3.169) may now be replaced by

$$K = \frac{1}{2}\left[1 + \frac{1}{2}\frac{a}{b} + \frac{3}{8}\frac{a^2}{b^2} - 0.363\,\frac{a^3}{b^3} + 0.731\,\frac{a^4}{b^4}\right], \tag{3.171}$$

and this result, presented in Figure 3.12, agrees excellently with the findings of the (limited) numerical calculations in the literature.

In the case of loading by a *bending moment M* from (3.37) for small values of the crack depth is obtained

$$\sigma_z \to 1.122\,\frac{4M}{\pi b^3}\,\frac{c^{\frac{1}{2}}}{(2r)^{\frac{1}{2}}}\left[1 + O\left(\frac{c}{b}\right)\right] \quad \text{for} \quad c/b \to 0\,, \tag{3.172}$$

where again no better error estimate is available at present. At the other end of the range, $c/b \to 1$, however, from (3.94)

$$\sigma_z \to \frac{3}{8}\,\frac{4M}{\pi a^3}\,\frac{a^{\frac{1}{2}}}{(2r)^{\frac{1}{2}}}\left[1 + O\left(\frac{a^5}{b^5}\right)\right] \quad \text{for} \quad a/b \to 0\,. \tag{3.173}$$

As before, the error estimate in equation (3.173) is obtained by removing the traction along the cylindrical surface $\rho = b$ in the infinite space problem.

These tractions are of order $M/b^3$, and their removal induces a curvature of the dam surface of order $M/Eb^4$, which vanishes at the center and is therefore of order $Ma/Eb^5$ in the dam area. The additional stresses in the dam are thus of order $a^5/b^5$ times the primary stresses (cf. [3, §116a]).

The *nondimensional stress intensity factor* $K$ in the *bending* case is defined by

$$\sigma_z \to K \frac{4M}{\pi a^3}\left(\frac{ac}{b}\right)^{\frac{1}{2}} \frac{1}{(2r)^{\frac{1}{2}}}, \qquad \left(k_1 = K \frac{4M}{\pi a^3}\left(\frac{ac}{b}\right)^{\frac{1}{2}}\right), \tag{3.174}$$

and the asymptotic formulae are

$$K \to 1.122\left[1+O\left(\frac{c}{b}\right)\right] \quad \text{for} \quad c/b \to 0,$$

$$K \to \frac{3}{8}\left[1+\frac{1}{2}\frac{a}{b}+\frac{3}{8}\frac{a^2}{b^2}+\frac{5}{16}\frac{a^3}{b^3}+\frac{35}{128}\frac{a^4}{b^4}+O\left(\frac{a^5}{b^5}\right)\right] \quad \text{for} \quad a/b \to 0. \tag{3.175}$$

The approximate formula is obtained by quintic interpolation in this case, and the result

$$K = \frac{3}{8}\left[1+\frac{1}{2}\frac{a}{b}+\frac{3}{8}\frac{a^2}{b^2}+\frac{5}{16}\frac{a^3}{b^3}+\frac{35}{128}\frac{a^4}{b^4}+0.531\frac{a^5}{b^5}\right] \tag{3.176}$$

is shown in Figure 3.12.

In view of the large number of exact data at the end $c/b=1$ (the function itself and no less than four derivatives), and observing that the last terms, obtained by interpolation to the value at $c/b=0$, have no large coefficient, it may be expected that the approximation (3.176) is quite accurate, even if the exact value of the derivative of $K$ in the bending case at $c/b=0$ is not available. Without question, the derivative at $a/b=1$ is obtained considerably more accurately from equation (3.176) than the derivative of $K$ in the tension case is obtained from (3.169), and the *exact* relation (3.170) between the derivatives in both cases leads to the more accurate approximation (3.171) in the tension case.

Finally, in the loading condition of a *torque* $T$ for small crack depth $c$ from equation (3.74)

$$\tau_{z\phi} \to \frac{2T}{\pi b^3}\frac{c^{\frac{1}{2}}}{(2r)^{\frac{1}{2}}}\left[1+O\left(\frac{c}{b}\right)\right] \quad \text{for} \quad c/b \to 0. \tag{3.177}$$

At the other end of the range of the parameter, $c/b \to 1$, from equation (3.96)

$$\tau_{z\phi} \to \frac{3}{8} \frac{2T}{\pi a^3} \frac{a^{\frac{1}{2}}}{(2r)^{\frac{1}{2}}} \left[1+O\left(\frac{a^5}{b^5}\right)\right] \quad \text{for} \quad a/b \to 0\,, \tag{3.178}$$

where the error estimate is obtained in the same way as in the bending case.

Define the *nondimensional stress intensity factor* $K$ in the torsion case by

$$\tau_{z\phi} \to K \frac{2T}{\pi a^3} \left(\frac{ac}{b}\right)^{\frac{1}{2}} \frac{1}{(2r)^{\frac{1}{2}}}, \qquad \left(k_3 = K \frac{2T}{\pi a^3} \left(\frac{ac}{b}\right)^{\frac{1}{2}}\right), \tag{3.179}$$

and the asymptotic formulae are

$$\begin{aligned} &K \to 1+O\left(\frac{c}{b}\right) \quad \text{for} \quad c/b \to 0\,, \\ &K \to \frac{3}{8}\left[1 + \frac{1}{2}\frac{a}{b} + \frac{3}{8}\frac{a^2}{b^2} + \frac{5}{16}\frac{a^3}{b^3} + \frac{35}{128}\frac{a^4}{b^4} + O\left(\frac{a^5}{b^5}\right)\right] \quad \text{for} \quad a/b \to 0\,. \end{aligned} \tag{3.180}$$

The approximate formula is again obtained by quintic interpolation (Figure 3.12)

$$K = \frac{3}{8}\left[1 + \frac{1}{2}\frac{a}{b} + \frac{3}{8}\frac{a^2}{b^2} + \frac{5}{16}\frac{a^3}{b^3} + \frac{35}{128}\frac{a^4}{b^4} + 0.208\,\frac{a^5}{b^5}\right], \tag{3.181}$$

and it is estimated to be accurate far better than one percent.

## References


[1] Sneddon, I. N. and Lowengrub, M., *Crack problems in the classical theory of elasticity.* New York, London (1969).

[2] Sadowski, M. A., Zweidimensionale Probleme der Elastizitätstheorie. *Zeitschrift für Angew. Math. und Mech.*, 8, S. 107 (1928).

[3] Muskhelishvili, N. I., *Some basic problems of the mathematical theory of elasticity.* Groningen, Netherlands, 1953. (translation from the Russian 1948).

[4] Koiter, W. T., On the flexural rigidity of a beam weakened by transverse saw cuts. *Proc. Royal Neth. Acad. of Sciences*, B59, 354–374 (1956).

[5] Koiter, W. T., Rectangular tensile sheet with symmetric edge cracks. *Journal of Appl. Mech.*, 32, p. 237 (1965).

[6] Koiter, W. T., Note on the stress intensity factors for sheet strips with cracks under tensile loads. Univ. of Techn.; Lab. of Eng. Mech. Rep. nr. 314, Delft, Netherlands (1965).

[7] Koiter, W. T., An infinite row of parallel cracks in an infinite elastic sheet. *Problems of Continuum Mechanics* (Muskhelishvili Ann. Vol.), page 246. Philadelphia (1961).

[8] Sneddon, I. N. and Berry, D. S., The classical theory of elasticity in: S. Flügge (editor), *Encyclopedia of physics*, Vol. VI, Elasticity and Plasticity, Berlin (1958).

[9] Sneddon, I. N., *Fourier transforms.* New York, London, 1951.

[10] Weinstein, A., On cracks and dislocations in shafts under torsion. *Quart. of Appl. Math.* X, p. 77 (1952).
[11] Boussinesq, J., Application des potentiels ..., Paris (1885).
[12] Lekkerkerker, J. G., The effective length of a cantilever beam with circular cross-section (in Dutch). *De Ingenieur*, 73, p. O 110 (1961).
[13] Lurje, A. I., *Räumliche Probleme der Elastizitätstheorie*, Berlin (1963) (translation from the Russian 1954.).
[14] Rooke, D. P., Width corrections in fracture mechanics. *Engng. Fracture Mech.* 1, p. 727 (1970).
[15] Isida, M., *Crack tip stress-intensity factors for the tension of an eccentrically cracked strip.* Lehigh Univ. Dept. of Mech. (1965).
[16] Feddersen, C. E., Discussion to: Plane strain crack toughness testing. *ASTM Spec. Tech. Publ. No.* 410, p. 77 (1967).
[17] Elst van, H. C., Crack-extension susceptibility of high-strength steels (in Dutch). *De Ingenieur*, 82, p. W 131 (1970).
[18] Formann, R. G. and Kobayashi, A. S., *Journal of Basic Eng.*, 86, p. 693 (1964).
[19] Sneddon, I. N. and Srivastav, R. P., The stress field in the vicinity of a Griffith crack in a strip of finite width. *Int. J. Engng. Sci.* 9, p. 479 (1971).
[20] Irwin, G. R., Liebowitz, H. and Paris, P. C., A mystery of fracture mechanics. *Engng. Fracture Mech.* 1, p. 235 (1970).
[21] Bowie, O. L., Rectangular tensile sheet with symmetric edge cracks. *Journ. Appl. Mech.* 31, 208–212 (1964).
[22] Bueckner, H. F., *ASTM-STP* 381 (1965) p. 82.
[23] Westergaard, H. M., Bearing pressures and cracks. *Journal of Appl. Mech.* 6, p. A-49 (1939).
[24] Koiter, W. T., An infinite row of collinear cracks in an infinite elastic sheet. *Ing. Arch.* 28, s. 169 (1959).
[25] Stallybrass, M. P., A semi-infinite crack perpendicular to the surface of an elastic half-plane. *Int. J. Engng. Sci.* 9, p. 133 (1971).
[26] Cross, B., Srawley, J. E. and Brown, W. F., Jr., N.A.S.A. Techn. Note D-2395 (1964).
[27] Cross, B. and Srawley, J. E., N.A.S.A. Techn. Note D-2603 and D-3092 (1965).
[28] Titchmarsh, E. C., *Theory of Fourier integrals.* Oxford, 1937.
[29] Erdélyi, A., Magnus, W., Oberhettinger, F. and Tricomi, F. G., *Tables of integral transforms.* 2 vols., New York, 1954.
[30] Collins, W. D., Some axially symmetric stress distributions in elastic solids containing penny-shaped cracks. Part I, *Proc. Royal Soc. London* A266, 359–386 (1962).
Part II, *Mathematika* 9, 25–37 (1962).
Part III, *Proc. Edinburgh Math. Soc.* 13, 69–78 (1962).
[31] Reissner, E. and Sagoci, H. F., Forced Torsional oscillations of an elastic half-space. *Journ. Appl. Phys.* 15, 652 (1944).
[32] Rice, J. R. and Levy, N., The part-through surface crack in an elastic plate. N.A.S.A. Techn. Rep. NGL 40-002-080/3 (1970).
[33] Koiter, W. T., Solution of some elasticity problems by asymptotic methods. *Proc. IUTAM Symposium on Applications of the Theory of Functions in Continuum Mechanics* (Tbilisi 1963). Nauka Publ. House, Moscow (1965), pp. 15–31.

*R. J. Hartranft and G. C. Sih*

# 4 *Alternating method applied to edge and surface crack problems*

## 4.1 Introduction

Over the past decade, numerous analytical solutions of crack problems have appeared in the open literature [1]. A great number of these solutions are concerned with idealized crack geometries in plane or axisymmetric elasticity. However, only a few problems involving the interaction of cracks with neighboring boundaries have been solved satisfactorily. In situations where the crack intersects a free edge or surface, the method of solution becomes much more difficult and requires special attention. The advent of computers has no doubt facilitated the numerical computation of stress distributions around cracks. Without them, many of the tedious calculations would not be attempted. The alternating method is one which intimately combines analytical results with the numerical calculations.

One of the requisites for solving any crack problem is to handle the stress singularities at the crack tips properly. This involves first of all a knowledge of the correct behavior of the stress singularity, which is a task normally accomplished by analytical means. Next, it is essential to preserve this singular behavior of the solution in the problem either by isolating it away from numerical computations or by treating it numerically with the utmost care. Generally speaking, the error committed near a singular point such as a crack tip or border will not be confined locally but will cause errors elsewhere as well. The same applies to corners or points in the elastic solid where high stress gradients are present. This point will be demonstrated in the present work on the surface crack problem in three dimensions.

What follows is a treatment of the alternating method as applied to solve edge crack problems in two-dimensions and surface crack problems in three-dimensions. Although the surface crack solution is not complete, it serves as a good example for illustrating the complexities and understanding required to treat problems of this type. The mechanics of the alternating method is in fact rather rudimentary. It is described in the work of Kantorovich and Krylov [2] who obtained the solutions to potential problems by using successive, iterative superposition of sequences of solutions. Their illustration involves two sequences of solutions, each sequence applying to a particular geometry. By the alternating superposition of the sequences, the solution for the region common to both geometries may be found. The method is called the Schwarz–Newmann Alternating Technique in their book. In this work it will be referred to as the alternating method, and it will be applied to solve elastic crack problems.

As an example of the alternating method, consider a simple problem with no singularities. Suppose the stresses in the quarter plane $x \geqslant 0$, $y \geqslant 0$ are to be found for the boundary conditions

$$\tau_{xy}(0, y) = 0 \qquad \tau_{xy}(x, 0) = 0$$
$$\sigma_x(0, y) = 0 \qquad \sigma_y(x, 0) = \sigma(x) .$$

One sequence of solutions, Sequence A, leads from the stress $\sigma_{yA}(x, 0) = q(x)$ on the half plane $y \geqslant 0$ to the stress $\sigma_{xA}(0, y) = p(y)$ in particular and to all other stresses in the quarter plane considered. The second, Sequence B, for $x \geqslant 0$ leads from $\sigma_{xB}(0, y) = p(y)$ to a solution yielding $\sigma_{yB}(x, 0) = q(x)$. The sequences, A and B, may be formed to yield the solution for the quarter plane common to both half planes.

From A, let $q^{(0)}(x) = \sigma(x)$, $x > 0$ and $q^{(0)}(x) = \sigma(-x)$, $x < 0$. For the first solution in sequence A,

$$\sigma_{yA}^{(1)}(x, 0) = q^{(0)}(x) \text{ yields } \sigma_{xA}^{(1)}(0, y) = p^{(1)}(y) \text{, say .}$$

For the first solution in sequence B,

$$\sigma_{xB}^{(1)}(0, y) = -p^{(1)}(y) \text{ yields } \sigma_{yB}^{(1)}(x, 0) = -q^{(1)}(x) \text{, say .}$$

Next in sequence A,

$$\sigma_{yA}^{(2)}(x, 0) = q^{(1)}(x) \text{ yields } \sigma_{xA}^{(2)}(0, y) = p^{(2)}(y) \text{, say}$$

and then in sequence B,

$$\sigma_{xB}^{(2)}(0, y) = -p^{(2)}(y) \text{ yields } \sigma_{yB}^{(2)}(x, 0) = -q^{(2)}(x) \text{, say .}$$

The sequences continue alternating this way. Both sequences apply to the quarter plane, and by superposition, the boundary values on the quarter plane are

$$\sigma_y(x, 0) = \sigma_{yA}^{(1)}(x, 0) + \sigma_{yB}^{(1)}(x, 0) + \sigma_{yA}^{(2)}(x, 0) + \sigma_{yB}^{(2)}(x, 0) + \dots$$

$$\sigma_x(0, y) = \sigma_{xA}^{(1)}(0, y) + \sigma_{xB}^{(1)}(0, y) + \sigma_{xA}^{(2)}(0, y) + \sigma_{xB}^{(2)}(0, y) + \dots$$

or

$$\sigma_y(x, 0) = q^{(0)}(x) - q^{(1)}(x) + q^{(1)}(x) - q^{(2)}(x) + \dots$$

$$\sigma_x(0, y) = p^{(1)}(y) - p^{(1)}(y) + p^{(2)}(y) - p^{(2)}(y) + \dots$$

Therefore, after superposing the first $n$ terms of each sequence, the boundary values are

$$\sigma_y(x, 0) = \sigma(x) - q^{(n)}(x)$$

$$\sigma_x(0, y) = 0 \, .$$

Additional solutions from the sequences are superposed until the residual stress, $q^{(n)}(x)$, on the $x$-axis is negligible compared to the applied stress $\sigma(x)$.

A different form of the alternating method was applied to the problem of an edge crack in a semi-infinite region by Irwin [3], who essentially used only the first few solutions of each sequence to estimate the stress intensity factor. Lachenbruch [4] included additional terms from each sequence to obtain a better estimate. Further study of the alternating method applied to this problem will be carried out in detail in Section 4.2. One of the two sequences of solutions involved there is for an infinite region containing a finite straight crack subjected to arbitrary normal stresses on its faces. The solution is written in integral form and may be evaluated for any given loads. In this case, the analytical result gives the singularity separately, and no additional special numerical treatment is required.

Recently, application of the alternating method to finite width strips cracked on one edge has been formulated by Brož [5]. The method of Section 4.2 could be extended to this case by using three sequences of solutions. Reference to such extensions will be made later on. However, Brož considers more complicated analytical problems involving only two sequences. One sequence considers a row of collinear equally spaced cracks of equal length, each loaded by the same arbitrary normal stresses on its faces. He gives the formulas for the stresses at the locations of the edges of the strip in terms of integrals which may be numerically evaluated. The other sequence gives a similar expression for the stress produced at the location of the crack in an uncracked strip loaded arbitrarily on its edges. He has included very few numerical results in [5], but a computer program for obtaining additional results could be easily written.

In a similar fashion, three-dimensional problems may also be solved by the alternating method. Smith et al. [6, 7, 8] have published a series of papers containing numerical results for the effect of interaction of cracks with free surfaces. One of the geometries considered [6] was a surface crack perpendicular to the surface of a half space. The crack is in the shape of half of a flat circular disk with the diameter on the surface. The other problems involved cases where the crack is a larger or smaller portion of the circular disk whose diameter parallel to the surface lies either above or below the surface. In [7] the crack intersects the free surface, and in [8] it is completely embedded in the half space. A generalization of [8] was later made by Shah and Kobayashi [9] in which the embedded crack is elliptical in shape with the minor axis perpendicular to the surface.

As in the plane problems, one sequence of solutions for the three-dimensional problems above is chosen to be an infinite body containing a crack. For the penny-shaped crack a fairly convenient solution is known for pressure on the crack of the form $r^n \cos m\theta$ where $r$ and $\theta$ are the usual polar coordinates. A less convenient solution for the elliptical crack is known for polynomial variations of the stress on the crack. But these solutions account only partially for the effect of the stress singularity. The other sequence used for the three-dimensional problems is made up of solutions of arbitrarily loaded half spaces. It may be seen in Section 4.3 that the stress applied to the half space must be singular. The previous solutions do not satisfactorily treat this effect, and hence the numerical results are drastically altered. The singularity on the surface of the half space may be taken care of analytically, and then one is faced with an additional singularity in the other sequence. That is, the stress applied to the crack is singular at two points of the crack front. No analytical way of handling this singularity could be found at present, but special precaution was taken numerically.

## 4.2 Edge crack problem

The determination of stress intensity factors for bodies containing cracks extending inward from a free edge has attained importance because such cracks are frequently found in structural members. The analytical model is usually based on the theory of elasticity in plane strain or generalized plane stress. The case of a crack perpendicular to the edge of a semi-infinite material (Figure 4.1) will be considered.

Irwin [3, 10] estimated the stress intensity factor, $k_1$, for the case of uniform pressure on the edge crack (or uniform tension at infinity perpendicular to the crack) and obtained a value of $k_1$ ten percent higher than that for an infinite plate with a crack of twice the length. He used two problems whose solutions are given in this section. The infinite plate with an arbitrarily

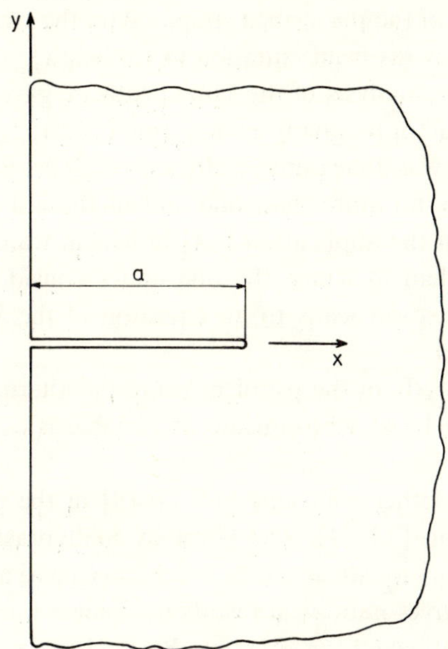


Figure 4.1. Edge-cracked plate.

pressurized crack is considered first. Then the half plate subjected to arbitrary normal stresses is superposed on the infinite cracked plate. Irwin used the superposition to obtain two simultaneous integral equations in two unknown functions and solved them by an iterative method which is equivalent to the alternating method presented here.

Later, Bueckner [11] reconsidered the problem and obtained a more accurate correction (13%)*. By superposing the two problems described

* Winne and Wundt [12] overlooked the correction of Bueckner when they used some of his other results on cracks in finite width strips (also contained in [11]). The review of Paris and Sih [1] points out the error in [12].

above, he obtained a singular integral equation for the deformed shape of the crack. The equation permitted straightforward integration for obtaining the pressure required to produce certain prescribed deformed shapes. A number of different deformed shapes and associated pressure distributions were obtained. A linear combination of these pressures was fitted to the actual uniform pressure by collocation (equating the actual pressure to the linear combination at a number of points equal to the number of undetermined coefficients). Thus the deformed shape of the crack was determined as a known linear combination of the prescribed shapes. And the stress intensity factor followed immediately (as from equation (4.16) below).

Koiter [13], in his analysis of the same problem, gave a refined value of the stress intensity factor involving a correction of 12.15%. The computation of a definite integral was done numerically with such precision that the factor above is in error by no more than one unit in the last digit. The definite integral results from the application [14] of Mellin transforms [15] to the quarter plane problem to which the one under consideration reduces by symmetry. The transform leads to an equation of the Wiener–Hopf type which is then solved.

A more general study of the problem using the alternating method, was made by Lachenbruch [4], who considered variable as well as uniform loads on the crack.

There have been other solutions [16, 17, 18] of the problem using the same technique as in [13, 14]. The latest by Stallybrass [18] obtains the stress intensity factors for stress on the crack varying as a power of distance from the edge. He gives numerical results for integer powers, 0, 1, 2, ..., 10, which could be combined if the stress on the crack is a polynomial function of that distance. Sneddon [19] also used the quarter plane formulation, but his technique of solving the problem (similar to the method of Section 4.2) leads to a Fredholm integral equation of the second kind. Two approximate methods of solving the integral equation are compared for the case of uniform stress on the crack, and both give very good accuracy.

The sections which follow contain the solutions of the two problems used in the alternating method. The alternating method is first presented in detail for the case of concentrated forces acting on the crack. The numerical results are then used to develop an integral expression for the stress intensity factor for arbitrary loads on the crack. The integral is evaluated for several examples which are contained in the Appendix. One of the examples is the problem of uniform pressure on the crack which has been considered by many others. A comparison of this result with those obtained by the various other investigators can be found in Table I.

TABLE I

*$k_1$ values for uniform tension in Figure* 4.1

| *Source* | $k_1/\sigma a^{\frac{1}{2}}$ |
|---|---|
| Irwin [3, 10] | 1.1 |
| Bueckner [11] | 1.13 |
| Koiter [13, 14] | 1.1215 |
| Wigglesworth [16] | 1.122 |
| Lachenbruch [4] | 1.1 |
| Stallybrass [18] | 1.1215 |
| Sneddon [19] | 1.1215 |
| Appendix | 1.1215* |

* Obtained by using equation (4.50).

*Infinite plate with a central crack.* The method of integral transforms [15] may be used to obtain a solution of the equations of elasticity general enough to satisfy the boundary conditions of this problem. Formulas from Chapter 9 of [15] for the special case of symmetry about the $x$- and $y$-axes may be written in integral form. For the present work, it is sufficient to note that the expressions satisfy all of the equations of elasticity by direct substitution. For the case of plane strain the displacements $u_x$ and $v_y$ in the $x$- and $y$-directions, respectively, are given by

$$\begin{aligned} \frac{E}{1+\nu} u_x &= -\int_0^\infty [(1-2\nu)-\alpha y]\, B(\alpha)\, \mathrm{e}^{-\alpha y} \sin \alpha x \, \mathrm{d}\alpha \\ \frac{E}{1+\nu} v_y &= \int_0^\infty [2(1-\nu)+\alpha y]\, B(\alpha)\, \mathrm{e}^{-\alpha y} \cos \alpha x \, \mathrm{d}\alpha \end{aligned} \tag{4.1}$$

where $E$ and $\nu$ are the modulus of elasticity and Poisson's ratio. $B(\alpha)$ is a function which will be determined by the boundary conditions. The stresses associated with the displacement field of equations (4.1) are

$$\sigma_x = -\int_0^\infty (1-\alpha y)\,\alpha B(\alpha)\, \mathrm{e}^{-\alpha y} \cos \alpha x \, \mathrm{d}\alpha$$

$$\sigma_y = -\int_0^\infty (1+\alpha y)\alpha B(\alpha) e^{-\alpha y} \cos \alpha x \, d\alpha$$

$$\tau_{xy} = -y \int_0^\infty \alpha^2 B(\alpha) e^{-\alpha y} \sin \alpha x \, d\alpha \,. \tag{4.2}$$

The solution given by equations (4.1) and (4.2) may also be used for generalized plane stress if $E$ is replaced by $(1+2\nu)E/(1+\nu)^2$ and $\nu$ by $\nu/(1+\nu)$. Equations (4.1) and (4.2) should be applied only for $y \geqslant 0$. For negative $y$, the conditions of symmetry about the $x$-axis should be used. The crack

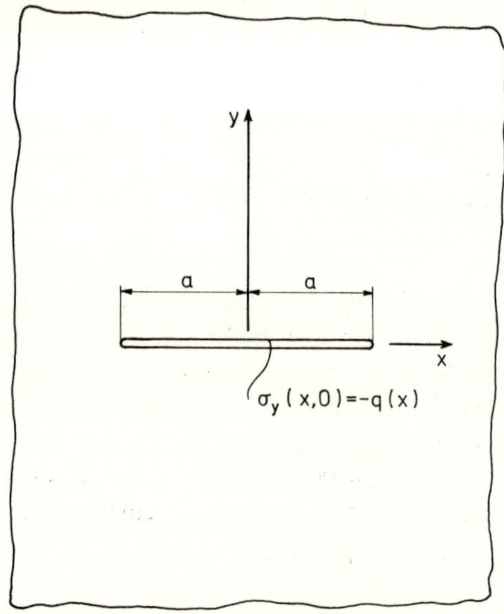


Figure 4.2. Cracked infinite plate.

problem of this Section is restricted to the case of loading by normal stresses on the crack which are symmetric about the $y$-axis. With this restriction, the problem has the symmetry required for application of equations (4.1) and (4.2). It may be noticed that the last of equations (4.2) satisfies the boundary condition $\tau_{xy}(x, 0) = 0$ on the crack and the $x$-axis of symmetry outside of the crack.

The boundary conditions which determine the value of $B(\alpha)$ are

$$\sigma_y(x, 0) = -q(x) \,, \quad |x| < a \qquad v_y(x, 0) = 0 \,, \quad |x| > a \tag{4.3}$$

where the first of equations (4.3) comes from the imposed pressure, $q(x)$, on the crack. The second of equations (4.3) results from symmetry about the $x$-axis.

The boundary conditions (4.3) require these special results from equations (4.1) and (4.2):

$$\frac{E}{1+\nu} v_y(x, 0) = 2(1-\nu)\int_0^\infty B(\alpha)\cos\alpha x\,\mathrm{d}\alpha$$

$$\sigma_y(x, 0) = -\int_0^\infty \alpha B(\alpha)\cos\alpha x\,\mathrm{d}\alpha\,. \tag{4.4}$$

Let

$$v(x) = \frac{E}{2(1-\nu^2)} v_y(x, 0) = \int_0^\infty B(\alpha)\cos\alpha x\,\mathrm{d}\alpha\,. \tag{4.5}$$

The Fourier inversion [15] of equation (4.5) gives

$$B(\alpha) = \frac{2}{\pi}\int_0^\infty v(x)\cos\alpha x\,\mathrm{d}x\,.$$

But by the second of equations (4.3), $v(x)=0$ for $x>a$. Therefore, $B(\alpha)$ becomes

$$B(\alpha) = \frac{2}{\pi}\int_0^a v(x)\cos\alpha x\,\mathrm{d}x\,. \tag{4.6}$$

Now, since $v(x)$ is a displacement, the representation

$$v(x) = \int_x^a \frac{\psi(t)\,t\,\mathrm{d}t}{(t^2-x^2)^{\frac{1}{2}}}, \qquad 0<x<a \tag{4.7}$$

may be used. Integration of equation (4.7) by parts shows the correct crack tip opening shape. If equation (4.7) is substituted into equation (4.6),

$$\begin{aligned} B(\alpha) &= \frac{2}{\pi}\int_0^a \cos\alpha x\,\mathrm{d}x\int_x^a \frac{\psi(t)\,t\,\mathrm{d}t}{(t^2-x^2)^{\frac{1}{2}}} \\ &= \frac{2}{\pi}\int_0^a \psi(t)\,t\,\mathrm{d}t\int_0^t \frac{\cos\alpha x}{(t^2-x^2)^{\frac{1}{2}}}\,\mathrm{d}x \\ &= \int_0^a \psi(t)J_0(\alpha t)\,t\,\mathrm{d}t\,. \end{aligned} \tag{4.8}$$ ⋆

⋆ The formula for $J_0(x)$, the Bessel function of order zero, on page 27 of [20] is used in the last step.

The first of equations (4.3) and the second of (4.4) give the equation

$$\int_0^\infty \alpha B(\alpha)\cos \alpha x\,\mathrm{d}\alpha = q(x)\,, \qquad 0<x<a \tag{4.9}$$

Integrating equation (4.9),

$$\int_0^\infty B(\alpha)\sin \alpha x\,\mathrm{d}\alpha = \int_0^x q(\xi)\mathrm{d}\xi\,, \qquad 0<x<a\,. \tag{4.10}$$

If equation (4.8) is substituted into equation (4.10) an integral equation for $\psi(t)$ results.

$$\int_0^\infty \sin \alpha x\,\mathrm{d}\alpha \int_0^a \psi(t) J_0(\alpha t)\,t\,\mathrm{d}t = \int_0^x q(\xi)\mathrm{d}\xi\,, \qquad 0<x<a$$

or

$$\int_0^a \psi(t)\,t\,\mathrm{d}t \int_0^\infty J_0(\alpha t)\sin \alpha x\,\mathrm{d}\alpha = \int_0^x q(\xi)\mathrm{d}\xi\,, \qquad 0<x<a \tag{4.11}$$

But*

$$\int_0^\infty J_0(\alpha t)\sin \alpha x\,\mathrm{d}\alpha = \begin{cases} 0 & t>x \\ (x^2-t^2)^{-\frac{1}{2}} & t<x\,. \end{cases}$$

Therefore, equation (4.11) reduces to

$$\int_0^x \frac{\psi(t)\,t\,\mathrm{d}t}{(x^2-t^2)^{\frac{1}{2}}} = \int_0^x q(\xi)\mathrm{d}\xi\,, \qquad 0<x<a\,. \tag{4.12}$$

Equation (4.12) may be solved as a special case of Abel's integral equation**

$$\int_0^x \frac{\phi(t)\,\mathrm{d}t}{(x^2-t^2)^{\frac{1}{2}}} = h(x)\,, \qquad 0<x<a \tag{4.13}$$

which has the solution

$$\phi(t) = \frac{2}{\pi}\left\{t\int_0^t \frac{h'(x)\,\mathrm{d}x}{(t^2-x^2)^{\frac{1}{2}}} + h(0)\right\}, \qquad 0<t<a\,. \tag{4.14}$$

Therefore, the solution of equation (4.12) is

$$\psi(t) = \frac{2}{\pi}\int_0^t \frac{q(x)\,\mathrm{d}x}{(t^2-x^2)^{\frac{1}{2}}}\,, \qquad 0<t<a\,. \tag{4.15}$$

* First formula on page 37 of [20].

** This form may be obtained from that on page 141 of [20] by a change of variables.

It is known that the opening displacement of the crack tip is given by (see equations (3.15) of [21])

$$v_y(x, 0) = \frac{2(1-\nu^2)}{E} k_1 (2r)^{\frac{1}{2}} + O(r) \tag{4.16}$$

for plane strain. In this result, $r = a - x$, and $k_1$ is the Mode I stress intensity factor. From equation (4.5)

$$v_y(x, 0) = \frac{2(1-\nu^2)}{E} v(x) \tag{4.17}$$

and from equation (4.7) for $x = a - r$,

$$v(x) = \int_{a-r}^{a} \frac{\psi(t) t \, dt}{[t^2 - (a-r)^2]^{\frac{1}{2}}} = \int_0^r \frac{a-p}{(r-p)^{\frac{1}{2}}} \frac{\psi(a-p) dp}{(2a-r-p)^{\frac{1}{2}}} . \tag{4.18}$$

In equation (4.18), when $r$ is small, $r$ and $p$ may be neglected in comparison with $a$, and then

$$\begin{aligned} v(x) &= \frac{a\psi(a)}{(2a)^{\frac{1}{2}}} \int_0^r \frac{dp}{(r-p)^{\frac{1}{2}}} + O(r) \\ &= (2r)^{\frac{1}{2}} (a^{\frac{1}{2}}) \psi(a) + O(r) . \end{aligned} \tag{4.19}$$

Thus, by comparing equations (4.16), (4.17), (4.19), the stress intensity factor

$$k_1 = a^{\frac{1}{2}} \psi(a)$$

is obtained. Hence, utilizing equation (4.15),

$$k_1 = \frac{2}{\pi} a^{\frac{1}{2}} \int_0^a \frac{q(x) dx}{(a^2 - x^2)^{\frac{1}{2}}} . \tag{4.20}$$

It is also necessary to evaluate the normal stress on the $y$-axis which, from equations (4.2), is

$$\sigma_x(0, y) = -\int_0^{\infty} (1 - \alpha y) \alpha B(\alpha) e^{-\alpha y} d\alpha . \tag{4.21}$$

From equations (4.8) and (4.15)

$$\begin{aligned} B(\alpha) &= \frac{2}{\pi} \int_0^a J_0(\alpha t) t \, dt \int_0^t \frac{q(x) dx}{(t^2 - x^2)^{\frac{1}{2}}} \\ &= \frac{2}{\pi} \int_0^a q(x) dx \int_x^a \frac{J_0(\alpha t)}{(t^2 - x^2)^{\frac{1}{2}}} t \, dt . \end{aligned}$$

Therefore, interchanging the order of integration in equation (4.21),

$$\sigma_x(0, y) = \frac{2}{\pi}\int_0^a f(x, y)q(x)\mathrm{d}x \tag{4.22}$$

where

$$f(x, y) = \int_x^a g(y, t)\frac{t\,\mathrm{d}t}{(t^2-x^2)^{\frac{1}{2}}} \tag{4.23}$$

where

$$g(y, t) = -\int_0^\infty (1-\alpha y)\alpha J_0(\alpha t)\mathrm{e}^{-\alpha y}\mathrm{d}\alpha\,.$$

The evaluation of $g(y, t)$ leads directly to

$$g(y, t) = y\frac{y^2-2t^2}{(t^2+y^2)^{\frac{5}{2}}}\,. \tag{4.24}$$

The integration of equation (4.23) may be accomplished with $g(y, t)$ given by equation (4.24). The result is

$$f(x, y) = \frac{y}{x^2+y^2}\left(\frac{a^2-x^2}{a^2+y^2}\right)^{\frac{1}{2}}\left[\frac{2y^2}{x^2+y^2} - \frac{2a^2+y^2}{a^2+y^2}\right]. \tag{4.25}$$

Finally, substituting equation (4.25) into (4.22), the integral

$$p(y) = \frac{2}{\pi}\frac{y}{(a^2+y^2)^{\frac{1}{2}}}\int_0^a q(x)\frac{(a^2-x^2)^{\frac{1}{2}}}{x^2+y^2}\left[\frac{2y^2}{x^2+y^2} - \frac{2a^2+y^2}{a^2+y^2}\right]\mathrm{d}x \tag{4.26}$$

for the desired stress,

$$\sigma_x(0, y) = p(y) \tag{4.27}$$

is obtained.

The above results apply also to the case of shear loading on the crack. In fact if the stress on the crack is

$$\tau_{xy}(x, 0) = -q(x)\,, \qquad |x| < a$$

where $q(x)=q(-x)$, then the Mode II stress intensity factor is

$$k_2 = \frac{2}{\pi}a^{\frac{1}{2}}\int_0^a \frac{q(x)\mathrm{d}x}{(a^2-x^2)^{\frac{1}{2}}} \tag{4.28}$$

and

$$\tau_{xy}(0, y) = p(y)$$

where $p(y)$ is given by equation (4.26).

*Edge-loaded semi-infinite plate.* The solution given by equations (4.1) and (4.2) applies to the case in which a half plane, $y \geqslant 0$ (Figure 4.3) is subjected to normal stresses

$$\sigma_y(x, 0) = p(x) \tag{4.29}$$

symmetric about the $y$-axis. It is found from the second of equations (4.2)

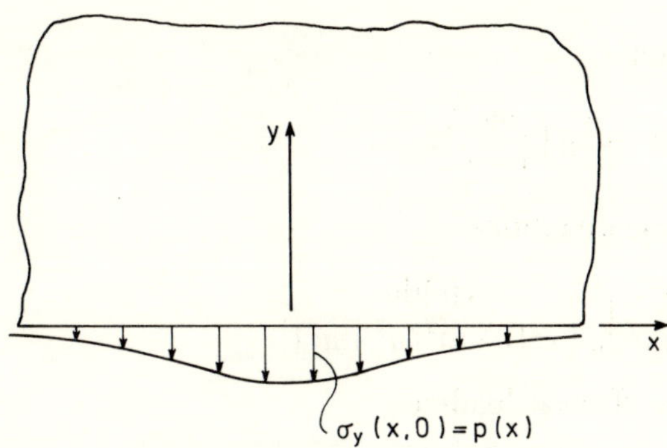


Figure 4.3. Edge-loaded half plane.

and the Fourier inversion theorem that

$$\alpha B(\alpha) = -\frac{2}{\pi}\int_0^\infty p(x)\cos \alpha x \, dx\,. \tag{4.30}$$

The first of equations (4.2) and equation (4.30) combine to give

$$\begin{aligned}\sigma_x(0, y) &= \frac{2}{\pi}\int_0^\infty (1-\alpha y)\alpha e^{-\alpha y} d\alpha \int_0^\infty p(x)\cos \alpha x \, dx \\ &= \frac{2}{\pi}\int_0^\infty p(x) dx \int_0^\infty (1-\alpha y)\alpha e^{-\alpha y}\cos \alpha x \, d\alpha \\ &= \frac{4}{\pi} y \int_0^\infty \frac{x^2 p(x) dx}{(x^2+y^2)^2}\,.\end{aligned} \tag{4.31}$$

A change of coordinates gives us that for a half plane, $x \geqslant 0$, subjected to tension on the edge,

$$\sigma_x(0, y) = p(y) \tag{4.32}$$

the desired stress is

$$\sigma_y(x,0) = \frac{4}{\pi} x \int_0^\infty \frac{y^2 p(y)\mathrm{d}y}{(x^2+y^2)^2}. \tag{4.33}$$

For convenience in the numerical work, introduce the new variable,

$$u = \frac{y}{a+y}$$

and the function,

$$t(u) = u^2 p(y) = u^2 p\left(\frac{au}{1-u}\right).$$

In terms of these quantities,

$$\sigma_y(x,0) = \frac{4}{\pi}\frac{x}{a} \int_0^1 \frac{t(u)\mathrm{d}u}{[x^2(1-u)^2/a^2+u^2]^2}. \tag{4.34}$$

Again, the case of shear loading

$$\tau_{xy}(0,y) = p(y) \tag{4.35}$$

on a half plane has a similar solution yielding

$$\tau_{xy}(x,0) = \frac{4}{\pi} x \int_0^\infty \frac{y^2 p(y)\mathrm{d}y}{(x^2+y^2)^2}. \tag{4.36}$$

*Iterative formulation of the problem.* The iterative approach may be thought about in two ways which are essentially equivalent. One way used by Irwin [3], involves the superposition of the preceding two solutions. The boundary conditions of the edge-cracked plate problem result in a pair of integral equations:

$$q(x) - \frac{4}{\pi} x \int_0^\infty \frac{y^2 p(y)\mathrm{d}y}{(x^2+y^2)^2} = \sigma(x), \qquad 0 \leqslant x \leqslant a$$

and

$$p(y) + \frac{2}{\pi}\frac{y}{(a^2+y^2)^{\frac{1}{2}}} \int_0^a \frac{q(x)(a^2-x^2)^{\frac{1}{2}}}{x^2+y^2}\left[\frac{2y^2}{x^2+y^2} - \frac{2a^2+y^2}{a^2+y^2}\right]\mathrm{d}x = 0, \qquad 0 \leqslant y < \infty$$

where $\sigma(x)$ is the stress applied to the edge crack. These equations may be solved by iteration by assuming some form for $q(x)$, say $q(x)=\sigma(x)$, and using this assumed form in the second equation to solve for $p(y)$. This value of $p(y)$ is then used in the first equation to obtain a corrected value of $q(x)$. Then a corrected value of $p(y)$ is found, and the procedure is continued until there is no significant change in $p(y)$ or $q(x)$. The last value of $q(x)$ gives the stress intensity factor according to equation (4.20).

A more satisfactory way, from a physical and computational point of view, of interpreting the iterative approach involves the alternating superposition of the previously developed solutions. Consider the details of the case,

$$\sigma(x)=P\delta(x-b)\,, \qquad 0<b<a$$

where $\delta(x)$ is the Dirac delta generalized function. The above expression represents a line force, $P$ (force per unit length in the $z$-direction), acting a distance $b$ from the free edge (Figure 4.4).

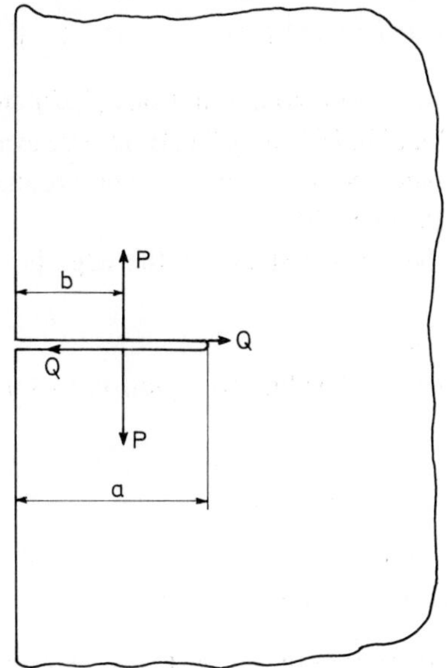


Figure 4.4. Concentrated forces on a edge crack.

Now refer to the infinite plate containing the crack as Problem A, and to the edge-loaded semi-infinite plate as Problem B. Subscripts of A or B will denote the problem with which the subscripted stress is associated.

Then the iterations consist of superimposing Problem A and Problem B. The procedure begins with Problem A.

In Problem A, let the stress on the crack be

$$\sigma_{yA}^{(0)}(x,0) = -q^{(0)}(x) = -\sigma(x) = -P\delta(x-b), \qquad x>0 \tag{4.37}$$

and require it to be even in $x$. That is, the crack is opened by four symmetrically located concentrated forces. The present simple form of $\sigma(x)$ allows

$$k_1^{(0)} = \frac{2}{\pi}\frac{Pa^{\frac{1}{2}}}{(a^2-b^2)^{\frac{1}{2}}}$$

and

$$\sigma_{xA}^{(0)}(0,y) = p^{(0)}(y)$$

where

$$p^{(0)}(y) = \frac{2P}{\pi}\frac{(a^2-b^2)^{\frac{1}{2}}}{(a^2+y^2)^{\frac{1}{2}}}\frac{y}{b^2+y^2}\left[\frac{2y^2}{b^2+y^2} - \frac{2a^2+y^2}{a^2+y^2}\right] \tag{4.39}$$

to be evaluated in closed form from equations (4.20) and (4.26). In the case of more complicated loading $k_1^{(0)}$ and $p^{(0)}(y)$ could be evaluated by numerical integration as in the succeeding iterations. The second part of the zeroth integration involves Problem B.

In Problem B, let the normal stress on the edge be

$$\sigma_{xB}^{(0)}(0,y) = -p^{(0)}(y).$$

This load gives the stress according to equation (4.33) as

$$\sigma_{yB}^{(0)}(x,0) = q^{(1)}(x)$$

where

$$q^{(1)}(x) = -\frac{4}{\pi}x\int_0^\infty \frac{y^2 p^{(0)}(y)\mathrm{d}y}{(x^2+y^2)^2}. \tag{4.40}$$

The superposition of these two problems gives a stress state which satisfies all of the boundary conditions of the edge-cracked plate except that the stress on the crack is

$$\sigma_y(x,0) = -\sigma(x) + q^{(1)}(x).$$

If $q^{(1)}(x)$ is not negligible compared to $\sigma(x)$, further iterations are required.

The first part of the first iteration requires the solution of Problem A for

$$\sigma_{yA}^{(1)}(x,0) = -q^{(1)}(x).$$

This leads to an additional contribution to the stress intensity factor of

$$k_1^{(1)} = \frac{2}{\pi}a^{\frac{1}{2}}\int_0^a \frac{q^{(1)}(x)\mathrm{d}x}{(a^2-x^2)^{\frac{1}{2}}}$$

and a stress on $x=0$ of

$$\sigma_{xA}^{(1)}(0,y) = p^{(1)}(y)$$

where

$$p^{(1)}(y) = \frac{2}{\pi}\frac{y}{(a^2+y^2)^{\frac{1}{2}}}\int_0^a \frac{q^{(1)}(x)(a^2-x^2)^{\frac{1}{2}}}{x^2+y^2}\left[\frac{2y^2}{x^2+y^2} - \frac{2a^2+y^2}{a^2+y^2}\right]\mathrm{d}x\,. \tag{4.41}$$

An application of Problem B for a load of

$$\sigma_{xB}^{(1)}(0,y) = -p^{(1)}(y)$$

gives

$$\sigma_{yB}^{(1)}(x,0) = q^{(2)}(y)$$

where

$$q^{(2)}(y) = -\frac{4}{\pi}x\int_0^\infty \frac{y^2p^{(1)}(y)\mathrm{d}y}{(x^2+y^2)^2}\,.$$

The superposition of these two parts of the first iteration gives the solution of the edge cracked plate with load on the crack,

$$\sigma_y^{(1)}(x,0) = -q^{(1)}(x)+q^{(2)}(x)\,.$$

After superposing the zeroth and first iterations, the stress on the crack is

$$\sigma_y(x,0) = -\sigma(x)+q^{(2)}(x)$$

and the stress intensity factor is

$$k_1 = \frac{2}{\pi}\frac{Pa^{\frac{1}{2}}}{(a^2-b^2)^{\frac{1}{2}}} + \frac{2}{\pi}a^{\frac{1}{2}}\int_0^a \frac{q^{(1)}(x)\mathrm{d}x}{(a^2-x^2)^{\frac{1}{2}}}\,.$$

The second iteration has the stresses

$$\sigma_{yA}^{(2)}(x,0) = -q^{(2)}(x) \qquad \sigma_{xA}^{(2)}(0,y) = p^{(2)}(y)$$
$$\sigma_{xB}^{(2)}(0,y) = -p^{(2)}(y) \qquad \sigma_{yB}^{(2)}(x,0) = q^{(3)}(x)$$

and the stress intensity factor associated with $q^{(2)}(x)$. The functions $p^{(2)}(y)$ and $q^{(3)}(x)$ are given by formulas similar to equations (4.40) and (4.41). After superposing this iteration on the others, we have the edge-cracked plate subjected to stress

$$\sigma_y(x,0) = -\sigma(x) + q^{(3)}(x)$$

on the crack. For this load the stress intensity factor is

$$k_1 = \frac{2}{\pi} a^{\frac{1}{2}} \left\{ \frac{P}{(a^2-b^2)^{\frac{1}{2}}} + \int_0^a [q^{(1)}(x)+q^{(2)}(x)] \frac{\mathrm{d}x}{(a^2-x^2)^{\frac{1}{2}}} \right\}.$$

The pattern of the iterations should be clear now.

For the computations, introduce the dimensionless variables

$$\xi = x/a \qquad u = y/(a+y) \qquad c = b/a$$

and the functions

$$s^{(n)}(\xi) = \frac{1}{P} (a^2-b^2)^{\frac{1}{2}} q^{(n)}(a\xi)$$

$$t^{(n)}(u) = \frac{u^2}{P} (a^2-b^2)^{\frac{1}{2}} p^{(n)} \left( \frac{au}{1-u} \right).$$

Then the iteration is from $t^{(0)}(u)$ to $s^{(1)}(\xi)$ to $t^{(1)}(u)$, etc. where

$$t^{(0)}(u) = \frac{2}{\pi} \frac{1-c^2}{c^2(1-u)^2+u^2} \frac{u^3(1-u)^2}{\{(1-u)^2+u^2\}^{\frac{1}{2}}} \times \left[ \frac{2u^2}{c^2(1-u)^2+u^2} - \frac{2(1-u)^2+u^2}{(1-u)^2+u^2} \right] \tag{4.42}$$

$$s^{(n)}(\xi) = -\frac{4}{\pi} \xi \int_0^1 \frac{t^{(n-1)}(u)\mathrm{d}u}{[\xi^2(1-u)^2+u^2]^2}, \qquad n \geqslant 1 \tag{4.43}$$

$$t^{(n)}(u) = \frac{2}{\pi} \frac{u^3(1-u)^2}{[(1-u)^2+u^2]^{\frac{1}{2}}} \int_0^1 \frac{s^{(n)}(\xi)(1-\xi^2)^{\frac{1}{2}}}{\xi^2(1-u)^2+u^2} \times \left[ \frac{2u^2}{\xi^2(1-u)^2+u^2} - \frac{2(1-u)^2+u^2}{(1-u)^2+u^2} \right] \mathrm{d}\xi, \qquad n \geqslant 1. \tag{4.44}$$

In terms of

$$s(\xi) = s^{(1)}(\xi) + s^{(2)}(\xi) + \dots$$

the stress intensity factor is given by

$$k_1 = \frac{2}{\pi}\frac{P(a^{\frac{1}{2}})}{(a^2-b^2)^{\frac{1}{2}}}\left[1+f\left(\frac{b}{a}\right)\right] \tag{4.45}$$

where

$$f(c) = \int_0^1 \frac{s(\xi)\mathrm{d}\xi}{(1-\xi^2)^{\frac{1}{2}}}\,.$$

The dependence of $f(c)$ on $c$ comes from $t^{(0)}(u)$.

The stress intensity factor for the edge-cracked plate subjected to shear forces (Figure 4.4) is given in terms of the same function $f(c)$ above as

$$k_2 = \frac{2}{\pi}\frac{Q(a^{\frac{1}{2}})}{(a^2-b^2)^{\frac{1}{2}}}\left[1+f\left(\frac{b}{a}\right)\right]\,. \tag{4.46}$$

*Numerical results and Green's function.* The numerical iteration of equations (4.42)–(4.44) is straightforward except when the concentrated forces are very near the edge of the plate. For $b/a \geqslant 0.2$ the integrals were evaluated using Simpson's rule with 250 subdivisions, and five iterations were used. Tests using more subdivisions and iterations showed that the accuracy of the results is about one percent. As $b/a$ was decreased, more subdivisions were required and it appeared that eight iterations were required. The results are shown in Figure 4.5. The correction factor plotted there is the fractional increase due to the edge of the stress intensity factor for an infinite plate with four symmetrically located normal or shear forces.

No values of the function $f(b/a)$ were obtained for $b/a < 0.05$, but the curve plotted in Figure 4.5 seems a reasonable extension of the computed points. The computed points are shown as circles, and the curve drawn through the points is

$$f(c) = (1-c^2)[0.2945 - 0.3912\,c^2 + 0.7685\,c^4 - 0.9942\,c^6 + 0.5094\,c^8]\,. \tag{4.47}$$

This function will be used in a Green's function analysis of the stress intensity factor for an edge crack subjected to some arbitrary distribution of pressure.

One reason for the difficulty in obtaining points on the curve for small values of $b/a$ is the singularity at the point of application of the concentrated load. As long as $b > 0$, the stress $\sigma_{xA}^{(0)}(0, y)$ (equation 4.39) has no singularity, and its removal by the half plane solution given previously presents no difficulty. But when $b=0$, the stress on the $y$-axis has a singularity at the origin which requires special treatment. A straight application of the alternating method to this case would, if the numerical analysis were exactly accurate, give a stress on the crack, $\sigma_{yB}^{(0)}(x, 0)$, with a singularity at the origin. Each step of the procedure would leave a stress singularity at the origin. This points up a fundamental difficulty associated with the alternating method in more general cases.

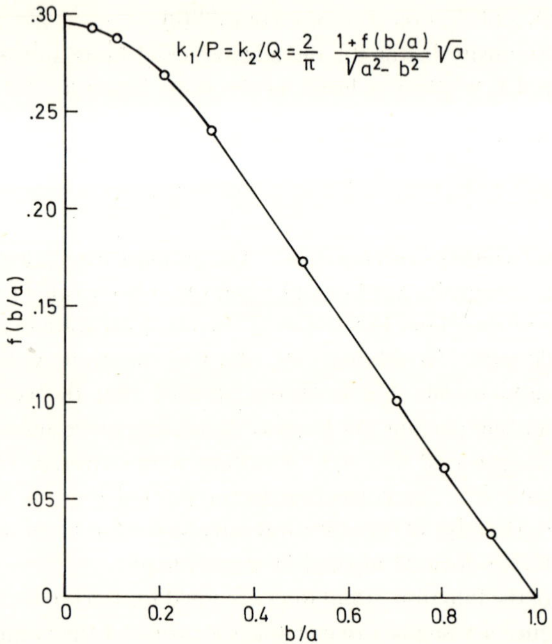


Figure 4.5. Stress intensity factor correction for concentrated forces on an edge crack.

To show the difficulty, recall that equations (4.1) and (4.2) give the solution for the half plane, $y>0$, loaded by normal stresses on the edge, $y=0$. It can be seen from equations (4.2) that

$$\sigma_x(x, 0) = \sigma_y(x, 0) .$$

That is, both normal stresses are the same at each point of the edge. Applying this result to equations (4.37) and those following for arbitrary stress

$$\sigma_{yA}^{(0)}(x, 0) = -q^{(0)}(x) = -\sigma(x)$$

on the crack, it is found that the stress, $\sigma_{xA}^{(0)}(0, y)$ satisfies

$$\sigma_{xA}^{(0)}(0, 0) = \sigma_{yA}^{(0)}(0, 0) \quad \text{or} \quad p^{(0)}(0) = -q^{(0)}(0).$$

Similarly, in the half plane problem,

$$\sigma_{yB}^{(0)}(0, 0) = \sigma_{xB}^{(0)}(0, 0) \quad \text{or} \quad q^{(1)}(0) = -p^{(0)}(0).$$

And so it would continue giving

$$q^{(0)} = -p^{(0)} = q^{(1)} = -p^{(1)} = q^{(2)} = -p^{(3)} = \dots$$

where each function is evaluated at the origin. Therefore,

$$q^{(n)}(0) = q^{(0)}(0) = \sigma(0).$$

After the $n$th iteration, the superposition of all steps gives the exact solution of the edge crack problem for stress on the crack given by

$$\sigma_y(x, 0) = -\sigma(x) + q^{(n)}(x).$$

And since

$$\sigma_y(0, 0) = -\sigma(0) + \sigma(0) = 0$$

the scheme does not converge to the desired solution. That is, $q^{(n)}(x)$ is not negligible compared to $\sigma(x)$ at $x=0$ at least. But no difficulty should be expected if the stress on the edge crack at the edge, $\sigma(0)$, is zero. And in the case of a concentrated force as in equation (4.37) $\sigma(0)=0$ as long as $b>0$.

In the case of shear loading, the same difficulty exists. The equality of the normal stresses on the edge and on the plane at right angles to the edge has its counterpart in the obvious statement about the shear stresses on the same planes. The same kind of details could be given for this case, but fundamentally the difficulty is that the shear stress on the crack at the intersection of the crack and edge must be equal to that on the edge unless the stress tensor is allowed to be non-symmetric.

Consider now the case of arbitrary stress on the edge crack,

$$\sigma_y(x, 0) = -\sigma(x), \qquad 0 < x < a \tag{4.48}$$

For a concentrated force, $dP = \sigma(x)dx$, located a distance $x$ from the edge, equation (4.45) gives the stress intensity factor

$$dk_1 = \frac{2}{\pi} a^{\frac{1}{2}} \frac{\sigma(x)dx}{(a^2 - x^2)^{\frac{1}{2}}} [1 + f(x/a)]. \tag{4.49}$$

The contribution of each part of the pressure distribution on the crack may be written as in equation (4.49). The total stress intensity factor is obtained by integrating,

$$k_1 = \frac{2}{\pi} a^{\frac{1}{2}} \left\{ \int_0^a \sigma(x) \frac{\mathrm{d}x}{(a^2-x^2)^{\frac{1}{2}}} + \int_0^a \sigma(x) \frac{f(x/a)\mathrm{d}x}{(a^2-x^2)^{\frac{1}{2}}} \right\}. \tag{4.50}$$

For shear loading, $\tau_{xy}(x, 0) = -\tau(x)$, $k_2$ is given by the same expression with $\sigma(x)$ replaced by $\tau(x)$. Note that the first term is the stress intensity factor for an infinite plate containing a crack of length $2a$ subjected to the same pressure (even in $x$). The second term is the correction due to the presence of the free edge. The results of integrating equation (4.50) for a number of pressure variations are contained in the Appendix.

The value of the stress intensity factor listed in Table I was obtained by application of equation (4.50). It may be seen to agree with the value calculated by Koiter [13]. In addition, the other edge crack results in the Appendix agree with those of Lachenbruch [4], though he published only two digit stress intensity factors. The result for linearly varying pressure agrees with that of Stallybrass [18].

## 4.3 Surface crack problem

One of the crack geometries that has received continual interest in fracture mechanics is that of a semi-elliptical crack whose major axis lies on a stress free surface. This configuration is analogous to the two-dimensional version of the edge crack problem discussed earlier except that no reliable method has yet been found to solve the three-dimensional problem. The major cause of difficulty lies in the lack of information at the points where the crack border intersects with the free surface. Sih [22] has discussed this point in detail in connection with the finite thickness crack problem. This difficulty is clearly evidenced by the variation among results [6, 9, 10] published on the semi-circular and semi-elliptical surface crack problems. In [6, 9], the maximum value of the stress-intensity factor was found to be on the free surface, whereas in [10] the maximum occurred at the utmost interior point of the crack front.

In order to demonstrate the sensitivity of the solution to the influence of the free surface the semi-circular crack problem is again treated in this section by the alternating method. With special care given to the stress state around the critical points where the crack border intersects with the free surface, a more accurate numerical solution was achieved. The results are strikingly different from those published in the literature [6, 7, 8, 9, 10, 23] and are given for each iteration to illustrate the rate of convergence. The construction of the two sequences used in the alternating method is described with the general solution given for one of the sequences. The half space solution required by the other sequence is then presented. Finally, the singularities which affect the numerical solution are discussed, and methods of treating them are given. And, concluding this Section, the numerical results are shown.

*Penny-shaped crack in an infinite body.* The basic solution used for the infinite body containing a penny-shaped crack (Figure 4.6) is obtained from

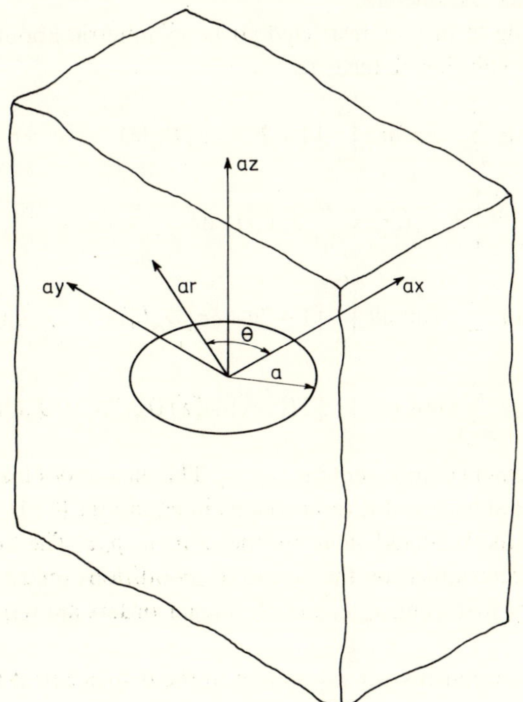


Figure 4.6. Penny-shaped crack in infinite body.

Muki [24]. His solution is valid for completely arbitrary loading of a semi-infinite solid. When the loading is restricted to normal stresses on the surface, his solution may be applied to the case of a penny-shaped crack subjected to arbitrary variation of pressure on its faces. The crack problem may be reduced by symmetry to a half space problem in which known normal stress acts inside a circular region and unknown normal stress (producing known displacement) acts outside this region.

A characteristic length is taken to be $a$, and all length coordinates are expressed as the product of $a$ and a dimensionless coordinate. The cylindrical coordinates $(ar, \theta, az)$ shown in Figure 4.6 involve the dimensionless variables $r$ and $z$. In the following equations, a quantity, $p_0$, which has dimensions of stress is shown explicitly so that all quantities appearing after the summation signs are dimensionless. In this Section $a$ will be the radius of the crack and $p_0$, the uniform pressure acting on it. In other applications they would designate some other parameters.

If the loading is further restricted to be symmetric about the $xz$-plane $(\theta=0)$, Muki's solution reduces to

$$2\mu u_r = -p_0 a \sum_{m=0}^{\infty} \cos m\theta \int_0^{\infty} [1-2\nu-\xi z] D_m(\xi) \times e^{-\xi z}\left[J_{m+1}(\xi r) - \frac{m}{\xi r} J_m(\xi r)\right] d\xi$$

$$2\mu v_\theta = -p_0 a \sum_{m=0}^{\infty} \sin m\theta \int_0^{\infty} [1-2\nu-\xi z] D_m(\xi) e^{-\xi z} \frac{m}{\xi r} J_m(\xi r) d\xi$$

$$2\mu w_z = p_0 a \sum_{m=0}^{\infty} \cos m\theta \int_0^{\infty} [2(1-\nu)+\xi z] D_m(\xi) e^{-\xi z} J_m(\xi r) d\xi \tag{4.51}$$

for the displacement components, $u_r$, $v_\theta$, $w_z$. The shear modulus and Poisson's ratio are denoted by $\mu$ and $\nu$, respectively, in equations (4.51). Note that the dimensions of the left-hand sides are the same as $p_0 a$. The functions $D_m(\xi)$ remain to be determined by the boundary conditions on $z=0$. The Bessel functions of the first kind, $J_m(\xi r)$, of all integer orders are required in equations (4.51).

The stress components corresponding to the displacement field (4.51) are given by

$$\sigma_r = p_0 \sum_{m=0}^{\infty} \cos m\theta \left\{ -\int_0^{\infty} (1-\xi z)\xi D_m(\xi) e^{-\xi z} J_m(\xi r) d\xi + \frac{1}{r}\int_0^{\infty} (1-2\nu-\xi z) D_m(\xi) e^{-\xi z}\left[\frac{m(m-1)}{\xi r} J_m(\xi r) + J_{m+1}(\xi r)\right] d\xi \right\}$$

$$\sigma_\theta = \quad p_0 \sum_{m=0}^{\infty} \cos m\theta \left\{ -2\nu \int_0^\infty \xi D_m(\xi) \mathrm{e}^{-\xi z} J_m(\xi r) \mathrm{d}\xi \right.$$

$$\left. - \frac{1}{r} \int_0^\infty (1-2\nu-\xi z) D_m(\xi) \mathrm{e}^{-\xi z} \left[ \frac{m(m-1)}{\xi r} J_m(\xi r) + J_{m+1}(\xi r) \right] \mathrm{d}\xi \right\}$$

$$\sigma_z = -p_0 \sum_{m=0}^{\infty} \cos m\theta \int_0^\infty (1+\xi z) \xi D_m(\xi) \mathrm{e}^{-\xi z} J_m(\xi r) \mathrm{d}\xi$$

$$\tau_{\theta z} = -p_0 \frac{z}{r} \sum_{m=0}^{\infty} m \sin m\theta \int_0^\infty \xi D_m(\xi) \mathrm{e}^{-\xi z} J_m(\xi r) \mathrm{d}\xi$$

$$\tau_{rz} = -p_0 \frac{z}{r} \sum_{m=0}^{\infty} \cos m\theta \int_0^\infty \xi D_m(\xi) \mathrm{e}^{-\xi z} [\xi r J_{m+1}(\xi r) - m J_m(\xi r)] \mathrm{d}\xi$$

$$\tau_{r\theta} = \quad p_0 \frac{1}{r} \sum_{m=0}^{\infty} m \sin m\theta \int_0^\infty (1-2\nu-\xi z) D_m(\xi) \mathrm{e}^{-\xi z}$$

$$\left[ J_{m+1}(\xi r) - \frac{m-1}{\xi r} J_m(\xi r) \right] \mathrm{d}\xi \,. \tag{4.52}$$

The shear stresses, $\tau_{\theta z}$ and $\tau_{rz}$, vanish on the plane $z=0$ as required by symmetry about the $xy$-plane. The boundary conditions of the penny-shaped crack problem which must be satisfied by appropriately choosing $D_m(\xi)$ are

$$\begin{aligned} \sigma_z &= -p_0 p(r, \theta) \quad \text{on } z=0 \text{ for } r<1 \\ w_z &= 0 \qquad\qquad\quad\;\, \text{on } z=0 \text{ for } r>1 \end{aligned} \tag{4.53}$$

where $p(r, \theta)$ is the dimensionless pressure acting on the faces of the crack. Since $p(r, \theta)$ is required to be symmetric about the $xz$-plane, it may be expanded in a cosine series as

$$p(r, \theta) = \sum_{m=0}^{\infty} K_m(r) \cos m\theta \tag{4.54}$$

where

$$K_0(r) = \frac{1}{\pi} \int_0^\pi p(r, \theta) \mathrm{d}\theta$$

$$K_m(r) = \frac{2}{\pi} \int_0^\infty p(r, \theta) \cos m\theta \, \mathrm{d}\theta \,, \qquad m \geqslant 1 \,. \tag{4.55}$$

The equations for $D_m(\xi)$ are obtained by using equations (4.53) in conjunction with (4.51), (4.52) and (4.54). They are

$$\begin{aligned} &\int_0^\infty \xi D_m(\xi) J_m(\xi r) \mathrm{d}\xi = K_m(r)\,, \qquad r<1 \\ &\int_0^\infty D_m(\xi) J_m(\xi r) \mathrm{d}\xi = 0\,, \qquad r>1 \end{aligned} \tag{4.56}$$

for each $m=0, 1, 2, \ldots$ . The procedure for solving these equations is analogous to that for the plane problem. The displacement-like function,

$$w_m(r) = \int_0^\infty D_m(\xi) J_m(\xi r) \mathrm{d}\xi \tag{4.57}$$

is introduced. Since by the second of equations (4.56), $w_m(r)=0$ for $r>1$, the Hankel inverse transform [15] of equation (4.57) is

$$D_m(\xi) = \xi \int_0^1 w_m(r) J_m(\xi r) \mathrm{d}r\,. \tag{4.58}$$

The displacement-like behavior of $w_m(\xi)$ is assured if it is represented in the form

$$w_m(r) = \frac{2}{\pi} r^m \int_r^1 g_m(t) \frac{t^{-m+1}\mathrm{d}t}{(t^2-r^2)^{\frac{1}{2}}}\,. \tag{4.59}$$

Use of equation (4.59) in equation (4.58) gives another representation of $D_m(\xi)$,

$$\begin{aligned} D_m(\xi) &= \frac{2}{\pi}\xi \int_0^1 J_m(\xi r) r^m \mathrm{d}r \int_r^1 g_m(t) \frac{t^{-m+1}\mathrm{d}t}{(t^2-r^2)^{\frac{1}{2}}} \\ &= \frac{2}{\pi}\xi \int_0^1 t^{-m+1} g_m(t) \mathrm{d}t \int_0^t r^{m+1} \frac{J_m(\xi r)\mathrm{d}r}{(t^2-r^2)^{\frac{1}{2}}} \\ &= \frac{2}{\pi}\xi \int_0^1 t^2 g_m(t) j_m(\xi t) \mathrm{d}t \end{aligned} \tag{4.60}$$ ★

where

$$j_m(\xi t) = \left(\frac{\pi}{2\xi t}\right)^{\frac{1}{2}} J_{m+\frac{1}{2}}(\xi t)$$

is the spherical Bessel function of the first kind of order $m$.

★ The formula at the top of page 32 of [20] may be used in the last step if the substitution, $r=t \sin\theta$ is made.

If the first of equations (4.56) is multiplied on both sides by $r^{m+1}$, it can be integrated with respect to $r$ to become

$$\int_0^\infty D_m(\xi) r^{m+1} J_{m+1}(\xi r) \mathrm{d}\xi = L_m(r), \qquad r<1 \tag{4.61}$$

where

$$L_m(r) = \int_0^r \eta^{m+1} K_m(\eta) \mathrm{d}\eta. \tag{4.62}$$

The equation which determines $g_m(t)$ is obtained by substituting equation (4.60) into (4.61):

$$\frac{2}{\pi}\int_0^\infty \xi J_{m+1}(\xi r)\mathrm{d}\xi \int_0^1 t^2 g_m(t) j_m(\xi t)\mathrm{d}t = r^{-m-1} L_m(r), \qquad r<1$$

or

$$\int_0^1 t^2 g_m(t)\mathrm{d}t \int_0^\infty \xi J_{m+1}(\xi r) j_m(\xi t)\mathrm{d}\xi = \frac{\pi}{2} r^{-m-1} L_m(r), \qquad r<1. \tag{4.63}$$

Knowing that*

$$\int_0^\infty \xi J_{m+1}(\xi r) j_m(\xi t)\mathrm{d}\xi = \begin{cases} 0, & t>r \\ \dfrac{t^m r^{-m-1}}{(r^2-t^2)^{\frac{1}{2}}}, & t<r \end{cases}$$

equation (4.63) may be written as

$$\int_0^r g_m(t)\frac{t^{m+2}\mathrm{d}t}{(r^2-t^2)^{\frac{1}{2}}} = \frac{\pi}{2} L_m(r), \qquad r<1. \tag{4.64}$$

By application of the solution, equation (4.14), of equation (4.13) to equation (4.64),

$$t^{m+2} g_m(t) = t\int_0^t \frac{L_m'(r)\mathrm{d}r}{(t^2-r^2)^{\frac{1}{2}}}$$

or, since equation (4.62) gives

$$L_m'(r) = r^{m+1} K_m(r)$$

$$g_m(t) = t^{-m-1}\int_0^t K_m(r)\frac{r^{m+1}\mathrm{d}r}{(t^2-r^2)^{\frac{1}{2}}}. \tag{4.65}$$

Changing the variable of integration in equation (4.65) from $r$ to $\zeta = r/t$ gives the form,

* Third formula on page 37 of [20] with $\mu = m$, $\nu = m+\frac{1}{2}$.

$$g_m(t) = \int_0^1 K_m(t\zeta) \frac{\zeta^{m+1} d\zeta}{(1-\zeta^2)^{\frac{1}{2}}} .$$

The stress intensity factor is most conveniently determined by examining the crack opening shape near $r=1$. From equations (4.51) and (4.57) the displacement for $z=0$ is

$$w_z = \frac{1-\nu}{\mu} p_0 a \sum_{m=0}^{\infty} w_m(r) \cos m\theta . \tag{4.66}$$

For $r=1-\rho$, equation (4.59) becomes

$$w_m(1-\rho) = \frac{2}{\pi}(1-\rho)^m \int_{1-\rho}^{1} g_m(t) \frac{t^{-m+1} dt}{(t^2-r^2)^{\frac{1}{2}}}$$

$$= \frac{2}{\pi}(1-\rho)^m \int_0^{\rho} g_m(1-\eta) \frac{(1-\eta)^{-m+1} d\eta}{(\rho-\eta)^{\frac{1}{2}}(2-\rho-\eta)^{\frac{1}{2}}} .$$

As $\rho \to 0$, $\rho$ and $\eta$ may be neglected in comparison with 1, and this leads to

$$w_m(1-\rho) = \frac{2}{\pi} g_m(1) \frac{1}{2^{\frac{1}{2}}} \int_0^{\rho} \frac{d\eta}{(\rho-\eta)^{\frac{1}{2}}} + O(\rho)$$

$$= \frac{2}{\pi} g_m(1)\,(2\rho)^{\frac{1}{2}} + O(\rho) .$$

Therefore, the equation (4.66) for the crack opening near the crack front becomes

$$w_z = \frac{2}{\pi} \frac{2(1-\nu^2)}{E} p_0(a^{\frac{1}{2}})(2a\rho)^{\frac{1}{2}} \sum_{m=0}^{\infty} g_m(1) \cos m\theta + O(\rho) . \tag{4.67}$$

The relationship between displacement and stress intensity factor (See equation (5.37) of [21]) is the same as in equation (4.16) for plane strain. In equation (4.16), note that $r$ is not dimensionless. Hence it is identified with $a\rho$ in equation (4.67). Thus, comparing equations (4.67) and (4.16),

$$k_p = \frac{2}{\pi} p_0(a^{\frac{1}{2}}) \sum_{m=0}^{\infty} g_m(1) \cos m\theta . \tag{4.68}$$

In equation (4.68), $k_p$ is the physically dimensioned stress intensity factor. The subscript $p$ is used to distinguish it from the dimensionless stress intensity factor

$$k(\theta) = \sum_{m=0}^{\infty} g_m(1)\cos m\theta = k_p \Big/ \left[\frac{2}{\pi}\, p_0(a^{\frac{1}{2}})\right] \tag{4.69}$$

which will be used in the numerical calculations.

The stress intensity factor may also be determined by calculating the singular part of the stress distribution. If equation (4.60) is integrated by parts,

$$D_m(\xi) = \frac{2}{\pi}\, g_m(1) j_{m+1}(\xi) - \frac{2}{\pi}\int_0^1 t^{m+2} j_{m+1}(\xi t)\,\frac{\mathrm{d}}{\mathrm{d}t}\,[t^{-m} g_m(t)]\,\mathrm{d}t$$

only the first term contributes to the stress singularities. The spherical Bessel functions may be written as the product of polynomials in $\xi^{-1}$ and trigonometric functions as e.g.

$$j_0(\xi) = \frac{\sin\xi}{\xi}, \quad j_1(\xi) = \frac{\sin\xi}{\xi^2} - \frac{\cos\xi}{\xi}.$$

The resulting integrals of the type

$$\int_0^{\infty} \xi^n \mathrm{e}^{-\xi z} J_m(\xi r)\sin\xi\,\mathrm{d}\xi$$

obtained from equations (4.52) may be evaluated from formulas based on the third one on page 33 of [20]. By differentiating and separating the real and imaginary parts of the formula referred to, results of the type

$$\int_0^{\infty} \mathrm{e}^{-\xi z} J_m(\xi r)\sin\xi\,\mathrm{d}\xi = \frac{r^{-m}}{(2\rho)^{\frac{1}{2}}}\sin\left(\frac{\phi}{2} + \frac{m\pi}{2}\right) + O(1)$$

are found. After these results are used, and the nonsingular terms discarded, the stress state may be written in the form

$$\sigma_r = \frac{2}{\pi}\, p_0 \frac{k(\theta)}{(2\rho)^{\frac{1}{2}}}\left[\cos\frac{\phi}{2} - \tfrac{1}{2}\sin\phi\,\sin\frac{3\phi}{2}\right] + O(1)$$

$$\sigma_\theta = 2\nu\,\frac{2}{\pi}\, p_0 \frac{k(\theta)}{(2\rho)^{\frac{1}{2}}}\left[\cos\frac{\phi}{2}\right] + O(1)$$

$$\sigma_z = \frac{2}{\pi}\, p_0 \frac{k(\theta)}{(2\rho)^{\frac{1}{2}}}\left[\cos\frac{\phi}{2} + \tfrac{1}{2}\sin\phi\,\sin\frac{3\phi}{2}\right] + O(1)$$

$$\tau_{rz} = -\frac{2}{\pi}\, p_0 \frac{k(\theta)}{(2\rho)^{\frac{1}{2}}}\left[\tfrac{1}{2}\sin\phi\,\cos\frac{3\phi}{2}\right] + O(1). \tag{4.70}$$

In equations (4.70), $\rho$ and $\phi$ are the dimensionless polar coordinates shown in Figure 4.7. The stresses $\tau_{\theta z}$ and $\tau_{r\theta}$, are nonsingular. It may be verified that the singular terms satisfy the condition of plane strain,

$$\sigma_\theta = \nu(\sigma_r + \sigma_z)$$

locally.

The present iterative approach requires the calculation of the stresses on the $yz$-plane. In general each of the stresses, $\sigma_x$, $\tau_{xy}$, $\tau_{xz}$, is not zero, but for the case in which the pressure on the crack is symmetric about the $yz$-plane,

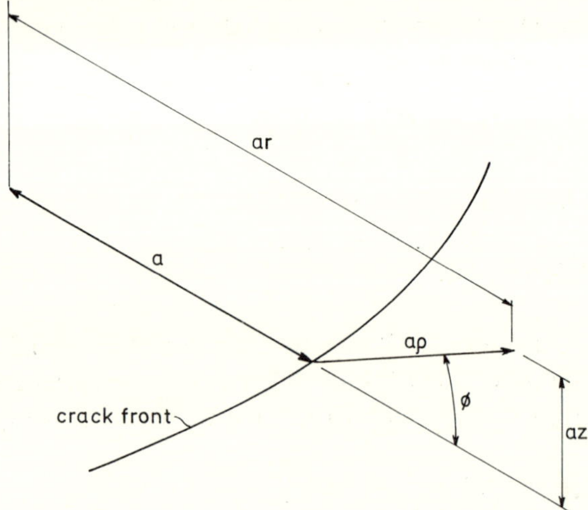


Figure 4.7. Local coordinates $(\rho, \phi)$ at crack front.

only the normal stress is not zero. It may be evaluated from equations (4.52) by

$$\sigma_x(0, y, z) = \sigma_\theta(y, \pi/2, z) .$$

It is helpful in the numerical analysis to separate the singular and non-singular parts of this stress as

$$\sigma_x = 2\nu \frac{2}{\pi} p_0 k\left(\frac{\pi}{2}\right) \left[(2\rho_1)^{-\frac{1}{2}} \cos\frac{\phi_1}{2} + (2\rho_2)^{-\frac{1}{2}} \cos\frac{\phi_2}{2}\right] + p_0 q(y, z) \quad (4.71)$$

where $(\rho_1, \phi_1)$ and $(\rho_2, \phi_2)$ are coordinates in the $yz$-plane as shown in Figure 4.9. The nonsingular function $q(y, z)$ is computed from the preceding two equations as long as the point $(0, y, z)$ is not close to the crack front. When $(0, y, z)$ nears one of the points $(0, \pm a, 0)$, the singular parts of equations (4.71) and (4.72) are equal, and $q(y, z)$ may be written in a form in which the singularities are analytically cancelled. From equations (4.52) and (4.71),

$$q(y,z) = -\frac{4v}{\pi} k\left(\frac{\pi}{2}\right)\left[(2\rho_1)^{-\frac{1}{2}}\cos\frac{\phi_1}{2} + (2\rho_2)^{-\frac{1}{2}}\cos\frac{\phi_2}{2}\right]$$
$$+ \sum_{m=0}^{\infty}(-1)^m\left\{-2v\int_0^{\infty}\xi D_{2m}(\xi)\mathrm{e}^{-\xi z}J_{2m}(\xi y)\mathrm{d}\xi\right.$$
$$-\frac{1}{y}\int_0^{\infty}(1-2v-\xi z)D_{2m}(\xi)\mathrm{e}^{-\xi z}\left[\frac{2m(2m-1)}{\xi y}J_{2m}(\xi y)\right.$$
$$\left.\left. + J_{2m+1}(\xi y)\right]\mathrm{d}\xi\right\}. \quad (4.72)$$

Use has been made of the fact that for symmetry of loading about the $yz$-plane, all $K$, $g$, $D$ with odd subscripts vanish.

To illustrate the lack of singularity in $q(y, z)$, the form valid for $z=0$ and $y \leqslant 1$ will be given. A similar expression can be written for $y \geqslant 1$, and at $y=1$, both give the same value of $q(y, z)$. The Fourier coefficients of $p(r, \theta)$ can be written as polynomials,

$$K_0(r) = p_c + \sum_{n=1}^{N} A_{1n}r^n$$
$$K_{2m}(r) = \sum_{n=1}^{N} A_{m+1,n}r^n, \qquad m \geqslant 1 \quad (4.73)$$

where $p_c = K_0(0) = p(0, \theta)$ is the dimensionless pressure at the center of the crack. For $K_{2m}(r)$ in this form, equation (4.65) gives

$$g_0(t) = p_c + \sum_{n=1}^{N} C_{1n}t^n$$
$$g_{2m}(t) = \sum_{n=1}^{N} C_{m+1,n}t^n, \qquad m \geqslant 1 \quad (4.74)$$

where

$$C_{mn} = I_{2m+n-1}A_{mn} \quad (4.75)$$

and

$$I_0 = \frac{\pi}{2}, \quad I_1 = 1, \quad I_k = \frac{k-1}{k} I_{k-2}. \tag{4.76}$$

The part of $D_0(\xi)$ resulting from the constant part of $g_0(t)$ may be evaluated as

$$\frac{2}{\pi} \xi \int_0^1 t^2 p_c j_0(\xi t) \mathrm{d}t = p_c j_1(\xi)$$

and the integrals of this part of $D_0(\xi)$ in equation (4.72) can be determined in closed form. The remaining part of $D_0(\xi)$ and the other $D_{2m}(\xi)$ can be found in terms of spherical Bessel functions and the sine integral function. In general, the remaining integrals in equation (4.72) are evaluated numerically, but when $z=0$ they can be explicitly obtained.

For $z=0$ and $y \leqslant 1$, the integrations in equation (4.72) result in

$$q(y, 0) = \frac{2}{\pi} p_c F_0(y, 0) + \frac{2}{\pi} \sum_{m=1}^{\infty} (-1)^m \sum_{n=1}^{N} I_{2m+n} C_{mn} y^n [1-2\nu + 2\nu(2m+n) + (1-2\nu)(2m-2)(2m-3) F_{mn}(y)] \tag{4.77}$$

where

$$F_{mn}(y) = \begin{cases} \dfrac{1 - \dfrac{I_{4m-4}}{I_{2m+n}} y^{2m-4+n}}{2m-4-n}, & n \neq 2m-4 \\[2ex] \ln \dfrac{y}{2} + \dfrac{1}{1 \cdot 2} + \dfrac{1}{3 \cdot 4} + \ldots + \dfrac{1}{(4m-5)(4m-4)}, & n = 2m-4 \end{cases}$$

A similar but more complicated result can be found for $y \geqslant 1$. Using the polar coordinates shown in Figure 4.9, the other function in equation (4.77) may be written

$$\begin{aligned} F_0(y, z) = 2\nu \Bigg[ & \frac{1}{(\rho_1 \rho_2)^{\frac{1}{2}}} \sin \frac{\phi_1 + \phi_2}{2} - (2\rho_1)^{-\frac{1}{2}} \cos \frac{\phi_1}{2} - (2\rho_2)^{-\frac{1}{2}} \cos \frac{\phi_2}{2} \Bigg] \\ & - \frac{1+2\nu}{2} \arcsin \left( \frac{2}{\rho_1 + \rho_2} \right) + \frac{1-2\nu}{2y^2} (\rho_1 \rho_2)^{\frac{1}{2}} \\ & \times \left[ \sin \frac{\phi_1 + \phi_2}{2} + z \cos \frac{\phi_1 + \phi_2}{2} \right] + \frac{z}{y^2 (\rho_1 \rho_2)^{\frac{1}{2}}} \\ & \times \left[ z \sin \frac{\phi_1 + \phi_2}{2} + (\rho_1 \rho_2 - 1) \cos \frac{\phi_1 + \phi_2}{2} \right]. \end{aligned} \tag{4.78}$$

For $z=0$, equation (4.78) reduces to

$$F_0(y, 0) = -\frac{\nu[2(y-1)]^{\frac{1}{2}}}{[2(y+1)]^{\frac{1}{2}}+y+1} - \frac{1+2\nu}{2}\arcsin\frac{1}{y} + \frac{1-2\nu}{2y}[y^2-1]^{\frac{1}{2}} \tag{4.79}$$

for $y \geqslant 1$. For $y<1$, $F_0(y, 0)$ is equal to the constant obtained from equation (4.79) by putting $y=1$.

Reiterating the important results, if $p(r, \theta)$ is symmetric about both the $xz$- and the $yz$-planes, then it can be written

$$p(r, \theta) = -\frac{1}{p_0}\sigma_z(x, y, 0) = \sum_{m=0}^{\infty} K_{2m}(r)\cos 2m\theta\,. \tag{4.80}$$

It follows that

$$g_{2m}(t) = \int_0^1 \zeta^{2m+1}\frac{K_{2m}(\zeta t)\mathrm{d}t}{(1-\zeta^2)^{\frac{1}{2}}} \tag{4.81}$$

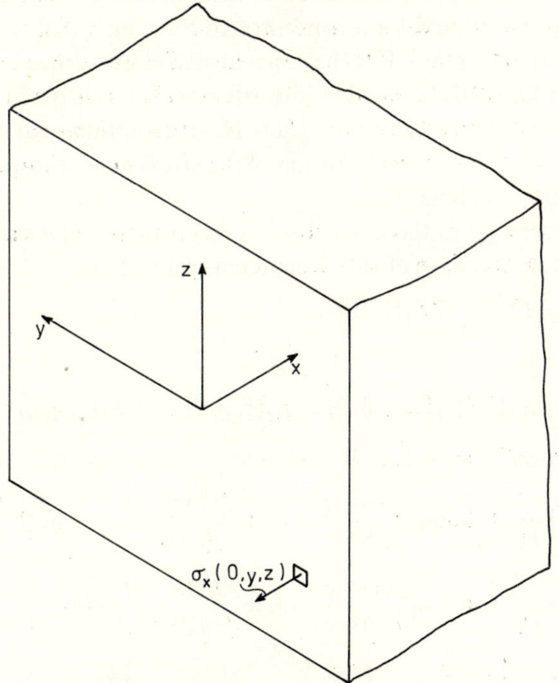


Figure 4.8. Half space ($x \geqslant 0$) loaded on surface ($x=0$).

can be computed to yield the stress intensity factor

$$k_p(\theta) = \frac{2}{\pi} p_0(a^{\frac{1}{2}}) \sum_{m=0}^{\infty} g_{2m}(1)\cos 2m\theta \tag{4.82}$$

and

$$D_{2m}(\xi) = \frac{2}{\pi}\xi \int_0^1 t^2 g_{2m}(t) j_{2m}(\xi t)\mathrm{d}t\,. \tag{4.83}$$

Once $D_{2m}(\xi)$ is computed, any of the stresses and displacements may be determined. In particular $\sigma_x(0, y, z)$ of equation (4.71) is obtained by calculating the function $q(y, z)$ from eq. (4.72).

*Surface loads on half space.* Consider the case of normal stresses $\sigma_x$ acting on the surface occupying the $yz$-plane as shown in Figure 4.8. For the case

$$\sigma_x(0, y, z) = p_0 q(y, z) \tag{4.84}$$

in which there is no stress singularity, there are a number of possible methods of solution. For example, the solution at the beginning of Section 4.3 could be applied or the solution for a concentrated force on a half space could be used as a Green's function. For the convenience of numerical calculation as in [6], the surface will be divided into squares sufficiently small in size so that the stress is nearly uniform on each. Next a solution provided by Love [25] is used to obtain the contribution of the stress $\sigma_x$ on each square to the stress $\sigma_z$ on the $xy$ plane.

Love's solution gives the normal stress $\sigma_z$ on the $xy$-plane due to a unit tensile stress on a square of side $b$ centered at $(0, \bar{y}, \bar{z})$ as

$$\sigma_z(x, y, 0) = F(|\bar{y}-y|\,,\, |\bar{z}|\,,\, |x|)$$

where

$$2\pi F(\xi, \eta, \zeta) = (1-2\nu)F_1(\xi, \eta, \zeta) - \zeta F_2(\xi, \eta, \zeta) - 2\nu F_3(\xi, \eta, \zeta) \tag{4.85}$$

and the functions $F_j$ $(j=1, 2, 3)$ stand for

$$F_1 = \tan^{-1}\frac{b-\xi}{b-\eta} + \tan^{-1}\frac{b+\xi}{b-\eta} - \tan^{-1}\frac{\zeta(b-\xi)}{a_1(b-\eta)} - \tan^{-1}\frac{\zeta(b+\xi)}{b_2(b-\eta)}$$

$$+ \tan^{-1}\frac{b-\xi}{b+\eta} + \tan^{-1}\frac{b+\xi}{b+\eta} - \tan^{-1}\frac{\zeta(b-\xi)}{d_4(b+\eta)} - \tan^{-1}\frac{\zeta(b+\xi)}{c_3(b+\eta)}$$

$$F_2 = \frac{b-\eta}{(b-\eta)^2+\zeta^2}\left[\frac{b-\xi}{a_1} + \frac{b+\xi}{b_2}\right] + \frac{b+\eta}{(b+\eta)^2+\zeta^2}\left[\frac{b-\xi}{d_4} + \frac{b+\xi}{c_3}\right]$$

$$F_3 = \tan^{-1}\frac{\zeta a_1}{(b-\xi)(b-\eta)} + \tan^{-1}\frac{\zeta b_2}{(b+\xi)(b-\eta)}$$

$$+ \tan^{-1}\frac{\zeta c_3}{(b+\xi)(b+\eta)} + \tan^{-1}\frac{\zeta d_4}{(b-\xi)(b+\eta)} - \begin{cases} 2\pi, & |\xi|<b,\ |\eta|<b \\ 0, & \text{otherwise} \end{cases} \tag{4.86}$$

with the following contractions

$$a_1^2 = (b-\xi)^2 + (b-\eta)^2 + \zeta^2$$
$$b_2^2 = (b+\xi)^2 + (b-\eta)^2 + \zeta^2$$
$$c_3^2 = (b+\xi)^2 + (b+\eta)^2 + \zeta^2$$
$$d_4^2 = (b-\xi)^2 + (b+\eta)^2 + \zeta^2 .$$

By taking advantage of symmetry, attention may be confined to the centers $(0, \bar{y}_i, \bar{z}_j)$ of the squares which lie in the first quadrant of the $yz$-plane. The effect of the other three symmetrically located quadrants is included in the result

$$\sigma_z(x, y, 0) = p_0 t(x, y)$$

where

$$t(x, y) = 2\sum_{i=1}^{N}\sum_{j=1}^{N} q(\bar{y}_i, \bar{z}_j)\left[F(|\bar{y}_i - y|, |\bar{z}_j|, |x|)\right.$$
$$\left. + F(|\bar{y}_i + y|, |\bar{z}_j|, |x|)\right] . \tag{4.87}$$

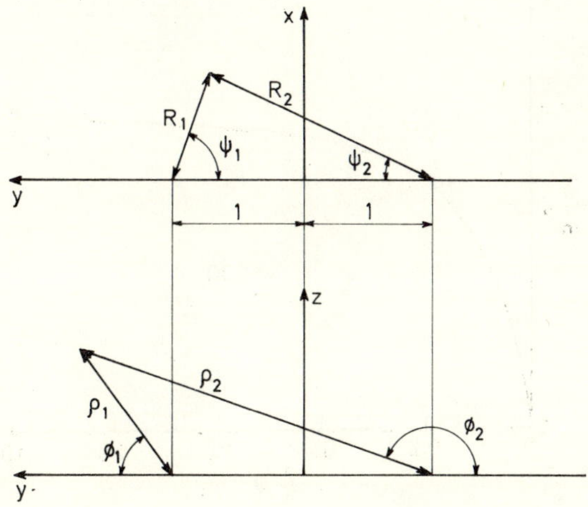


Figure 4.9. Auxiliary coordinates for singular solutions.

The symmetry of the stress on the $yz$-plane insures that there will be no shear stresses on the $xy$-plane.

When the stress on the $yz$-plane is singular, it is desirable to seek a closed form solution. For the case of interest, from equation (4.71),

$$\sigma_x(0, y, z) = \frac{4v}{\pi} p_0 k \left(\frac{\pi}{2}\right) \left[(2\rho_1)^{-\frac{1}{2}} \cos \frac{\phi_1}{2} + (2\rho_2)^{-\frac{1}{2}} \cos \frac{\phi_2}{2}\right] \tag{4.88}$$

such a solution can be found by applying Muki's half-space solution, equations (4.51) and (4.52), to each term. The result is

$$\sigma_z(x, y, 0) = \frac{4v}{\pi} p_0 k \left(\frac{\pi}{2}\right) [(2R_1)^{-\frac{1}{2}} S(\psi_1) + (2R_2)^{-\frac{1}{2}} S(\psi_2)] \tag{4.89}$$

where the dimensionless coordinates, $\rho$, $\phi$, $R$, $\psi$ are shown in Figure 4.9. The function $S(\psi)$ is in the form of an infinite series involving hypergeometric functions. Rather than give the formulas, the results are shown graphically in Figure 4.10. An excellent fit of the exactly computed points of the curve is obtained by the approximation,

$$\begin{aligned} S(\psi) = \psi^{\frac{1}{2}} [&0.344 + 1.851\psi - 11.637\psi^2 + 34.545\psi^3 \\ &- 61.286\psi^4 + 68.124\psi^5 - 47.644\psi^6 + 20.321\psi^7 \\ &- 4.823\psi^8 + 0.488\psi^9] . \end{aligned} \tag{4.90}$$

This expression is valid only for $0 \leqslant \psi \leqslant \pi/2$ and for Poisson's ratio, $v = 0.30$.

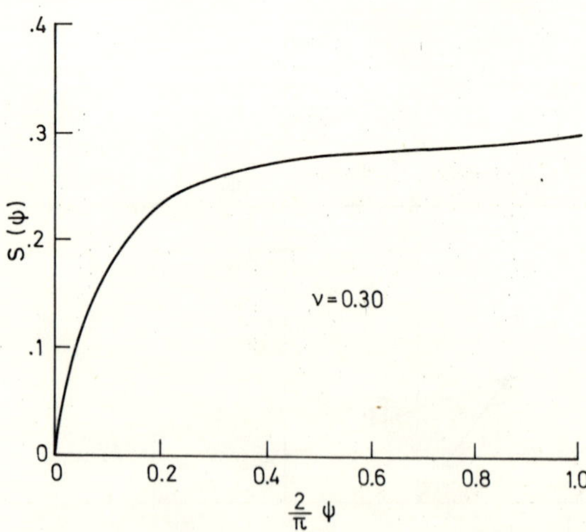


Figure 4.10. Function for equation (4.89) for singular normal stress on half-space.

It should be noted that there are singularities in $\sigma_z(x, y, 0)$ at the points where the crack front intersects the free surface. The numerical difficulties due to the singularity will be discussed in more detail subsequently.

*Iterative formulation.* The solutions of the two preceding problems will be superposed alternately to solve the case of a uniformly pressurized half penny-shaped crack perpendicular to a free surface (Figure 4.11). The proce-

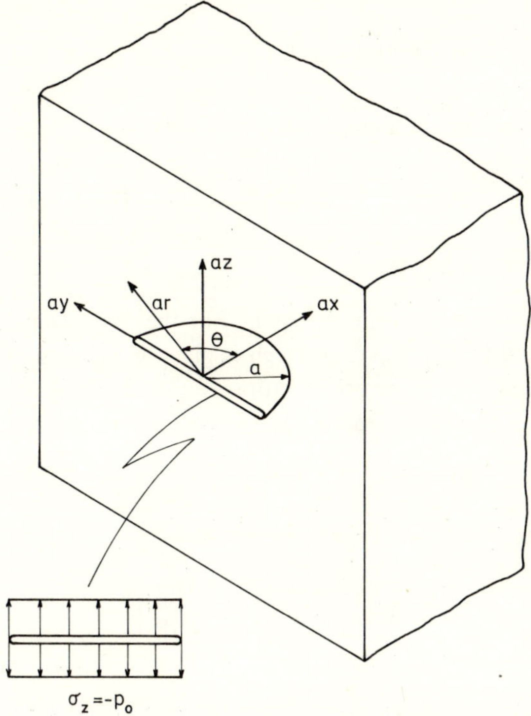


Figure 4.11. Half-penny surface crack.

dure begins with the well known solution [26] for uniform pressure of a penny-shaped crack in an infinite medium. This solution gives stresses on the *yz*-plane which are removed by superposing the half space solution for equal and opposite stresses on the *yz*-plane. As a result of the half space solution, there is a residual normal stress on the crack in addition to the uniform pressure. This is eliminated by superposing the solution for pressure on the crack equal to the residual stress. Again, stresses on the *yz*-plane are produced which are removed by application of the half space solution. The cycle of superpositions is continued until the residual stress on the crack is negligible compared to the uniform pressure.

As outlined, the first step in the procedure requires the solution of the penny-shaped crack problem for the case

$$\sigma_{zA}^{(0)}(x, y, 0) = -p_0 . \tag{4.91}$$

The crack solution above may be used to find the stress intensity factor,

$$k_p^{(0)} = \frac{2}{\pi} p_0(a^{\frac{1}{2}}) \quad \text{or} \quad k^{(0)}(\theta) = 1 \tag{4.92}$$

and the stress on the $yz$-plane,

$$\sigma_{xA}^{(0)}(0, y, z) = \frac{4\nu}{\pi} p_0 k^{(0)}\left(\frac{\pi}{2}\right)\left[(2\rho_1)^{-\frac{1}{2}}\cos\tfrac{1}{2}\phi_1 + (2\rho_2)^{-\frac{1}{2}}\cos\tfrac{1}{2}\phi_2\right] + p_0 q^{(0)}(y, z) . \tag{4.93}$$

See Figure 4.9 for the coordinates, $\rho$, $\phi$. For the case of uniform pressure, $q^{(0)}(y, z)$ is equal to $(2/\pi)F_0(y, z)$ which is given in equation (4.78).

The second half of the zeroth iteration is used to eliminate the stress on the free surface. The stress applied to the surface is

$$\sigma_{xB}^{(0)}(0, y, z) = \sigma_{xA}^{(0)}(0, y, z) \tag{4.94}$$

which gives the result (See Figure 4.9 for $R$, $\psi$).

$$\sigma_{zB}^{(0)}(x, y, 0) = \frac{4\nu}{\pi} p_0 k^{(0)}\left(\frac{\pi}{2}\right)\left[\frac{S(\psi_1)}{(2R_1)^{\frac{1}{2}}} + \frac{S(\psi_2)}{(2R_2)^{\frac{1}{2}}}\right] + p_0 t^{(1)}(x, y) \tag{4.95}$$

where $t^{(1)}(x, y)$ is determined by applying the method described earlier. It depends only on $q^{(0)}$, and not on the singular part of $\sigma_{xA}^{(0)}$.

After the second half of this iteration is subtracted from the first half, the surface will be free of stress, but the stress on the crack will be

$$\sigma_z(x, y, 0) = -p_0 - \sigma_{zB}^{(1)}(x, y, 0) \tag{4.96}$$

rather than constant. An additional iteration is used to eliminate the stress $\sigma_{zB}^{(0)}$.

For the first iteration, the stress on the penny-shaped crack is taken to be

$$\sigma_{zA}^{(1)}(x, y, 0) = \sigma_{zB}^{(0)}(x, y, 0)$$

$$= \frac{4v}{\pi} p_0 k^{(0)} \left(\frac{\pi}{2}\right) \left[\frac{S(\psi_1)}{(2R_1)^{\frac{1}{2}}} + \frac{S(\psi_2)}{(2R_2)^{\frac{1}{2}}}\right] + p_0 t^{(1)}(x, y) . \qquad (4.97)$$

Because of the singularity in the first term of equation (4.97), care must be taken in the numerical procedure. The description of the numerical method will be given in the following section. The contribution of the two parts of $\sigma_{zA}^{(1)}$ are computed separately and the desired quantities expressed as

$$k^{(1)}(\theta) = \frac{4v}{\pi} k^{(0)} \left(\frac{\pi}{2}\right) k_S(\theta) + k_N^{(1)}(\theta) \qquad (4.98)$$

$$q^{(1)}(y, z) = \frac{4v}{\pi} k^{(0)} \left(\frac{\pi}{2}\right) q_S(y, z) + q_N^{(1)}(y, z) \qquad (4.99)$$

$$\sigma_{xA}^{(1)}(0, y, z) = \frac{4v}{\pi} p_0 k^{(1)} \left(\frac{\pi}{2}\right) [(2\rho_1)^{-\frac{1}{2}} \cos \tfrac{1}{2}\phi_1 + (2\rho_2)^{-\frac{1}{2}} \cos \tfrac{1}{2}\phi_2]$$

$$+ p_0 q^{(1)}(y, z) \qquad (4.100)$$

where $k_S(\theta)$ and $p_0 q_S(y, z)$ are the dimensionless stress intensity factor and nonsingular part of $\sigma_x(0, y, z)$, respectively, due to stress on the crack of

$$\sigma_z(x, y, 0) = p_0 \left[\frac{S(\psi_1)}{(2R_1)^{\frac{1}{2}}} + \frac{S(\psi_2)}{(2R_2)^{\frac{1}{2}}}\right] . \qquad (4.101)$$

Similarly, $k_N^{(1)}(\theta)$ and $p_0 q_N^{(1)}(y, z)$ stem from the stress $p_0 t^{(1)}(x, y)$ on the crack.

The effects of these two terms are kept separate into the second half of the iteration. The stress on the half space is taken to be

$$\sigma_{xB}^{(1)}(0, y, z) = \sigma_{xA}^{(1)}(0, y, z) .$$

This results in the residual stress on the crack,

$$\sigma_{zB}^{(1)}(x, y, 0) = \frac{4v}{\pi} p_0 k^{(1)} \left(\frac{\pi}{2}\right) \left[\frac{S(\psi_1)}{(2R_1)^{\frac{1}{2}}} + \frac{S(\psi_2)}{(2R_2)^{\frac{1}{2}}}\right] + p_0 t^{(2)}(x, y) \qquad (4.102)$$

where

$$t^{(2)}(x, y) = \frac{4v}{\pi} k^{(0)} \left(\frac{\pi}{2}\right) t_S(x, y) + t_N^{(2)}(x, y) . \qquad (4.103)$$

In a way similar to the definitions of $q_S$ and $q_N^{(1)}$, $t_S$ and $t_N^{(2)}$ are defined to be the stress $\sigma_z(x, y, 0)$ due to the stress $\sigma_x(0, y, z)$ equal to $q_S$ and $q_N^{(1)}$, respectively.

Again, the second half of the iteration when subtracted from the first half frees the surface of stresses. Adding the first iteration to the zeroth cancells the residual stress $\sigma_{z\mathrm{B}}^{(0)}$ and gives

$$\sigma_z(x, y, 0) = -p_0 - \sigma_{z\mathrm{B}}^{(1)}(x, y, 0)\,.$$

Further iterations are added until the stress on the crack is reduced to the applied stress $-p_0$ to within some arbitrary tolerance. The additional computations which must be made for each iteration are straightforward. For the second iteration, the dimensionless intensity factor, $k_N^{(2)}(\theta)$, and stress $\sigma_x(0, y, z) = p_0 q_N^{(2)}(y, z)$ due to stress on the crack of $\sigma_z(x, y, 0) = p_0 t^{(2)}(x, y)$ are obtained. Then the half space solution gives $\sigma_z(x, y, 0) = p_0 t_N^{(3)}(x, y)$ due to $\sigma_x(0, y, z) = p_0 q_N^{(2)}(y, z)$. The contribution of this iteration to the dimensionless stress intensity factor is

$$k^{(2)}(\theta) = \frac{4\nu}{\pi}\, k^{(1)}\left(\frac{\pi}{2}\right) k_S(\theta) + k_N^{(2)}(\theta)$$

and the stress on the crack is

$$\sigma_z(x, y, 0) = -p_0 - \sigma_{z\mathrm{B}}^{(2)}(x, y, 0)$$

where

$$\sigma_{z\mathrm{B}}^{(2)} = \frac{4\nu}{\pi}\, p_0 k^{(2)}\left(\frac{\pi}{2}\right)\left[\frac{S(\psi_1)}{(2R_1)^{\frac{1}{2}}} + \frac{S(\psi_2)}{(2R_2)^{\frac{1}{2}}}\right] + p_0 t^{(3)}(x, y)$$

$$t^{(3)}(x, y) = \frac{4\nu}{\pi}\, k^{(1)}\left(\frac{\pi}{2}\right) t_S(x, y) + t_N^{(3)}(x, y)\,.$$

For the $n$th iteration, the functions, $k_S(\theta)$, $q_S(y, z)$, and $t_S(x, y)$, are used again in conjunction with the results of the previous iteration, $k^{(n-1)}(\theta)$ and $t^{(n)}(x, y)$. The current technique gives the new functions, $k_N^{(n)}(\theta)$ and $q_N^{(n)}(y, z)$, which are, respectively, the dimensionless stress intensity factor and the nonsingular part of $\sigma_x(0, y, z)$ produced by application of the stress $\sigma_z(x, y, 0) = t^{(n)}(x, y)$ to the crack. The half space solution of this section gives $t_N^{(n+1)}(x, y)$, the stress $\sigma_z(x, y, 0)$ produced by $\sigma_x(0, y, z) = q_N^{(n)}(y, z)$ acting on the surface. The new addition to the dimensionless stress intensity factor is

$$k^{(n)}(\theta) = \frac{4\nu}{\pi}\, k^{(n-1)}\left(\frac{\pi}{2}\right) k_S(\theta) + k_N^{(n)}(\theta)$$

and the residual stress on the crack is

$$\sigma_{zB}^{(n)}(x, y, 0) = \frac{4\nu}{\pi} p_0 k^{(n)} \left(\frac{\pi}{2}\right) \left[\frac{S(\psi_1)}{(2R_1)^{\frac{1}{2}}} + \frac{S(\psi_2)}{(2R_2)^{\frac{1}{2}}}\right] + p_0 t^{(n+1)}(x, y)$$

where the term carried into the next iteration, if necessary, is

$$t^{(n+1)}(x, y) = \frac{4\nu}{\pi} k^{(n-1)} \left(\frac{\pi}{2}\right) t_S(x, y) + t_N^{(n+1)}(x, y) .$$

The alternating method applied in this Section is modified by the separate treatment of the singular part of the stress (equation (4.93)) applied to the half space. The solution for the half space subjected to the singular stress is given by the first part of equation (4.95). The resulting stress which must be removed from the crack, equation (4.97), has two parts which are kept separate. The singular part is the same in each iteration except for the coefficient involved. Therefore, the numerical solution for a penny-shaped crack subjected to the singular stress (tensile) of equation (4.101) is determined to as high a degree of accuracy as possible. The contribution of this solution to each iteration is contained in the functions with a subscript, $S$. The remaining contribution, from the rest of equation (4.97) gives the terms denoted by the subscript, $N$. The parts are combined as in equations (4.98) to (4.100) to give the total solution of the penny-shaped crack problem for this iteration. The solution is treated by the same method in each successive iteration.

The next section contains a description of the numerical details and the difficulties associated with the stress singularity of equation (4.97).

*Numerical treatment of singularities.* For the case of uniform pressure on the crack, the first step requiring numerical analysis is the clearing of the stress, equation (4.93) from the surface in the second half of the zeroth iteration. For convenience the problem is a half space with tensile stresses,

$$\sigma_x(0, y, z) = p_0 q(y, z)$$

on the surface. The function $q(y, z)$ can be any of the nonsingular functions of the first line of Table II. The solution for the singular part of the stress $\sigma_x(0, y, z)$ is given analytically by equations (4.88) to (4.90).

The surface of the half space is divided into a number of regions as shown in Figure 4.12. Each region is then subdivided into a number of squares, larger squares being used where the stress becomes small and less rapidly varying. The values of $q^{(0)}(y, z)$ at the centers of each of the squares are

TABLE II

*Results of half space solution—nonsingular stress*

| | | | $n \geqslant 1$ |
|---|---|---|---|
| surface stress $\sigma_x(0, y, z)/p_0 =$ | $q^{(0)}(y, z)$ | $q_S(y, z)$ | $q_N^{(n)}(y, z)$ |
| results in $\sigma_z(x, y, 0)/p_0 =$ | $t^{(1)}(x, y)$ | $t_S(x, y)$ | $t_N^{(n+1)}(x, y)$ |

computed from the closed form expression, equation (4.78). Because of the lengthy calculations needed to obtain the values of the other $q$'s, they were computed at fewer points and the six point bivariate interpolation formula was used to get the values at the remaining points. The numbers in parentheses in Figure 4.12 give the number of values of $y$ and the number of values of

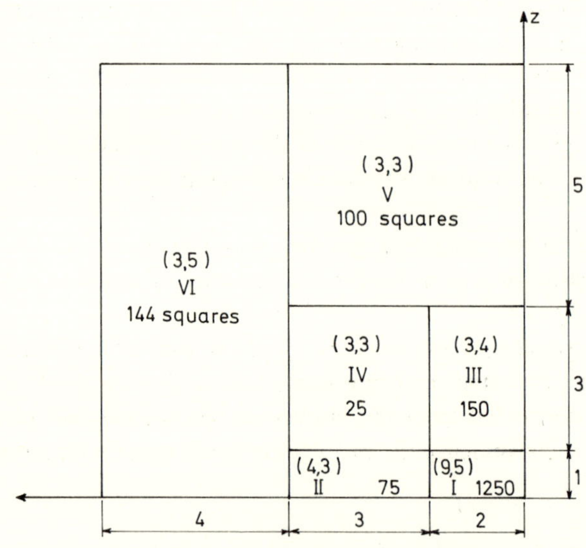


Figure 4.12. Grid on surface of half-space.

$z$, respectively, used to obtain the grid points at which $q$ was calculated exactly. After the values of $q(y, z)$ have been computed at the center of each square, the solution for the nonsingular stresses on a half space in this section is used directly to obtain the values of the stress on the $xy$-plane denoted by

$$\sigma_z(x, y, 0) = p_0 t(x, y) \tag{4.104}$$

where the $t$'s corresponding to the various $q$'s are shown in Table II.

The values of $t(x, y)$ are computed at the 36 points $(0.2m, 0.2n)$, $m, n = 0, 1, \ldots, 5$. The six point bivariate interpolation formula is used to obtain intermediate values of $t(x, y)$. This completes the description of the half space problem's numerical solution.

The solution of the penny-shaped crack problem is needed for stresses on the crack which contain no singularities.

$$\sigma_z(x, y, 0) = p_0 t(x, y) .$$

The function $t(x, y)$ may represent any of the $t$'s in Table III. In a case of non-uniform pressure on the crack, it could also represent the initial stress on the crack, $t^{(0)}(x, y)$. (In this example, $t^{(0)}(x, y) = -1$). As indicated above, $t(x, y)$ is known at 36 points and interpolation gives its value at any other point on the crack.

The solution for the case of a penny-shaped crack requires first the computation of the Fourier coefficients of

$$p(r, \theta) = -t(x, y)$$

which are

$$\begin{aligned} K_0(r) &= \frac{2}{\pi}\int_0^{\frac{1}{2}\pi} p(r, \theta)\, d\theta \\ K_{2m}(r) &= \frac{4}{\pi}\int_0^{\frac{1}{2}\pi} p(r, \theta)\cos 2m\theta\, d\theta\,, \qquad m \geqslant 1 \\ K_{2m+1}(r) &= 0\,, \qquad m \geqslant 0\,. \end{aligned} \tag{4.105}$$

The first 40 $K$'s are each computed for 30 values of $r$ using a technique given by Filon [27]. His technique gives as good accuracy with 100 divisions of the range of integration as the usual Simpson's rule does with 500.

The coefficients of a five term polynomial approximation to each $K$ are determined by the method of least squares. This gives

$$\begin{aligned} K_0(r) &= p_c + \sum_{n=1}^{5} A_{1n} r^n \\ K_{2m}(r) &= \sum_{n=1}^{5} A_{m+1,n} r^n\,, \qquad m \geqslant 1 \end{aligned} \tag{4.106}$$

where $A_{mn}$, $m = 1, 2, \ldots, 40$; $n = 1, 2, \ldots, 5$ are known constants and $p_c = p(0, \theta)$.

The polynomial form of the $K$'s enable $g_m(t)$ to be determined also in polynomial form as in equation (4.74),

$$g_0(t) = p_c + \sum_{n=1}^{5} C_{1n} t^n$$

$$g_{2m}(t) = \sum_{n=1}^{5} C_{m+1,n} t^n , \qquad m \geqslant 1 \tag{4.107}$$

where $C_{mn}$ is given by equations (4.75) and (4.76).

The stress intensity factor may be obtained simply from equations (4.69) and (4.107). The integrals of equation (4.60) which give $D_m(\xi)$ may be evaluated in closed form in terms of spherical Bessel functions and the sine integral function. Recurrence relations between the integrals are used for the numerical calculations. The details are lengthy but straightforward. Finally, the integrals in equation (4.72) involving $D_m(\xi)$ must be evaluated to obtain the function $q(y, z)$ of equation (4.72) which gives the nonsingular part of $\sigma_x(0, x, z)/p_0$. These integrals are evaluated by Simpson's rule using intervals of length 0.25 and integrating over segments of length 5.0 until the integral over the last segment is small compared to the total integral. As pointed out in the first part of this section, $q(y, z)$ is computed for various

TABLE III

*Stress intensity factors and stresses on yz-plane for various loads on crack*

| | | | $n \geqslant 1$ |
|---|---|---|---|
| stress on crack $\sigma_z(x, y, 0)/p_0 =$ | eq. (4.101)$/p_0$ | $t^{(0)}(x, y) = -1$ | $t^{(n)}(x, y)$ |
| produces stress $\sigma_x(0, y, z)/p_0$ whose nonsingular part is = | $q_S(y, z)$ | $q^{(0)}(y, z) = \frac{2}{\pi} F_0(y, z)$ (see eq. (4.78)) | $q_N^{(n)}(y, z)$ |
| produces stress intensity factor $k_p(\theta)\Big/\frac{2}{\pi} p_0 a^{\frac{1}{2}} =$ | $k_S(\theta)$ | $k^{(0)}(\theta) = 1$ | $k_N^{(n)}(\theta)$ |

For $n \geqslant 2$, $t^{(n)}(x, y) = \frac{4\nu}{\pi} k^{(n-2)}\left(\frac{\pi}{2}\right) t_S(x, y) + t_N^{(n)}(x, y)$ (See Table II)

For $n \geqslant 1$, $k^{(n)}(\theta) = \frac{4\nu}{\pi} k^{(n-1)}\left(\frac{\pi}{2}\right) k_S(\theta) + k_N^{(n)}(\theta)$

points of the $yz$-plane. The notation used for the results of this part of the numerical analysis is summarized in Table III.

The singular stress on the crack was given in equation (4.89). For the numerical analysis consider the dimensionless form,

$$\sigma_z(x, y, 0) = \frac{S(\psi_1)}{(2R_1)^{\frac{1}{2}}} + \frac{S(\psi_2)}{(2R_2)^{\frac{1}{2}}}.$$

Note that since $\sigma_z > 0$, the singular part of the load on the crack gives a negative stress intensity factor. Further, since larger tensile stresses occur near the intersection of the crack front with the surface, the stress intensity factor should be expected to be larger in magnitude there. Thus, when combined with the other contributions to the stress intensity factor, this will cause a decrease in the total as the surface is approached.

Because of the singularity, it would be advantageous to have a closed form solution for this part of the problem. But since none could be found, it was decided to use the same numerical method given for nonsingular loads. It is evident that more terms and finer mesh sizes are required. Limitations on computer time forced the use of the same number of points in $r$ and the same number of terms in the $r$ expansion as above. More terms were included in the Fourier series, and smaller subdivisions were used in the integration. It was found that $K_m(1)$ decreased very slowly as $m$ increased. For $m = 400$, $K_m(1)$ was of the order of $10^{-2}$, $K_m(0.967)$, of $10^{-3}$, $K_m(0.933)$, of $10^{-4}$, and for $r < 0.5$, $K_m(r)$ was of the order $10^{-5}$. For $m = 1600$ these figures decreased by a factor of 10. Each time additional terms were included in the Fourier series, the magnitude of $k_S(\theta)$, the dimensionless stress intensity factor due to the singular pressure on the crack, increased near $\theta = \pi/2$ and remained the same elsewhere. Finally $k_S(\theta)$ was calculated using 841 terms in the cosine series.

The function $q_S(y, z)$ which gives the nonsingular part of the stress $\sigma_x(0, y, z)$ due to the load on the crack was computed using 241 terms in the Fourier series. This part of the numerical solution required more than half of the computer time used. But note that the same function is used again in each iteration.

In reference [6] which also treats this problem there is no separation of singular and nonsingular terms in the numerical analysis. The half space problem utilizes stresses at points within Regions I and III of Figure 4.11. Outside these regions, the stress is approximated by zero. In [6], the stress is computed at 80 points within Region I, as compared with 45 in this analysis. The additional points should not be expected to account accurately for the singularities.

To obtain the stresses in the regions mentioned above and to obtain the stress intensity factor for a given stress on the penny-shaped crack in [6], the stress on the crack is computed for five values of $r$ and 19 values of $\theta$. Weddle's rule is applied to the integrals for $K_m(r)$ for $m=0, 2, 4, \ldots, 10$. Simpson's rule using, say, 40 values of $\theta$ should be as accurate. $K_m(r)$ is

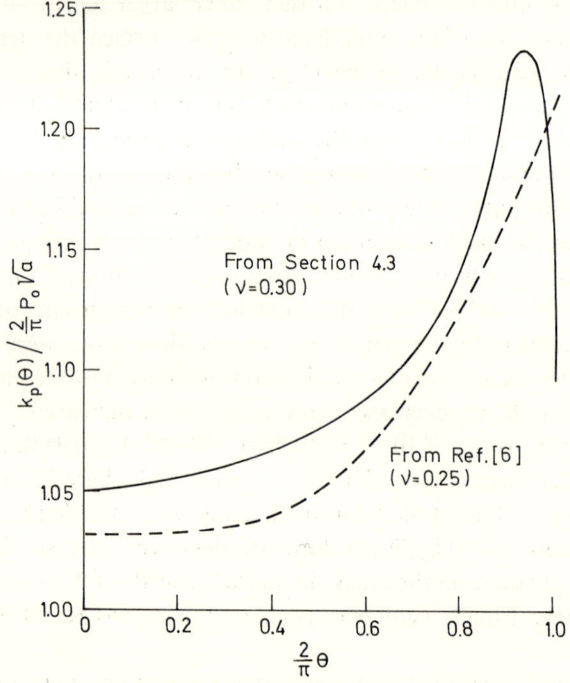


Figure 4.13. Stress intensity factor for Figure 4.11.

obtained in [6] as a fifth order polynomial in $r$ by forcing the polynomial to pass through the computed points. In this work, $K_m(r)$ is known at 30 values of $r$, and the fifth-order polynomial is obtained by least squares curve fitting.

*Discussion of numerical results.* The result of the analysis described above is the curve in Figure 4.13 which shows the variation of the stress intensity factor along the crack front. The deepest point of the crack is $\theta=0$. Initially, moving away from this point, there is a gradual increase in stress intensity factor. Then about 20° from the surface there is a rapid increase. Following this a more rapid decrease occurs in the last 3°. Note that the stress intensity factor tends to zero at the surface, a result discussed in several previous papers [22, 28, 29]. It appears that an increase in numerical accuracy would further decrease the value at the surface toward zero. However, the lack of an exact solution of the problem of a crack loaded by the stress of equation (4.101) is

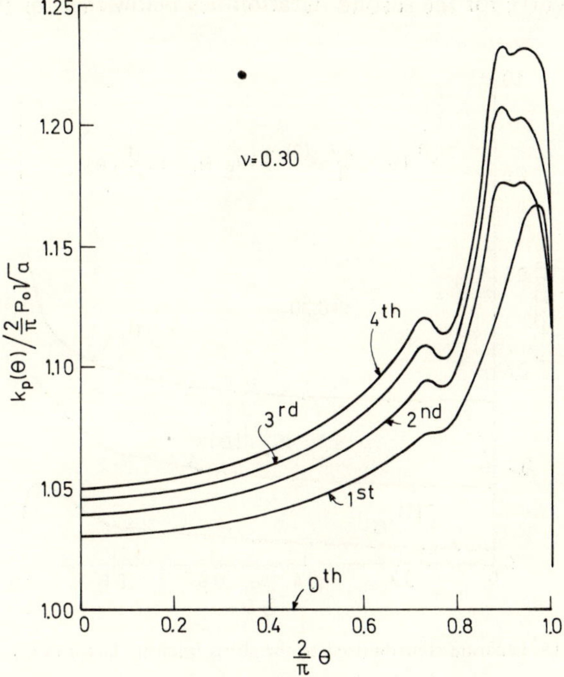


Figure 4.14. Successive iterations for stress intensity factor.

reflected by an accumulation of numerical inaccuracies as the number of iterations is increased.

It should be clear that the results in Figure 4.13 are by no means the exact solution of the problem but are presented as the best approximation available. The trend of the curves differs substantially from those obtained in [6] using basically the same numerical procedure. The results of [6], shown as a dashed line in Figure 4.13, would be expected to differ slightly from the present results because they were obtained for a Poisson's ratio of 0.25 and the present ones, 0.30. The apparent reason for the qualitative difference between the results is the differing treatments of the singularities. As background information some intermediate results will be given.

The curve of Figure 4.13 is the result after the fourth iteration. The stress intensity factors after each iteration are shown in Figure 4.14. Each new iteration raises the stress intensity factor, but the amount of increase decreases as the number of iterations increases, with one exception. Near the surface the curve for the second iteration lies below that for the first.

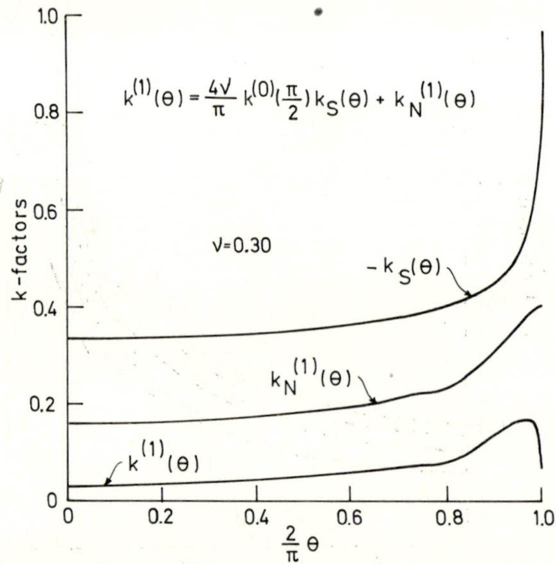


Figure 4.15. Separate contributions to the stress intensity factor in first iteration.

As described earlier, the contribution of each iteration to the stress intensity factor consists of two parts. The two contributions for the first iteration are shown in Figure 4.15. The function $k_S(\theta)$ is used with each iteration as the stress intensity factor for the singular pressure on the crack. The curve for $k_N^{(1)}(\theta)$ comes from the nonsingular part of the stress being removed from the crack by this iteration. The combination of the two designated $k^{(1)}(\theta)$ is what its iteration adds to the previous total stress intensity factor. (In this case, $k(\theta)=k^{(0)}(\theta)=1$.)

## 4.4 Future applications—semi-elliptical crack

The problem of a flat elliptical crack lying completely within a half space and perpendicular to the surface has already been attempted in [9]. The results in [9] for the case in which the major axis of the ellipse lies in the surface (Figure 4.16) are labeled as approximate in recognition of the greater accuracy required due to the intersection of the crack front with the surface. The

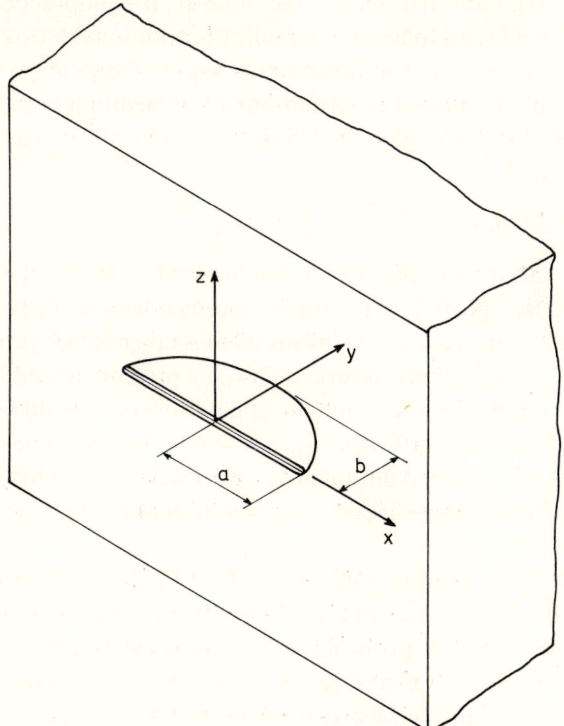


Figure 4.16. Semi-elliptical surface crack.

troublesome singularities encountered in the alternating method which were discussed in detail in Section 4.3 are present in the semi-elliptical surface crack problem. Such considerations are neglected in [9], but other difficulties remain. The major difficulty is that the solution for the elliptical crack in an infinite medium is extremely complicated.

For the penny-shaped crack problem, the pressure on the crack needs only to permit a cosine series expansion of the form of equations (4.73) and

(4.80). And the inclusion of more terms in the series is easily handled because of the nature of the solution. However, the solution for the elliptical crack requires the pressure to be given as a polynomial

$$p(x, y) = \sum_{n=0}^{N} \sum_{m=0}^{M} a_{mn} x^m y^n . \tag{4.108}$$

It is possible to work out the solution for any finite number of terms of equation (4.108), although, for a large number, it is impractical. Thus far, no way has been found to include the effect of additional terms in (4.108) by a recurrence scheme. The additional terms require the solution for the lower order terms and additional contributions combined in a complicated way.

As a result of the difficulty, the elliptical crack problem was restricted in [9] to be of the form

$$p(x, y) = a_{00} + a_{01} y + a_{20} x^2 + a_{02} y^2 + a_{21} x^2 y + a_{03} y^3 . \tag{4.109}$$

The six constants in this equation are determined by a least squares technique of fitting equation (4.109) to a number of known values of $p(x, y)$. The odd powers of $x$ are omitted because the problem is taken to be symmetric about the $yz$-plane. For the elliptical surface flaw, the pressure would also be made symmetric about the $xz$-plane, and the odd powers of $y$ would be eliminated. Thus, only three constants, $a_{00}$, $a_{20}$, $a_{02}$, would be available for the least square fitting necessary. Additional terms are clearly needed, but the 200 terms used in the penny-shaped crack solution (40 in $\theta$, 5 in $r$) cannot be approached.

As suggested by Kantorovich and Krylov [2], the Schwarz–Neumann Alternating Technique has wide applicability. It can be applied to other three-dimensional crack problems of interest such as the finite thickness crack problem. The crack may be taken to intersect the plate at any angle. Such problems would involve combined modes of crack extension even though the plate is loaded in simple tension.

There is some arbitrariness in the choice of the two sequences of problems involved in the alternating method. For example the stress, $\sigma(x)$, on the edge crack was extended symmetrically to negative values of $x$ in the zeroth iteration. It would have been possible to make the stress skew-symmetric about the $y$-axis instead. The two sequences of problems would then be quite different. Similarly, in the surface crack problem, the initial pressure on the penny-shaped crack could be taken to be skew-symmetric about the $yz$-plane. The problems which make up the two sequences would then have different characteristics. In general, the various alternatives should be studied to determine whether one provides a better representation of the essential properties of the problem.

It is conceivable that in some applications the alternating method will involve three or more sequences of solutions. In general, one sequence would be expected to be necessary for each part of the boundary of the material. If there were three parts of the boundary and three solutions, the sequence would be formed similarly to the examples presented here. Each solution would produce some undesirable residual stresses on the other two boundaries. Thus, each iteration would use each solution in turn to remove the residual stresses produced by the other two solutions in their previous applications.

## 4.5 Appendix

Stress intensity factors for various particular loads on the edge crack are listed here. For other loads the general expression of equation (4.50) may be used. In addition, results for the penny-shaped crack in an infinite body are given, including some recently published stress intensity factors for non-axisymmetrically loaded cracks.

### EDGE CRACKS

*Concentrated forces.* For concentrated normal and shearing forces applied to the crack as shown in Figure 4.4, the stress intensity factors are

$$k_1 = \frac{2}{\pi} P(a^{\frac{1}{2}}) \frac{1+f(b/a)}{(a^2-b^2)^{\frac{1}{2}}}$$

$$k_2 = \frac{2}{\pi} Q(a^{\frac{1}{2}}) \frac{1+f(b/a)}{(a^2-b^2)^{\frac{1}{2}}}$$

where $f(b/a)$ is given by equation (4.47) and Figure 4.5.

*Linear stress.* In this case, the stress on the crack varies linearly from zero at the edge to twice the average stress at the crack tip (Figure 4.17). The stress intensity factors in terms of the average stresses are

$$k_1 = 1.3660\ \sigma_a(a^{\frac{1}{2}})$$

$$k_2 = 1.3660\ \tau_a(a^{\frac{1}{2}})\,.$$

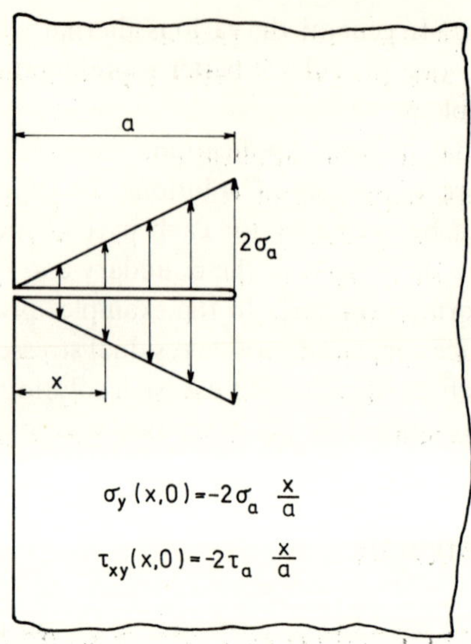


Figure 4.17. Linearly loaded edge crack.

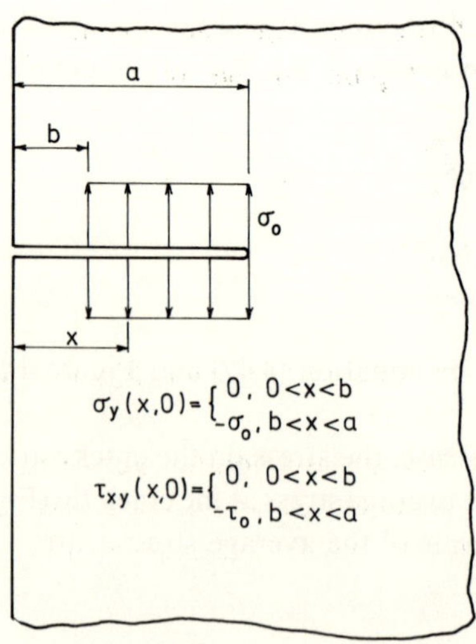


Figure 4.18. Partially loaded edge crack.

*Uniform stress near crack tip.* As shown in Figure 4.18, the crack is free of stress on the part of its faces adjacent to the edge and is subjected to uniform stresses on the other part up to the crack tip. The stress intensity factors may be expressed as

$$k_1 = \frac{2}{\pi} \arccos(b/a)\sigma_0(a^{\frac{1}{2}})[1+h_1(b/a)]$$

$$k_2 = \frac{2}{\pi} \arccos(b/a)\,\tau_0(a^{\frac{1}{2}})[1+h_1(b/a)]$$

where $h_1(b/a)$ is shown in Figure 4.19. These results may also be expressed in terms of the average stresses on the crack,

$$\sigma_a = (1-b/a)\sigma_0 \qquad \tau_a = (1-b/a)\tau_0 \,.$$

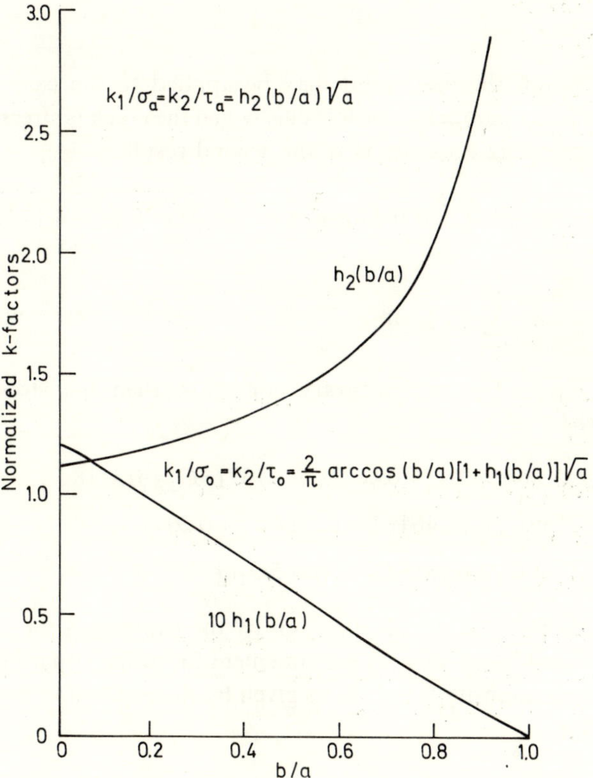


Figure 4.19. Correction factors for partially loaded edge crack.

The stress intensity factors

$$k_1 = h_2(b/a)\sigma_a(a^{\frac{1}{2}})$$
$$k_2 = h_2(b/a)\,\tau_a(a^{\frac{1}{2}})$$

then show the expected increase when the same total load is applied nearer the crack tip. See Figure 4.19.

*Uniform stress* (See Table I). In the case of uniform shear loading, the stress where the crack intersects the edge must be zero. See also the numerical discussion in Section 4.2 of a similar restriction on the normal stress when the alternating method is used directly. These restrictions do not affect the use of the previous example for very small values of $b/a$. As $b/a \to 0$ it is found that

$$k_1 = 1.1215\,\sigma_0\,(a^{\frac{1}{2}})$$
$$k_2 = 1.1215\,\tau_0\,(a^{\frac{1}{2}})\,.$$

By superposition the first result may be applied to the case of uniform tension at infinity parallel to the free edge when the crack is stress free. There is no corresponding application of the second result.

*Polynomial stress* [18]. If the form of the stress in Figure 4.17 is not linear, but is given by

$$\sigma_y(x, 0) = -\sigma_0 \sum_{n=0}^{10} C_n(x/a)^n$$

where $C_0, C_1, \ldots, C_{10}$ are arbitrary constants, then the stress intensity factor is [18]

$$k_1 = \sigma_0(2a)^{\frac{1}{2}}[0.7930\,C_0 + 0.4829\,C_1 + 0.3716\,C_2 + 0.3118\,C_3 + 0.2735\,C_4 + 0.2464\,C_5 + 0.2260\,C_6 + 0.2099\,C_7 + 0.1968\,C_8 + 0.1858\,C_9 + 0.1765\,C_{10}]\,.$$

If there is shear loading $\tau_{xy}(x, 0)$ given by the expression for $\sigma_y(x, 0)$, but with $\sigma_0$ replaced by $\tau_0$, then by the statement following equation (4.50), the Mode II stress intensity factor, $k_2$, is given by the expression for $k_1$ with $\sigma_0$ replaced by $\tau_0$.

**PENNY-SHAPED CRACKS**

*General result.* The pressure applied to the crack in Figure 4.6 is in the form

$$\sigma_z = -p_0 \sum_{m=0}^{\infty} K_m(r) \cos m\theta\,, \qquad z=0\,,\ r \leqslant 1$$

where $r$ is the distance from the $z$-axis divided by the crack radius, $a$. The resulting Mode I stress intensity factor is

$$k_1 = \frac{2}{\pi}\, p_0(a^{\frac{1}{2}}) \sum_{m=0}^{\infty} g_m(1) \cos m\theta$$

where

$$g_m(t) = \int_0^1 K_m(t\zeta)\, \frac{\zeta^{m+1}\,\mathrm{d}\zeta}{(1-\zeta^2)^{\frac{1}{2}}}\,.$$

These formulas apply only when the stress on the crack is symmetric about a plane containing the $z$-axis (the $\theta=0$ plane). For the special case of axisymmetric loading the only nonzero $K_m(r)$ is $K_0(r)$, and the equations simplify considerably.

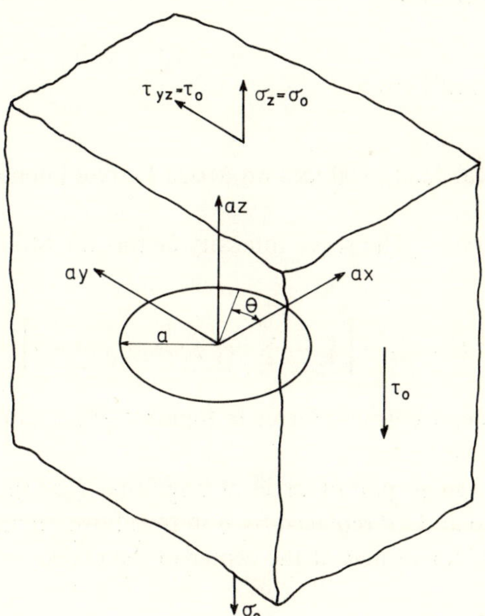


Figure 4.20. Penny-shaped crack in an infinite body under uniaxial tension and uniform shear.

*Uniform tension.* By superposition, the stress intensity factor due to the uniaxial tension as shown in Figure 4.20 may be determined by considering the same geometry, but with uniform pressure,

$$\sigma_z = -\sigma_0 , \quad z = 0 , \ r < 1$$

on the crack and zero stress at infinity. For this problem the parameters of the preceding case become

$$p_0 = \sigma_0 , \ K_0(r) = 1 , \ K_m(r) = 0 , \qquad m \geqslant 1 .$$

Therefore, the stress intensity factor is [26].

$$k_1 = \frac{2}{\pi} \sigma_0 (a^{\frac{1}{2}}) .$$

*Uniform shear.* The solution in [30] for constant shearing stresses on the faces of a penny-shaped crack may be used to obtain the stress intensity factors for Figure 4.20.

$$k_2 = \frac{1}{2-\nu} \frac{4}{\pi} \tau_0 (a^{\frac{1}{2}}) \sin \theta$$

$$k_3 = \frac{1-\nu}{2-\nu} \frac{4}{\pi} \tau_0 (a^{\frac{1}{2}}) \cos \theta .$$

The uniform shear load produces no Mode I stress intensity factor.

*Concentrated forces.* The stress intensity factors for Modes II and III are zero, and [6]

$$k_1 = \frac{2}{\pi} \frac{2P}{\pi a^2} a^{\frac{1}{2}} (1 - r_1^2)^{-\frac{1}{2}} \left[ \tfrac{1}{2} + \sum_{m=1}^{\infty} r_1^m \cos m\theta_1 \cos m\theta \right] .$$

The two pairs of concentrated forces in Figure 4.21 act a distance of $ar_1$ out on the rays $\theta = \pm \theta_1$.

In the case of a single pair of forces of magnitude $F$ on the crack, $\theta$ may be put equal to zero and $2P$ replaced by $F$ in the above expression. If, further, the single pair of forces acts at the center of the crack,

$$k_1 = \frac{1}{\pi} \frac{F}{\pi a^2} a^{\frac{1}{2}} .$$

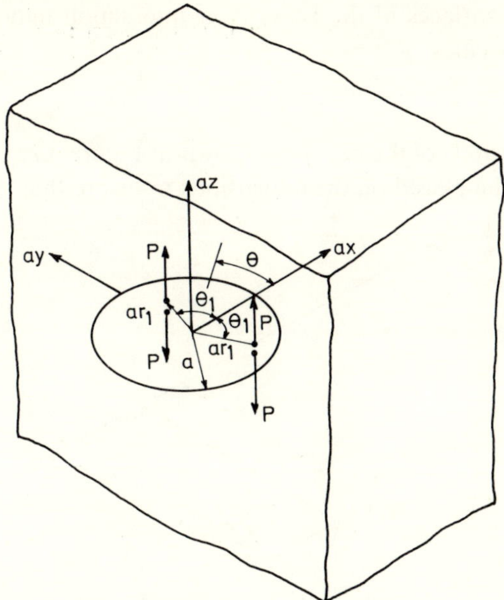


Figure 4.21. Concentrated forces on a penny-shaped crack.

*Linear stress.* In the case of normal stress varying linearly with $x = r\cos\theta$, i.e.,

$$\sigma_{zz} = -p_0 r \cos\theta\,, \quad z = 0\,, \ r < 1$$

the stress varies from zero on the $y$-axis to a maximum of $p_0$ at the points where the crack front intersects the $x$-axis of Figure 4.6. The only non-zero $K_m(r)$ is

$$K_1(r) = r$$

and the stress intensity factor [6],

$$k_1 = \frac{4}{3\pi} p_0(a^{\frac{1}{2}}) \cos\theta$$

is easily obtained.

*Bending stress.* The cases of uniform tension and linear stress may be superposed to obtain the solution of the bending of a beam containing a small penny-shaped crack in a transverse section. If the crack is far enough from the lateral surfaces of the beam their interaction may be neglected. Reference [6] specifies

$$d > 3a \qquad c - b > 3a$$

where the dimensions of the beam are shown in Figure 4.22. An additional restriction must be placed on the dimensions to insure that the crack faces

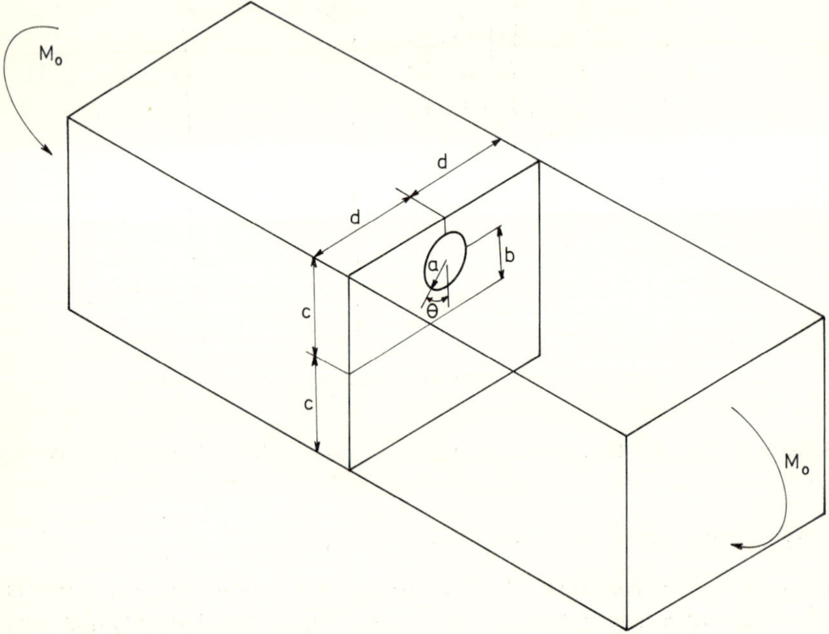


Figure 4.22. Penny-shaped crack in a cross-section of a beam under pure bending.

are not compressed together, for the solution would be invalid if that happened. This restriction,

$$b > \tfrac{2}{3}a$$

simply makes the stress intensity factor positive at all points of the crack front. In terms of the moment of inertia, $I = 2dc^3/3$, of the cross-section, the stress produced in an uncracked beam is

$$\sigma_{zz} = \frac{M_0 a}{I}(b/a - r\cos\theta), \qquad z = 0, \; r < 1 .$$

By superposition, the stress intensity factor,

$$k_1 = \frac{2}{\pi}\frac{M_0 a}{I} a^{\frac{1}{2}}(b/a-(2/3)\cos\theta)\,.$$

may be found by considering the beam to be loaded only by stresses on the faces of the crack equal and opposite to those produced in the uncracked beam.

### SURFACE CRACK

The geometry of the half penny-shaped crack at the surface of a half space is shown in Figure 4.11. For the case of uniform tension at infinity perpendicular to the crack (or uniform pressure on the crack faces), the stress intensity factor is given in Figure 4.13.

## References

[1] Paris, P. C. and Sih, G. C., *Fracture Toughness Testing and Its Applications*, ASTM STP 381, p. 30 (1965).
[2] Kantorovich, L. V. and Krylov, V. I., *Approximate Methods of Higher Analysis*, Interscience, New York (1964).
[3] Irwin, G. R., *The Crack-Extension Force for a Crack at a Free Surface Boundary*, U. S. Naval Research Laboratory Report 5120 (1958).
[4] Lachenbruch, A. H., *J. Geophys. Res.*, 66, p. 4273 (1961).
[5] Brož, P., *Acta Technical Čsav*, 15, p. 724 (1970).
[6] Smith, F. W., Kobayashi, A. S. and Emery, A. F., *J. Appl. Mech.*, 34, p. 947 (1967).
[7] Smith, F. W. and Alavi, M. J., *Proc. First Int. Conf. Pres. Ves. Tech.*, Delft, The Netherlands (1969).
[8] Smith, F. W. and Alavi, M. J., *Engineering Fracture Mechanics*, (to appear).
[9] Shah, R. C. and Kobayashi, A. S., *Int. J. Fracture Mechanics*, (to appear).
[10] Irwin, G. R., *J. Appl. Mech.*, 29, p. 651 (1962).
[11] Bueckner, H. F., *Boundary Problems in Differential Equations*, University of Wisconsin Press, Madison, Wis. (1960).
[12] Winne, D. H. and Wundt, B. M., *Trans. ASME*, 80, p. 1643 (1958).
[13] Koiter, W. T., discussion of paper by Bowie, *J. Appl. Mech.*, 32, p. 237 (1965).
[14] Koiter, W. T., *Proc. Kon. Ned. Akad. Wet. (B)*, 59, p. 354 (1956).
[15] Sneddon, I. N., *Fourier Transforms*, McGraw-Hill, New York, 1951.
[16] Wigglesworth, L. A., *Mathematika*, 4, p. 76 (1957).
[17] Doran, H. E. and Buchwald, V. T., *J. Inst. Maths. Applics.*, 5, p. 91 (1969).
[18] Stallybrass, M. P., *Int. J. Engr. Sci.*, 8, p. 351 (1970).
[19] Sneddon, I. N. and Das, S. C., *Int. J. Engr. Sci.*, 9, p. 25 (1971).
[20] Magnus, W. and Oberhettinger, F., *Formulas and Theorems for the Functions of Mathematical Physics*, Chelsea, New York (1949).

[21] Sih, G. C. and Liebowitz, H., (Mathematical Theories of Brittle Fracture), *Fracture*, 2, Academic Press, New York (1968).
[22] Sih, G. C., *Int. J. Fract. Mech.*, 7, p. 39 (1971).
[23] Thresher, R. W. and Smith, F. W., *ASME Paper No.* 71-*APMW*-6, *J. Appl. Mech.*, in press.
[24] Muki, R., *Progress in Solid Mechanics*, Vol. 1, North-Holland Publishing Co., Amsterdam, p. 401 (1960).
[25] Love, A. E. H., *Phil. Trans. Roy. Soc.* (*A*), 228, p. 377 (1929).
[26] Sneddon, I. N., *Proc. Roy. Soc.* (*A*), 187, p. 229 (1946).
[27] Filon, L. N. G., *Proc. Roy. Soc. Edin.*, 49, p. 38 (1928–29).
[28] Hartranft, R. J. and Sih, G. C., *J. Math. Mech.*, 19, p. 123 (1969).
[29] Sih, G. C., Williams, M. L. and Swedlow, J. L., AFML-TR-242, Air Force Materials Laboratory, Wright-Patterson Air Force Base, Dayton, (November, 1966).
[30] Segedin, C. M., *Proc. Camb. Phil. Soc.*, 47, p. 396 (1951).

*Hans F. Bueckner*

# 5 Field singularities and related integral representations

## 5.1 Introduction

This chapter is concerned with certain stress fields in isotropic elastic bodies. Assumed will be the laws of linear elasticity where the stresses and strains are related to one another by Hooke's law. Emphasis will be placed on field singularities due to cracks. Moreover the analysis will be confined to fields of plane strain and to special cases of axial symmetry. The results for plane strain apply to cases of plane stress, if the familiar approximation is admitted whereby a field of plane strain can be interpreted as one of plane stress for different elastic parameters.

The stresses in an elastic body can be caused by surface tractions, by body forces and by a temperature field. Some examples pertaining to a hollow cylinder of inner radius $a$, outer radius $b$ and with a sharp radial notch as shown in Figure 5.1 will be considered. Cartesian rectangular coordinates $x$, $y$, $z$ as well as polar coordinates defined by $x = r\cos\theta$, $y = r\sin\theta$ are used.

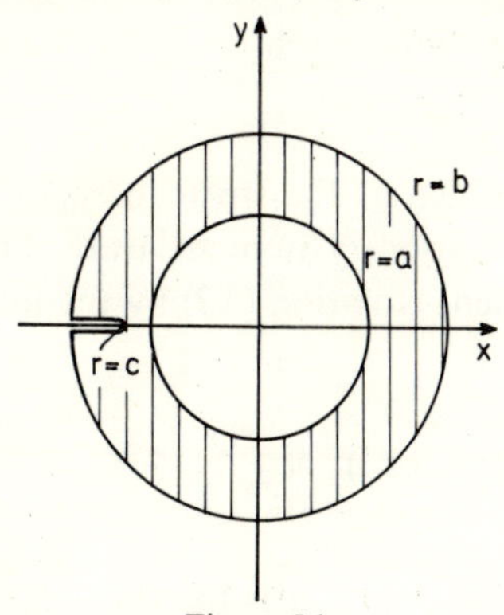


Figure 5.1

Assume that the origin is at the center of the cross-section of the cylinder and that the notch is defined by $y=\pm 0$, $a<c\leqslant -x\leqslant b$ in cartesian and by $\theta=\pm\pi$, $c<r\leqslant b$ in polar coordinates. Plane strain will be assumed, so that all field quantities depend on $x$ and $y$ only; there is no displacement in the direction of the axis of the cylinder. This leads to two displacements $u$ in $x$- and $v$ in $y$-direction. The stresses of interest are $\sigma_x$, $\sigma_y$ and $\tau_{xy}$ since $\tau_{xz}=0$, $\tau_{yz}=0$ and $\sigma_z=\nu(\sigma_x+\sigma_y)$ where $\nu$ denotes Poisson's parameter. Let the cylinder be subjected to the following three cases of loading:

(1) A uniform pressure $p$ acts on the inner cylinder $r=a$; all other parts of the surface are free from traction.

(2) The cylinder rotates with angular velocity $\omega$ around its axis; the surface is free from traction.

(3) A steady state temperature field $T=T_0 \log(b/r)$ exists in the specimen. The surface is free from traction.

It is to be understood that the faces of the notch are included in the surface.

All three cases, different as they seem to be, have one phenomenon in common. At the root $x=-c$, $y=0$ of the notch the stresses are unbounded. If local polar coordinates $d$, $t$ are introduced in accordance with

$$x+c=d\cos t\ ;\quad y=d\sin t\,, \tag{5.1}$$

the stresses can be represented asymptotically ($d\to 0$) as follows:

$$\sigma = \frac{k}{(2d)^{\frac{1}{2}}}\cdot\cos\tfrac{1}{2}t\cdot f(t)$$

with

$$\begin{aligned} f(t)&=1-\sin\tfrac{1}{2}t\cdot\sin\tfrac{3}{2}t \quad &&\text{for}\quad \sigma=\sigma_x\\ f(t)&=1+\sin\tfrac{1}{2}t\cdot\sin\tfrac{3}{2}t \quad &&\text{for}\quad \sigma=\sigma_y\\ f(t)&=\sin\tfrac{1}{2}t\cdot\cos\tfrac{3}{2}t \quad &&\text{for}\quad \sigma=\tau_{xy}\,. \end{aligned} \tag{5.2}$$

Here $k$ is a constant, known as the stress intensity factor; more precisely $k=k_1$ in terms of contemporary nomenclature. Disregarding rigid body motion we can supplement equation (5.2) by the following asymptotic law for the displacements:

$$\begin{aligned} u &= \frac{k}{2\mu}\left(\frac{d}{2}\right)^{\frac{1}{2}}\cdot\cos\tfrac{1}{2}t\,(\kappa-\cos t)\quad \kappa=3-4\nu\\ v &= \frac{k}{2\mu}\left(\frac{d}{2}\right)^{\frac{1}{2}}\cdot\sin\tfrac{1}{2}t\,(\kappa-\cos t);\ \mu \text{ is the shear modulus}\,. \end{aligned} \tag{5.3}$$

The factor $k$ is the same in equations (5.2) and (5.3) but it will take different values for all three cases in general. A derivation of equations (5.2), (5.3) will be given in the next section.

The solutions to the problems (1), (2), (3) are well-known and can be found in many text books of elasticity if the cylinder is *without* notch. The fields have axial symmetry in the sense that the stresses $\sigma_r$, $\sigma_\theta$, $\tau_{r\theta}$ depend on $r$ only; moreover $\tau_{r\theta} \equiv 0$ while $\sigma_r$ and $\sigma_\theta$ are regular analytic functions of $r$ for $a \leqslant r \leqslant b$. Setting $\sigma_\theta = g(r)$ we list the various functions $g$. They are

$$g(r) = \frac{pa^2}{b^2 - a^2} \cdot \left(1 + \frac{b^2}{r^2}\right)$$

for problem (1);

$$g(r) = \frac{3+\nu}{8}\,\rho\omega^2 \left(a^2 + b^2 + \frac{a^2 b^2}{r^2} - \frac{1+3\nu}{3+\nu}\,r^2\right) \tag{5.4}$$

($\rho$ denotes the mass density) [1]

for problem (2) and

$$g(r) = -\frac{1+\nu}{1-\nu}\,\mu\alpha T_0 \left\{\log\frac{b}{a} \cdot \frac{a^2(b^2 + r^2)}{r^2(b^2 - a^2)} - 1 + \log\frac{b}{r}\right\}$$

($\alpha$ denotes the thermal expansion coefficient) [13].

for problem (3).

Since the theory of elasticity is linear, the field in the notched cylinder can be divided into two parts: The regular field which appears under the same loading conditions in the *unnotched* specimen, and a corrective field due to the presence of the notch. The regular field having no singularity at the root of the notch, we see at once that the field quantities of the corrective field prevail over those of the regular one at the root; therefore $k$ is completely determined by the corrective field alone. The latter is easily characterized. It is the field in the notched cylinder caused by tractions distributed over the two faces of the notch in accordance with

$$\sigma_y = -g(r)\,, \quad \tau_{xy} = 0 \text{ on both faces}\,. \tag{5.5}$$

No other tractions appear; centrifugal forces and temperature distribution for problems (2) and (3) respectively are absent. Thus all three problems are reduced to the same type (5.5), so far as the field singularity at $x=-c$, $y=0$ is concerned.

The idea of reduction (suggested in [1]) applies to other specimens as well. In the realm of linear elasticity any crack or notch problem can be reduced to one where the external load appears in the form of tractions distributed over the faces of the crack. This reduction is valuable not only for reasons of a systematic analysis; in many cases it facilitates the analysis. Even where an exact analysis is difficult or next to impossible on geometric grounds alone, the reduction permits the engineer to compare different load systems with one another by comparing the systems of tractions as they appear on the faces of the crack. If, for instance, two different cylinder problems possess approximately the same function $g(r)$ in equation (5.5) then the value of $k$ is practically the same for both of them.

From here on all further analysis is confined to cases where the specimen's surface alone is loaded by tractions, the faces of the crack included. In this context it is said that the tractions on the crack faces are of "induced" nature if at opposite points of the faces the traction vectors are opposite in direction but equal in magnitude; this characterizes the outcome of the reduction and applies in particular to (5.5).

## 5.2 Analysis by field continuation

Consider states of plane strain and refer to $z$ as the complex variable $z=x+iy$. Let some domain $A$ be in the $z$-plane, bounded by a closed curve $B$ as shown in Figure 5.2. In the absence of body forces, any state of plane strain in $A$ can be expressed by the formulas of Muskhelishvili [9] as follows:

$$2\mu w = \kappa\phi(z) - z\cdot\overline{\phi'(z)} - \overline{\psi(z)}\ ; \quad w = u + iv \tag{5.6}$$

$$\sigma_x = \mathrm{Re}\,[\phi'(z) + \overline{\phi'(z)} - z\cdot\overline{\phi''(z)} - \overline{\psi'(z)}] \tag{5.7}$$

$$\sigma_y = \mathrm{Re}\,[\phi'(z) + \overline{\phi'(z)} + z\cdot\overline{\phi''(z)} + \overline{\psi'(z)}] \tag{5.8}$$

$$\tau_{xy} = \mathrm{Im}\,[\bar{z}\phi''(z) + \psi'(z)]\,. \tag{5.9}$$

Here

$$\mu = E/2(1+\nu)\,, \quad \kappa = 3-4\nu\ ; \tag{5.10}$$

$u$, $v$ are the displacements in $x$-direction and $y$-direction respectively. The functions $\phi''(z)$, $\psi'(z)$ are holomorphic in $A$. If $A$ is bounded and simply connected, then $\varphi$, $\psi$ are also holomorphic in $A$. If $\phi$, $\psi$ yield a stress field $\sigma_x, \sigma_y, \tau_{xy}$ then all other pairs $\phi^* = \phi + \mathrm{i}\alpha z + \beta$, $\psi^* = \psi + \gamma$ with arbitrary complex constants $\beta$, $\gamma$ and an arbitrary real constant $\alpha$ yield the same stress field. The constants $\alpha$, $\kappa\beta - \gamma$ represent the family of rigid body motions. It should be emphasized, that $\phi$, $\psi$ depend on the choice of the coordinate system $x$, $y$. If the variable $z$ is changed to

$$\zeta = \varepsilon(z - a)\,, \quad |\varepsilon| = 1 \tag{5.11}$$

then the pair of functions

$$\phi_1(\zeta) = \varepsilon\phi(z)\,, \quad \psi_1(\zeta) = \bar{\varepsilon}[\psi(z) + \bar{a}\phi'(z)] \tag{5.12}$$

yields the same field of stresses and displacements. It can be observed that

$$\bar{\zeta}\phi_1'(\zeta) + \psi_1(\zeta) = \bar{\varepsilon}[\bar{z}\phi'(z) + \psi(z)]\,. \tag{5.13}$$

If d$s$ is an oriented arc-element in $A$, then the components $X$ in $x$-direction, $Y$ in $y$-direction of the traction which attacks the elastic material to the left of d$s$ are given by the formula

$$Z\,\mathrm{d}s = (X + \mathrm{i}\,Y)\mathrm{d}s = -\mathrm{i}\,\mathrm{d}P \quad \text{(Figure 5.2)} \tag{5.14a}$$

where

$$P = P(z) = \phi(z) + z\overline{\phi'(z)} + \overline{\psi(z)}\,. \tag{5.14b}$$

The function $P(z)$ is not analytic in general. Note that equation (5.6) can also be written in the form

$$2\mu w = (\kappa + 1)\phi(z) - P(z)\,.$$

The strain energy density $W$ per unit length can be expressed in the form

$$\begin{aligned} 2EW &= (1 - \nu^2)(\sigma_x^2 + \sigma_y^2) + 2(1 + \nu)(\tau_{xy}^2 - \sigma_x\sigma_y) \\ &= 8(1 + \nu)(1 - 2\nu)(\mathrm{Re}\,\phi')^2 + 2(1 + \nu)|\bar{z}\phi''(z) + \psi'(z)|^2\,. \end{aligned} \tag{5.15}$$

Of practical importance are those states of plane strain for which the integral

$$H = \int_A W\,\mathrm{d}x\,\mathrm{d}y \tag{5.16}$$

exists. This means that finite energy (per unit length) is stored in $A$. In most engineering applications only such states are encountered. Yet this investigation will lead to special fields of infinite energy in $A$, even if $A$ is of bounded area. For this reason the distinction $H<\infty$ and $H=\infty$ has to be kept in mind.

Domain $A$ shows a crack with faces $C^+, C^-$; the root is at the origin, and the crack runs along the negative real axis. Dealing with the field near $z=0$,

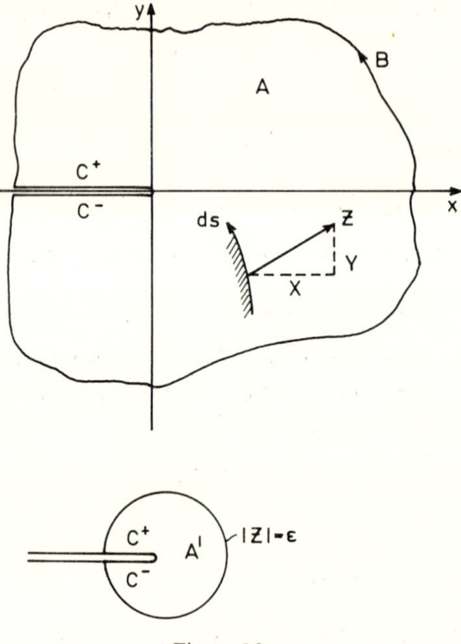


Figure 5.2

we prefer to describe it not with the aid of $\phi$, $\psi$ but rather with $\phi$ and $\rho$, where $\rho(z)=\psi(z)+z\phi'(z)$. Consequently the formulas take the form

$$2\mu w=\kappa\phi(z)-\overline{\rho(z)}+(\bar{z}-z)\overline{\phi'(z)}\,, \tag{5.17}$$

$$\sigma_x \;=\operatorname{Re}\,[\phi'(z)+2\overline{\phi'(z)}-\overline{\rho'(z)}+(\bar{z}-z)\overline{\phi''(z)}] \tag{5.18}$$

$$\sigma_y \;=\operatorname{Re}\,[\phi'(z)+\overline{\rho'(z)}-(\bar{z}-z)\overline{\phi''(z)}]\,, \tag{5.19}$$

$$\tau_{xy} \;=-\operatorname{Im}\,[\phi'(z)-\rho'(z)+(z-\bar{z})\phi''(z)]\,. \tag{5.20}$$

Take now a field in $A$ responding to boundary tractions on $B$ and in particular to tractions of an induced nature on some part of the faces $C^+$, $C^-$ near $z=0$; more precisely assume that the latter are given by

$$\sigma_y + i\tau_{xy} = f(x) = \sum_{n=0}^{\infty} f_n x^n \; ; \quad -\varepsilon < x \leqslant 0 \; ; \text{ on } C^+, C^- \tag{5.21}$$

where $f(x)$ is a converging power series for the indicated interval; due to the induced nature $f(x)$ is the same on both faces. (Note that this is a restriction; one could prescribe tractions of a different kind on the faces.) The coefficients $f_n$ can be complex-valued. The number $\varepsilon > 0$ can be as small as one wishes. In this context we take a region $A'$ consisting of all those points $0 < |z| < \varepsilon$ which are either in $A$ or on the crack faces (Figure 5.2). Without loss of generality it can be assumed that the boundary of $A'$ consists of the circle $|z| = \varepsilon$ and of two segments on $C^+$, $C^-$. We assert the

*Theorem.* If the field has finite energy in $A'$ and if $\phi'$, $\rho'$ are continuous in $A'$, then the functions $\phi$, $\rho$ admit expansions

$$\phi(z) = \sum_{k=0}^{\infty} a_k z^{\frac{1}{2}k} \; ; \quad \rho(z) = \sum_{k=0}^{\infty} b_k z^{\frac{1}{2}k}$$

valid for $|z| < \varepsilon$. Moreover $b_n = \bar{a}_n$ for $n =$ odd and $mb_{2m} = -m\bar{a}_{2m} + f_{m-1}$ else.

A few explanatory remarks about the theorem are perhaps in order.

(1) Although inessential we define $z^{\frac{1}{2}}$ by its main branch (positive for positive real $z$).

(2) Since $z = 0$ is not in $A'$ the assumptions of the theorem do not refer to continuity of $\phi'$, $\rho'$ at $z = 0$. They do refer to continuity on the segments of $C^+$, $C^-$ which belong to $A'$. This is the relevant point of the assumption since $\phi'$, $\rho'$ are holomorphic, hence continuous in the interior of $A'$.

(3) Disregarding the condition (5.21) one can construct fields in $A'$ of the more general form

$$\phi(z) = \sum_{n=-\infty}^{\infty} a_n z^{\frac{1}{2}n}, \quad \rho(z) = \sum_{n=-\infty}^{\infty} b_n z^{\frac{1}{2}n} \tag{5.22}$$

where all $a_n$, $b_n$ are different from zero for all negative $n$. Such fields can have finite energy in $A'$ and they can even be of such a nature that all field quantities are continuous in $A'$ with the point $z = 0$ added. An example is

$$\phi = \rho = z^2 \exp(-z^{-\frac{1}{2}}) = \sum_{k=-\infty}^{4} \frac{(-1)^k}{(4-k)!} z^{\frac{1}{2}k} . \tag{5.23a}$$

On $C^+$, $C^-$ we find $\tau_{xy} = 0$ and

$$\sigma_y = 2\,\mathrm{Re}\,\phi' = 4x\cos|x|^{-\frac{1}{2}} - |x|^{\frac{1}{2}}\sin|x|^{-\frac{1}{2}} = f(x)\,. \tag{5.23b}$$

The function $f(x)$, although continuous in $-1 \leqslant x \leqslant 0$ with $f(0)=0$, is *not* of type (5.21). For this example it is essential that $z^{\frac{1}{2}}$ is defined as the main branch of the square root of $z$.

If the Laurent expansions (5.22) in powers of $z^{\frac{1}{2}}$ are such that

$$a_n = 0\,,\; b_n = 0 \quad \text{for} \quad n < -p\,, \quad p \text{ a positive integer}\,,$$

then the functions $\phi$, $\rho$ have dominant terms of the form

$$\phi(z) \approx a_{-p} z^{-\frac{1}{2}p}\,, \quad \rho(z) \approx b_{-p} z^{-\frac{1}{2}p} \text{ if } a_{-p} \neq 0\,,\; b_{-p} \neq 0\,,$$

and the behavior of the field near $z=0$ is determined by them. It is easy to show that the field has infinite energy. This does not make the field unrealistic if one excludes some neighborhood of zero. The case $p=1$ is quite important, and its consideration and use will be discussed in the next section. In the case of the expansions shown in the theorem the energy is finite. This is well-known.

(4) The theorem will be proved in two steps. In the first it will be shown that $\phi$, $\rho$ admit expansions of the general form (5.22). In the second step it will be proved, on the grounds of finite energy, that $a_n$, $b_n$ must vanish for negative $n$. In view of the example (5.23) this will be neither trivial nor simple. In the proof below use will be made of the fact that the sum of two fields, each of finite energy, is also of finite energy.

*Proof of theorem* (*Step* 1). By way of equations (5.19) and (5.20), it is found that

$$\sigma_y + i\tau_{xy} = \bar{\phi}' + \rho' = f(x) \text{ on } C^+, C^-\,. \tag{5.24}$$

A field of finite energy complying with equation (5.24) and the theorem is given by $\rho(z) = F(z)$, $\phi(z) \equiv 0$ where $F'(x) = f(x)$. Since this is not the most general field, an additional field of finite energy must be found for which the tractions vanish on $C^+$, $C^-$. This amounts to proving the theorem for the case $f(x) \equiv 0$, and the condition

$$\bar{\phi}' + \rho' = 0 \text{ on } C^+, C^-$$

must be satisfied. Here one must bear in mind that neither $\phi'$ nor $\rho'$ will show the same boundary value on opposite points of the crack. Whenever boundary tractions vanish on a straight line segment of the boundary, it is indicated that the field can be "analytically" continued across and beyond the segment. The use of the functions $\phi$, $\rho$ rather than of $\phi$, $\psi$ has advantages in this respect. The combinations $g(z)=\rho'-\phi'$ and $h(z)=\mathrm{i}(\rho'+\phi')$ are even better. By virtue of the above relation between $\phi'$ and $\rho'$ the new functions take real values on $C^+$, $C^-$. Set $t=\mathrm{i}z^{\frac{1}{2}}$ and map $A'$ into the complex $t$-plane. The mapping is 1:1 and Figure 5.3 shows the image $A_t$ of $A'$. $A_t$ is a semi-circular

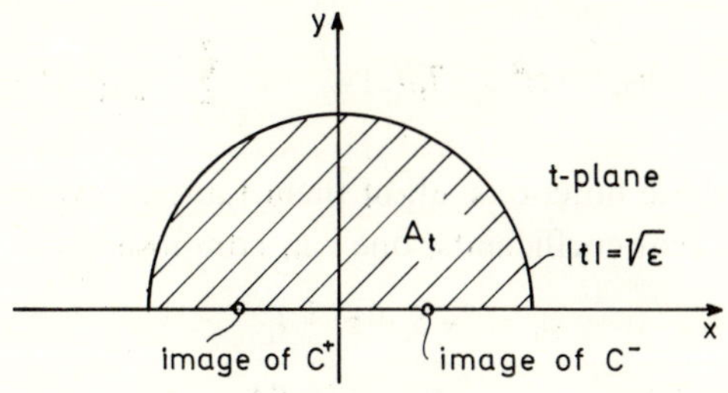


Figure 5.3

disk in the upper half of the $t$-plane, the radius being $\varepsilon^{\frac{1}{2}}$. $C^+$, $C^-$ are taken into intervals on the real $t$-axis, separated by $t=0$; the latter point does not belong to $A_t$. The functions $G(t)=g(z)$, $H(t)=h(z)$ are real on the real $t$-axis in $A_t$. Since by assumption they approach the boundary values on the real $t$-axis for $t\neq 0$ continuously, both $G$ and $H$ can be analytically continued into the lower $t$- half-plane by means of the formulas

$$G(t)=\overline{G(\bar{t})}\,,\quad H(t)=\overline{H(\bar{t})}\,. \tag{5.25a}$$

This implies Laurent expansions

$$G(t)=\sum_{n=-\infty}^{\infty} g_n t^n\,,\quad H(t)=\sum_{n=-\infty}^{\infty} h_n t^n\,,\ \text{valid for } 0<|t|<\varepsilon^{\frac{1}{2}}\,. \tag{5.25b}$$

The coefficients $g_n$, $h_n$ are real. Returning now to $g$, $h$ and back to $\phi'$, $\rho'$ we obtain the expansions:

$$2\rho'(z)=\sum_{n=-\infty}^{\infty}(g_n-\mathrm{i}h_n)\mathrm{i}^n\cdot z^{\frac{1}{2}n}\,;\quad 2\phi'(z)=-\sum_{n=-\infty}^{\infty}(g_n+\mathrm{i}h_n)\mathrm{i}^n z^{\frac{1}{2}n}\,. \tag{5.26a}$$

For $n\neq -2$ coefficients $a_{n+2}$, $b_{n+2}$ can be defined by

$$(n+2)a_{n+2}=(g_n+\mathrm{i}h_n)(\mathrm{i})^n\,,\quad (n+2)b_{n+2}=(g_n-\mathrm{i}h_n)(\mathrm{i})^n\,. \tag{5.26b}$$

The $a_k$, $b_k$ of the theorem are those defined by equations (5.26b). It can be seen that the coefficient rule $b_k = -\bar{a}_k$ for even $k$ and $b_k = \bar{a}_k$ for odd $k$, stated in the theorem, holds in the more general case (5.26b).

*Proof* of *theorem* (*Step* 2). It is useful to introduce the functions

$$\begin{aligned} S_{\mathrm{I}}(s) &= -\tfrac{1}{2}\sum_{k=-\infty}^{\infty} g_{2k}(-z)^k\,; \quad T_{\mathrm{I}}(z) = -\tfrac{1}{2}\sum_{k=-\infty}^{\infty} h_{2k-1}(-z)^k \\ S_{\mathrm{II}}(z) &= -\tfrac{1}{2}\sum_{k=-\infty}^{\infty} h_{2k}(-z)^k\,; \quad T_{\mathrm{II}}(z) = \tfrac{1}{2}\sum_{k=-\infty}^{\infty} g_{2k-1}(-z)^k\,. \end{aligned} \tag{5.27}$$

With the aid of these functions, all of them Laurent expansions in integral powers of $z$ with real coefficients, one can compose

$$\begin{aligned} \phi'_{\mathrm{I}} &= S_{\mathrm{I}} + z^{-\frac{1}{2}} T_{\mathrm{I}}\,, & \rho'_{\mathrm{I}} &= -S_{\mathrm{I}} + z^{-\frac{1}{2}} T_{\mathrm{I}} \\ \phi'_{\mathrm{II}} &= \mathrm{i}(S_{\mathrm{II}} + z^{-\frac{1}{2}} T_{\mathrm{II}})\,, & \rho'_{\mathrm{II}} &= \mathrm{i}(S_{\mathrm{II}} - z^{-\frac{1}{2}} T_{\mathrm{II}})\,. \end{aligned} \tag{5.28a}$$

The functions $\phi'$, $\rho'$ can be split as follows

$$\phi' = \phi'_{\mathrm{I}} + \phi'_{\mathrm{II}}\,, \quad \rho' = \rho'_{\mathrm{I}} + \rho'_{\mathrm{II}}\,. \tag{5.28b}$$

The split gives rise to two stress fields. The field described by $\phi'_{\mathrm{I}}$, $\rho'_{\mathrm{I}}$ will be referred to as mode I and the other one as mode II. For mode I each of the stresses $\sigma_x$, $\sigma_y$ takes the same value at conjugate complex points $z$, $\bar{z}$; $\tau_{xy}$ changes sign. For mode II $\sigma_x$, $\sigma_y$ switch sign as we change from $z$ to $\bar{z}$, while $\tau_{xy}$ stays unchanged. If one considers a subdomain $A''$ in $A'$, symmetrically shaped with respect to the $x$-axis and outside of some neighborhood of $z=0$, the two modes are orthogonal to one another in the sense that the mixed energy of them in $A''$ vanishes. The energy of the sum of the two fields equals the sum of their individual energies. Since it is assumed that the field generated by the pair $\phi$, $\rho$ has finite energy in $A'$, each of the two modes must have finite energy in $A'$. It therefore suffices to prove for each mode that its finite energy implies that the functions $S$, $T$ are ordinary power series in $z$ and not genuine Laurent expansions. This will be done in detail for mode I. For the sake of easier writing the subscript I will be omitted and we set

$$\phi' = S + z^{-\frac{1}{2}} T\,, \quad \rho' = -S + z^{-\frac{1}{2}} T\,. \tag{5.29}$$

Going back to equation (5.15) we see that $\bar{z}\phi'' + \psi' = (\bar{z} - z)\phi'' + \rho' - \phi' = -2\mathrm{i}(y\phi'' - \mathrm{i}S)$. The condition of bounded energy yields the existence of $U^2 < \infty$ such that

$$\iint_{A'} (\mathrm{Re}\,\phi')^2\,\mathrm{d}x\,\mathrm{d}y \leqslant U^2\,, \qquad \iint_{A'} |y\phi'' - \mathrm{i}S|^2\,\mathrm{d}x\,\mathrm{d}y \leqslant U^2\,. \tag{5.30}$$

In order to apply these inequalities one must use certain results from the theory of analytic functions. The reader will find them in the form of six lemmas, derived and compiled in Appendix I. The function $F$ of the lemmas is to be identified with $\phi'$ and the function $G$ of the lemmas 4, 6 with $-\mathrm{i}S$. The lemmas imply that the composition $y^2\phi''(z)$ is bounded in $A'$ and that the Laurent expansion of $S(z)$ is of the form

$$S(z) = \frac{c}{z} + S_0(z)\,, \quad S_0 = \text{regular power series}\,. \tag{5.31}$$

Consequently $y^2(z^{-\frac{1}{2}}\,T(z))' = y^2 z^{-\frac{1}{2}}\,(T'(z) - \frac{1}{2z}\,T(z))$ is bounded in $A'$. By way of lemma 5 one has

$$T'(z) - \frac{1}{2z}\,T(z) = \sum_{n=-1}^{\infty} d_n z^n\,, \tag{5.32}$$

whence

$$T(z) = z^{\frac{1}{2}} \sum_{n=-1}^{\infty} \frac{d_n}{n+\frac{1}{2}}\,z^{n+\frac{1}{2}} = \sum_{n=-1}^{\infty} \frac{d_n}{n+\frac{1}{2}}\,z^{n+1}\,. \tag{5.33}$$

This makes $T(z)$ a regular power series in $z$. From equations (5.29), (5.31) it follows that

$$\phi' \approx \frac{c}{z}, \quad \rho' \approx -\frac{c}{z} \text{ as } z \to 0\,. \tag{5.34}$$

But such functions are incompatible with finite energy unless $c=0$. Thus it has been proved that $S$ and $T$ of equations (5.29) are regular power series. Integration of $\phi', \rho'$ to $\phi, \rho$ leads to the expansions of the theorem with the coefficients $a_k$, $b_k$ defined by equations (5.26b). This concludes the proof for mode I. For mode II it can be conducted in similar vein. The lemmas are so formulated that they cover the case of mode II as well. Further details will not be given.

A few consequences of the theorem will now be mentioned. Since the asymptotic behavior of the field is determined by the functions $\phi(z) = az^{\frac{1}{2}}$, $\rho(z) = \bar{a}z^{\frac{1}{2}}$, it is easy to find the behavior of the field quantities near $z=0$. If $a$ is real, the relations shown in equations (5.21), (5.3) govern the behavior. A real coefficient $a$ pertains to mode I; reference to that mode was already made in section 5.1. For mode II the coefficient $a$ is purely imaginary. The asymptotic relations (well-known) are, with $z = r\mathrm{e}^{\mathrm{i}\theta}$, $\theta = 2t$,

$$\sigma_x = \frac{k_2}{(2r)^{\frac{1}{2}}}\left[-\sin t\,(2+\cos t\cos 3t)\right]$$

$$\tau_{xy} = \frac{k_2}{(2r)^{\frac{1}{2}}}\left[\cos t\,(1-\sin t\sin 3t)\right]$$

$$\sigma_y = \frac{k_2}{(2r)^{\frac{1}{2}}}\left[\sin t\cos t\cos 3t\right] \tag{5.35}$$

$$u = \frac{k_2}{2\mu}\left(\frac{r}{2}\right)^{\frac{1}{2}}\left[\sin t\,(\kappa+1+2\cos^2 t)\right]$$

$$v = \frac{k_2}{2\mu}\left(\frac{r}{2}\right)^{\frac{1}{2}}\left[-\cos t\,(\kappa-1-2\sin^2 t)\right]$$

(see also [8a] with respect to equations (5.1), (5.3) and (5.35)). In the general case of a complex coefficient $a$, the stress field may consist of both mode I and mode II with intensity factors $k_1$, $k_2$ given by

$$k_1 - \mathrm{i}k_2 = 2^{\frac{1}{2}}\cdot a\,. \tag{5.36}$$

Two particular limit relations are noteworthy. If $Z=\sigma_y+\mathrm{i}\,\tau_{xy}$ on the positive real axis then

$$2^{\frac{1}{2}}\cdot x^{\frac{1}{2}}(-\mathrm{i}Z)\to k \text{ as } x\to +0\,. \tag{5.37}$$

If $\Delta w = w(x^+) - w(x^-)$, $x^+$, $x^-$ opposite points on the crack faces $C^+$, $C^-$ respectively, then the "complex crack opening" $\Delta$ satisfies

$$(-x)^{-\frac{1}{2}}\Delta w \to \frac{2\cdot 2^{\frac{1}{2}}(1-\nu)}{\mu}k\,; \quad x\to -0\,. \tag{5.38}$$

From here on $k$ will be referred to as the complex intensity factor. A field whose behavior complies with the theorem will be referred to as a regular field with a *normal* singularity at $z=0$.

The theorem can be generalized. Instead of a straight line crack one can assume an analytic crack ($C^+$, $C^-$ have the shape of an analytic curve). It has been shown in [5] that the expansions of the theorem stay verbally valid if the crack leaves the origin in the direction of the negative $x$-axis and if the crack is load free in some neighborhood of the origin.

The functions $\phi$, $\rho$ can be analytically continued, as already mentioned. They become well-defined functions on the Riemann surface of $z^{\frac{1}{2}}$. Consequently $A'$ has a counterpart $A''$ on the other sheet, and $\phi$, $\rho$ define a regular field with a normal singularity in $A''$ as well. If for example one takes mode II then it follows from equation (5.38) that the functions $\phi$ and $\rho$ exchange their roles, as one goes from $A'$ to $A''$. This is a feature which should be useful in crack analysis.

This section will be concluded with a special problem for which the method of analytic field continuation yields the solution in closed form. The elastic domain is the upper half of the $z$-plane, and the $x$-axis is its boundary. Along the interval $-1<x<1$, referred to as $L^+$, constant traction is prescribed. On the remaining part $L$ of the $x$-axis the displacements are zero, i.e., $u=v=0$ (Figure 5.4). The problem is to find a field of finite energy in the

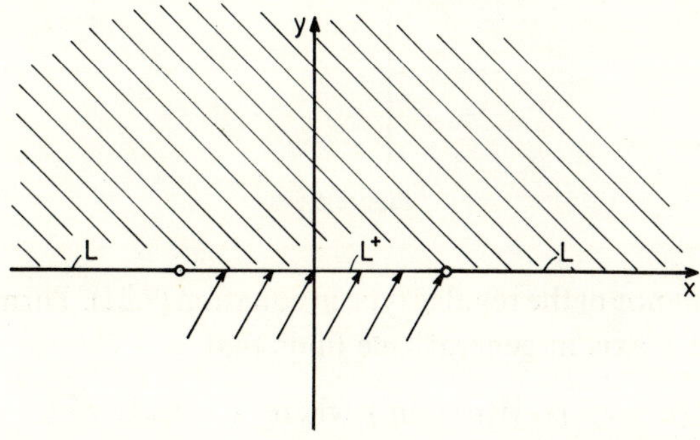


Figure 5.4

domain. The condition on $L$ suggests that the field can be continued into the lower half-plane. In order to show this, let the functions $\phi$, $\rho$ be replaced by functions $p(z)$, $q(z)$ shown in

$$2\kappa\phi = p - iq\,, \quad 2\rho = p + iq\,; \tag{5.39}$$

hence for points of the real axis

$$4\mu w = p - \bar{p} - i(q - \bar{q})\,. \tag{5.40}$$

The condition on $L$ implies that $p$ and $q$ take real values there. Consequently $p$, $q$, well determined by equations (5.39) in the upper half-plane, can be continued into the lower one by means of the formulas

$$p(z) = \overline{p(\bar{z})}\,, \quad q(z) = \overline{q(\bar{z})}\,. \tag{5.41}$$

The functions $p$, $q$, hence the functions $\phi$, $\rho$ define a field in the whole plane, cut along the interval $-1<x<1$ on the boundary. In other words: a field in the domain of the so-called Griffith-crack has been obtained. The faces of the crack will be distinguished by $L^+$ and $L^-$ (Figure 5.5). It will turn out that the field in the larger domain shows tractions on $L^-$. These tractions are

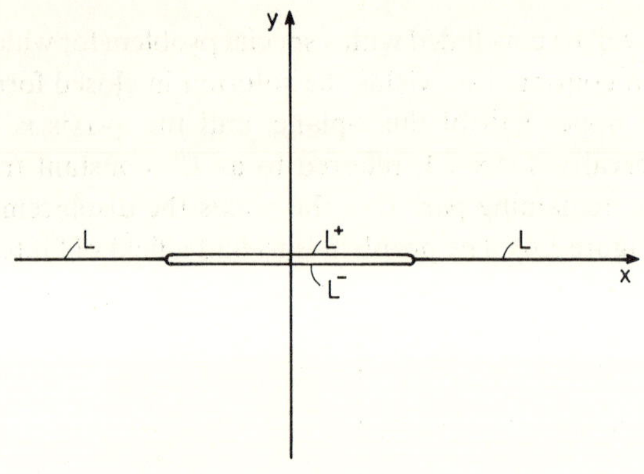


Figure 5.5

neither constant nor of the regular type in equation (5.21). Turning now to the stresses on the $x$-axis in general, one finds that

$$\sigma_y + i\tau_{xy} = \tfrac{1}{2}(p' + \lambda\bar{p}') + \tfrac{1}{2}i(q' + \lambda\bar{q}') \text{ where } \lambda = 1/\kappa\,. \tag{5.42}$$

Now, let $\sigma_y$ and $\tau_{xy}$ be constant along $L^+$. Introduce the related constant and real coefficients

$$a = \frac{2\sigma_y}{1+\lambda}, \quad b = \frac{2\tau_{xy}}{1+\lambda}. \tag{5.43}$$

The boundary conditions on $L^+$ can be given the form

$$\left.\begin{aligned}(1+\lambda)\,\mathrm{Re}\,(p'-a)-(1-\lambda)\,\mathrm{Im}\,(q'-b)=0\\ (1-\lambda)\,\mathrm{Im}\,(p'-a)+(1+\lambda)\,\mathrm{Re}\,(q'-b)=0\,.\end{aligned}\right\}\ \text{on } L^+ \tag{5.44}$$

From this it follows that

$$\mathrm{Re}\,(p'-a)\,\mathrm{Im}\,(p'-a)+\mathrm{Re}\,(q'-b)\,\mathrm{Im}\,(q'-b)=0\,, \tag{5.45}$$

and therefore the function

$$R(z)=(p'-a)^2+(q'-b)^2 \tag{5.46}$$

takes real values on $L^+$. Since $R(z)$ follows the law $R(z)=\overline{R(\bar{z})}$, $R(z)$ takes the same real value on opposite points of the crack, and $R(z)$ is seen to be regular analytic in the whole $z$-plane with the exception of the endpoints of the crack where it will display the singularities of a meromorphic function. Now introduce the condition that $p$, $q$ are *holomorphic at infinity*. This implies that $R(z)$ has the same behavior. This circumstance and the condition of finite energy of the field in the upper half-plane suggest that $R(z)$ is a rational function of the form

$$R(z) = \frac{c_0+c_1 z+c_2 z^2}{z^2-1}. \tag{5.47}$$

The coefficients $c_0$, $c_1$, $c_2$ are to be real. Since $p$, $q$ admit expansions of the form

$$p = p_0 + \frac{p_1}{z} + \ldots, \quad q = q_0 + \frac{q_1}{z} + \ldots \text{ at } z = \infty, \tag{5.48}$$

one can express the $c_k$ in terms of the coefficients in equation (5.48). This leads to

$$c_2 = a^2+b^2, \quad c_1 = 0; \qquad c_0 = -a^2-b^2+2p_1 a+2q_1 b. \tag{5.49}$$

Assume now a solution $A(z)$, $B(z)$ of the functional equation

$$A^2(z)+B^2(z) = R(z), \tag{5.50}$$

where $A$, $B$ are to be holomorphic in the slot-domain of Figure 5.5, the point $z=\infty$ included, while the faces of the crack are to be disregarded. It is also required that $A(z)$, $B(z)$ follow the law (5.41), i.e. they have to be real on $L$ and to take conjugate complex values at conjugate complex points $z$, $\bar{z}$ in general. Then (at least locally) the most general solution to equation (5.46) can be given the form

$$p'(z) = a+A\cos S-B\sin S; \quad q'(z) = b+A\sin S+B\cos S, \tag{5.51}$$

where $S$ is some analytic function. To find some "simple" pair $A$, $B$ satisfying equation (5.50), introduce the function

$$T(z) = \left(\frac{z-1}{z+1}\right)^{\frac{1}{2}}, \tag{5.52}$$

with the square-root so taken that $T(\infty)=1$. This function satisfies the conditions of regularity as well as the law (5.41), demanded of $A$ and $B$. Setting up

$$A(z) = a_1 T+a_2/T; \quad B = b_1 T+b_2/T, \tag{5.53}$$

with constant and real coefficients $a_1, a_2, b_1, b_2$, we find that

$$A^2+B^2=(a_1^2+b_1^2)\frac{z-1}{z+1}+(a_2^2+b_2^2)\frac{z+1}{z-1}+2a_1a_2+2b_1b_2\,. \tag{5.54}$$

The right hand side is of type (5.47). Note that $A$ and $B$ take purely imaginary values on $L^+, L^-$. The next step is to find a suitable function $S=S(z)$ following the law (5.41) and making it possible to satisfy the condition (5.44) on $L^+$. Since $A$, $B$ are imaginary on $L^+$, the boundary condition can be given the form

$$(A+\mathrm{i}B)(\mathrm{e}^{\mathrm{i}S}-\lambda\,\mathrm{e}^{\mathrm{i}\bar{S}})=0 \ \text{ on } \ L^+\ ; \tag{5.55a}$$

multiplying this relation by $\mathrm{e}^{-\mathrm{i}S}$, we see that

$$1-\lambda\exp(2\ \mathrm{Im}\ S)=0 \ \text{ on } \ L^+\ , \tag{5.55b}$$

and $S$ must show a constant imaginary part on $L^+$. A suitable function $S$ is

$$S(z)=\gamma\log\frac{z-1}{z+1}\ ;\quad \gamma=-\tfrac{1}{2}\log\lambda=\tfrac{1}{2}\log\kappa\,. \tag{5.56}$$

The log is to be so taken that $S(\infty)=0$. The function $S$ is holomorphic in the slot-domain, the point $z=\infty$ included. The same holds for cos $S$ and sin $S$. As regards the approach to $z=1$, $z=-1$ let

$$\frac{z-1}{z+1}=d\cdot\mathrm{e}^{\mathrm{i}t},\quad d>0\,. \tag{5.57}$$

Then one can write

$$\begin{aligned}\mathrm{e}^{\mathrm{i}S}&=\mathrm{e}^{-\gamma t}\,(\cos(\gamma\log d)+\mathrm{i}\sin(\gamma\log d))\\ \mathrm{e}^{-\mathrm{i}S}&=\mathrm{e}^{\gamma t}\,\ \ (\cos(\gamma\log d)-\mathrm{i}\sin(\gamma\log d))\end{aligned} \tag{5.58}$$

and this shows that cos $S$, sin $S$ stay bounded as $d\to 0$ or $d\to\infty$.

On $L^+$, $L^-$ $S, T$ can be represented in the form

$$T=\pm\mathrm{i}|T|\ \text{ with } |T|=\left(\frac{1-x}{1+x}\right)^{\frac{1}{4}}\ ;\ S=2\gamma\log|T|\pm\gamma\pi\mathrm{i}\ ; \tag{5.59}$$

upper sign for $L^+$, lower sign for $L^-$.

It remains to deal with the coefficients $a_k$, $b_k$ in equations (5.53). To this end impose on $p'$, $q'$ by equation (5.51) the conditions that their expansions at $z=\infty$ begin, due to equations (5.48), with the term in $z^{-2}$. This yields two conditions for either function. If one uses the following expansions at infinity

$$\begin{aligned} &T(z) = 1 - 1/z + \ \dots, \quad 1/T(z) = 1 + 1/z + \ \dots \\ &S(z) = -2\gamma/z + \ \dots, \quad \cos S = 1 + \ \dots, \quad \sin S = -2\gamma/z + \ \dots \end{aligned} \tag{5.60}$$

which enter the conditions with the explicitly written terms only, then one obtains:

$$\begin{aligned} &a_1 + a_2 + a = 0, \quad -a_1 + a_2 + 2\gamma(b_1 + b_2) = 0 \\ &b_1 + b_2 + b = 0, \quad -2\gamma(a_1 + a_2) - b_1 + b_2 = 0 \end{aligned} \tag{5.61}$$

with the solution

$$\begin{aligned} a_1 &= -\tfrac{1}{2}(a + 2\gamma b) \ ; \quad a_2 = -\tfrac{1}{2}(a - 2\gamma b) \\ b_1 &= \ \tfrac{1}{2}(2\gamma a - b) \ ; \quad b_2 = -\tfrac{1}{2}(2\gamma a + b), \end{aligned} \tag{5.62}$$

whence

$$A = -\tfrac{1}{2}a(T + 1/T) - \gamma b(T - 1/T) \ ; \quad B = -\tfrac{1}{2}b(T + 1/T) + \gamma a(T - 1/T). \tag{5.63}$$

The solution so determined satisfies all conditions and in particular the one of finite energy. The energy in the whole slot-domain is bounded. To check the tractions along the $x$-axis, we observe that

$$\sigma_y + \mathrm{i}\tau_{xy} = \tfrac{1}{2}(1 + \lambda)(a + \mathrm{i}b) + (A + \mathrm{i}B)(\mathrm{e}^{\mathrm{i}S} - \lambda\,\mathrm{e}^{\mathrm{i}\bar{S}}) \ \text{ for } L^+, L^- \tag{5.64a}$$

$$\sigma_y + \mathrm{i}\tau_{xy} = \tfrac{1}{2}(1 + \lambda)(a + \mathrm{i}b) + (A + \mathrm{i}B)\,\mathrm{e}^{\mathrm{i}S} \ \text{ on } L, \tag{5.65}$$

where

$$\begin{aligned} \mathrm{e}^{\mathrm{i}S} - \lambda\,\mathrm{e}^{\mathrm{i}\bar{S}} &= \mathrm{e}^{\mathrm{i}S}(1 - \lambda \exp(2 \operatorname{Im} S)) = 0 \ \text{ on } L^+ \\ &= (1 - \lambda^2)\,\mathrm{e}^{\mathrm{i}S} \ \text{ on } L^-. \end{aligned} \tag{5.64b}$$

This makes it clear that the tractions on $L^-$ are neither constant nor of type (5.21). From equation (5.65) it follows that the stresses do not follow the asymptotic laws of mode I and II. Although the field exhibits singularities at $z = \pm 1$, they are not the normal ones described in the theorem.

The conditions (5.44) can also be expressed as integral equations. Setting

$$p'(z) = \frac{1}{\pi}\int_{-1}^{1} \frac{f(t)\mathrm{d}t}{t - z}, \quad q'(z) = \frac{1}{\pi}\int_{-1}^{1} \frac{g(t)\mathrm{d}t}{t - z} \ ; \ f(t),\ g(t) \text{ real} \tag{5.66}$$

we make use of the well-known properties of Cauchy integrals according to which

$$\begin{aligned} &\operatorname{Im} p'(x^+) = f(x), \quad \operatorname{Im} q'(x^+) = g(x)\ ; \quad x^+ \in L^+ \\ &\operatorname{Re} p'(x) = Kf, \quad \operatorname{Re} q'(x) = Kg \ \text{ with} \\ &Kf = \mathrm{CPV}\, \frac{1}{\pi} \int_{-1}^{1} \frac{f(t)\,\mathrm{d}t}{t-x}, \ \text{etc}. \end{aligned} \tag{5.67}$$

Consequently equations (5.44) become equivalent with

$$\begin{aligned} &(1-\lambda) f + (1+\lambda) Kg = (1+\lambda) b \\ &(1+\lambda) Kf - (1-\lambda) g = (1+\lambda) a . \end{aligned} \tag{5.68}$$

This represents a system of two simultaneous* integral equations, each with the integral operator $K$. The equations are coupled; they have to be solved for the two unknown functions $f$, $g$; equations (5.68) are a special case of the more general system

$$f + \tau Kg = \beta, \quad g - \tau Kf = \alpha\ ; \quad \tau = \frac{1+\lambda}{1-\lambda} \tag{5.69}$$

where $\alpha$, $\beta$ are functions over the interval $-1 < x < 1$. These functions can represent a traction distribution more general than the one of constant tractions along $L^+$. One can convert the system into one equation for say $f$ alone. Substituting $g$ into the first equation one obtains

$$f + \tau^2 K^2 f = \beta + \tau K\alpha . \tag{5.70}$$

This has the formal appearance of a Fredholm equation, but the appearance is misleading. What accounts for the results of Fredholm's theory is the circumstance that the integral operator $K^2$ is compact. This is not the case in equation (5.70) where $K$, in the language of functional analysis, is bounded but certainly not compact. This implies phenomena of solvability and of solutions which are outside of Fredholm's theory. One could develop a theory of equations (5.69) or (5.70) in analogy to the procedure developed in [4], and it should be possible to use the already known solution for the case (5.68) in order to construct the solution in the general case. This ends the application of the method of field continuation for this section. Another example will be discussed in section 5.4.

* Such equations have been commonly referred to in the open literature as a system of "dual integral equations".

## 5.3 Fundamental fields and weight functions

Assume a simply connected domain $A$ with an open crack, extending from $z=0$ along part of the negative $x$-axis. Figure 5.6 shows such a configuration. Returning now to the general expansions (5.22) we single out the special case

$$\phi_s = a \cdot z^{-\frac{1}{2}}, \quad \rho_s = \bar{a} \cdot z^{-\frac{1}{2}}, \tag{5.71}$$

where the functions $\phi_s$, $\rho_s$ are well-defined in $A$. Now consider the associated field in the large. The field displays no tractions along the faces of the crack

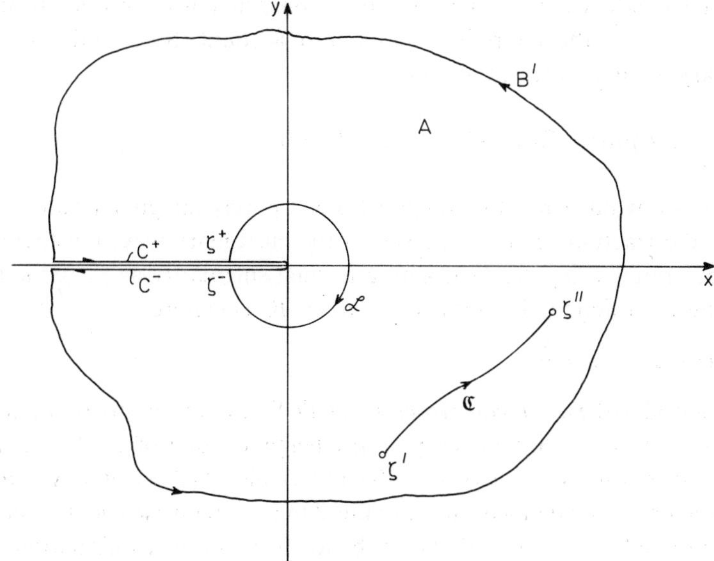


Figure 5.6

but there will be tractions on the remaining part $B'$ of the boundary $B$. These tractions are well-behaved if $B'$ is smooth enough. They are also self-equilibrated. In order to verify the latter it is necessary to show that (see [9])

$$\int_{B'} \mathrm{d}P = 0, \quad \mathrm{Re} \int_{B'} \bar{z} \cdot \mathrm{d}P = 0. \tag{5.72}$$

By (5.14b) and due to $\rho = \psi + z\phi'$ it follows that

$$P = \phi + \bar{\rho} + (z - \bar{z})\bar{\phi}'. \tag{5.73}$$

Evidently $P$ vanishes on the crack. Since $B'$ goes from the endpoint of $C^-$ to the endpoint of $C^+$ the first integral in equation (5.72) is indeed zero. Integrate the second integral by parts and the condition becomes ($P=0$ at the endpoints applied)

$$\operatorname{Re}\int_{B'}\bar{z}\,\mathrm{d}P=-\operatorname{Re}\int_{B'}P\,\mathrm{d}\bar{z}=-\operatorname{Re}\int_{B'}\bar{P}\,\mathrm{d}z\,.$$

But

$$\int_{B'}\bar{P}\cdot\mathrm{d}z=\int_{B'}(\rho+\bar{\phi})\mathrm{d}z+\int_{B'}(\bar{z}-z)\mathrm{d}\phi=\int_{B'}(\rho+\phi)\mathrm{d}z+\int_{B'}(\bar{\phi}\,\mathrm{d}z-\phi\,\mathrm{d}\bar{z})$$

where the integral over $(\bar{z}-z)\cdot\mathrm{d}\phi$ has been converted by integration by parts with $\bar{z}-z=0$ at the endpoints taken into account. Since $\bar{\phi}\,\mathrm{d}z-\phi\,\mathrm{d}\bar{z}$ is imaginary, it suffices to show that

$$\operatorname{Re}\int_{B'}(\phi+\rho)\mathrm{d}z=\operatorname{Re}(a+\bar{a})\cdot\int_{B'}z^{-\frac{1}{2}}\mathrm{d}z=0\,.$$

But this is obvious since the integral takes a purely imaginary value.

Now the tractions can be relieved in the customary way. Let therefore $\phi_r$, $\rho_r$ determine a regular field with a normal singularity at $z=0$, the field opposing the tractions induced by $\phi_s$, $\rho_s$ on $B$. Therefore

$$\phi=\phi_s+\phi_r\,,\quad \rho=\rho_s+\rho_r \tag{5.74}$$

defines a field with *no boundary tractions at all.* Due to the nature of equations (5.71) the field has infinite energy in any neighborhood of $z=0$. This field and also its strong singularity at $z=0$ will be called *fundamental.* All fundamental fields in $A$ so defined can be reduced to essentially two fields: One for $a=1$ and one for $a=\mathrm{i}$. All others can be found by linear combination with constant real combination coefficients.

The construction of a fundamental field makes it clear that an expansion of type (5.22) holds with $a_n=0$, $b_n=0$ for $n<-1$; the coefficients follow the rule $b_n=\bar{a}_n$ for $n=$odd and $b_n=-\bar{a}_n$ for $n=$even.

The fundamental fields permit one to calculate the stress intensity factor of a given regular field in a simple manner. Let the pair $\phi_1$, $\rho_1$ denote a given regular field with a normal singularity of complex intensity $k$ at $z=0$, and $\phi_2$, $\rho_2$ denote a fundamental field with coefficient $a$ in equations (5.71). The field quantities will be distinguished by subscripts 1, 2 respectively. In what follows the words "field" and "state" will be used synonymously.

Take now an oriented, piecewise smooth curve $\mathfrak{C}$ in $A$, going from some point $\zeta'$ to another point $\zeta''$. Along $\mathfrak{C}$ the two states exhibit tractions which attack the material to the left. Consider the case where the tractions of state $\phi_k, \rho_k$ accomplish work through the displacements of state $\phi_m, \rho_m$. This work is

$$W_{km} = \int_{\mathfrak{C}} (X_k u_m + Y_k v_m) \mathrm{d}s = \mathrm{Re} \int_{\mathfrak{C}} (X + \mathrm{i}\, Y)_k (u - \mathrm{i}\, v)_m \mathrm{d}s$$

$$= -\mathrm{Re}\,\mathrm{i} \int_{\mathfrak{C}} \bar{w}_m \mathrm{d}P_k = \mathrm{Im} \int_{\mathfrak{C}} \bar{w}_m \mathrm{d}P_k \; ; \qquad k, m = 1, 2 \, . \tag{5.75}$$

In particular, consider

$$W^*(\mathfrak{C}) = W_{12} - W_{21} = \mathrm{Im} \int_{\mathfrak{C}} (\bar{w}_2 \mathrm{d}P_1 - \bar{w}_1 \mathrm{d}P_2), \tag{5.76}$$

which may be written as

$$2\mu W^*(\mathfrak{C}) = \mathrm{Im} \int_{\mathfrak{C}} (\bar{P}_1 \mathrm{d}P_2 - \bar{P}_2 \mathrm{d}P_1) + (\kappa + 1)\, \mathrm{Im} \int_{\mathfrak{C}} (\bar{\phi}_2 \mathrm{d}P_1 - \bar{\phi}_1 \mathrm{d}P_2). \tag{5.77}$$

Integration by parts leads to

$$\mathrm{Im} \int_{\mathfrak{C}} (\bar{P}_1 \mathrm{d}P_2 - \bar{P}_2 \mathrm{d}P_1) =$$

$$= \mathrm{Im}\, \bar{P}_1 P_2 \Big|_{\zeta'}^{\zeta''} - \mathrm{Im} \int_{\mathfrak{C}} (\bar{P}_2 \mathrm{d}P_1 + P_2 \mathrm{d}\bar{P}_1) = \mathrm{Im}\, \bar{P}_1 P_2 \Big|_{\zeta'}^{\zeta''}, \tag{5.78}$$

$$\mathrm{Im} \int_{\mathfrak{C}} (\bar{\phi}_2 \mathrm{d}P_1 - \bar{\phi}_1 \mathrm{d}P_2) =$$

$$= \mathrm{Im}\, (\bar{\phi}_2 P_1 - \bar{\phi}_1 P_2) \Big|_{\zeta'}^{\zeta''} + \mathrm{Im} \int (P_2 \overline{\mathrm{d}\phi_1} - P_1 \overline{\mathrm{d}\phi_2}) \, . \tag{5.79}$$

Furthermore

$$\int_{\mathfrak{C}} (P_2 \overline{\mathrm{d}\phi_1} - P_1 \overline{\mathrm{d}\phi_2}) = \int_{\mathfrak{C}} (\phi_2 \overline{\mathrm{d}\phi_1} - \phi_1 \overline{\mathrm{d}\phi_2}) \; +$$

$$+ \int_{\mathfrak{C}} (z - \bar{z}) \overline{(\phi_2' \mathrm{d}\phi_1 - \phi_1' \mathrm{d}\phi_2)} - \int_{\mathfrak{C}} \bar{g}(z) \mathrm{d}\bar{z} \, , \tag{5.80}$$

where $g(z) = \rho_1 \phi_2' - \rho_2 \phi_1'$. The second integral in equation (5.80) has a vanishing integrand; the first one, in analogy to equation (5.78), can be given the form

$$\operatorname{Im}\int_{\mathfrak{C}}(\phi_2 \mathrm{d}\bar{\phi}_1-\phi_1 \mathrm{d}\bar{\phi}_2)=-\operatorname{Im}\phi_1\bar{\phi}_2\Big|_{\zeta'}^{\zeta''}. \tag{5.81}$$

Putting all the pieces together we obtain

$$2\mu W^*(\mathfrak{C})=(\kappa+1)\operatorname{Im}\int_{\mathfrak{C}} g(z)\mathrm{d}z+\operatorname{Im}R\Big|_{\zeta'}^{\zeta''} \tag{5.82}$$

with

$$R=\bar{P}_1 P_2+(\kappa+1)[\bar{\phi}_2(P_1-\phi_1)-\bar{\phi}_1 P_2].$$

Two special cases of $\mathfrak{C}$ will be considered. If $\mathfrak{C}$ is closed then $R$ does not contribute and

$$2\mu W^*(\mathfrak{C})=\operatorname{Im}(\kappa+1)\int_{\mathfrak{C}} g(z)\mathrm{d}z\ ;\quad \mathfrak{C}\text{ is closed}. \tag{5.83}$$

The integral over $g(z)$ in equation (5.83) does not necessarily vanish, if $A$ were not simply connected. However if $\mathfrak{C}$ is a *Jordan* curve whose interior belongs to $A$, then by *Cauchy*'s integral theorem $W^*(\mathfrak{C})=0$. This is a special case of *Betti*'s theorem, according to which the work done by the tractions of state (1) through the displacements of state (2) equals the work which the tractions of state (2) accomplish through the displacements of state (1).

Next let $\mathfrak{C}$ be a closed path $\mathfrak{C}'$, consisting of a circle of sufficiently small radius $r$ around the origin ($\mathscr{L}$ in Figure 5.6), of the faces outside of $\mathscr{L}$ and of $B'$. The circle goes from a point $\zeta^+$ on $C^+$ to the opposite point $\zeta^-$ on $C^-$. The path $\mathfrak{C}'$ is to bound a simply connected subdomain of $A$, and this can be achieved if $r>0$ is small enough. Since part of the path is on the boundary $B$ of $A$, it is assumed that the relevant field quantities are continuous in the closure of the subdomain so that the preceding formulas stay applicable. $\mathfrak{C}'$ is so oriented that the interior of the subdomain is to the left. Since $W^*(\mathfrak{C}')=0$ it follows that

$$W^*(\mathscr{L})=-W^*(B'') \tag{5.84}$$

where $B''$ consists of $B'$ and of $C^+$, $C^-$ outside of $\mathscr{L}$; in other words: $B''$ is that part of $B$ which lies outside of $\mathscr{L}$. On $B''$ the fundamental field has no tractions. Hence

$$-W^*(B'')=-\operatorname{Im}\int_{B''}\bar{w}_2 \mathrm{d}P_1\,. \tag{5.85a}$$

If the tractions of state (1) along the crack are bounded and continuous, then $W^*(B'')$ goes to a limit as $r \to 0$, namely

$$\lim_{r\to 0} W^*(B'') = \operatorname{Im} \int_B \bar{w}_2 \, \mathrm{d}P_1 \, . \tag{5.85b}$$

An opposing limit must exist for $W^*(\mathscr{L})$. For small $r$ the behavior of this integral depends on the dominant terms in the expansions (5.22). These are

$$\phi_1 \approx \frac{k}{2^{\frac{1}{2}}} z^{\frac{1}{2}}, \quad \rho_1 \approx \frac{\bar{k}}{2^{\frac{1}{2}}} z^{\frac{1}{2}} ; \qquad \phi_2 \approx a z^{-\frac{1}{2}}, \quad \rho_2 \approx \bar{a} z^{-\frac{1}{2}} . \tag{5.86}$$

Turning first to the term $R$ in the formula (5.82) for the case $\mathfrak{C} = \mathscr{L}$, we observe that $P_2 = 0$ can be assumed along $B$ and in particular on the crack. Indeed $P_2$ has to be a constant on the boundary. By adding a suitable constant to $\phi_2$, which does not essentially change the field (it amounts to rigid body motion) one can normalize $P_2$ to zero. Since $P_1 - \varphi_1 = \bar{\rho}_1$ on the crack, $R = (\kappa+1)\overline{\phi_2 \rho_1}$. As $r \to 0$, the product $\phi_2 \rho_1$ goes to a finite limit, the same for both sides of the crack. Hence $R$ does not contribute to the limit of $W^*(\mathscr{L})$. It remains to deal with $g(z)$. On the basis of equations (5.86), it is found that

$$z g(z) \to -\frac{k\bar{a} + \bar{k}a}{2 \cdot 2^{\frac{1}{2}}} \text{ as } z \to 0 \, . \tag{5.87}$$

Consequently

$$\int_{\mathscr{L}} g(z) \mathrm{d}z \to \tfrac{1}{2} \cdot 2^{\frac{1}{2}} \pi \cdot (k\bar{a} + \bar{k}a) \cdot \mathrm{i} \text{ as } r \to 0 \, , \tag{5.88}$$

$$\lim_{r\to 0} W^*(\mathscr{L}) = 2^{\frac{1}{2}} \pi \, \frac{\kappa+1}{4\mu} (k\bar{a} + \bar{k}a) \, . \tag{5.89}$$

The coefficient $a$ can be normalized by

$$a = \frac{2\mu}{\kappa+1} \varepsilon \text{ where } |\varepsilon| = 1 \, . \tag{5.90a}$$

In this case the asymptotic behavior of $w_2$ takes the form

$$w_2 \approx \varepsilon z^{-\frac{1}{2}} \text{ as } z \to 0 \, . \tag{5.90b}$$

A combination of (5.84), (5.85) and (5.89) leads to the integral formula

$$\mathrm{Re}(k\bar{\varepsilon}) = \mathrm{Im}\,\frac{1}{2^{\frac{1}{2}}\cdot\pi}\int_B \overline{w_2}\,\mathrm{d}P_1 = -\frac{1}{2^{\frac{1}{2}}\cdot\pi}\int_B (X_1 u_2 + Y_1 v_2)\,\mathrm{d}s \tag{5.91}$$

where $s$ denotes the arclength on the boundary. In general two fundamental fields are needed in order to determine a complex $k$. If the boundary displacements of the case $\varepsilon = -1$ are denoted by $u_2^{(r)}$, $v_2^{(r)}$ and those of the case $\varepsilon = \mathrm{i}$ by $u_2^{(i)}$ and $v_2^{(i)}$, then

$$\begin{aligned} k_1 &= \quad \mathrm{Re}\,k = \frac{1}{2^{\frac{1}{2}}\cdot\pi}\int_B (X_1 u_2^{(r)} + Y_1 v_2^{(r)})\,\mathrm{d}s \\ k_2 &= -\mathrm{Im}\,k = \frac{1}{2^{\frac{1}{2}}\cdot\pi}\int_B (X_1 u_2^{(i)} + Y_1 v_2^{(i)})\,\mathrm{d}s\,. \end{aligned} \tag{5.92}$$

As one can see, $k_1$, $k_2$ appear as weighted averages over the boundary tractions $X_1$, $Y_1$. For this reason the boundary displacements $u_2$, $v_2$ of the fundamental states will be referred to as *weight functions.*

The first formula becomes particularly simple for a domain $A$, symmetrically shaped with respect to the $x$-axis and loaded along the crack by normal tractions of induced nature, as it was indicated in the examples associated with Figure 5.1. The integrals in equations (5.92) will have contributions from the crack faces only; moreover $C^+$, $C^-$ contribute equally. Let the crack extend from $x = -l$ to $x = 0$ and set (along $C^+$) $Y_1 = p(x)$, $v_2 = m(x)$. The latter is the normal displacement on $C^+$ and $m(x)$ is so normalized that asymptotically $m(x) = (-x)^{\frac{1}{2}}$ near $x = 0$. Under these circumstances $k_1$ becomes

$$k_1 = \frac{2^{\frac{1}{2}}}{\pi}\int_{-l}^{0} p(s)\,m(s)\,\mathrm{d}s\,. \tag{5.93a}$$

Formulas (5.92) are not changed if one superimposes a rigid body motion on either field. Such a motion for the fundamental field adds nothing to the integral since the tractions of state (1) are naturally self-equilibrated. Therefore the normalization $P_2 = 0$ on $B$ is no limitation of the validity of formulas (5.92).

While it was assumed that the crack leaves the origin in the direction of the negative $x$-axis, the basic results are not confined to such a configuration. Equations (5.92) can be easily adjusted to a situation where $\zeta$ by equation (5.11) replaces $z$. To take an example, reference is made to Figure 5.1. Here (5.93a) applies in the form

$$k_1 = \frac{2^{\frac{1}{2}}}{\pi} \cdot \int_{-b}^{-c} g(x)\, m(x)\, \mathrm{d}x\,, \tag{5.93b}$$

with $m(x)$ related to the fundamental field whose fundamental singularity is at $z = -c$ rather than $z = 0$. Expansions (5.22) take the modified form

$$\phi(z) = \sum_{n=-1}^{\infty} a_n (z+c)^{\frac{1}{2}n}\,, \quad \rho(z) = \sum_{n=-1}^{\infty} b_n (z+c)^{\frac{1}{2}n} \tag{5.22a}$$

for the fundamental fields; for fields with a normal singularity at $z = -c$ the coefficients $a_{-1}$, $b_{-1}$ vanish.

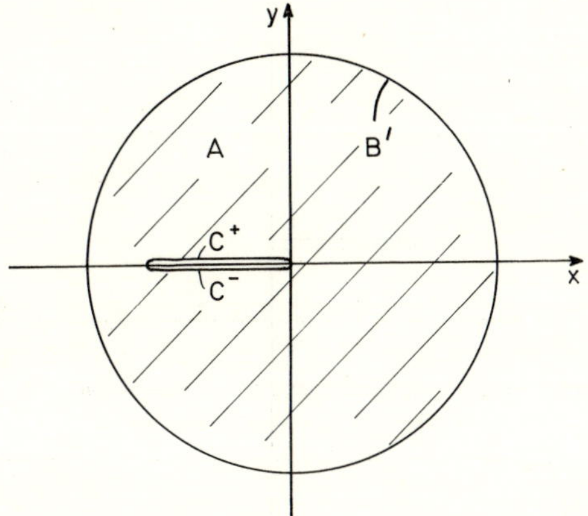


Figure 5.7

So far only the configuration of an open crack has been considered. Let now the crack be inside, as shown in Figure 5.7. Let the crack be again on the negative $x$-axis, extending from $x = -l$ to $x = 0$. Let the domain $A$ become simply connected if the crack were removed. Set

$$\phi_s(z) = a\left(\frac{l+z}{lz}\right)^{\frac{1}{2}}, \quad \rho_s(z) = \bar{a}\left(\frac{l+z}{lz}\right)^{\frac{1}{2}}. \tag{5.94}$$

These functions are well defined in $A$. The associated field has the asymptotic behavior of equations (5.71) as $z \to 0$; it induces no tractions along the faces of the crack. But there are boundary tractions on $B'$. As before they can be relieved by an additional field, regular and with normal singularities at $z = 0$, $z = -l$. Let $\phi_r$, $\rho_r$ denote the functions of that field. Now compose

$$\phi = \phi_s + \phi_r\,, \quad \rho = \rho_s + \rho_r \tag{5.95}$$

in analogy to equations (5.74). Again the associated field will be called fundamental with a fundamental singularity at $z=0$. It has a normal singularity at $z=-l$; it has no boundary tractions on $C^+, C^-, B'$. With insignificant modifications, the procedure marked by the formulas (5.75)–(5.89) can be repeated. In the end it turns out that the formulas (5.92) are valid as before.

All of this can be generalized. The reader will find the extension to analytic cracks in [5]; the same reference deals with loads on the crack more general than the one given by equations (5.21). General multiply connected and also infinite domains $A$ are also covered. But this is not the end of it. One can extend the definition of the fundamental field as well as its applications to

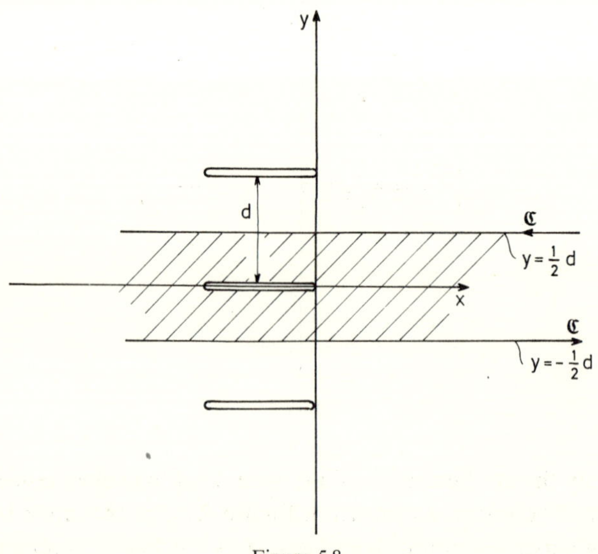


Figure 5.8

the case, where boundary conditions on $B$, pertaining to tractions, are at least partially changed into geometric conditions on the displacements. In view of the implications of the half-plane problem treated in section 5.2 one must exclude a neighborhood of the root of the crack from geometric conditions. Finally one can deal with fields which have more than one fundamental singularity. Assume for instance a cascade of cracks as shown in Figure 5.8. Let $d$ be the pitch of the cascade and assume that all cracks are identically loaded, the load consisting of normal tractions of induced nature. The stress field will have period $d$ with respect to $y$; the displacements can be expected to stay bounded as $|x|\to\infty$, for the same approach let the stresses go to zero.

Under these circumstances one should look for a fundamental field of the same features, the fundamental singularities to be at the points $z=ndi$, $n$ running through all integers. Take now the period strip $-\frac{1}{2}d\leqslant y\leqslant\frac{1}{2}d$. If $\mathfrak{C}$ consists of the two straight lines $y=\pm\frac{1}{2}d$, then $v=0$, $\tau_{xy}=0$ on $\mathfrak{C}$ and $W^*(\mathfrak{C})=0$. Hence equations (5.92), applied to the strip, stay valid with $B$ replaced by the two crack faces in the strip.

We turn to the means of finding weight functions. To begin with, equations (5.94) can be identified as the exact description of fundamental states for the domain of the Griffith crack, the crack to occupy the interval $-l<x<0$ of the real axis. Hence equation (5.93) takes the special form

$$k_1 = \frac{2^{\frac{1}{2}}}{\pi}\int_{-l}^{0}\left(\frac{l+x}{|x|\,l}\right)^{\frac{1}{2}} p(x)\mathrm{d}x\,. \tag{5.96}$$

The same formula was given by Barenblatt [10] who derived the value of $k$ from a direct representation of the actual field responding to the tractions on the crack. As $l\to\infty$ the domain becomes that of the $z$-plane cut along the negative $x$-axis. The functions (5.94) take the form (5.71), and the latter are (obvious anyway) the functions of the fundamental fields. The weight function is $m=|x|^{-\frac{1}{2}}$.

In order to obtain weight functions in other cases one can resort to at least three different methods:

(1) Compute the functions $\phi_r$, $\rho_r$. This is a common problem of elasticity; any method suitable to deal with the problem of finding the field for prescribed traction on the boundary can be employed, in particular the methods of collocation, of finite elements and those which bear the names of Ritz and Galerkin. Note that the effort is not greater than that of finding $k$ for a particular problem, while the result permits us to compute $k$ for all other problems by way of equations (5.92).

(2) Parameter differentiation with respect to $c$ in (5.22a) can be used. Assume here a fixed load on the boundary with a fixed function $f(x)$ in equation (5.21) prescribed. Look for a regular field with a normal singularity at $x=-c$ and consider the approach $c\to 0$. In accordance with the theorem one writes

$$\phi(z,c) = \sum_{n=1}^{\infty} a_n(c)(z+c)^{\frac{1}{2}n}\,; \qquad \rho(z,c) = \sum_{1}^{\infty} b_n(c)(z+c)^{\frac{1}{2}n}\,. \tag{5.97a}$$

Differentiation with respect to $c$ leads to a field with vanishing boundary tractions, while

$$\frac{\partial \phi}{\partial c} = \tfrac{1}{2} a_1(c)(z+c)^{-\frac{1}{2}} + \phi_r$$

$$\frac{\partial \rho}{\partial c} = \tfrac{1}{2} b_1(c)(z+c)^{-\frac{1}{2}} + \rho_r \tag{5.97b}$$

indicate that $\partial\phi/\partial c$ and $\partial\rho/\partial c$ at $c=0$ represent a fundamental field with the singularity at $z=0$. As an example consider the problem of finding $m(x)$ for the half-plane $-1<x$ with an edge crack from $x=-1$ to $x=0$ on the real axis (Figure 5.9).

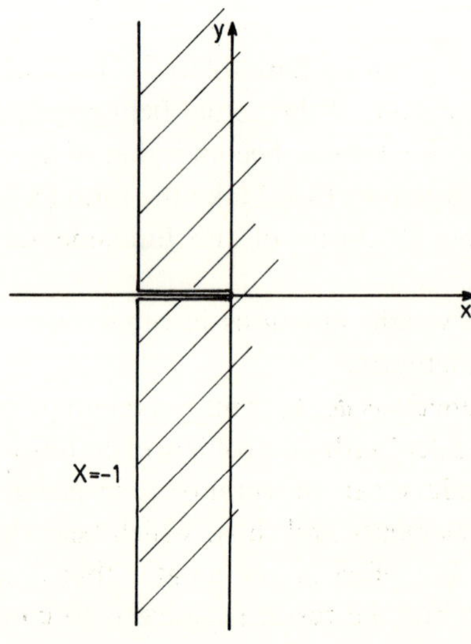


Figure 5.9

Let the crack be under constant pressure $p \equiv 1$ and assume the normal displacement $v(x)$ on $C^+$ to be known for this very case. Then

$$v(x, c) = (1-c)\,v\left(\frac{x+c}{1-c}\right) \tag{5.98}$$

for the case that the root is at $x=-c$ rather than $x=0$, the crack to be again under constant pressure $p\equiv 1$. Equation (5.98) is due to the circumstances that the two crack configurations are geometrically and dynamically similar. The procedure of differentiation leads to

$$\frac{\partial v(x,c)}{\partial c} = (1+x)\,v'(x)-v(x) \ \text{ at } \ c=0\ ; \tag{5.99}$$

consequently

$$\eta m(x) = (1+x)\,v'(x)-v(x) \tag{5.100}$$

with a normalizing factor $\eta$ such that $m(x)\approx |x|^{-\frac{1}{2}}$ near $x=0$.

If one writes $v(x)=(-x)^{\frac{1}{2}}\cdot q_0(x)$ then

$$m(x) = \frac{(1-x)q_0+2x(1+x)q_0'}{q_0(0)(-x)^{\frac{1}{2}}}\,. \tag{5.101}$$

In earlier numerical computations $q_0(x)$ was approximated by a polynomial of degree 3; a corresponding approximation to $m(x)$ can be found in Appendix II.

(3) In some cases and at least for the computation of $m(x)$ integral equations play an important and natural role. The relevant features of this method will be shown for the case of the Griffith crack with $l=1$. It is useful to assume that the normal displacement $v(x)$ on $C^+$ of the crack is known under the assumption that the latter is under normal tractions $p(x)$ such that $\sigma_y=-p(x)$ on both faces. Assuming $\phi(z)=\rho(z)$ and also $\phi(z)=\overline{\phi(\bar{z})}$ we ascertain from the very beginning that $\tau_{xy}=0$ on the crack. Furthermore

$$p(x) = -\frac{\mathrm{d}}{\mathrm{d}x}\,2\,\mathrm{Re}\,\phi \ \text{ on the faces}\,. \tag{5.102}$$

The displacement becomes

$$2\mu w = \kappa\phi-\bar{\phi} \ \text{ on } \ C^+,\,C^- \tag{5.103a}$$

whence

$$v(x) = \frac{\kappa+1}{2\mu}\,\mathrm{Im}\,\phi \ \text{ on } \ C^+\,. \tag{5.103b}$$

Prescribing $v(x)$ thus amounts to prescribing $\mathrm{Im}\,\phi=\pi q(x)$ on $C^+$. This determines $\phi(z)$ as the Cauchy integral

$$\phi(z) = \int_{-1}^{0} \frac{q(t)\,dt}{t-z}. \tag{5.104}$$

This leads to the equation

$$p(x) = -\frac{d}{dx}\,\mathrm{CPV}\int_{-1}^{0} \frac{2q(t)\,dt}{t-x}; \tag{5.105}$$

It has nontrivial solutions in the homogeneous case $p \equiv 0$, and they furnish the two weight functions associated with the crack, namely one with the singularity of the corresponding fundamental field at $z=0$ and one with the singularity at $z=-1$. It is well known that $R(t)=[-t(1+t)]^{\frac{1}{2}}$ satisfies the equation

$$\mathrm{CPV}\int_{-1}^{0} \frac{dt}{R(t)(t-x)} = 0 \text{ for } -1<x<0. \tag{5.106}$$

Using this relation one can easily show that the functions

$$R_1(t) = \left(\frac{1+t}{-t}\right)^{\frac{1}{2}}, \quad R_2(t) = 1/R_1(t) \tag{5.107}$$

satisfy

$$\mathrm{CPV}\int_{-1}^{0} \frac{R_1(t)\,dt}{t-x} = \pi; \quad \mathrm{CPV}\int_{-1}^{0} \frac{R_2(t)\,dt}{t-x} = -\pi, \tag{5.108}$$

and therefore $R_1$, $R_2$ are nontrivial solutions of (5.105) for $p \equiv 0$.

Set now

$$g(x) = \int_{-1}^{x} p(t)\,dt.$$

Then integration of equation (5.105) with respect to $x$ yields

$$\mathrm{CPV}\int_{-1}^{0} \frac{q(t)\,dt}{t-x} = -\tfrac{1}{2}(g(x)+c_0) = h(x) \tag{5.109}$$

where $c_0$ is some integration constant. The commonly known solution to equation (5.109) is

$$q(t) = \frac{1}{\pi^2 R(t)}\left(c_1 + \mathrm{CPV}\int_{-1}^{0} \frac{R(x)h(x)\,dx}{t-x}\right). \tag{5.110}$$★

★ From here on the sign CPV will be omitted.

Here $c_1$ is an arbitrary constant. One can use the constants $c_0$, $c_1$ in order to impose two conditions on $q(t)$. One finds the following limit relations:

$$\lim_{t\to 0}\int_{-1}^{0}\frac{R(x)h(x)}{t-x}\mathrm{d}x = \int_{-1}^{0} R_1(x)h(x)\mathrm{d}x = h_1$$
$$-\lim_{t\to -1}\int_{-1}^{0}\frac{R(x)h(x)}{t-x}\mathrm{d}x = \int_{-1}^{0} R_2(x)h(x)\mathrm{d}x = h_2\,. \tag{5.111}$$

Now if one wishes to impose the conditions $q(0)=q(-1)=0$, then it follows that $c_1+h_1=0$, $c_1-h_2=0$ and thus $h_1+h_2=0$ which in turn is equivalent with

$$\int_{-1}^{0}\frac{h(x)\mathrm{d}x}{R(x)} = 0\,. \tag{5.112}$$

This yields exactly one value of $c_0$, and this determines $h_1$, $h_2$ as well as $c_1$. There is then exactly one solution to equation (5.105) with $q(0)=q(-1)=0$. This solution leads to a regular state with normal singularities at $z=0$, $z=-1$. This result can also be reinterpreted in the following way: Let $q(t)=q^*(t)$ be some particular solution to equation (5.105); then the general solution is $q(t)=q^*(t)+d_1R_1(t)+d_2R_2(t)$, and one can fix the arbitrary constants $d_1$, $d_2$ in order to impose the conditions that $q(t)$ vanish at the endpoints.

In more general situations one meets integral equations of the more general type

$$p(x) = \frac{\mathrm{d}}{\mathrm{d}x}\int_{a}^{b} L(x,t)q(t)\mathrm{d}t \tag{5.113}$$

where $L(x, t)$ is a singular kernel, similar in nature to the one in equation (5.105). The phenomena, associated with the homogeneous case $p(x)\equiv 0$, are the same as those shown for equation (5.105). Because of the importance of such equations for the calculation of weight functions we shall devote all of the next section to integral equations.

This section will be concluded with some remarks on the function $m(x)$. Is this function always positive? The answer is affirmative for the two examples above, and in general $m(x)>0$ for at least small $|x|$. Linked to this question and perhaps even more important is this one: Let $m_1(x)$, $m_2(x)$ pertain to two domains $A_1$, $A_2$ of which $A_1$ is contained in $A_2$ as shown in Figure 5.10. Let the cracks be open, and let the one of $A_1$ be part of the crack of $A_2$. One would expect $m_1(x)\geqslant m_2(x)$ for all $x$ on the crack of

$A_1$, since in the case of loading the crack interval of $A_1$ by pressure one would think that the notch of $A_1$ would open up more than the one in $A_2$. Now a partial answer to the question will be given. In accord with the theorem the

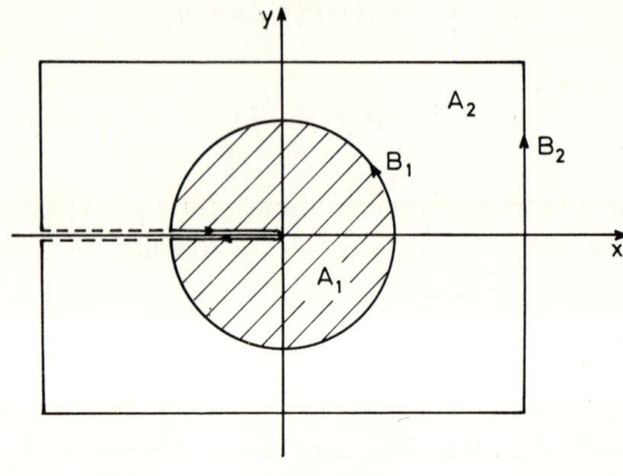


Figure 5.10

two weight functions have expansions in odd powers of $(-x)^{\frac{1}{2}}$,

$$m_k(x) = |x|^{-\frac{1}{2}} + c_k|x|^{\frac{1}{2}} + \dots \text{ for small } |x| \; ; \tag{5.114}$$

the leading terms of the associated functions are given by

$$\phi, \rho = -cz^{-\frac{1}{2}} + c_k z^{\frac{1}{2}} + \dots \; ; \qquad c = \frac{2\mu}{\kappa+1} . \tag{5.115}$$

Now the difference state defined by $\phi = \phi_2 - \phi_1$, $\rho = \rho_2 - \rho_1$ has a normal singularity at $z=0$ with $k_1 = 2^{\frac{1}{2}} \cdot (c_2 - c_1)$. The use of formula (5.92) gives

$$2\pi(c_1 - c_2) = -\int_{B_1} [(X_2 - X_1)u_1 + (Y_2 - Y_1)v_1] \mathrm{d}s \tag{5.116}$$

where $X_2$, $Y_2$ is the traction, induced by field $(\phi_2, \rho_2)$ on the boundary of $A_1$; $X_1$ and $Y_1$ vanish on the boundary of $A_1$. Let now $u_2$, $v_2$ denote the displacements of field (2) on $B_1$. Then equation (5.116) can be written in the form

$$\begin{aligned} 2\pi(c_1 - c_2) = & \int_{B_1} [(X_2 - X_1)(u_2 - u_1) + (Y_2 - Y_1)(v_2 - v_1)] \mathrm{d}s \\ & + \int_{B_1} [(X_1 - X_2)u_2 + (Y_1 - Y_2)v_2] \mathrm{d}s . \end{aligned} \tag{5.117}$$

In this representation the first integral is twice the energy of the difference field in $A_1$ while the second integral is twice the energy of state (2) in the domain $A_2 - A_1$ (note that state (2) has no tractions on the boundary of $A_2$). Altogether it is found that

$$c_1 - c_2 > 0\,, \tag{5.118}$$

which means that at least for small $|x|$ the function $m_1$ is larger than the function $m_2$. If one takes for $A_2$ the whole plane cut along the negative $x$-axis then $c_2 = 0$ and hence $c_1 > 0$.

If this result can be made more general then one could construct upper and lower bounds for $m(x)$ by bracketing the domain $A$ between domains with well-known weight functions. The result (5.118) can be extended towards other configurations, in particular those with interior cracks.

## 5.4 Integral equations for various configurations

Adhering to mode I we shall discuss certain integral equations of type (5.105). Let us begin with the derivation of two special ones, pertaining to an array of radial cracks. Figure 5.11 shows a set of $n$ cracks $C_0, C_1, \ldots, C_{n-1}$; $C_0$

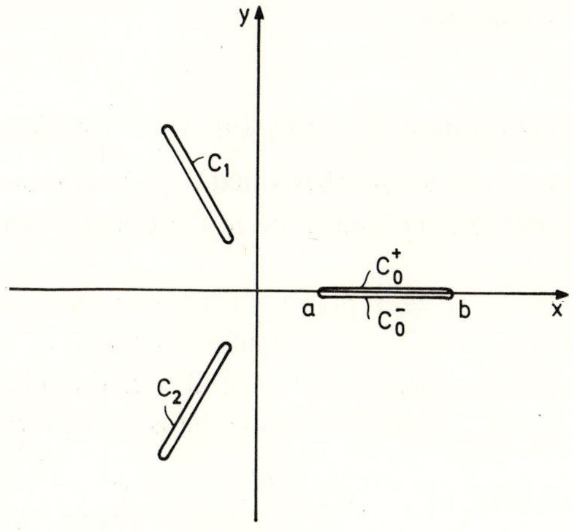


Figure 5.11

is on the real $x$-axis extending from $x = a$ to $x = b$, where $0 < a < b < \infty$. $C_k$ is obtained by rotating $C_0$ around the origin in the counterclockwise sense through the angle $2\pi k/n$. All cracks are assumed to be under normal pressure, and the pressure distribution along a crack is the same for all of them. The condition for $C_0$ can be written in the usual form

$$\sigma_y = -g(x)\ ;\tau_{xy} = 0 \text{ on } C_0^+ \text{ and } C_0^- . \tag{5.119}$$

The functions $\phi$, $\psi$ are particularly suited to analyze the situation. In order to express the same way of loading all cracks, it is necessary to write

$$\phi(\varepsilon z) = \varepsilon\phi(z)\ ;\quad \psi(\varepsilon z) = \bar{\varepsilon}\psi(z)\ ;\quad \varepsilon = \exp(2\pi i/n)\,. \tag{5.120}$$

This yields for the complex displacement

$$w(\varepsilon z) = \varepsilon w(z)\,. \tag{5.121}$$

Relation (5.121) is necessary to obtain the desired state of deformation in the plane. Since $C_0$ is on the real axis, one has, in addition to (5.120)

$$\phi(z) = \overline{\phi(\bar{z})}\,,\qquad \psi(z) = \overline{\psi(\bar{z})}\,. \tag{5.122}$$

The functions, $\phi$, $\psi$ are required to be holomorphic at infinity. Introduce the function $\omega(z) = z\phi' - \phi + \psi$, a function also following the law (5.121) but not (5.120). Then the conditions (5.119) can be expressed in the following form:

$$-g(x) = \frac{d}{dx}\,\mathrm{Re}\,(2\phi + \omega)\,, \tag{5.123a}$$

and

$$\tau_{xy} = \mathrm{Im}\,(\bar{z}\phi'' + \psi') = \mathrm{Im}\,(z\phi'' + \psi') = \mathrm{Im}\,\omega' = 0 \text{ on } C_0\,, \tag{5.123b}$$

which leads to the working hypothesis that $\omega(z)$ be real on $C_0$, continuous across the crack and thus holomorphic in the domain of Figure 5.11 with $C_0$ removed.

The functions $\phi$, $\psi$ can be represented by the Cauchy integral formula where the integration pertains to all cracks. By virtue of equations (5.120) the integration can be reduced to an integral over $C_0$ only and finally, due to (5.122), to one on $C_0^+$ alone. The outcome is of the form

$$\phi(z) = \int_a^b K_1(z,t)\,h_1(t)\,dt\,,\qquad K_1 = \frac{nzt^{n-2}}{t^n - z^n} = \sum_{k=0}^{n-1} \frac{\varepsilon^{2k}}{t\varepsilon^k - z} \tag{5.124}$$

$$\psi(z) = \int_a^b K_2(z,t)\,h_2(t)\,dt\,,\qquad K_2 = \frac{nz^{n-1}}{t^n - z^n} = \sum_{k=0}^{n-1} \frac{1}{t\varepsilon^k - z}\,. \tag{5.125}$$

Here the functions $\pi h_1$, $\pi h_2$ represent the imaginary parts of $\phi$, $\psi$ respectively on $C_0^+$. The reality and regularity of $\omega$ on points of $C_0$ permits us to set from now on

$$h_1(t) = q(t), \quad h_2(t) = q(t) - tq'(t). \tag{5.126}$$

Formulas (5.124), (5.125) are restricted to points $z$ not on a crack. If $z$ is on a crack face the Plemelj formulas on Cauchy integrals must be used.

Relations (5.126) imply in particular

$2\mu v(x) = (\kappa + 1) q(x)$ for the normal displacement $v$ on $C_0^+$,

while for points $z$ not on a crack:

$$\psi(z) = \int_a^b K_2^*(z, t) q(t) \mathrm{d}t \text{ with } K_2^* = K_2 + \frac{\partial}{\partial t}(tK_2). \tag{5.127}$$

Equation (5.127) is due to integration by parts with respect to the $q'$-term in (5.125) with $q(a) = q(b) = 0$ assumed. Similarly one can write

$$z\phi'(z) - \phi(z) = \int_a^b K_1(z, t)(tq' - q)\mathrm{d}t = -\int_a^b K_1^*(z, t) q(t) \mathrm{d}t$$

with (5.128)

$$K_1^* = K_1 + \frac{\partial}{\partial t}(tK_1).$$

(5.123a) can be given the equivalent form

$$-g(x) = \frac{\mathrm{d}}{\mathrm{d}x}\left\{2\int_a^b K_1 q \,\mathrm{d}t + \int_a^b (K_2 - K_1)(q - tq')\mathrm{d}t\right\}. \tag{5.129}$$

While $K_1$ is a singular integral operator the difference $K_2 - K_1$ takes the form of a smooth function of $x$ and $t$, namely

$$K_2 - K_1 = K_0 = -nx\,\frac{x^{n-2} - t^{n-2}}{x^n - t^n}. \tag{5.130}$$

Therefore the use of equations (5.127), (5.128) can be extended to points of $C_0$ inasmuch as $K_0$ is concerned. Introduce $K_0^* = K_2^* - K_1^*$, and equation (5.129) can be written as

$$-g(x) = \frac{\mathrm{d}}{\mathrm{d}x}\int_a^b [2K_1(x, t) + K_0^*(x, t)]\, q(t) \mathrm{d}t. \tag{5.131}$$

This is an integral equation for $q$ if $g(x)$ is prescribed. It is slightly more general than the type (5.105). For $n = 1$ the equations are the same. The above equation was derived to determine a regular state with normal singularities only. Obviously equation (5.131) can also be used to deal with the homogeneous case $g \equiv 0$. If one lifts the restriction that $q$ vanish at the endpoints two kinds of weight functions are obtained: One associated with singularities at $|z| = a$ and one for singularities at $|z| = b$ simultaneously on all cracks.

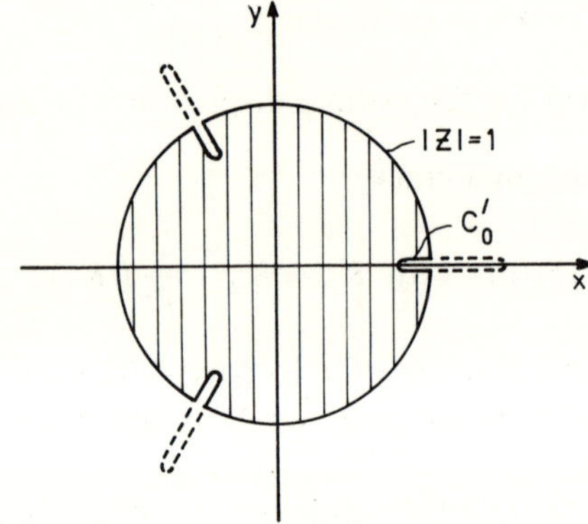


Figure 5.12

Let $ab = 1$ and modify the problem above by confining the cracks to the unit disk $|z| = 1$ (Figure 5.12). Conditions (5.119), (5.120), (5.122) shall stay the same; $g(x)$ is of course restricted to the parts $C_k'$ (of the original cracks $C_k$) in the disk. The disk boundary $|z| = 1$ is to be free from traction. Because of this circumstance one can expect that the field can be analytically continued into the domain of Figure 5.11. It will turn out that certain allowances must be made. To be cautious the continued field functions $\phi$, $\psi$ will not be required to be holomorphic at infinity. Now consider $P$ on the circular boundary where

$$P = \bar{\phi} + \bar{z}\phi' + \psi = \bar{\phi} + z^{-1}\phi'(z) + \psi(z) \text{ since } \bar{z} = z^{-1} . \tag{5.132}$$

Furthermore, as a consequence of equation (5.122),

$$\overline{\phi(z)} = \phi(\bar{z}) = \phi(1/z) \text{ on } |z| = 1 . \tag{5.133}$$

Since $P = 0$ can be assumed without essential loss of generality, the continuation rules

$$\phi(1/z) = -\psi(z) - \phi'(z)/z \, ; \quad \psi(1/z) = -\phi(z) - z\phi'(1/z) . \tag{5.134}$$

can be derived from equations (5.132) and (5.133). These formulae permit us to continue the field into the exterior of the disk. The function $\omega(z)$ above takes the form

$$\omega(z) = (z - 1/z)\,\phi'(z) - \phi(z) - \phi(1/z) . \tag{5.135}$$

As before this function has to be holomorphic in the domain (of the disk this time) without $C_0'$. On the latter it is real-valued. Relations (5.126), restricted to $C_0'$, describe as before the relation between $q$ (the imaginary part of $\phi$) and the imaginary part of $\psi$ on $C_0^+$. A new feature is that the regularity of $\phi$, $\psi$ at $z=0$ imply the behavior $\phi(z)\approx -\phi'(0)z$ as $z\to\infty$.

A slight deviation from the procedure of the first problem is made by setting up right away

$$\phi(z) = \int_a^1 M(z, t)q(t)\mathrm{d}t \; ; \quad \psi(z) = \int_a^1 N(z, t)q(t)\mathrm{d}t \, . \tag{5.136}$$

Note that both equations pertain to $q(t)$. $M$, $N$ are assumed to behave like $K_1$, $K_2^*$ respectively inside the disk. Consequently $M$ must have simple and $N$ double poles at the points $t<1$ of $C_0$. They will also have poles on the cracks $C_k$ outside the disk; otherwise $\phi$, $\psi$ would be regular for $|z|>1$ which is not in accord with the rules (5.134). The relation

$$N(z, t) = -M(1/z, t) - M'(z, t)/z \, , \tag{5.137}$$

is specified in accordance with (5.134); the prime denotes differentiation with respect to $z$. Introducing the analogue of $\omega$ as

$$\Omega(z, t) = (z-1/z)M'(z, t) - M(z, t) - M(1/z, t) \tag{5.138}$$

we require that this function be regular at any point $z=t$ on $C_0'$. Due to the representation (5.124) the function $M$ is constructed by adding a special rational function of $z$ to $K_1$:

$$M(z, t) = n\left[\frac{zt^{n-2}}{t^n - z^n} + z\,\frac{A(t)}{1-t^n z^n} + z\,\frac{B(t)}{(1-t^n z^n)^2} + cz\right]. \tag{5.139}$$

This function has simple poles at the points $z=\varepsilon^k t$ as required; at the points $z=\varepsilon^k/t$ allowance must be made for double poles since otherwise $N(z, t)$ might not have a double pole at $z=t$. But more than double poles are not admissible. An attempt is now made to find functions $A(t)$, $B(t)$ and a constant $c$, such that $\Omega$ is regular as required. The term $cz$ is necessary in order to enforce regularity of $N(z, t)$ at $z=0$. The following formulas list pertinent terms with respect to the point $t$ on $C_0'$:

$$\begin{aligned} M/n &= \frac{zt^{n-2}}{t^n - z^n} + \dots \; ; \quad M'/n = \frac{t^{n-2}}{t^n - z^n} + n\,\frac{z^n t^{n-2}}{(t^n - z^n)^2} + \dots \\ M(1/z, t)/n &= -A\,\frac{z^{n-1}}{t^n - z^n} + B\,\frac{z^{2n-1}}{(t^n - z^n)^2} + \dots \, . \end{aligned} \tag{5.140}$$

It must be required that

$$(t^n - z^n)[(z-1/z)M'(z,t) - M(z,t) - M(1/z,t)] \to 0 \text{ as } z \to t. \tag{5.141}$$

This condition yields at once

$$B = n\frac{t^2-1}{t^2} \tag{5.142}$$

as a necessary condition to remove the double pole of $\Omega$ at $z=t$. The unknown $A$ can now be so determined that the simple pole left over is also destroyed. This leads to

$$A = 2 - n + (n+1)t^{-2}. \tag{5.143}$$

Finally $c$ is found as

$$c = -\frac{1+t^2}{t^2}. \tag{5.144}$$

The verification of equations (5.143) and (5.144) is an elementary exercise left to the reader. The functions $M$ and $N$ satisfy all conditions imposed above. This is obvious. It remains to write down the analogue of equation (5.105) in explicit form. The result is

$$-g(x) = \frac{\mathrm{d}}{\mathrm{d}x}\left\{\int_a^1 [2M(x,t) + \Omega(x,t)]\,q(t)\,\mathrm{d}t\right\}. \tag{5.145}$$

That $q(x)$ has the meaning (5.126) follows from the already demonstrated behavior of the component $K_1$ of $M$.

The equation leads to a regular field with a normal singularity at $z=a$ on $C_0$ if one imposes the condition $q(a)=0$. If one lifts the condition one obtains weight functions as solutions of the homogeneous case $g\equiv 0$. In order to find the normalized weight function one should set $q(t)=(t-a)^{-\frac{1}{2}}+q_0(t)$. This leads to an inhomogeneous equation for $q_0(t)$, the latter to be solved under the condition $q_0(a)=0$. The numerical computation of $q_0(t)$ can follow any method designed for solving singular integral equations. Formula (5.93b) is applicable with the integration to go over the interval $(a, 1)$. This yields immediately the value of $k_1$ for any of the $n$ cracks. The same applies to the configuration of Figure 5.11.

While $ab=1$ was assumed for simplicity, the pertinent equations of the disk problem can be easily transformed to any disk radius $r=(ab)^{\frac{1}{2}}$. In addition one can apply a shift transformation. Both steps will be combined.

We employ in particular a transformation from $z$ to $\zeta = r(z-1)$. Prescribe some $d > 0$, $\alpha < 0$ and set $r = nd/2\pi$ and $a = 1 - \alpha/r$. If $n \to \infty$ then Figure 5.12 changes into Figure 5.13 of the $\zeta$-plane. The limit procedure can be applied to the integral equation for $q$ and leads to an equation which links $q$ and $g$ to one another. This equation governs the cascade of edge cracks of Figure 5.13.

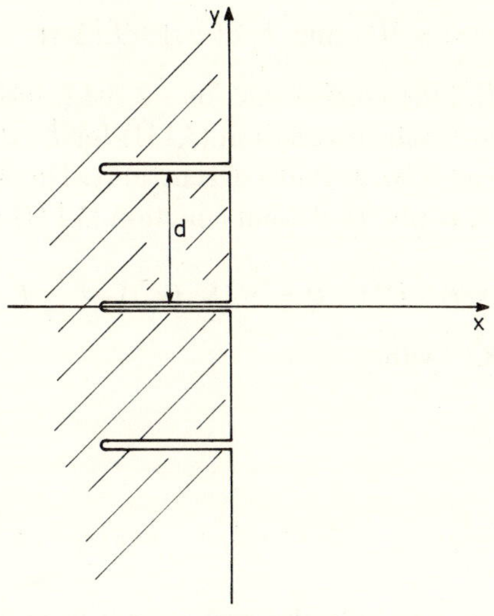


Figure 5.13

A similar procedure can be applied to Figure 5.11, so as to obtain the cascade of Figure 5.8 as well as the associated integral equation. The latter will be pursued in some detail. Now set $a = 1 - l/r$, $b = 1$; $l$ takes the place of $\alpha$; the definition of $\zeta$ and $r$ stay unchanged. The limit procedure yields precisely the cascade of Figure 5.8.

Now if one deals quite generally with equations of the type (5.113), which can be rewritten in the form

$$g(x) = \frac{\mathrm{d}}{\mathrm{d}x} \int_a^b L(x, t)\, q(t)\, \mathrm{d}t\,, \tag{5.146}$$

where $q(t)$ is related to the normal displacement on a crack by equation (5.126) and where $g(x)$ describes the normal traction, then any transformation of the type

$$x = \delta_0 + \delta_1 \xi\ ;\quad t = \delta_0 + \delta_1 \tau\ ;\quad \delta_1 > 0 \tag{5.147}$$

leads to an integral equation of the form

$$\tilde{g}(\xi) = \frac{\mathrm{d}}{\mathrm{d}\xi}\int_{\alpha}^{\beta} \tilde{L}(\xi, \tau)\tilde{q}(\tau)\mathrm{d}\tau\,, \tag{5.148}$$

where $\alpha$, $\beta$ are related to $a$, $b$ by equations (5.147). If one wishes to retain the meaning of $q(t)$ in equation (5.126), then one must set

$$q(t) = \tilde{q}(\tau)\,,\quad \delta_1 g(x) = \tilde{g}(\xi) \text{ and } \delta_1 L(x, t) = \tilde{L}(\xi, \tau) \tag{5.149}$$

in order to make $\tilde{g}(\xi)$ the normal traction on the transformed crack.

Applying the last result to equation (5.131) for $l > 0$ which can be any number, one can set $d = \pi$ without loss of generality. This leads to $x = 1 + 2\xi/n$, $t = 1 + 2\tau/n$. Hence it is observed from equation (5.131) that

$$L(x, t) = -2K_1(x, t) - K_0^*(x, t) = -2K_2(x, t) - t\frac{\partial}{\partial t}K_0(x, t)\,, \tag{5.150}$$

$$\tilde{L}(\xi, \tau) = 2T_n - 4S_n \quad \text{with} \tag{5.151}$$

$$S_n = \frac{x^{n-1}}{t^n - x^n}\,,$$

$$T_n = xt\frac{\partial}{\partial t}\frac{x^{n-2} - t^{n-2}}{x^n - t^n}\,; \qquad x = 1 + 2\xi/\mathrm{n}\,,\ t = 1 + 2\tau/n\,.$$

Dealing with $S_n$ first we apply the well-known formula $\mathrm{e}^s = \lim(1 + s/n)^n$ as $n \to \infty$. This yields at once

$$\lim_{n\to\infty} S_n = S = \frac{\mathrm{e}^{2\xi}}{\mathrm{e}^{2\tau} - \mathrm{e}^{2\xi}} = \tfrac{1}{2}\,\mathrm{ctgh}\,(\tau - \xi) - \tfrac{1}{2}\,. \tag{5.152}$$

More attention should be given to

$$T_n = (1 + 2\xi/n)(1 + 2\tau/n)\frac{\partial}{\partial\tau}\left\{\tfrac{1}{2}n\frac{(1 + 2\xi/n)^{n-2} - (1 + 2\tau/n)^{n-2}}{(1 + 2\xi/n)^n - (1 + 2\tau/n)^n}\right\}. \tag{5.153}$$

This is identical with

$$T_n = (1 + 2\xi/n)(1 + 2\tau/n)\frac{\partial}{\partial\tau}Q \quad \text{with}$$

$$Q = \tfrac{1}{2}n\frac{(1 + 2\xi/n)^n[1 + 4\tau/n + 4(\tau/n)^2] - (1 + 2\tau/n)^n[1 + 4\xi/n + 4(\xi/n)^2]}{[(1 + 2\xi/n)^n - (1 + 2\tau/n)^n](1 + 2\xi/n)^2(1 + 2\tau/n)^2}$$

$$= \tfrac{1}{2}n + 2\frac{\tau(1 + 2\xi/n)^n - \xi(1 + 2\tau/n)^n}{(1 + 2\xi/n)^n - (1 + 2\tau/n)^n} + O(n^{-1})\,; \quad \xi \neq \tau\,.$$

Here the additive constant $\frac{1}{2}n$ is destroyed by the differentiation $\partial/\partial\tau$ and for the explicitly written middle term one can go to the limit as in equation (5.152). Consequently

$$\lim_{n\to\infty} T_n = T = \frac{\partial}{\partial\tau}\frac{\tau e^{2\xi}-\xi e^{2\tau}}{e^{2\xi}-e^{2\tau}} = \frac{\partial}{\partial\tau}[\xi+\tau-(\tau-\xi)\operatorname{ctgh}(\tau-\xi)]\,. \qquad (5.154)$$

Altogether

$$\tilde{L}(\xi,\tau) = -2\left[\operatorname{ctgh}(\tau-\xi)-2+\frac{\partial}{\partial\tau}(\tau-\xi)\operatorname{ctgh}(\tau-\xi)\right]. \qquad (5.155)\star$$

and equations (5.148), (5.155) constitute the integral equation for the cascade of Figure 5.8 if one identifies the coordinates $x$, $y$ in the future with $\xi$, $\eta$ respectively.

In a similar vein $\tilde{L}(\xi,\tau)$ for the cascade of edge cracks can be found. The steps are basically the same and naturally more laborious. This goal will not be pursued in this chapter. A few remarks are still in order. For $n=1$ the cascade of edge cracks reduces to a single edge crack as shown in Figure 5.9. Here the $\tilde{L}$ is obtained by letting $r\to\infty$, while $d$ plays no role. The outcome is known, and with respect to Figure 5.9 the operator $L$ in equation (5.146) takes the form

$$L(x,t) = \frac{-2}{t-x}+\frac{2}{t+x+2}+\frac{4(1+x)}{(t+x+2)^2}-\frac{8(1+x)^2}{(t+x+2)^3}\,. \qquad (5.156)\star\star$$

A reasonable approximation to the weight function for this case can be found in Appendix II. It was obtained by the method of parameter differentiation. The method employed to attack the problem of the disk with radial notches is equally applicable to the "inverse" geometry, i.e. to the plane with a circular hole, from which radial notches enter the elastic domain. For the case of just one such notch (Figure 5.14) the operator $L$ is

$$L(x,t) = -\frac{2}{t-x}+\frac{2x}{tx-1}-\frac{x^2-1}{x^2}\frac{d}{dx}\left[x^2(x^2-1)\frac{d}{dx}\left(\frac{x}{tx-1}\right)\right]-\frac{2(x^2-1)}{xt^2}\,. \qquad (5.157)\star\star$$

Before leaving the subject of radial notches one should observe that equation (5.145) and the special cases derived from it are equally valid if the

⋆ In view of the $\xi$-differentiation in equation (5.148) the constant term can be dropped.

⋆⋆ The integral equations were first published by Wigglesworth [11] and independently by the author [2] who also used them in connection with the $k_1$-computation for a notched rotor [3].

cracks are *inside*, i.e. not running up to the edge. Thus if the integration in (5.147) is restricted to the interval $(a, c)$ with $a < c < 1$, a relation is obtained expressing the normal displacement in terms of the normal traction along radial slots of which the one on the $x$-axis has endpoints $a$ and $c$. In this case there are two weight functions, one of them with a fundamental singularity at $x = a$ and another one at $x = c$. Consider now the case of a half-plane with

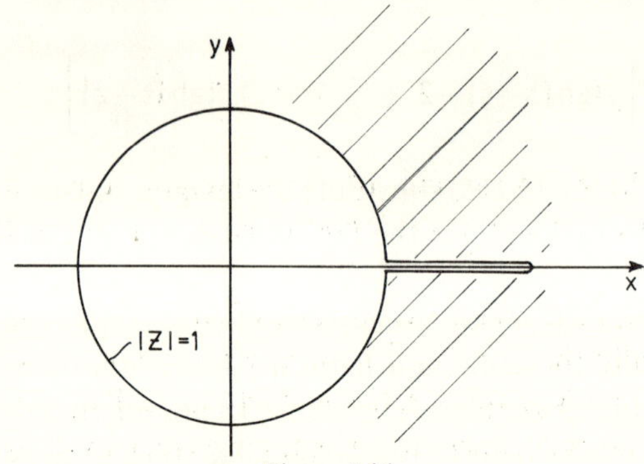


Figure 5.14

an inner crack. Let the half-plane be bounded by $x = 0$ and notched from $x = 1$ to the right up to infinity. The integral equation is

$$g(x) = \frac{\mathrm{d}}{\mathrm{d}x} \int_1^{\infty} L(x, t)\, q(t)\, \mathrm{d}t \; ;$$

$$L = -\frac{2}{t-x} + \frac{2}{t+x} + \frac{4x}{(t+x)^2} - \frac{8x^2}{(t+x)^3} \, .$$

In the homogeneous case we set $q(t) = m(t)$. Introduce new variables $\xi = 1/x$, $\tau = 1/t$, then the following relations emerge:

$$\tfrac{1}{2} L(x, t) = \tau^2 \left[ \frac{-\xi}{\tau(\xi - \tau)} + \frac{\xi}{\tau(\xi + \tau)} + \frac{2\xi}{(\xi + \tau)^2} - \frac{4\xi\tau}{(\xi + \tau)^3} \right]$$

$$\frac{-\xi}{\tau(\xi - \tau)} = \frac{-1}{\xi - \tau} - \frac{1}{\tau}$$

$$\frac{\tau\xi}{(\xi + \tau)^3} = \frac{\xi}{(\xi + \tau)^2} \left( 1 - \frac{\xi}{\xi + \tau} \right)$$

$$\frac{\xi}{\tau(\xi + \tau)} = \frac{-1}{\xi + \tau} + \frac{1}{\tau} \, .$$

With the aid of them we find at once that

$$\tau^2 L(x, t) = -L(\xi, \tau) .$$

Consequently the homogeneous equation changes into

$$0 = \frac{\mathrm{d}}{\mathrm{d}\xi} \int_0^1 L(\xi, \tau) m(1/\tau) \mathrm{d}\tau ,$$

and the weight function for the half-plane with an infinitely deep notch is simply

$m(t) = m^*(1/t)$, $m^*(t)$ = weight function for half-plane with edge notch.

This section will be concluded with a description of the integral equation for the notched bar. Details have been already published [6].

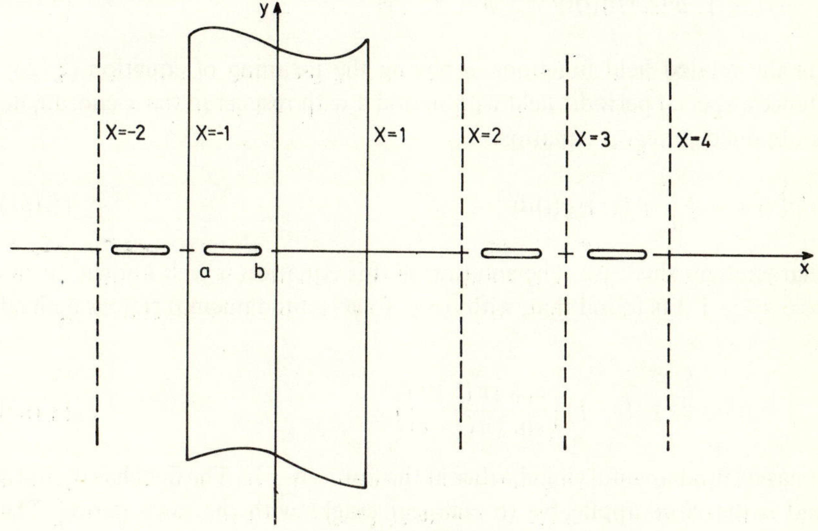


Figure 5.15

Let the bar occupy the domain of the strip $-1 \leqslant x \leqslant 1$ of the complex $z$-plane. The crack lies on the real axis, and occupies the interval $a \leqslant x \leqslant b$ such that $-1 \leqslant a < b < 1$. The case $a = -1$ refers to an edge crack. The operator $L(x, t)$ is representable in the following form:

$$L(x, t) = -2\phi(x, t) - \frac{\partial}{\partial t} K(x, t) \tag{5.158}$$

where $\phi(x, t)$ refers to the values on the crack interval of the analytic function

$$\phi(z,t)=\frac{\pi}{4}\left[\cot\frac{\pi}{4}(t-z)+\cot\frac{\pi}{4}(2-t-z)\right]=\frac{\frac{\pi}{2}\cos\frac{\pi}{2}z}{\sin\frac{\pi}{2}t-\sin\frac{\pi}{2}z}\;;\; a<t<b\,. \tag{5.159}$$

The functions $\phi(z, t)$ and $\psi(z, t)=\phi-z\phi'$ are closely related to the row of collinear cracks obtained from the one in the bar by certain shifts and reflections, as indicated in Figure 5.15. The shifts are by $4n$ in the $x$-direction, $n$ running through all integers, and the reflections are with respect to the lines $x=2n+1$. If all the cracks so obtained are loaded by normal tractions $g(t)$ where $g(t+4n)=g(t)$ and $g(2-t)=g(t)$ then

$$\phi(z)=\int_a^b \phi(z,t)\,q(t)\,\mathrm{d}t\,,\qquad \psi=\phi-z\phi' \tag{5.160}$$

are the related field functions, $q$ having the meaning of equation (5.126). Hence a special periodic field with period 4 with respect to the $x$-coordinate is obtained. Integral equation

$$g(x)=-2\int_a^b \phi(x,t)\,q(t)\,\mathrm{d}t \tag{5.161}$$

characterizes this case. The solution of this equation is well known. In the case $a=-1$ it is found that, with $\rho=\psi+z\phi'$, a fundamental state is defined by

$$\phi=\rho=\cos\tfrac{1}{4}\pi(z-b)\left(\frac{\cos\frac{1}{4}\pi(z+b)}{\sin\frac{1}{4}\pi(z-b)}\right)^{\frac{1}{2}}. \tag{5.162}$$

It has its fundamental singularities at the points $b+4n$. The field has period 4 and is therefore applicable to collinear cracks with the same period. The corresponding weight function is listed in Appendix II.

The function $K(x, t)$ in equation (5.158) is obtained by relieving the tractions of the field (5.160) on the bar boundaries $|x|=1$ with the crack ignored. This process does not change $v(x)$ on the upper side of the crack. In order to relieve the tractions a Fourier-transform can be employed. The outcome is

$$K(x,t)=\int_0^\infty\left\{\frac{H_1(x,\lambda)\,H_1(t,\lambda)}{P_1(\lambda)\sinh^2\lambda}+\frac{H_2(x,\lambda)\,H_2(t,\lambda)}{P_2(\lambda)\cosh^2\lambda}\right\}\lambda\,\mathrm{d}\lambda \tag{5.163}$$

with

$$P_1(\lambda) = \sinh 2\lambda + 2\lambda, \qquad P_2(\lambda) = \sinh 2\lambda - 2\lambda$$

$$H_1(t, \lambda) = (1-t)\sinh(1+t)\lambda - (1+t)\sinh(1-t)\lambda$$
$$= -2\sinh^2\lambda \frac{\partial}{\partial\lambda}\left(\frac{\sinh t\lambda}{\sinh\lambda}\right)$$

$$H_2(t, \lambda) = (1-t)\sinh(1+t)\lambda + (1+t)\sinh(1-t)\lambda$$
$$= -2\cosh^2\lambda \frac{\partial}{\partial\lambda}\left(\frac{\cosh t\lambda}{\cosh\lambda}\right).$$

These functions have the $t$-derivatives

$$\frac{\partial}{\partial t} H_1(t, \lambda) = -2\sinh^2\lambda \frac{\partial}{\partial\lambda}\left(\frac{\lambda\cosh t\lambda}{\sinh\lambda}\right)$$

$$\frac{\partial}{\partial t} H_2(t, \lambda) = -2\cosh^2\lambda \frac{\partial}{\partial\lambda}\left(\frac{\lambda\sinh t\lambda}{\cosh\lambda}\right).$$

The integral (5.163) can be reduced to integrals which depend on one parameter only. Indeed one may write

$$\begin{aligned} 2H_k(x, \lambda) H_k(t, \lambda) = {} & (1-x)(1-t)[\cosh(2+x+t)\lambda - \cosh(x-t)\lambda] + \\ & + (1+x)(1+t)[\cosh(2-x-t)\lambda - \cosh(x-t)\lambda] + \\ & + \varepsilon(1-x)(1+t)[\cosh(2+x-t)\lambda - \cosh(x+t)\lambda] + \\ & + \varepsilon(1+x)(1-t)[\cosh(2-x+t)\lambda - \cosh(x+t)\lambda], \end{aligned} \tag{5.164}$$

where $\varepsilon = -1$ for $k = 1$ and $\varepsilon = 1$ for $k = 2$.

It is now obvious that it suffices to deal with the integrals

$$Q_1(\tau) = \int_0^\infty \frac{(\cosh\tau\lambda - 1)\lambda\,\mathrm{d}\lambda}{P_1(\lambda)\sinh^2\lambda}, \quad Q_2(\tau) = \int_0^\infty \frac{(\cosh\tau\lambda - 1)\lambda\,\mathrm{d}\lambda}{P_2(\lambda)\cosh^2\lambda}, \tag{5.165}$$

where $|\tau| < 4$. For practical reasons it is preferable to represent $K(x, t)$ by means of

$$4F_1(\tau) = \int_0^\infty \left(\frac{1}{P_1(\lambda)\sinh^2\lambda} - 8\mathrm{e}^{-4\lambda}\right)(\cosh\tau\lambda - 1)\lambda\,\mathrm{d}\lambda,$$

$$4F_2(\tau) = \int_0^\infty \left(\frac{1}{P_2(\lambda)\cosh^2\lambda} - 8\mathrm{e}^{-4\lambda}\right)(\cosh\tau\lambda - 1)\lambda\,\mathrm{d}\lambda, \tag{5.166}$$

$$F^*(\tau) = 2\int_0^\infty \mathrm{e}^{-4\lambda}(\cosh\tau\lambda - 1)\lambda\,\mathrm{d}\lambda = \frac{1}{(\tau-4)^2} + \frac{1}{(\tau+4)^2} - \frac{1}{8}.$$

Since, for $\lambda \to \infty$,

$$\frac{1}{P_1(\lambda)\sinh^2\lambda} - 8\,e^{-4\lambda} = O(\lambda e^{-6\lambda}), \quad \frac{1}{P_2(\lambda)\cosh^2\lambda} - 8\,e^{-4\lambda} = O(\lambda e^{-6\lambda}) \tag{5.167}$$

the integrals offer the advantage that $F_1(\tau)$, $F_2(\tau)$ are holomorphic for $|\tau| < 6$. Tables for the functions $F_1$, $F_2$ can be found in Appendix II. The tables also show the values of two polynomials $G_1$, $G_2$ for comparison. These have the form

$$G_k = a_k\tau^2 + b_k\tau^4 + c_k\tau^6\ ; \qquad k = 1, 2\,. \tag{5.168}$$

It appears that $G_k$ is an acceptable approximation to $F_k$, at least for engineering applications.

The function $K(x, t)$, due to (5.163) is easily expressed with the aid of the functions $F_1$, $F_2$ and $F^*$.

Writing

$$K(x, t) = K_1(x, t) + K_2(x, t) + K^*(x, t) \tag{5.169}$$

where $K_1$, $K_2$, $K^*$ represent the contributions of $F_1$, $F_2$, $F^*$ respectively, we obtain

$$\begin{aligned}
\tfrac{1}{2}K_1(x, t) = {} & (1-x)(1-t)[F_1(2+x+t) - F_1(x-t)] - \\
& - (1-x)(1+t)[F_1(2+x-t) - F_1(x+t)] + \\
& + (1+x)(1+t)[F_1(2-x-t) - F_1(x-t)] - \\
& - (1+x)(1-t)[F_1(2-x+t) - F_1(x+t)]\,,
\end{aligned}$$

$$\begin{aligned}
\tfrac{1}{2}K_2(x, t) = {} & (1-x)(1-t)[F_2(2+x+t) - F_2(x-t)] + \\
& + (1+x)(1+t)[F_2(2-x-t) - F_2(x-t)] + \\
& + (1-x)(1+t)[F_2(2+x-t) - F_2(x+t)] + \\
& + (1+x)(1-t)[F_2(2-x+t) - F_2(x+t)]\,,
\end{aligned}$$

$$\begin{aligned}
\tfrac{1}{4}K^*(x, t) = {} & (1-x)(1-t)[F^*(2+x+t) - F^*(x-t)] + \\
& + (1+x)(1+t)[F^*(2-x+t) - F^*(x-t)]\,.
\end{aligned} \tag{5.170}$$

It was already suggested in connection with Figure 5.12 how to solve the associated integral equation. The computation of the weight function is reduced to the computation of a function $q_0(t)$ which itself determines a regular state with a normal singularity. As for $q_0$ any method designed to deal with singular integral equations is reasonable. Since the theorem on the expansions of $\phi$, $\rho$ implies expansions of type (5.114), more specifically (5.114) in odd powers of $|x|^{\frac{1}{2}}$ (if the root of the crack is at $x=0$), it is useful to set up

$$q_0(t)=|t|^{\frac{1}{2}}Q(t) \qquad (5.171)$$

where $Q(t)$ is an ordinary power series which converges best where it is most needed, namely at the root of the crack. $Q(t)$ will even converge in any disk which can be placed inside the elastic domain without exceeding the elastic domain without the crack. Thus the disk radius is $b+1$ in the case of the notched bar. At least for $b \leqslant 0$ the series $Q(t)$ must converge along the crack, the point $z=-1$ possibly excluded. Even for $b>0$ one can, by slight modification, assure convergence if one replaces $x$ by $\xi$, where $\zeta=\xi+i\eta$ is related to $z$ by the simple transformation

$$\zeta=\frac{z-b}{1-bz}\,. \qquad (5.172)$$

If instead of $Q(t)$ one uses a power series in $\xi$, then convergence will hold along the whole crack. This justifies more than enough the approach (5.171). For practical reasons one will approximate $Q$ by a polynomial and determine its coefficients by any procedure of satisfying the integral equation, in particular by collocation or least squares. In the present computation a low order polynomial has always been sufficient. The theorem and equation (4.114) explain why the approach has to be successful.

## 5.5 Special fields in three dimensions

Here the attention will be centered on how to generalize the concept and use of fundamental fields, as they were defined for states of plane strain. To begin with consider a "simple" generalization of the Griffith crack. Cartesian rectangular coordinates $x$, $y$, $z$ will be adopted. We assume a crack of domain $C$ in the $xy$-plane. Let its contour be $C'$. The crack faces are distinguished as $C^+$, $C^-$ as shown in Figure 5.16. Let the crack be under the action of normal tractions of induced nature such that

$$\sigma_z=-g(x,y) \text{ on } C \text{ as well as } \tau_{xz}=\tau_{yz}=0\,. \qquad (5.173)$$

The normal displacement on $C^+$ will be denoted by $w(x,y)$. As in the case of the Griffith crack, where $\phi\equiv\rho$ was found, only one harmonic function

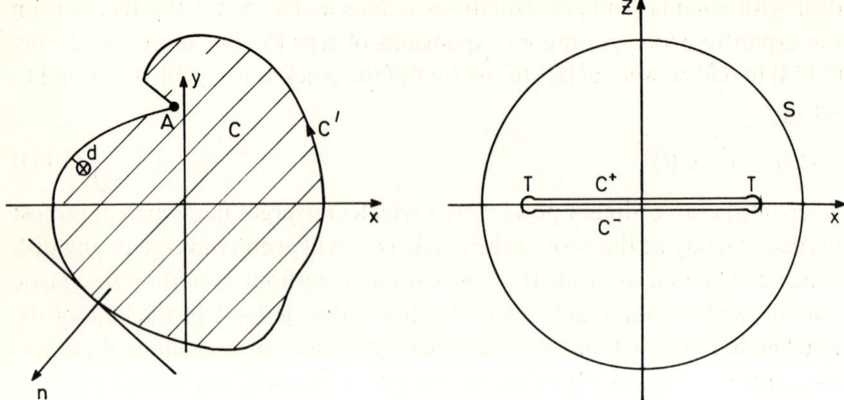


Figure 5.16

$G(x, y, z)$ will suffice to describe the field. This is the Boussinesq–Papkovich potential. With its aid the displacements and the stresses can be expressed as follows:

$$\begin{aligned} u &= -zG_{xz} - (1-2\nu)G_x \\ v &= -zG_{yz} - (1-2\nu)G_y \\ w &= -zG_{zz} + 2(1-\nu)G_z \end{aligned} \tag{5.174}$$

and

$$\begin{aligned} \sigma_x &= -2\mu[(zG_{xx})_z + 2\nu G_{yy}]\,; \quad & \tau_{yz} &= -2\mu zG_{yzz} \\ \sigma_y &= -2\mu[(zG_{yy})_z + 2\nu G_{xx}]\,; \quad & \tau_{zx} &= -2\mu zG_{xzz} \\ \sigma_z &= -2\mu[zG_{zzz} - G_{zz}]\,; \quad & \tau_{xy} &= -2\mu[zG_{xyz} + (1-2\nu)G_{xy}], \end{aligned}$$

where

$$\nabla^2 G = G_{xx} + G_{yy} + G_{zz} = 0$$

and subscripts of $G$ indicate partial derivatives. $G$ will be set up in the form of a single layer potential, the layer distributed over $C$ with a continuous density function $f(x, y)$; more precisely

$$G(x, y, z) = -\frac{1}{4\pi(1-\nu)} \int_C f(\xi, \eta) R^{-1} \cdot \mathrm{d}\xi\, \mathrm{d}\eta \tag{5.175}$$

with

$$R^2 = z^2 + (x-\xi)^2 + (y-\eta)^2 .$$

$G$ is regular harmonic outside of the crack domain. The shearing stresses $\tau_{xy}$, $\tau_{yz}$ vanish along the whole plane $z=0$. The displacement $w$ vanishes on $z=0$ outside of $C$. On $C^+$ it is given by

$$w = 2(1-\nu)G_z = f. \tag{5.176}$$

On $C^-$ the opposite value is obtained. Furthermore

$$\sigma_z = 2\mu G_{zz} = -2\mu\Delta G \text{ where } \Delta = \frac{\partial^2}{\partial x^2} + \frac{\partial^2}{\partial y^2}. \tag{5.177}$$

Since the limit of $G$ exists as a point $(x, y)$ of $C$ is approached along the $z$-axis, the limit being the same for both sides and determined by simply setting $z=0$ in equation (5.175), one can write (see also [16])

$$g(x, y) = -\Delta \frac{\mu}{2\pi(1-\nu)} \iint_C f(\xi, \eta)\left[(x-\xi)^2 + (y-\eta)^2\right]^{-1} \mathrm{d}\xi\, \mathrm{d}\eta. \tag{5.178}$$

Here equations (5.173) have been fed in. It is clear that once $g$ is given, equation (5.178) should furnish the associated displacement functions $f(x, y)$. Equation (5.178) is expected to have precisely one solution, $f$, if the condition $f=0$ is satisfied on the contour $C'$. With this solution there should go a field of finite energy in the cracked space if the contour of $C$ is sufficiently regular and the function $g(x, y)$ at least bounded and continuous on $C$. If one lifts the restriction $f=0$ on the boundary $C'$ of $C$ then the homogeneous case $g\equiv 0$ should yield infinitely many linearly independent solutions $f$, each defining a "fundamental" field. If one assumes a smooth contour $C'$ and $f(x, y)$ to show the behavior

$$f(x, y) = d^{\frac{1}{2}} \cdot F(x, y), \quad F \text{ continuous on } C + C' \tag{5.179}$$

where $d$ is the smallest distance of the point $(x, y)$ from $C'$ (Figure 5.16) then one can show that the field generated by equation (5.175) is essentially one of plane strain in the neighborhood of a point of $C'$, the state referring to the plane $(n, z)$ where $n$ is the outer normal to that point. Moreover, if the point $(x, y)$ on $C'$ is approached along the normal $n$ from outside or inside, the old asymptotic relations hold, in particular

$$\sigma_z = \frac{k_1}{(2d)^{\frac{1}{2}}} \quad \text{for the approach from outside } C \tag{5.180}$$

$$w = k_1 \frac{\kappa+1}{2\mu}\left(\frac{d}{2}\right)^{\frac{1}{2}} \quad \text{for the approach from inside the crack}. \tag{5.181}$$

The factor $k_1$ is of course related to $F(x, y)$:

$$k_1 \frac{\kappa+1}{2\mu} = 2^{\frac{1}{2}} \cdot F(x, y).$$

Finally it can be stated that our field is locally of the nature of the mode I deformation of plane strain, with a stress intensity factor $k_1$ varying along the contour $C'$ of the crack. Introducing the arclength $s$ along $C'$, one can set $k_1 = k(s)$. The singularity is of the "normal" type. All of this depends very much on the regularity of $C'$. If one would consider the point $A$ of the crack domain of Figure 5.16, the field behavior might deviate, perhaps drastically, from the one near a normal singularity. Turning now to fundamental fields, one can define them as follows: Any field, regular outside $C$, will be called fundamental, if its local behavior is governed by a state of plane strain due to

$$\phi(\zeta) = a\zeta^{-\frac{1}{2}}, \quad \rho(\zeta) = \bar{a}\zeta^{-\frac{1}{2}}, \quad \zeta = n + iz \tag{5.182}$$

in analogy to equations (5.71); $n + iz$ replaces the former complex variable $z$. One must allow the coefficient $a$ to depend on $s$, the arclength on $C'$ so that $a = a(s)$. The field shall have no tractions on $C$. In addition to this it is required that the stresses of the fundamental field go strongly enough to zero as the distance from $C$ goes to infinity. The displacements should stay bounded for the same approach. Now another look will be given to the formulas (5.92). They are based on Betti's theorem and the behavior of the related fields near a singularity. All of this is applicable to the new situation. Let us take a domain in the space, bounded by a sphere $S$ around the origin, of sufficiently large radius to contain $C$, and by the surface of a torus, surrounding $C'$. We set up Betti's theorem for this domain and go to the limit in double respect. We let the radius of $S$ grow to infinity, and in this context we assume now quite precisely, that the energy of the fundamental state as well as that of the regular one outside of $S$ is bounded, and that the outside energy goes to zero as $S$ approaches infinity. We also shrink the torus $T$ into $C'$, representing the work integrals on $T$ in harmony with the local behavior of plane strain as postulated above. In the limit we obtain the formula:

$$\frac{\kappa+1}{2\mu} \int_{C'} k(s)\, a(s)\, ds = \frac{-2^{\frac{1}{2}}}{\pi} \int_C M(x, y)\, g(x, y)\, dx\, dy \tag{5.183}$$

where $M(x, y)$ is the normal displacement of the singular field on $C^+$. This formula does not determine $k(s)$ locally. In order to do this an infinity of fundamental fields is needed. For practical reasons, a finite number of them, large enough as to draw conclusions from the corresponding integrals (5.183) on the local behavior, has to be found. As in the case of plane strain, one must provide weight functions $M(x, y)$ together with the associated $a(s)$. To some extent the three methods listed in section 3 appear to admit a generalization. Here reference will be made to parameter differentiation. Assume that the crack is loaded by constant normal pressure and that we possess the displacement $w(x, y)$ on $C^+$ of the regular field that responds to the pressure. If the crack is extended by dilatation, then $G(x, y, z)$ should change into $c^{-2}G(cx, cy, cz)$ with $c$ as dilatation factor. Differentiation with respect to $c$ at $c=1$ leads to

$$H(x, y, z) = -2G(x, y, z) + xG_x + yG_y + zG_z \tag{5.184}$$

as the potential of a fundamental field. The displacement on $C^+$ of $H$ is given by equation (5.174) and leads to

$$M(x, y) = xw_x + yw_y - w \ . \tag{5.185}$$

The function $a(s)$ is in proportion to $k(s)$, more precisely

$$a(s) = \alpha k(s) \, , \quad \alpha = -\tfrac{1}{4} \cdot 2^{\frac{1}{2}} \, . \tag{5.186}$$

Without normalization the integral in equation (5.183) has the integrand $\alpha k^2(s)$. The integral so obtained is essentially the energy release rate with respect to a growing dilatation factor $c$. One can normalize $a(s)$ such that

$$\int_{C'} a(s)\,\mathrm{d}s = 1 \, . \tag{5.187}$$

In this case equation (5.183) yields an average value of $k(s)$ on $C'$. That the integral equation (5.178) should yield weight functions was already mentioned.

For the axially symmetric case of a penny-shaped crack, introduce cylindrical coordinates $\theta$, $r$, $z$ in accordance with $r \cos\theta = x$, $r \sin\theta = y$ and assume that the crack domain is the unit disk $r \leqslant 1$. We consider first fundamental fields of axial symmetry; thus we may write $w(x, y) = w(r)$, $g(x, y) = g(r)$ and $M(x, y) = M(r)$. Thus the functions $a(s)$, $k(s)$ become constant, and equation (5.183) changes into

$$\frac{\kappa+1}{2\mu}\, ak_1 = \frac{-2^{\frac{1}{2}}}{\pi} \int_0^1 M(r)\, g(r)\, r\,\mathrm{d}r \, . \tag{5.188}$$

To determine the function $M(r)$, consideration is given to the potential $G$ associated with the crack loaded by *constant pressure*. In cylindrical coordinates the relations (5.174) become a special case of

$$u = (2\nu-1)P_r - zP_{rz} - Q_r\ ;\quad w = 2(1-\nu)P_z - zP_{zz} - Q_z$$

$$\sigma_r = -2\mu\left[\frac{2\nu}{r}P_r + (zP_{rr})_z + Q_{rr}\right]\ ;\quad \sigma_\theta = -2\mu\left[2\nu P_{rr} + \left(\frac{z}{r}P_r\right)_z + \frac{1}{r}Q_r\right]$$

$$\sigma_z = 2\mu(P_{zz} - zP_{zzz} - Q_{zz})\ ;\quad \tau_{rz} = -2\mu[zP_{rzz} + Q_{rz}]$$

$$\nabla^2 P = P_{rr} + \frac{1}{r}P_r + P_{zz} = 0\ ;\quad \nabla^2 Q = 0\,. \tag{5.189}$$

Here $u$ is the displacement in radial direction, while $w$ retains its old meaning. $P(r, z)$ stands for $G(x, y, z)$. $Q$ is an additional harmonic function which we have to resort to farther below. Right now $Q \equiv 0$ for the penny-shaped crack. Relation (5.184) takes the form

$$H(r, z) = -2P + DP\,,\quad D = r\frac{\partial}{\partial r} + z\frac{\partial}{\partial z}\,. \tag{5.190}$$

Assuming that $P$ is known, we possess in $H$ the potential of a fundamental field of axial symmetry. If for the case of a constant pressure

$$w(r) \approx 2^{\frac{1}{2}}\cdot(1-r)^{\frac{1}{2}} \ \text{ as } r\to 1 \tag{5.191}$$

then the displacement $M(r)$ due to equation (5.190) will be $M(r) = rw' - w$ with

$$M \approx -(\tfrac{1}{2})^{\frac{1}{2}}(1-r)^{-\frac{1}{2}} \ \text{ as } r\to 1\,.$$

Now the constancy of the pressure applied leads one to believe that other relations of $M(r)$ to $w(r)$ may exist. Looking for a different one, one constructs

$$H^* = D(D-1)P + P_{zz} - 2P\,. \tag{5.192}$$

This again is a harmonic function. The double differentiations and equation (5.191) indicate that the singularity of the field is even stronger than fundamental, which is indeed the case with respect to $D(D-1)P$. $P_{zz}$ has been added in order to reduce the singularity to a fundamental one, and the term $-2P$ will soon justify its existence. Using $\nabla^2 P = 0$, one finds explicitly

$$H^* = -(1-r^2+z^2)\left(P_{rr} + \frac{1}{r}P_r\right) + 2rzP_{rz} - 2P - rP_r\,.$$

In view of equations (5.189) $w$ is in proportion to $P_z$ on $z=0$, while $P_{zz} = \text{const.}$

expresses constancy of pressure. This leads at once to

$$H^*_{zz}=0 \text{ on } C\,, \tag{5.193}$$

and all indications are that $H^*$ represents a fundamental field, the normal displacement on $C^+$ being

$$w^*=(r^2-1)\left(w''+\frac{1}{r}w'\right)+rw'-2w\,. \tag{5.194}$$

Equations (5.191) and (5.194) imply $w^*=O(w)$ as $r\to 1$. Consequently $H^*$ has a normal singularity only. Since it has no tractions on $C$ we infer $H^*=0$, $w^*=0$ and

$$(r^2-1)\left(w''+\frac{1}{r}w'\right)+rw'-2w=\frac{\mathrm{d}}{r\,\mathrm{d}r}\left[r(r^2-1)w'-r^2w\right]=0\,.$$

This equation can be transformed into a hypergeometric one,

$$t(1-t)\ddot{w}+(1-3t/2)\dot{w}-\tfrac{1}{2}w=-\tfrac{1}{2}\frac{\mathrm{d}}{\mathrm{d}t}\left[2t(t-1)\dot{w}-tw\right]=0\,;\; r^2=t \tag{5.195}$$

dots referring to $t$-differentiation. The only solution regular at $t=0$ is

$$w=c(1-t)^{\frac{1}{2}}\,,\quad c=\text{const}\,. \tag{5.196}$$

In view of equation (5.191) we have $c=1$. The associated function $M(r)$ is, by equation (5.190),

$$M(r)=-(1-r^2)^{-\frac{1}{2}}\,. \tag{5.197}$$

Formula (5.188) for $k_1$ takes the simple form

$$k_1=\frac{2}{\pi}\int_0^1(1-r^2)^{-\frac{1}{2}}g(r)r\,\mathrm{d}r\,. \tag{5.198}$$

Both equation (5.196) and (5.198) are very well known⋆. If anything is new, it is the method of derivation, showing the power of the concept of a fundamental state. Here the integral formula (5.198) was derived without doing more than solving a second order linear ordinary differential equation. Additional conclusions can be drawn from equation (5.196). The fields defined by $H=P_x$, $H=P_y$ must also be singular fields. They are not of axial

⋆ Barenblatt [10] states that he found equation (5.698), the derivation based on the application of the method of Fourier–Hankel transforms, developed by I. N. Sneddon for solving axi-symmetrical problems of elasticity.

symmetry. Since $P_x + iP_y = e^{i\theta} P_r$, one finds that the associated normal displacement on the crack takes the form $M(x, y) = -r(1-r^2)^{-\frac{1}{2}} \cdot e^{i\theta}$. This is a weight function which permits to determine the first Fourier coefficients of the function $k_1(s)$ by virtue of (5.183), if the penny-shaped crack is under pressure of non-axisymmetric distribution. More generally one can introduce weight functions of the type $M(x, y) = M_n(r) \cdot e^{ni\theta}$ where $n$ denotes an integer. For the preceding cases one has $M_0 = M$ by (5.197) and $M_1 = rM$. One can show that $M_n = r^n M$ in general; the derivation can be based on induction by $n$ and on the technique of applying the operators $\partial/\partial x$, $\partial/\partial y$, $D$ to already obtained potentials of the type $H = H_n(r, z) \cdot e^{in\theta}$. These functions in turn will permit us to analyze $k_1$ $(s)$ for any state of mode I deformation of the penny-shaped crack.

The case of an infinitely deep outer notch of axial symmetry, characterized by $r \geqslant 1$ as the domain for $C$ in the $xy$-plane will be now considered. Figure

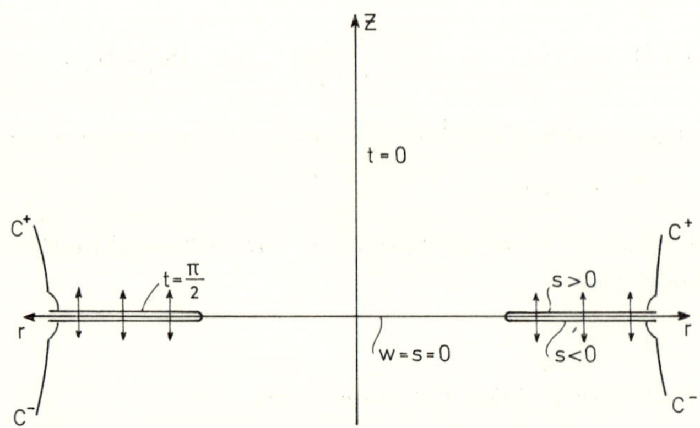


Figure 5.17a. Space with an axisymmetric infinitely deep notch.

5.17a indicates the necessary details of geometry and loading. A normal load is specified by

$$\sigma_z = -g(r)\ ; \quad 1 \leqslant r \leqslant r_1\,. \tag{5.199}$$

To be mentioned first is the case $g \equiv 0$ which has a nontrivial solution with a normal singularity at $r=1$, with finite field energy while the stresses go to zero at infinity. For the same field and the same approach the displacements approach constant limits, $u^*$ for $u$ and $\pm w^*$ for $w$ (the signs characterize upper and lower half-space). The potential of the field is defined by

$$P_0 = -\log \cosh s - \log(1+\cos t) + \cos t + z \arcsin(\tanh s) \tag{5.200}$$

where

$$\sinh(s+\mathrm{i}t) = z + \mathrm{i}r \ ; \quad -\infty < s < \infty \ ; \quad 0 \leqslant t \leqslant \tfrac{1}{2}\pi \, .$$

Details of the above equation are shown in Figure 5.17a. The field's normal singularity at $r=1$ has stress intensity factor $k_1 = 1$. More precisely one has

$$\sigma_z = (1-r^2)^{-\frac{1}{2}} \text{ on the cross section } r \leqslant 1 \, , \ z = 0 \, . \tag{5.201}$$

The normal displacement on $C^+$ is

$$w = w^* \arcsin\left(1 - \frac{1}{r^2}\right)^{\frac{1}{2}} \ ; \ w^* = \frac{2}{\pi\mu}(1-\nu) \, . \tag{5.202}\star$$

As $r \to \infty$, $w \to w^*$ on $C^+$. The stresses are of the order of $O(R^{-2})$ with

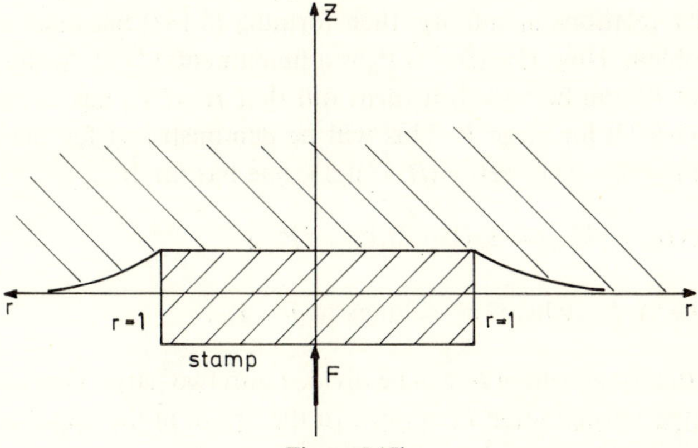


Figure 5.17b

$R = (r^2 + z^2)^{\frac{1}{2}}$ for large $R$. The very existence of $P_0$ as a field of finite energy with no tractions on $C$ is quite interesting. It means that the solution to our problem of prescribing $\sigma_z = -g(r)$ will not be unique unless we impose conditions on the solution. In similar vein the class of fundamental fields is ambiguous to the extent that the field of $P_0$ can be added to any fundamental field without changing the character of the latter. Before we discuss this in more detail, $P_0$ will be reinterpreted. If, instead of the displacement field $(u, w)$, one considers $(-u, w^* - w)$ for the upper half-space, then we meet the well-known indentation problem, indicated in Figure 5.17b. This makes it

⋆ This can be related to the second solution of the equation $r^{-1}\mathrm{d}/\mathrm{d}r[r(r^2-1)w' - r^2 w] = 0$. The latter is $w = (r^2-1)^{\frac{1}{2}} \arcsin r^{-1} - 1$ with $w - rw' = -\arcsin r^{-1} = -\frac{1}{2}\pi + \arccos r^{-1}$.

at once clear that the field $P_0$ has finite energy, equal to the work done by the stress of equation (5.201) through the displacement $w^*$. Note that the tractions of indentation due to (5.201) are not in equilibrium. There exists a resultant normal force $F=2\pi$. Returning to the actual problem, one can impose conditions in many ways. For instance, it is possible to prescribe that the displacements vanish at infinity. We can also—and this is our preference—prescribe that the half-space $z \geqslant 0$ be under tractions (reactionary ones for $r<1$ and impressed ones for $r>1$) which are in equilibrium with one another. This is what one ordinarily expects to happen, if the tractions $g(r)$ are beiny applied. In this very case the stresses at infinity are of the order $O(R^{-3})$ rather than of $O(R^{-2})$. With bounded displacements assumed, one can see that the energy of deformation can be expressed as the work of the applied tractions $g(r)$ through the normal displacements $w(r)$ on $C$. If the stresses and displacements of a fundamental field are to show the same order relations at infinity, then formula (5.183) becomes applicable to our problem. Now $H=(D-2)\,P_0$ is a fundamental field. In this case any multiple of $P_0$ can be added. It turns out that $H=DP_0$ has stresses of the order of $O(R^{-3})$ for large $R$. This will be demonstrated for the stress $\sigma_z$. Using $(Df)_z=(D+1)f_z$, $zDf=(D-1)(zf)$, one has for $H$

$$\begin{aligned}\sigma_z &= 2\mu(H_{zz}-zH_{zzz}) = 2\mu(D+2)(P_{zz}-zP_{zzz})\\ &= (D+2)\sigma_z^* \,, \quad \text{where} \quad \sigma_z^* \text{ belongs to } P=P_0 \,.\end{aligned} \tag{5.203}$$

For $z \geqslant 0$ the stress field of $P_0$ can be divided into two parts: One caused by a concentrated normal force $F$, applied to the origin of the upper half-space, and the other one caused by *self-equilibrated* tractions acting on $r \leqslant 1$. For the latter the stresses are of the order $O(R^{-3})$, while the stresses in response to the concentrated force have the order $O(R^{-2})$, in particular

$$\sigma_z = \text{const. } z^3 R^{-5} \tag{5.204}$$

But this term, as one can easily see, is destroyed by the operator $(D+2)$. Since $H=DP_0$ then is the right fundamental field, it is found that the associated $M(r)$ takes the form

$$M(r) = rw'(r) + w(r) \,. \tag{5.205}$$

This leads to the following formula for $k_1$ caused by the tractions $g(r)$:

$$k_1 = \frac{2}{\pi} \int_1^{r_1} r \left[ \arcsin \left(1 - \frac{1}{r^2}\right)^{\frac{1}{2}} + \frac{1}{(-1+r^2)^{\frac{1}{2}}} \right] g(r) \mathrm{d}r \,. \tag{5.206}$$

The last example is the notched cylinder (Figure 5.18) of infinite length and unit radius. The notch is in the plane $z=0$ and occupies the ring $0 < a \leqslant r \leqslant 1$. It shall be under normal traction of induced nature with $\sigma_z = -g(r)$ specified as usual. In order to find a stress field in the cylinder use must be made of the more general formulas (5.189). We shall derive an integral equation linking the

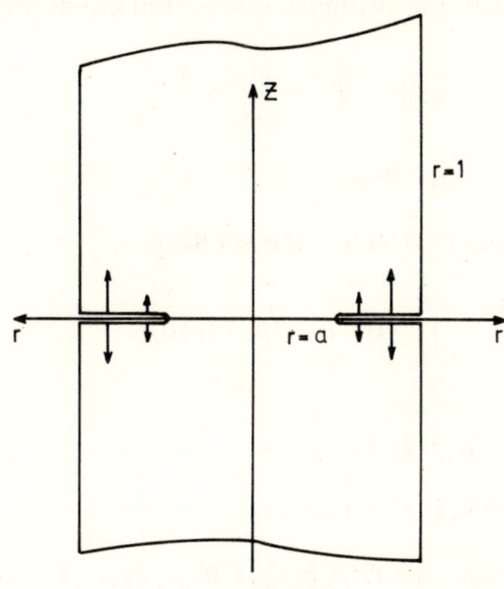


Figure 5.18

displacement $w(r)$ along the notch to the traction $g(r)$. This will be done in analogy to how we dealt with the notched bar in section 5.4. The function $w(r)$ will be specified along the notch and even beyond $r=1$ up to infinity. More precisely the special potential

$$P = P_1 = \frac{1}{4\pi(1-\nu)} \int_0^{\infty} \int_0^{2\pi} \frac{w(\rho)\rho \cdot \mathrm{d}\theta \cdot \mathrm{d}\rho}{R} \,;$$

$$R^2 = z^2 + r^2 - 2r\rho \cos\theta + \rho^2 \tag{5.207}$$

is introduced. $w(r)$, $w'(r)$ are assumed to belong to $L_1$ $(0, \leqslant)$; $w$ shall be continuous everywhere and vanish for $0 \leqslant r \leqslant a$. The field created by $P_1$ alone has $w(r)$ as normal displacement on the upper side $C^+$ of the notch. The next step is to find two functions $P_2$, $Q$, both regular harmonic in the unnotched cylinder such that the pair $P = P_1 + P_2, Q$ induces no tractions on the cylindrical surface $r = 1$. To this end Fourier-transforms will be applied leading to

$$\phi_k(r, \lambda) = \int_0^\infty P_k(r, z) \cos z\lambda \, dz \; ; \; \psi(r, \lambda) = \int_0^\infty Q(r, z) \cos z\lambda \, dz$$

$$P_k(r, z) = \frac{2}{\pi} \int_0^\infty \phi_k(r, \lambda) \cos z\lambda \, d\lambda \; ; \; Q(r, z) = \frac{2}{\pi} \int_0^\infty \psi(r, \lambda) \cos z\lambda \, d\lambda \, . \tag{5.208}$$

The transforms satisfy the ordinary differential equations:

$$(1) \quad \mathfrak{B}\phi_1 = \frac{w(r)}{2(1-\nu)} \text{ where } \mathfrak{B} = \frac{\partial^2}{\partial r^2} + \frac{1}{r}\frac{\partial}{\partial r} - \lambda^2 \; ;$$

$$(2) \quad \mathfrak{B}\phi_2 = 0 \; ; \qquad (3) \quad \mathfrak{B}\psi = 0 \, . \tag{5.209}$$

The first of equations (5.209) has the solution

$$\phi_1(r, \lambda) = -\frac{1}{2(1-\nu)} \int_0^\infty G(r, \rho; \lambda) \rho \cdot w(\rho) \, d\rho \tag{5.210}$$

where

$$G(r, \rho; \lambda) = I_0(\rho\lambda) K_0(r\lambda) \text{ for } \rho < r \; ;$$

$$= I_0(r\lambda) K_0(\rho\lambda) \text{ for } \rho > r \, .$$

We observe that $G(r, \rho; \lambda) = G(\rho, r; \lambda)$; $G(r, \rho; \lambda)$ is a Green's function of the operator $\mathfrak{B}$ and satisfies the customary conditions.

Formula (5.210) could have been directly derived from (5.207). One finds

$$\int_0^\infty R^{-1} \cdot \cos z\lambda \, dz = K_0[\lambda(r^2 + \rho^2 - 2r\rho \cos\theta)^{\frac{1}{2}}] = \sum_{n=0}^{\infty} G_n(r, \rho; \lambda) \cos n\theta \tag{5.211}$$

where $G_n(r, \rho; \lambda) = G_n(\rho, r; \lambda)$ and for $\rho < r$:

$$G_0 = I_0(\rho\lambda) K_0(r\lambda) \; ; \; G_n = 2I_n(\rho\lambda) K_n(r\lambda) \text{ for } n \geqslant 1 \, .$$

Since $z$ appears in $R^{-1}$ only, the integrations in equation (5.207) can be applied directly to the transform of $R^{-1}$. This must lead to $\phi_1$. The $\theta$-integration destroys all terms but $G_0$. This leads to equation (5.210) in another way.

Some properties of $G(r, \rho; \lambda)$ will now be listed:

$$\frac{\partial G}{\partial \rho} = \lambda I_1(\rho\lambda) K_0(r\lambda) \text{ for } \rho < r$$

and

$$\frac{\partial G}{\partial \rho} = -\lambda K_1(\rho\lambda) I_0(r\lambda) \text{ for } \rho > r \tag{5.212}$$

$$\frac{2}{\pi}\int_0^\infty G(r,\rho;\lambda)\cos z\lambda\,\mathrm{d}\lambda = [z^2+(r+\rho)^2]^{-\frac{1}{2}}\cdot F\left(\tfrac{1}{2},\tfrac{1}{2},1;\frac{4r\rho}{z^2+(r+\rho)^2}\right). \tag{5.213}$$

Using formulas on hypergeometric functions, in particular those by Gauss, one can rewrite equation (5.213). Introduce

$$\cosh t = \tfrac{1}{2}(s+1/s) \text{ where } 0 \leqslant s \leqslant 1 \;;\quad \cosh t = \frac{z^2+r^2+\rho^2}{2r\rho}.$$

Then

$$\frac{z^2+(r+\rho)^2}{4r\rho} = (1+s)^2/4s \;;\; [z^2+(r+\rho)^2]^{-\frac{1}{2}} = \left(\frac{s}{r\rho}\right)^{\frac{1}{2}}\cdot\frac{1}{1+s}.$$

Due to Gauss:

$$F\left(\tfrac{1}{2},\tfrac{1}{2};1;\frac{4s}{(1+s)^2}\right) = (1+s)F(\tfrac{1}{2},\tfrac{1}{2};1;s^2)$$

so that finally

$$\frac{2}{\pi}\int_0^\infty G(r,\rho;\lambda)\cos z\lambda\,\mathrm{d}\lambda = \left(\frac{s}{r\rho}\right)^{\frac{1}{2}}\cdot F(\tfrac{1}{2},\tfrac{1}{2};1;s^2).$$

At this point, the function

$$\begin{aligned} M_1(r,\rho) &= \frac{\pi\rho^2}{2r}F\left(\tfrac{1}{2},\tfrac{3}{2};2;\frac{\rho^2}{r^2}\right); \qquad \rho < r \\ M_1(r,\rho) &= \frac{\pi r^2}{2\rho}F\left(\tfrac{1}{2},\tfrac{3}{2};2;\frac{r^2}{\rho^2}\right), \qquad \rho > r. \end{aligned} \tag{5.214}$$

is introduced for subsequent use. Integration by parts of equation (5.210) yields, with $0 = \mathfrak{B}_\rho G$ used,

$$\phi_1(r,\lambda) = \frac{1}{2(1-\nu)\lambda^2}\left[\int_0^\infty \frac{\partial G}{\partial \rho}\cdot\rho w'(\rho)\mathrm{d}\rho - w(r)\right]. \tag{5.215}$$

In this context the function

$$\chi(r,\lambda) = \frac{1}{2(1-\nu)\lambda^2}\left[\int_0^1 \frac{\partial G}{\partial \rho}\cdot \rho w'(\rho)\,\mathrm{d}\rho - w(r)\right] \tag{5.216}$$

is introduced. The difference $\phi_1 - \chi$ must take the form

$$\phi_1 - \chi = A_1(\lambda) I_0(r\lambda) \text{ for } r<1\,, \tag{5.217}$$

with some coefficient $A_1$. Moreover

$$2(1-\nu)\mathfrak{B}\chi = w(r) \text{ for } r<1\,. \tag{5.218}$$

At this juncture the following compositions of Bessel functions will be introduced, where, if the argument is not explicitly written, $\lambda$ alone is that argument:

$$D(\lambda) = I_0^2 - I_1^2 - 2(1-\nu) I_1^2/\lambda^2 \tag{5.219}$$

$$\left.\begin{aligned} E(r,\lambda) &= r\frac{I_1(r\lambda)}{I_1} \\ F(r,\lambda) &= \frac{\partial E(r,\lambda)}{r\partial\lambda} = r\frac{I_0(r\lambda)}{I_1} - \frac{I_0\cdot I_1(r\lambda)}{I_1^2} \\ H^*(r,\lambda) &= \frac{1}{r}\cdot\frac{\partial}{\partial r}(rF(r,\lambda)) \\ H^*(r,\lambda)I_1^2 &= H(r,\lambda) = [2I_1 - \lambda I_0]\cdot I_0(r\lambda) + r\lambda I_1 I_1(r\lambda)\,. \end{aligned}\right\} \tag{5.220}$$

Here the identity

$$\frac{\partial}{\partial r}[rI_1(r\lambda)] = r\lambda I_0(r\lambda) \tag{5.221}$$

has been used. The functions $D$, $E$, $F$, $H$ are regular analytic in $\lambda$ in some neighborhood of $\lambda=0$. $D$ is even in $\lambda$; $E$, $H$ are even in $\lambda$ as well as in $r$, and $F$ is odd in both $\lambda$ and $r$.

The conditions that $\sigma_r$ and $\tau_{rz}$ must vanish on $r=1$ can be expressed as

$$2\nu P_r + (zP_{rr})_z + Q_{rr} = 0\,, \qquad r=1 \tag{5.222}$$

and

$$zP_{rzz} + Q_{rz} = 0\,, \qquad r=1 \tag{5.223}$$

respectively. Equations (5.222) and (5.223) are now subjected to a Fourier transform, (5.222) to the cos-type and (5.223) to the sin-type. In this process the factor $z$ or a $z$-derivative under the integral sign can be easily taken care of; the latter can be treated by integration by parts with respect to the derivative, and the former amounts to a $\lambda$-differentiation. Example:

$$\int_0^\infty (zP_{rr})_z \cos z\lambda \, dz = zP_{rr} \cos z\lambda \Big|_0^\infty + \lambda \int_0^\infty zP_{rr} \sin z\lambda \, dz$$

and (5.224)

$$\int_0^\infty zP_{rr} \sin z\lambda \, dz = -\frac{\partial}{\partial\lambda} \int_0^\infty P_{rr} \cos z\lambda \, dz \, .$$

If one treats the transforms in this way, then equation (5.222) leads to

$$\psi'' + 2\nu\phi' - \lambda \frac{\partial}{\partial\lambda} \phi'' = 0 \, . \tag{5.225}$$

Primes refer to $r$-derivatives, to be taken before one sets $r=1$ for (5.225). In similar vein

$$\int_0^\infty zP_{rzz} \sin z\lambda \, dz = -\frac{\partial^2}{\partial r \, \partial\lambda} \int_0^\infty P_{zz} \cos z\lambda \, dz \; ; \tag{5.226}$$

but $P_{zz} = -P_{rr} - P_r/r$; using this and equations (5.209) we find that

$$\int_0^\infty zP_{rzz} \sin z\lambda \, dz = \frac{\partial^2}{\partial r \, \partial\lambda} (\phi'' + \phi'/r) =$$

$$= \frac{\partial^2}{\partial r \, \partial\lambda} \left[ \lambda^2 \phi + \frac{w}{2(1-\nu)} \right] = \frac{\partial}{\partial\lambda} (\lambda^2 \phi') \, ,$$

where again the primes denote $r$-differentiation. Finally condition (5.223) becomes

$$-\lambda\psi' + \frac{\partial}{\partial\lambda} (\lambda^2 \phi') = 0 \, . \tag{5.227}$$

In the process of relieving the tractions (induced by $P_1$) on the cylindrical surface, the nature of $w(r)$ outside of the cylinder cannot influence the final outcome. What it contributes to the tractions is compensated again; it therefore suffices to set from now on:

$$\phi(r, \lambda) = \chi(r, \lambda) + A(\lambda) I_0(r\lambda) \, . \tag{5.228}$$

We shall write $\psi(r, \lambda)$ in the form

$$\psi(r, \lambda) = (C(\lambda) + \lambda A'(\lambda))\, I_0(r\lambda)\,. \tag{5.229}$$

The quantities $A(\lambda)$, $C(\lambda)$ can be determined on the basis of the conditions (5.225) and (5.227). To this end observe that

$$\begin{aligned} \phi(1, \lambda) &= \chi(1, \lambda) + AI_0\,; & \psi(1, \lambda) &= (C + \lambda A')I_0 \\ \phi'(1, \lambda) &= \chi'(1, \lambda) + \lambda AI_1\,; & \psi'(1, \lambda) &= (C + \lambda A')\lambda I_1 \\ \phi''(1, \lambda) &= \chi''(1, \lambda) + \lambda^2 AI^*\,; & \psi''(1, \lambda) &= (C + \lambda A')\lambda^2 I^*\,;\ I^* = I_0 - I_1/\lambda\,. \end{aligned} \tag{5.230}$$

As before a Bessel function written without argument refers to $\lambda$ alone. With $\chi'(1, \lambda) = N_1(\lambda)$, $\chi''(1, \lambda) = N_2(\lambda)$, the conditions take the form

$$\begin{aligned} A[\lambda^2 I_0 + \lambda^3 I_1 - 2\nu\lambda I_1] + \lambda C[-\lambda I_0 + I_1] &= 2\nu N_1 - \lambda N_2' \\ A[\lambda I_0 + 2I_1] \qquad - CI_1 &= -N_1' + 2N_1/\lambda\,. \end{aligned} \tag{5.231}$$

The integral representation of $\chi$ implies

$$\begin{aligned} 2(1-\nu)N_1 &= -\int_a^1 I_1(\rho\lambda)\, K_1(\lambda)\, \rho w'(\rho)\, \mathrm{d}\rho \\ 2(1-\nu)N_2 &= \int_a^1 I_1(\rho\lambda)\, [\lambda K_0(\lambda) + K_1(\lambda)]\, \rho w'(\rho)\, \mathrm{d}\rho\,. \end{aligned} \tag{5.232}$$

Postponing the integration in equations (5.232), we set

$$N_1 = -I_1(\rho\lambda)\, K_1(\lambda)\,; \quad N_2(\lambda) = I_1(\rho\lambda)\, [\lambda K_0(\lambda) + K_1(\lambda)]\,. \tag{5.233}$$

Equations (5.231) are solved for $A$, $C$ with $N_1$, $N_2$ given by (5.233). It is useful to split $A$, $C$ as follows:

$$A = A^* - N_1/\lambda I_1\,, \quad C = C^* + N_1'/I_1 - N_1 I_0/I_1^2\,. \tag{5.234}$$

It turns out that $A^*$, $C^*$ can be written in the form

$$A^* = -\frac{I_1\, \omega}{\lambda^2 D}, \quad C^* = -\frac{\lambda I_0 + 2I_1}{\lambda^2 D}\, \omega \tag{5.235}$$

with

$$\omega = -\frac{\partial E(\rho, \lambda)}{\rho\, \partial\lambda} = -\frac{\rho I_0(\rho\lambda)}{I_1} + \frac{I_1(\rho\lambda) I_0}{I_1^2}\,. \tag{5.236}$$

It remains to put all the pieces together and to retransform. Since only an expression for the normal traction along the notch is needed, it is not necessary to go through all the steps of retransforming. Prescribed is

$$g(r) = 2\mu(Q_{zz} - P_{zz}) = \frac{2\mu}{r}\frac{\partial}{\partial r}[rP_r - rQ_r] \text{ on } C. \tag{5.237}$$

Consequently

$$g(r) = \frac{4\mu}{\pi r}\frac{\partial}{\partial r}\int_0^\infty r\left\{\frac{\partial\chi}{\partial r} + \lambda A^*(\lambda) I_1(r\lambda) - N_1 I_1(r\lambda)/I_1 - \right.$$
$$\left. - \lambda[C^*(\lambda) + \lambda A^{*\prime}(\lambda)] I_1(r\lambda)\right\} \mathrm{d}\lambda. \tag{5.238}$$

Dealing with the derivative $A^{*\prime}(\lambda)$ by integration by parts one obtains

$$g(r) = \frac{4\mu}{\pi r}\frac{\partial}{\partial r}\int_0^\infty r\left\{\frac{\partial\chi}{\partial r} + \lambda A^*[2I_1(r\lambda) + r\lambda I_0(r\lambda)] - \right.$$
$$\left. - \lambda C^* I_1(r\lambda) - N_1 I_1(r\lambda)/I_1\right\} \mathrm{d}\lambda. \tag{5.239}$$

The contribution of $\chi_r$ to the integral (5.239) will be denoted by $g_1$; the remainder is $g_2$ which can be written in two parts as $g_2 = g_{21} + g_{22}$. The following formulae pertain to the various contributions. As for $g_1$ it is noted that

$$-2(1-\nu)\chi_r = \int_0^r I_1(\rho\lambda) K_1(r\lambda)\rho w'(\rho)\mathrm{d}\rho +$$
$$+ \int_r^1 I_1(r\lambda) K_1(\rho\lambda)\rho w'(\rho)\mathrm{d}\rho, \tag{5.240}$$

$$\int_0^\infty I_1(\rho\lambda) K_1(r\lambda)\mathrm{d}\lambda = \tfrac{1}{4}\pi\frac{\rho}{r^2} F\left(\tfrac{1}{2}, \tfrac{3}{2}; 2; \frac{\rho^2}{r^2}\right) = \frac{M_1(r, \rho)}{2r\rho}, \quad \rho < r, \tag{5.241}$$

(see [15], Vol. 2, 7.14.2, p. 93, formula (35)). Consequently

$$g_1(r) = -\frac{\mu}{\pi(1-\nu)}\cdot\frac{\partial}{r\partial r}\int_a^1 M_1(r, \rho) w'(\rho)\mathrm{d}\rho. \tag{5.242}$$

The other terms are

$$g_{21}(r) = \frac{2\mu}{\pi(1-\nu)}\frac{\mathrm{d}}{r\mathrm{d}r}\int_a^1\int_0^\infty \frac{I_1(r\lambda) I_1(\rho\lambda)}{I_1} K_1 \cdot r\rho w'(\rho)\mathrm{d}\lambda\,\mathrm{d}\rho \tag{5.243}$$

$$g_{22}(r) = \frac{2\mu}{\pi(1-\nu)}\frac{\mathrm{d}}{r\mathrm{d}r}\int_a^1\int_0^\infty \frac{I_1^2}{D} F(r, \lambda) F(\rho, \lambda) r\rho w'(\rho)\mathrm{d}\rho. \tag{5.244}$$

Defining the kernels

$$M_{21} = -2\int_0^\infty E(r,\lambda)E(\rho,\lambda)I_1K_1\,\mathrm{d}\lambda$$
$$M_{22} = -2\int_0^\infty \frac{I_1^2}{D}F(r,\lambda)F(\rho,\lambda)r\rho\cdot\mathrm{d}\lambda \tag{5.245}$$

and $M_2 = M_{21}+M_{22}$, $M=M_1+M_2$, one can write

$$g(r) = -\frac{\mu}{\pi(1-\nu)}\frac{\mathrm{d}}{r\,\mathrm{d}r}\int_a^1 M(r,\rho)w'(\rho)\mathrm{d}\rho\,. \tag{5.246}$$

This is the integral relation between the normal displacement $w(r)$ and the traction $g(r)$ along the upper side of the notch. In contrast to the integral equation for the notched bar, the derivative of $w$ is under the integral sign. This was done in order to exhibit the symmetry of the kernels $M_1$, $M_2$, $M_{21}$, $M_{22}$. By integration by parts, equation (5.246) can easily be changed into one involving $w(r)$ only. This would yield

$$g(r) = \frac{\mu}{\pi(1-\nu)}\frac{\mathrm{d}}{r\,\mathrm{d}r}\int_a^1 \frac{\partial M(r,\rho)}{\partial\rho}w(\rho)\mathrm{d}\rho\,. \tag{5.247}$$

The homogeneous form of this equation should be used in order to find the weight function.

In connection with the integration by parts it should be pointed out that $F(1,\lambda)=0$, and therefore $M_{22}(r,1)=0$. It is also obvious that $M_1(r,1)+M_{21}(r,1)=0$. This justifies equation (5.247). Equation (5.247) differs in quantitative respect only from the equation for the notched bar. For $|r-\rho|\ll 1$ the behavior of the kernel is dictated by that of $M_1$. From the properties of the hypergeometric function it follows that

$$\frac{\partial M_1}{\partial\rho} \approx \frac{2\rho^2}{r^2-\rho^2}\,. \tag{5.248}$$

It can also be shown that the asymptotic behavior of $\partial M/\partial\rho$ as $r,\rho\to 1$ is precisely that of the kernel $L(x,t)$ for the notched half-plane, as $x$, $t$ approach the edge. Numerical computations based on equation (5.246) were carried out in 1964. These were made for the case of constant pressure $g\equiv 1$, and intensity factors $k_1$ were computed for values of a ranging from $\frac{1}{2}$ to 1. Appendix II contains some data about those computations.

Future work should analyze the configuration of Figure 5.16 much more. This will amount to study in detail the phenomena associated with the integral equation (5.178).

## 5.6 Appendix I: Some lemmas on Taylor and Laurent expansions

*Lemma* 1. Let the power series

$$F(z) = \sum_{n=0}^{\infty} c_n(z-z_0)^n$$

have radius $2R > 0$ of convergence. Let $D$ denote the disk of radius $R$ centered on $z_0$. Then

$$J_r = \iint_D (\mathrm{Re}\ F)^2\,\mathrm{d}x\,\mathrm{d}y = \pi R^2\left\{(\mathrm{Re}\ c_0)^2 + \sum_{n=1}^{\infty} |c_n|^2 \frac{R^{2n}}{2n+2}\right\}$$

$$J_i = \iint_D (\mathrm{Im}\ F)^2\,\mathrm{d}x\,\mathrm{d}y = \pi R^2\left\{(\mathrm{Im}\ c_0)^2 + \sum_{n=1}^{\infty} |c_n|^2 \frac{R^{2n}}{2n+2}\right\} \tag{5.249}$$

$$J = \iint_D |F^2|\,\mathrm{d}x\,\mathrm{d}y = \pi R^2\left\{|c_0|^2 + 2\sum_{n=1}^{\infty} |c_n|^2 \frac{R^{2n}}{2n+2}\right\}.$$

*Proof.* It suffices to prove the first relation; the others follow from it. Introduce coordinates $r$, $\theta$ defined by $z-z_0 = r\,\mathrm{e}^{\mathrm{i}\theta}$ and use $\mathrm{d}x\mathrm{d}y = r\mathrm{d}r\mathrm{d}\theta$. Re $F$ is an ordinary Fourier-series in $\theta$. Use the orthogonality of the Fourier components and integrate. This leads immediately to the first relation (5.249).

*Lemma* 2. From equations (5.249) it follows that

$$|F(z_0)| = |c_0| \leqslant \frac{J^{\frac{1}{2}}}{R\pi^{\frac{1}{2}}}\,;\quad |F'(z_0)| = |c_1| \leqslant \frac{2J_r^{\frac{1}{2}}}{R^2\pi^{\frac{1}{2}}}. \tag{5.250}$$

*Proof.* Clear.

*Lemma* 3. Let $F(z)$ be holomorphic in the domain $\Delta$ of the unit disk $|z| < 1$, cut along the negative real axis (Figure 5.19). Assume

$$\iint_{\Delta} (\mathrm{Re}\ F)^2\,\mathrm{d}x\,\mathrm{d}y \leqslant U^2 < \infty\,. \tag{5.251}$$

Then

(a) $|y^2 F'(z)| \leqslant p_0 U$ for $|z| < \frac{1}{2}$; $z \in \Delta$,

(b) $\iint_D y^2 |F'(z)|^2\,\mathrm{d}x\,\mathrm{d}y \leqslant p_1 U^2$ for $|z_0| < \frac{1}{4}$, $z_0 \in \Delta$,

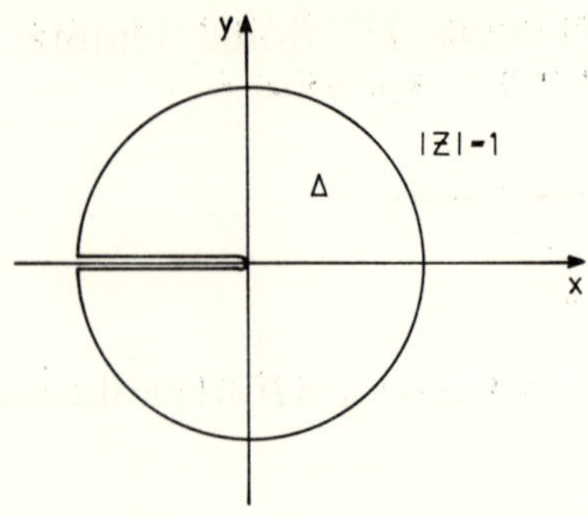


Figure 5.19

where $2R$ is the minimum distance of any point $z_0$ from the negative real axis and where $D$ denotes the disk of radius $R$ centered at $z_0$; $p_0$, $p_1$ are positive constants not depending on $F(z)$.

*Proof.* As a domain, $\Delta$ is an open set; therefore $R>0$; furthermore if $z_0=x_0+\mathrm{i}\,y_0$ we have $2R=|z_0|$ for $x_0\geqslant 0$ and $2R=|y_0|$ for $x_0<0$.

Lemmas 1, 2 stay valid if one takes for $2R$ any positive number not exceeding the radius of convergence. The disk $D$ is in $\Delta$.

Equations (5.250) can be applied with $J_r$ replaced by $U^2$. Set $p_0=2/\pi^{\frac{1}{2}}$; since $R\geqslant|y_0|$, it follows from equation (5.250) that $|y_0^2F'(z_0)|\leqslant|R^2F'(z_0)|\leqslant p_0U$, which proves assertion (a).

As for (b) apply $y^2|F'(z)|^2\leqslant p_0^2U^2/R^2$. Set $p_1=p_0^2\pi=4$, and (b) is satisfied.

*Lemma* 4. In addition to $F(z)$ of lemma 3 let another holomorphic function $G(z)$ in $\Delta$ be given. Compose $\omega=yF'(z)+G(z)$ and assume besides condition (5.251)

$$\iint_\Delta |\omega^2|\,\mathrm{d}x\,\mathrm{d}y\leqslant U^2 \tag{5.252}$$

with the same $U$ as in (5.251). Then

$$|RG(z_0)|\leqslant p_2U \quad \text{with} \quad p_2=\left(\frac{10}{\pi}\right)^{\frac{1}{2}} \quad \text{for} \quad |z_0|<\tfrac{1}{4},\ z_0\in\Delta\,.$$

*Proof.* $G(z)=\omega-yF'$; $|G(z)|^2\leqslant 2|\omega|^2+2|yF'(z)|^2$. From lemma 3(b) with $p_1=4$ and from (5.252) it follows that

$$\iint_D |G|^2\,\mathrm{d}x\,\mathrm{d}y\leqslant 10U^2\ ; \tag{5.253}$$

application of equation (5.250) with $J$ replaced by $10U^2$ proves the assertion of lemma 4.

*Lemma* 5. Let the Laurent series

$$H_0(z) = \sum_{m=-\infty}^{\infty} h_m z^m$$

converge for $0<|z|<1$.

Let $0<a<1$. Consider in $\Delta$ a function $H(z)$ which is either ($\alpha$) $H(z)\equiv H_0(z)$ or ($\beta$) $H(z)=z^{-\frac{1}{2}}H_0(z)$.

Assume that $|y^n H(z)|\leqslant C'$ for all $z\in\Delta$, $|z|<a$ with some constant $C'$ and $n$ a positive integer. Then $h_m=0$ for all $m<-n$; moreover $h_{-n}=0$ in the case ($\beta$).

*Proof.* In either case $|y^n H_0(z)|\leqslant C'$. Take the circle $|z|=r$ $(0<r<a)$ and form the integral

$$J(n, p) = \oint y^n z^p H_0(z)\frac{\mathrm{d}z}{z}, \quad p \text{ a positive integer}$$

over that circle. The preceding inequality yields $J(n, p)\leqslant 2\pi C' r^p$ and $J(n, p)\to 0$ as $r\to 0$. On the other hand one can evaluate $J(n, p)$ by using $2\mathrm{i}\,y=z-\bar{z}$ and $\bar{z}=r^2/z$. Consequently

$$(2\mathrm{i})^n J(n, p) = 2\pi\mathrm{i}\,h_{-n-p}+r^2 q(r^2)$$

where $q(r^2)$ is some polynomial of $r^2$ whose coefficients depend on the $h_k$ only, $k$ running from $-n-p+1$ to $-p+1$. But this is compatible with $J(n, p)\to 0$ only if $h_{-n-p}$ vanishes. The dominant term of $H_0$ as $z\to 0$ is $h_{-n}z^{-n}$. In the case ($\beta$) we have $y^n H(z)\approx h_{-n}(y/z)^n z^{-\frac{1}{2}}$; this contradicts the existence of $C'$ unless $h_{-n}=0$. This concludes the proof.

*Lemma* 6. Let $G(z)=H(z)$ where $G$ refers to lemma 4 and $H$ to lemma 5. Then $h_m=0$ for $m<-1$ and in addition $h_{-1}=0$ in the case ($\beta$).

*Proof.* By lemma 4 we can assume $|yG(z)|\leqslant p_2 U$ valid for $|z|<\frac{1}{4}$. This makes lemma 5 applicable for $n=1$, and lemma 6 follows.

Note that all lemmas 3–6 stay valid if one replaces $\Delta$ by a similar domain of disk radius $p>0$; the limitations $|z|<1$, $|z|<\frac{1}{2}$, $|z|<\frac{1}{4}$ are to be replaced by $|z|<p$, $|z|<\frac{1}{2}p$, $|z|<\frac{1}{4}p$.

## 5.7 Appendix II: Weight functions and stress-intensity factors

### WEIGHT FUNCTIONS

*Collinear cracks of period* $2\pi$. If $-c<x<c$ and its translations by $2\pi n$ along the $x$-axis describe the cracks and if all cracks are equally loaded by normal tractions then $\rho=\phi=(\sin z+\sin c)(\sin^2\frac{1}{2}z-\sin^2\frac{1}{2}c)^{-\frac{1}{2}}$ describe a fundamental field of period $2\pi$ with a normal singularity at $x=-c$ and a fundamental one at $x=c$. Formula (5.162) is an adaptation to the configuration discussed in section 5.4. The weight function with respect to $(-c, c)$ is in normalized form

$$m(x) = \frac{\sin x+\sin c}{2[(\cos x-\cos c)\sin c]^{\frac{1}{2}}}\ ; \quad \text{see equation (5.93)}\,.$$

*Edge crack in the half-plane, infinitely deep notch in the half-plane and notched bar.* For the half-plane $m(x)$ was found by means of equation (5.101); for the infinitely deep notch the relation $m(x)=m^*(1/x)$ was used where $m^*(x)$ is the weight function for half-plane with edge notch. For the notched bar the related integral equation was numerically solved. The following approximation to $m(x)$ of formula (5.93) covers all three cases. It is useful to replace $x$ at a generic notch by the distance $t$ of that point from the root of the notch. Figure 5.20 shows the relevant geometric quantities. Furthermore we write

$$m(x) = M(t)\,,\quad p(x) = P(t) \tag{5.254}$$

whence

$$k_1 = \frac{2^{\frac{1}{2}}}{\pi}\int_0^h M(t)\,P(t)\,dt\,. \tag{5.255}$$

The weight function may be represented with engineering accuracy (1% error) in the convenient form of the following approximation:

$$M(t) = t^{-\frac{1}{2}}(1+m_1 t/h+m_2(t/h)^2)\ ; \tag{5.256}$$

here $m_1$, $m_2$ are constant coefficients; i.e., they do not depend on the variable $t/h$. They depend however on the ratio

$$r = h/W \tag{5.257}$$

and this dependency is representable in the form of the following approximation:

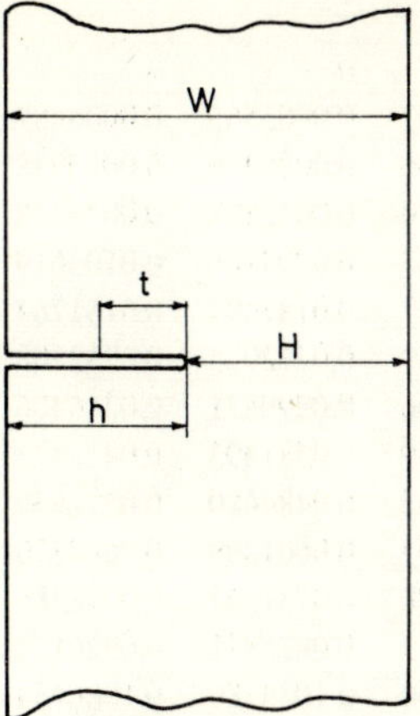


Figure 5.20

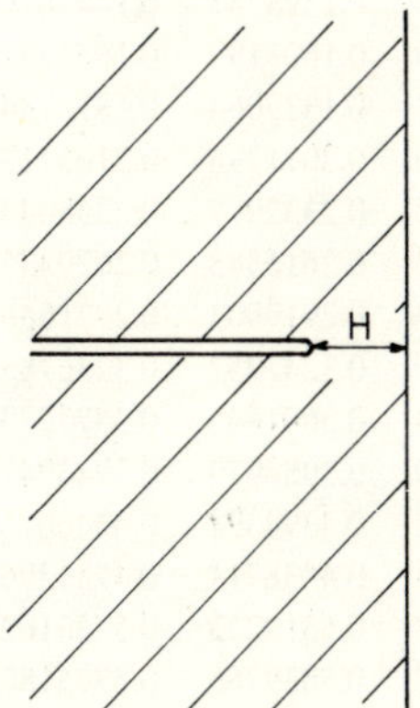


Figure 5.21

$$m_1 = A_1 + B_1 r^2 + C_1 r^6 , \quad m_2 = A_2 + B_2 r^2 + C_2 \cdot r^6 , \tag{5.258}$$

where $A_1 = 0.6147 \quad B_1 = 17.1844 \quad C_1 = \ 8.7822$

$A_2 = 0.2502 \quad B_2 = \ 3.2889 \quad C_2 = 70.0444 .$

These approximations are useful for $0 \leqslant r \leqslant 0.5$. They cover the case of the edge crack in the half-plane ($H = \infty$) so that

$m_1 = A_1 , \quad m_2 = A_2$ for edge crack in half-plane,

As shown in section 5.3, the information for an edge crack in a half-plane yields the weight function for what seems to be an entirely different case; namely the case of an infinitely deep notch in a half-plane (Figure 5.21) where $H$ is bounded while $h = \infty$. In this case one has

TABLE I *The functions* $F_1$, $F_2$ (5.166) *and* $G_1$, $G_2$ (5.168)

| $\tau$ | $F_1$ | $G_1$ | $F_2$ | $G_2$ |
|---|---|---|---|---|
| 0 | 0 | 0 | 0 | 0 |
| 0.1 | 0.0001087 | 0.0000607 | 0.0005865 | 0.0006487 |
| 0.2 | 0.0004346 | 0.0002446 | 0.0023476 | 0.0025938 |
| 0.3 | 0.0009772 | 0.0005570 | 0.0052888 | 0.0058328 |
| 0.4 | 0.0017355 | 0.0010065 | 0.0094186 | 0.0103619 |
| 0.5 | 0.0027079 | 0.0016047 | 0.0147494 | 0.0161764 |
| 0.6 | 0.0038925 | 0.0023659 | 0.0212975 | 0.0232708 |
| 0.7 | 0.0052869 | 0.0033066 | 0.0290831 | 0.0316406 |
| 0.8 | 0.0068878 | 0.0044449 | 0.0381303 | 0.0412823 |
| 0.9 | 0.0086912 | 0.0057995 | 0.0484680 | 0.0521949 |
| 1.0 | 0.0106924 | 0.0073895 | 0.0601299 | 0.0643810 |
| 1.1 | 0.0128854 | 0.0092328 | 0.0731547 | 0.0778486 |
| 1.2 | 0.0152628 | 0.0113453 | 0.0875871 | 0.0926124 |
| 1.3 | 0.0178161 | 0.0137400 | 0.1034780 | 0.1086954 |
| 1.4 | 0.0205344 | 0.0164252 | 0.1208851 | 0.1261315 |
| 1.5 | 0.0234049 | 0.0194038 | 0.1398743 | 0.1449670 |
| 1.6 | 0.0264121 | 0.0226710 | 0.1605199 | 0.1652631 |
| 1.7 | 0.0295371 | 0.0262134 | 0.1829064 | 0.1870984 |
| 1.8 | 0.0327570 | 0.0300070 | 0.2071295 | 0.2105717 |
| 1.9 | 0.0360443 | 0.0340152 | 0.2332977 | 0.2358043 |
| 2.0 | 0.0393653 | 0.0381872 | 0.2615345 | 0.2629435 |
| 2.1 | 0.0426795 | 0.0424559 | 0.2919803 | 0.2921656 |
| 2.2 | 0.0459373 | 0.0467355 | 0.3247957 | 0.3236788 |
| 2.3 | 0.0490782 | 0.0509194 | 0.3601645 | 0.3577273 |
| 2.4 | 0.0520284 | 0.0548780 | 0.3982979 | 0.3945947 |
| 2.5 | 0.0546976 | 0.0584558 | 0.4394394 | 0.4346075 |
| 2.6 | 0.0569745 | 0.0614693 | 0.4838714 | 0.4781396 |
| 2.7 | 0.0587219 | 0.0637037 | 0.5319223 | 0.5256162 |
| 2.8 | 0.0597701 | 0.0649107 | 0.5839766 | 0.5775182 |
| 2.9 | 0.0599079 | 0.0648053 | 0.6404863 | 0.6343865 |
| 3.0 | 0.0588718 | 0.0630625 | 0.7019865 | 0.6968269 |
| 3.1 | 0.0563308 | 0.0593147 | 0.7691149 | 0.7655151 |
| 3.2 | 0.0518675 | 0.0531480 | 0.8426360 | 0.8412010 |
| 3.3 | 0.0449518 | 0.0440990 | 0.9234743 | 0.9247149 |
| 3.4 | 0.0349063 | 0.0316513 | 1.0127565 | 1.0169718 |
| 3.5 | 0.0208599 | 0.0152320 | 1.1118682 | 1.1189778 |
| 3.6 | 0.0016822 | −0.0057922 | 1.2225289 | 1.2318352 |
| 3.7 | −0.0241040 | −0.0321188 | 1.3468959 | 1.3567486 |
| 3.8 | −0.0584490 | −0.0645129 | 1.4877056 | 1.4950309 |
| 3.9 | −0.1039549 | −0.1038120 | 1.6484688 | 1.6481096 |
| 4.0 | −0.1641229 | −0.1509298 | 1.8337606 | 1.8175327 |

$G_k = a_k\tau^2 + b_k\tau^4 + c_k\tau^6$

$a_1 = 0.006057$ $b_1 = 0.001486$ $c_1 = -0.0001534$

$a_2 = 0.06487$ $b_2 = -0.0007273$ $c_2 = 0.0002358$

$$M(t) = (Hs)^{-\frac{1}{2}}(1 + A_1 s + A_2 s^2) \text{ with } s = \frac{t}{t+H}. \tag{5.259}$$

Suppose that $P(t)$ is given or can be approximated by a polynomial

$$P(t) = \sum_{k=0}^{n} C_k (t/h)^k. \tag{5.260}$$

In this case the integration (5.255) can be carried out to closed form and

$$k_1 = 2\frac{(2h)^{\frac{1}{2}}}{\pi} \sum_{k=0}^{n} C_k \left[\frac{1}{2k+1} + \frac{m_1}{2k+3} + \frac{m_2}{2k+5}\right]. \tag{5.261}$$

This formula promises easy engineering applications. It takes but a slide rule in order to compute $k_1$ from equation (5.261).

For the notched bar a comparison of $k_1$-data, obtained by the method of weight functions, and of those available from ASTM 410 and NASA TN D-2395 was made. The data refer to bending and tension. The agreement is excellent; details can be found in [6].

Table I pertains to the integral equation for the notched bar and presents the functions $F_1$, $F_2$ as well as their approximations $G_1$, $G_2$. More details about the integral equation can be found in [6].

## STRESS-INTENSITY FACTORS

*Infinite plane with a circular hole and a radial edge notch.* The data pertaining to the case of tractions induced by centrifugal forces can be found in [13].

*Notched cylinder under tension.* The values of $k_1$ for various ratios of notch length to radius were computed on the basis of equation (5.246) in [7]. In this context approximations for the $r$-derivatives of $M_1$ and $M_2$ were developed; care was taken to preserve the singular character of $M$ at $r=\rho$ and its special character at $r=1$, $\rho=1$.

From the exact form of $M_1$ it follows that its $r$-derivative is a function of the ratio $x=\rho/r$ alone; more precisely

$$\frac{\partial M_1}{\partial r} = -\tfrac{1}{2}\pi x^2(1-x^2)^{-1} F(\tfrac{1}{2}, \tfrac{1}{2}, 2; x^2) \quad \text{for} \quad x<1$$

$$\frac{\partial M_1}{\partial r} = \pi x (x^2-1)^{-1} F(-\tfrac{1}{2}, \tfrac{1}{2}, 1; x^{-2}) \quad \text{for} \quad x>1.$$

The approximation is of the form

$$\frac{\partial M_1}{\partial r} = \tfrac{1}{2}\pi \frac{r^2}{\rho^2 - r^2} F(x) \quad \text{with}$$

$$F(x) = \frac{4}{\pi} + 1.1965(x^2-1) - 0.0500(x^2-1)^2/x + \frac{1}{\pi}(1-x^2)\log|1-x^2| \,.$$

$F(x)$ is accurate by better than 1% for $0.5 \leqslant x \leqslant 1.5$.

The function $M_2$ or rather its $r$-derivative can be presented in the form

$$\frac{\partial M_2}{\partial r} = -\frac{s^2+4st-t^2+Z}{(s+t)^3}\rho \quad \text{with} \quad s=1-r\,,\ t=1-\rho$$

where $Z=o(s^2+t^2)$ as $s,\ t\to 0$. $Z$ was computed on the basis of the integral representation of $M_2$; thereafter an approximation was developed. This is of the form

$$\tilde{Z} = 0.9(s+t)^4 + \tfrac{1}{2}(s+t)^3 \log\frac{1}{s+t} + c_0(s)(s+t)^3 + c_1(s)\,s(s+t)^2$$
$$+ c_2(s)\,s^2(s+t) + c_3(s)\,s^3\,,$$

with the coefficients

$$c_0 = 0.59 - 12.10\,s\,, \quad c_1 = 7.73 + 13.20\,s + 15.00\,s^2$$

$$c_2 = -14.13 - 0.94\,s - 30.00\,s^2\;; \quad c_3 = 5.86 - 1.94\,s + 15.00\,s^2\,.$$

TABLE II

*The functions $X$ and $\tilde{X}$ ($\nu = 0.3$)*

| | $r=0.5$ | $r=0.6$ | $r=0.7$ | $r=0.8$ | $r=0.9$ | $r=1$ |
|---|---|---|---|---|---|---|
| $\rho=0.5$ | 1.092 | 1.108 | 1.090 | 0.982 | 0.644 | −0.307 |
| | 1.038 | 1.069 | 1.063 | 0.966 | 0.639 | −0.307 |
| $\rho=0.6$ | 1.349 | 1.391 | 1.400 | 1.302 | 0.887 | −0.530 |
| | 1.331 | 1.373 | 1.381 | 1.282 | 0.869 | −0.546 |
| $\rho=0.7$ | 1.618 | 1.705 | 1.774 | 1.736 | 1.280 | −0.936 |
| | 1.634 | 1.718 | 1.782 | 1.739 | 1.282 | −0.936 |
| $\rho=0.8$ | 1.879 | 2.039 | 2.225 | 2.362 | 2.033 | −1.782 |
| | 1.915 | 2.078 | 2.266 | 2.409 | 2.096 | −1.713 |
| $\rho=0.9$ | 2.065 | 2.320 | 2.697 | 3.250 | 3.774 | −4.316 |
| | 2.092 | 2.359 | 2.753 | 3.333 | 3.923 | −4.084 |
| $\rho=1$ | 1.957 | 2.216 | 2.658 | 3.545 | 6.166 | |
| | 1.957 | 2.214 | 2.326 | 3.549 | 5.896 | |

Table II shows a comparison of two functions $X$, $\tilde{X}$ where

$$X = \frac{s^2+4st-t^2+Z}{2r(s+t)^3}$$

and $\tilde{X}$ the approximation obtained by replacing $Z$ by $\tilde{Z}$. The upper numbers refer to $X$, the lower ones to $\tilde{X}$.

For prescribed constant $g(r)$ the integral equation was solved by an "inverse" method, according to which $w(r)$ was set up in the form

$$w(r) = \gamma_0(r-a)^{\frac{1}{2}} + \gamma_1(r-a)^{1+\frac{1}{2}} + \gamma_2(r-a)^{2+\frac{1}{2}} + \gamma_3(r-a)^{3+\frac{1}{2}}$$

with unknown coefficients $\gamma_k$. For a function of this type the integral in equation (5.246) can be obtained in closed form, if one uses the approximations to $\partial M/\partial r$. The coefficients $\gamma_k$ were determined by collocation in order to match the prescribed $g(r)$ at four special points of the interval $(a, 1)$. Matching was within $1\%$. The values of $k_1$ were derived from $\gamma$. The related quantity $f_0 = k_1 h^{-\frac{1}{2}}$, $h = 1-a$, is shown in the "Table of coefficients" below.

If one wishes to use equation (5.247) in the homogeneous form for the computation of the weight function, then the approximation to the $r$-derivative of $M$ can be used by simply exchanging $r$ and $\rho$. Weight functions have not yet been computed; integral equation and approximation to its kernel are presented in order to stimulate the computation of related weight functions by those who are interested.

If one had the weight function $M(r)$, normalized to $M(r) \approx (r-a)^{-\frac{1}{2}}$, then

$$k_1 = \frac{2^{\frac{1}{2}}}{\pi a}\int_a^1 M(r)g(r)r\,\mathrm{d}r\,. \tag{5.262}$$

For deep notches one could perhaps use formula (5.206) for approximation, so that, after adaptation of it to the new notch,

$$k_1 \approx k_1^* = \frac{2}{\pi a}a^{-\frac{1}{2}}\int_a^1 [\arcsin(1-a^2/r^2)^{\frac{1}{2}} + a(r^2-a^2)^{-\frac{1}{2}}]\,g(r)r\,\mathrm{d}r\,.$$

For $g \equiv 1$ the integration yields

$$k_1^* = \frac{1}{\pi a}a^{-\frac{1}{2}}[\arccos a + a(1-a^2)^{\frac{1}{2}}], \tag{5.263}$$

and in particular

$$k_1^*(1-a)^{-\frac{1}{2}} = \tfrac{4}{3} + \frac{1}{\pi}3^{\frac{1}{2}} \approx 1.884 \text{ for } a = \tfrac{1}{2}\,.$$

Table III gives the value of $k_1(1-a)^{-\frac{1}{2}}$ as 1.92; this value and 1.88 are sufficiently close together, and the use of $k_1 \approx k_1^*$ for cases $a < \frac{1}{2}$ appears to be justified. As for $a \geqslant \frac{1}{2}$ an approximation to the weight function can be found

TABLE III

*Values of coefficients*

| $a$ | 0.5 | 0.6 | 0.7 | 0.8 | 0.9 |
|---|---|---|---|---|---|
| $-c_0$ | 0.960 | 0.708 | 0.531 | 0.392 | 0.259 |
| $c_1$ | 0.881 | 0.608 | 0.462 | 0.396 | 0.417 |
| $-c_2$ | 0.768 | 0.441 | 0.245 | 0.087 | 0.230 |
| $-c_3$ | −0.148 | 0.258 | 0.926 | 2.930 | 17.168 |
| $f_0$ | 1.920 | 1.583 | 1.371 | 1.240 | 1.158 |
| $\tilde{f}_0$ | 1.923 | 1.580 | 1.374 | 1.242 | 1.160 |
| $-10\,f_1$ | 4.589 | 3.435 | 2.610 | 2.020 | 1.610 |
| $-10\,\tilde{f}_1$ | 4.620 | 3.337 | 2.563 | 2.061 | 1.716 |
| $10\,f_2$ | 2.000 | 0.997 | 0.415 | 0.089 | −0.089 |
| $10\,\tilde{f}_2$ | 1.995 | 1.002 | 0.419 | 0.070 | −0.118 |
| $100\,f_3$ | −1.927 | 2.332 | 4.708 | 5.980 | 6.629 |
| $100\,\tilde{f}_3$ | −1.868 | 2.256 | 4.649 | 6.024 | 6.661 |

by parameter differentiation of $w(r)$ with respect to $a$. For this reason Table III gives an approximation to $w(r, a)$ which can be differentiated with respect to $a$.

*Details of computation for the case $g(r) \equiv 1$ and $\nu = 0.3$.* With

$$C = \frac{\kappa + 1}{2(2^{\frac{1}{2}})\mu},$$

$w(r)$ can be written in the form

$$w(r) = -C(2^{\frac{1}{2}}) \sum_{n=0}^{3} c_n(r-a)^n .$$

Thus $k_1 = -2^{\frac{1}{2}} c_0$. If one introduces $h = 1-a$, $x = (r-a)/h$ then $w(r)$ takes the form

$$w(r) = Chf_0(a)x^{\frac{1}{2}} \cdot [1 + f_1(a)x + f_2(a)x^2 + f_3(a)x^3],$$

where $f_0 = k_1 h^{-\frac{1}{2}}$, $f_n = h^n c_n / c_0$ for $n = 1, 2, 3$. Approximations $\tilde{f}_n(a)$ to $f_n(a)$ are:

$$\tilde{f}_0 = 1.080 + 0.281\,A + 0.040\,a^{10}\,, \quad A = (1-a^2)/a^2$$

$$\tilde{f}_1 = -0.147 - 0.105\,A$$

$$\tilde{f}_2 = -0.047 + 0.082\,A + 0.030\,a^6$$

$$\tilde{f}_3 = -0.6\,f_2 + 0.056 + 0.015\,A\,.$$

TABLE IV

*Normal tractions responding to approximate $w(r)$*

| $10\,x$ | 1 | 2 | 3 | 4 | 5 | 6 | 7 | 8 | 9 |
|---|---|---|---|---|---|---|---|---|---|
| $a = 0.5$ | 1.000 | 1.005 | 1.004 | 1.000 | 0.998 | 1.000 | 1.005 | 1.010 | 1.000 |
| $a = 0.6$ | 1.000 | 1.003 | 1.002 | 1.000 | 0.999 | 1.000 | 1.003 | 1.006 | 1.000 |
| $a = 0.7$ | 1.000 | 1.002 | 1.001 | 1.000 | 0.999 | 1.000 | 1.002 | 1.005 | 1.000 |
| $a = 0.8$ | 1.000 | 1.002 | 1.001 | 1.000 | 0.999 | 1.000 | 1.002 | 1.004 | 1.000 |
| $a = 0.9$ | 1.000 | 1.002 | 1.001 | 1.000 | 0.999 | 1.000 | 1.002 | 1.004 | 1.000 |

The values of $k_1$ and of related quantities such as $f_0$ were first published in [8b]. It should be emphasized that the $k_1$-values agree with computations by Zahn [12].

## References

[1] Bueckner, H. F., The Propagation of Cracks and the Energy of Elastic Deformation, *Transactions of the ASME*, pp. 1225–1230 (1958).

[2] Bueckner, H. F., *Some Stress Singularities and Their Computation by Means of Integral Equations, Boundary Problems in Differential Equations*, Madison 1960, The University of Wisconsin Press, pp. 215–230.

[3] Bueckner, H. F. and Giaever, I., The Stress Concentration of a Notched Rotor Subjected to Centrifugal Forces, *ZAMM* 46, pp. 265–273 (1966).

[4] Bueckner, H. F., On a Class of Singular Integral Equations, *Journal of Mathematical Analysis and Applications*, Vol. 14, No. 3 (1966) pp. 392–426.

[5] Bueckner, H. F., A Novel Principle for the Computation of Stress Intensity Factors, *ZAMM* 50, pp. 529–545 (1970).

[6] Bueckner, H. F., Weight Functions for the Notched Bar, *ZAMM*, 51, pp. 97–109 (1971).

[7] Bueckner, H. F., *The Stress Concentration of a Notched Cylinder under Tension.* Report

DF-64-LS-85, Oct. 1964, Class I. Large Steam Turbine-Generator Department, General Electric Co., Schenectady, N. Y.

[8] a. Paris, Paul C. and Sih, George C., *Stress Analysis of Cracks. Fracture Toughness Testing and its Applications.* ASTM Special Technical Publication No. 381; pp. 30–81.
b. Bueckner, H. F., Discussion (of the preceding paper), pp. 82–83.

[9] Muskhelishvili, N. I., *Singular Integral Equations*, Groningen, Holland. 1953, P. Noordhoff Ltd., chapter 11.

[10] Barenblatt, G. I., Mathematical theory of equilibrium cracks. *Advances in Applied Mathematics*, Academic Press, New York and London, Vol. 7, pp. 55–129, 1962.

[11] Wigglesworth, L. A.
a. Stress distribution in a notched plate. *Mathematika*, Vol. 4, 1957, pp. 76–96.
b. Stress relief in a cracked plate. *Mathematika*, Vol. 5, No. 9, 1958, pp. 67–81.

[12] Zahn, John J., Stress Intensity Factors for a Sharply Notched Infinite Cylinder under Tension. *Developments in Mechanics*, Vol. 3, part 1, p. 91. John Wiley & Sons, Inc. 1965.

[13] Melan, Ernst and Parkus, Heinz, *Warmespannungen* (p. 32), Wien, Springer-Verlag, 1953.

[14] Oberhettinger, Fritz, *Tabellen Zur Fourier Transformation*, p. 3. Springer-Verlag, 1957.

[15] California Institute of Technology, Bateman Manuscript Project. McGraw-Hill Book Company, Inc. 1954.

[16] Panasyuk, V. V., *Limiting equilibrium of brittle solids with fractures* (p. 184). Translation from Russian, distributed in 1971 by Management Information Services, P. O. Box 5129, Detroit, Michigan 48236.

*I. N. Sneddon*

# 6 *Integral transform methods*

## 6.1 Introduction

Among the earliest papers leading to a revival of interest in crack problems in the classical theory of elasticity were those in which the solution of the relevant boundary value problem was obtained by a systematic use of the theory of integral transforms. In this connection reference can be made to [1], [2]. Interest in this form of solution has been maintained over the years and the present chapter is an attempt to present the essentials of this method.

It is emphasized here that it is only the essentials of this approach which are described here. Several topics of interest are omitted, for example, the behavior under shear of solids containing cracks receives only a brief mention and there is no discussion of either torsion problems or of problems concerning cracks situated at the interface of two different media. There has certainly been a great deal of work done on these topics by the use of integral transform techniques but the techniques used are basically the same as those employed in the simpler problems discussed here, in which it is assumed that the deformation produced in the elastic body arises from the application of pressure to the crack surfaces. (By a simple superposition argument it is easily seen that the classical problem of the behavior of such a body under uniaxial tension can be reduced to a particular case—that in which the pressure is constant—of this form of problem.

## 6.2 The Griffith crack problem as a mixed boundary value problem

To determine the stress field in the $xy$-plane due to the application of a symmetric pressure $p(x)$ to the faces of the Griffith crack specified by the relations $|x| \leqslant c$, $y=0$, it is sufficient to calculate the stress field in the half-plane $y \geqslant 0$ when the boundary $y=0$ is subjected to the boundary conditions

$$\sigma_{yy}(x, 0) = -p(x), \qquad 0 \leqslant x \leqslant c,$$
$$u_y(x, 0) = 0, \qquad x > c,$$
$$\sigma_{xy}(x, 0) = 0, \qquad x \geqslant 0,$$

which can be deduced easily by considering the symmetry of the problem. It should be observed that by "symmetric" in this case it is meant that the prescribed function $p(x)$ is an *even* function of $x$ and also that the pressure applied to the face $y=0-$ of the crack is identical to that applied to the opposite face $y=0+$. The usual assumptions of the classical theory of elasticity are made; in particular it is assumed that the deformed surface of the crack does not differ from the undeformed one in any marked degree so that the pressure is assumed to be applied to the boundary $y=0$ and not to the faces of the deformed crack.

An elegant solution of this half-plane problem has been given by Lowengrub [3] using a method similar to that devised by Sneddon [4] for the case of a circular crack in an infinite elastic solid in three-dimensional space. Let the Fourier cosine and sine transforms be denoted by

$$\mathscr{F}_c[f(\xi, y); \xi \to x] = \left(\frac{2}{\pi}\right)^{\frac{1}{2}} \int_0^\infty f(\xi, y) \cos(\xi x) d\xi$$

$$\mathscr{F}_s[f(\xi, y); \xi \to x] = \left(\frac{2}{\pi}\right)^{\frac{1}{2}} \int_0^\infty f(\xi, y) \sin(\xi x) d\xi .$$

Thus the displacement field may be written in the form $\boldsymbol{u}=(u_x, u_y, 0)$ where the components $u_x$ and $u_y$ are given respectively by the equations

$$u_x(x, y) = -2^{\frac{1}{2}} \cdot (1+\nu)/\pi^{\frac{1}{2}} E \mathscr{F}_s[(1-2\nu-\xi y)\psi(\xi) e^{-\xi y}; \xi \to x] \tag{6.1}$$
$$u_y(x, y) = 2^{\frac{1}{2}} \cdot (1+\nu)/\pi^{\frac{1}{2}} E \mathscr{F}_c[(2-2\nu+\xi y)\psi(\xi) e^{-\xi y}; \xi \to x] \tag{6.2}$$

in which $E$ denotes Young's modulus and $\nu$ denotes Poisson's ratio. This is the solution of the equations for *plane strain*; if the solution for the case of plane stress is desired it may be obtained by means of the familiar change in the numerical values of the elastic constants. The corresponding expressions for the components of the stress tensor are readily obtained from the stress-strain relations for an isotropic solid. From these expressions it can be deduced immediately that $\sigma_{xy}(x, 0)$ vanishes for all real values of $x$ and that

$$\sigma_{yy}(x, 0) = -\left(\frac{2}{\pi}\right)^{\frac{1}{2}} \frac{d}{dx} \mathscr{F}_s[\psi(\xi);\ x].$$

The prescribed boundary conditions will, therefore, be satisfied if the function $\psi(\xi)$ can be determined as the solution of the "dual" integral equations

$$\left(\frac{2}{\pi}\right)^{\frac{1}{2}} \frac{d}{dx} \mathscr{F}_s[\psi(\xi); x] = p(x), \qquad 0 \leqslant x \leqslant c, \tag{6.3}$$

$$\mathscr{F}_c[\psi(\xi); x] = 0, \qquad x > c. \tag{6.4}$$

Now express the unknown function $\psi(\xi)$ in terms of another function $g(t)$ by the integral

$$\psi(\xi) = \int_0^c t\, g(t)\, J_0(\xi t)\, dt, \tag{6.5}$$

in which $J_0$ denotes the Bessel function of the first kind of order zero. It can be easily deduced from elementary properties of Bessel functions that whatever the function $g(t)$ is—provided, of course, that the integral (6.5) exists—the equation (6.4) is automatically satisfied. Similarly it can be shown that, if $0 \leqslant x < c$,

$$\mathscr{F}_c[\psi(\xi); x] = \left(\frac{2}{\pi}\right)^{\frac{1}{2}} \int_x^c \frac{t\, g(t)\, dt}{(t^2 - x^2)^{\frac{1}{2}}}. \tag{6.6}$$

In a similar way we can show that

$$\mathscr{F}_s[\psi(\xi); x] = \left(\frac{2}{\pi}\right)^{\frac{1}{2}} \int_0^{\min(x,c)} \frac{t\, g(t)\, dt}{(x^2 - t^2)^{\frac{1}{2}}} \tag{6.7}$$

and then deduce from equation (6.3) that the unknown function $g(t)$ is the solution of the Abel integral equation

$$\frac{2}{\pi} \frac{d}{dx} \int_0^x \frac{t\, g(t)\, dt}{(x^2 - t^2)^{\frac{1}{2}}} = p(x), \qquad 0 < x < c. \tag{6.8}$$

It is elementary that the solution of this equation is

$$g(t) = \int_0^t \frac{p(x)\, dx}{(t^2 - x^2)^{\frac{1}{2}}}, \qquad 0 < t < c.$$

In terms of this form of the function $g(t)$ the stress component $\sigma_{yy}(x, 0)$ and the displacement component $u_y(x, 0)$ can be expressed by the pair of equations

$$\sigma_{yy}(x, 0) = -\frac{2}{\pi}\frac{\mathrm{d}}{\mathrm{d}x}\int_0^c \frac{t\,g(t)\mathrm{d}t}{(x^2-t^2)^{\frac{1}{2}}}, \qquad x > c, \tag{6.9}$$

$$u_y(x, 0) = \frac{4(1-\nu^2)}{\pi E}\int_x^c \frac{t\,g(t)\mathrm{d}t}{(t^2-x^2)^{\frac{1}{2}}}, \qquad 0 \leqslant x \leqslant c. \tag{6.10}$$

If the function $g(t)$ is differentiable it can be deduced by performing an integration by parts in the integral on the right-hand side of equation (6.9) that when $x > c$,

$$\sigma_{yy}(x, 0) = \frac{2x g(c)}{\pi(x^2-c^2)^{\frac{1}{2}}} + x\int_0^c \frac{g'(t)\mathrm{d}t}{(x^2-c^2)^{\frac{1}{2}}},$$

from which it follows that

$$\lim_{x\to c+} (x-c)^{\frac{1}{2}}\sigma_{yy}(x, 0) = \frac{2}{\pi}g(c)c^{\frac{1}{2}}. \tag{6.11}$$

From the explicit expression for $g(t)$ it can be deduced immediately that in this case the right-hand side of equation (6.11) may be replaced by the expression

$$\frac{2}{\pi}c^{\frac{1}{2}}\int_0^c \frac{p(x)\mathrm{d}x}{(c^2-x^2)^{\frac{1}{2}}},$$

so that the stress intensity factor can be calculated directly once the variation of pressure on the crack faces is prescribed.

To calculate the elastic energy $W$ expended in forming the crack, use is made of the formula

$$W = 2\int_0^c p(x)u_y(x, 0)\mathrm{d}x. \tag{6.12}$$

If we substitute the form (6.10) for $u_y(x, 0)$ and use the formula

$$\int_0^c p(x)\mathrm{d}x\int_x^c \frac{t\,g(t)\mathrm{d}t}{(t^2-x^2)^{\frac{1}{2}}} = \int_0^c t\,g(t)\mathrm{d}t\int_0^t \frac{p(x)\mathrm{d}x}{(t^2-x^2)^{\frac{1}{2}}}$$

for the change in the order of the integrations, and recall the formula for $g(t)$ we find that

$$W = \frac{8(1-\nu^2)}{\pi E}\int_0^c t[g(t)]^2\mathrm{d}t. \tag{6.13}$$

For example, in the case in which $p(x)=p_0$, a constant, it is easily shown that $g(t)=\pi p_0/2$ and hence that the stress intensity factor is $p_0 a^{\frac{1}{2}}$ and that the crack energy is given by the formula

$$W = \frac{\pi(1-\nu^2)p_0^2 c^2}{E}$$

which is Inglis' expression for the crack energy in the case of a crack opened out by the application to its faces of a constant pressure.

It is useful, at this point, to recall the steps in this solution since they are common to many of the approaches which involve integral transforms:

(1) Replacement of the crack problem by a mixed boundary value problem for a half-space;
(2) Use of an integral transform solution of the equations of elasticity involving an unknown function $\psi$ to derive dual integral equations for this function;
(3) Formulation of an integral representation of $\psi$ in terms of a new unknown function $g$ in order to reduce the dual integral equations for $\psi$ to a single integral equation for $g$.
(4) Calculation of the quantities of physical interest—the stress intensity factor and the crack energy—in terms of the function $g$.

It will be seen that the same procedure is applicable in more complicated situations. In general, however, the integral equation in stage (3) is not an equation of Abel type with a closed form solution but a Fredholm integral equation of the second kind whose solution can be obtained numerically.

## 6.3 Griffith cracks with more complicated geometries

*Griffith crack in a strip with stress-free edges.* There are two cases of a crack in an infinite strip to be considered: that in which the crack is parallel to the edges of the strip and that in which it is perpendicular.

First consider the case shown in Figure 6.1 (Cf. [5]). Now suppose that a crack of length $2c$, situated symmetrically in a strip of width $2b$, is opened up by the application of a symmetric pressure $p(x)$ to its faces. It is easily seen that the problem of determining the stress field in the strip is equivalent to determining the stress field in the strip $-\infty < x < +\infty$, $0 \leqslant y \leqslant b$, subject to the boundary conditions

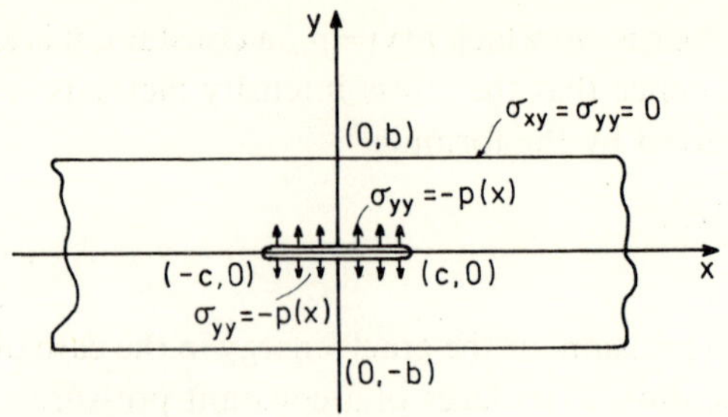

Figure 6.1

$$\sigma_{yy}(x,0) = -p(x)\,, \quad 0<|x|<c\,, \tag{6.14}$$
$$u_y(x,0) = 0\,, \quad |x|>c\,, \tag{6.15}$$
$$\sigma_{xy}(x,0) = 0\,, \quad -\infty < x < +\infty\,, \tag{6.16}$$
$$\sigma_{xy}(x,b) = u_y(x,b) = 0\,, \quad -\infty < x < +\infty\,. \tag{6.17}$$

If $p(x)$ is an even function of $x$, it is readily shown that the displacement field

$$\begin{aligned}\mu u_x(x,y) = &-(\pi/2)^{\frac{1}{2}}\mathscr{F}_s[\psi(\xi)\{U_1(\xi b)\cosh(\xi y) - \xi y U_2(\xi b)\sinh(\xi y)\};\ \xi\to x]\\ &+\tfrac{1}{2}(\pi/2)^{\frac{1}{2}}\mathscr{F}_s[\psi(\xi)\{(1-2\nu)\sinh(\xi y)+\xi y\cosh(\xi y)\};\ \xi\to x]\,,\end{aligned}$$

$$\begin{aligned}\mu u_y(x,y) = &\ (\pi/2)^{\frac{1}{2}}\mathscr{F}_c[\psi(\xi)\{U_3(\xi b)\sinh(\xi y) + \xi y U_2(\xi b)\cosh(\xi y)\};\ \xi\to x]\\ &+(\pi/2)^{\frac{1}{2}}\mathscr{F}_c[\psi(\xi)\{(1-\nu)\cosh(\xi y)-\tfrac{1}{2}\xi y\cosh(\xi y)\};\ \xi\to x]\,,\end{aligned}$$

where $\mu$ denotes the rigidity modulus $E/2(1+\nu)$ and the functions $U_1$, $U_2$, $U_3$ are defined by the equations

$$U_1(t) = \{t^2+(1-2\nu)\sinh^2 t\}/(2t+\sinh 2t)\,,$$
$$U_2(t) = \sinh^2 t/(2t+\sinh 2t)\,,$$
$$U_3(t) = \{t^2-2(1-\nu)\sinh^2 t\}/(2t+\sinh 2t)$$

satisfies the conditions (6.16) and (6.17). For this displacement field the following equations are obtained:

$$\sigma_{yy}(x,0) = -(\pi/2)^{\frac{1}{2}}\frac{\mathrm{d}}{\mathrm{d}x}\mathscr{F}_s[\psi(\xi)\{1-M(\xi b)\};\ x]\,,$$

$$u_y(x,0) = \frac{2(1-\nu^2)}{E}\left(\frac{\pi}{2}\right)^{\frac{1}{2}}\mathscr{F}_c[\psi(\xi);\ x]\,,$$

where the function $M$ is defined by the equation

$$M(t) = \{2t(t+1)+1-e^{-2t}\}/(2t+\sinh 2t).$$

The remaining two conditions (6.14), (6.15) are satisfied if the function $\psi(\xi)$ is the solution of the dual integral equations

$$\left(\frac{\pi}{2}\right)^{\frac{1}{2}} \mathscr{F}_s[\psi(\xi)\{1-M(\xi b)\};\, x] = p(x), \quad 0 \leqslant x \leqslant c, \tag{6.18}$$

$$\mathscr{F}_c[\psi(\xi);\, x] = 0, \qquad x > c. \tag{6.19}$$

Equation (6.19) is again satisfied by the representation (6.5). Substituting this form into equation (6.18) we obtain the relation

$$\frac{2}{\pi}\frac{d}{dx}\int_0^x \frac{t\,g(t)dt}{(x^2-t^2)^{\frac{1}{2}}} \frac{2}{\pi}\frac{d}{dx}\int_0^x M(\xi b)\psi(\xi)\cos(\xi x)d\xi = p(x), \quad 0 \leqslant x \leqslant c. \tag{6.20}$$

If the operator $I$ is defined by the equation

$$If(x) = \frac{2}{\pi}\frac{d}{dx}\int_0^x \frac{tf(t)dt}{(x^2-t^2)^{\frac{1}{2}}}$$

then it is easily shown that its inverse is defined by the equation

$$I^{-1}h(t) = \int_0^t \frac{h(x)dx}{(t^2-x^2)^{\frac{1}{2}}}.$$

Applying $I^{-1}$ to both sides of equation (6.20) a Fredholm integral equation of the second kind we find:

$$g(t) - \int_0^c sg(s)M(s,t)ds = \hat{p}(t)$$

which determines the auxiliary function $g(t)$; the kernel of the equation is defined by the equation

$$M(s,t) = \int_0^\infty \xi M(\xi b) J_0(\xi s) J_0(\xi t) d\xi,$$

and the right hand side by the equation

$$\hat{p}(t) = I^{-1}p(t) = \int_0^t \frac{p(x)dx}{(t^2-x^2)^{\frac{1}{2}}}. \tag{6.21}$$

From equation (6.12) and the relation

$$u_y(x,0)=\frac{2(1-\nu^2)}{E}\int_x^c \frac{t\,g(t)\mathrm{d}t}{(t^2-x^2)^{\frac{1}{2}}}$$

it is found that

$$W=\frac{4(1-\nu^2)}{E}\int_0^c t\,\hat{p}(t)\,g(t)\mathrm{d}t\,. \tag{6.22}$$

Lowengrub solved the integral equation for the case in which $c \ll b$ and $p(x)=p_0$, a constant and derived the formula

$$W=\frac{\pi p_0(1-\nu^2)}{E}\,c^2\left\{1+\mu_1(c/b)^2+\mu_2(c/b)^4+\mu_3(c/b)^6+O(c^8/b^8)\right\} \tag{6.23}$$

where $\mu_1=1.1749$, $\mu_2=0.1351$, $\mu_3=9.3291$.

*Infinite row of parallel cracks.* Also considered in [5] is the closely related problem of determining the distribution of stress in an infinite (two-dimensional) medium in which there is an infinite row of identical cracks spaced at equal intervals. (Cf. Figure 6.2). In this case the components of the displacement field are defined by the equations

$$\mu u_x(x,y)=-\tfrac{1}{2}\mathscr{F}_s\left[\left\{\frac{(1-2\nu)\cosh\xi(b-y)}{\sinh\xi b}+\right.\right.$$
$$\left.\left.-\frac{\xi b\cosh\xi y}{\sinh^2\xi b}-\frac{\xi y\sinh\xi(b-y)}{\sinh\xi b}\right\}\psi(\xi):\ \xi\to x\right]$$

$$\mu u_y(x,y)=\ \tfrac{1}{2}\mathscr{F}_c\left[\left\{\frac{2(1-\nu)\sinh\xi(b-y)}{\sinh\xi b}+\right.\right.$$
$$\left.\left.-\frac{\xi b\sinh\xi y}{\sinh^2\xi b}+\frac{\xi y\cosh\xi(b-y)}{\sinh\xi b}\right\}\psi(\xi):\ \xi\to x\right].$$

These expressions are formulated to ensure that the boundary conditions

$$\sigma_{xy}(x,0)=0\,, \qquad -\infty<x<\infty$$
$$\sigma_{xy}(x,\pm b)=u_y(x,\pm b)=0\,, \qquad -\infty<x<\infty$$

are satisfied. By means of them the displacement field is calculated in the band $|y|\leqslant b$; that in any other part of the solid can be obtained by a simple symmetry argument. From these equations it is easily shown that

$$\sigma_{yy}(x,0)=-\frac{\mathrm{d}}{\mathrm{d}x}\mathscr{F}_s[\{1+H(\xi b)\}\psi(\xi);\,x]$$

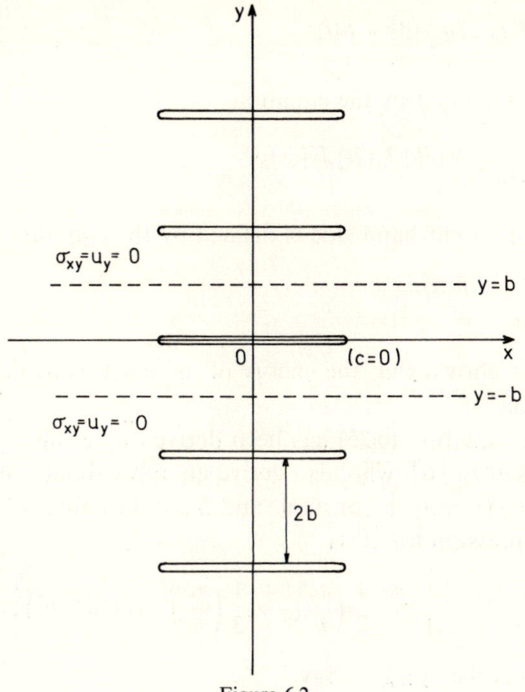


Figure 6.2

where the function $H$ is defined by the equation

$$H(t) = (t + e^{-t} \sinh t) \operatorname{cosech}^2 t, \tag{6.24}$$

so that the boundary conditions

$$\sigma_{yy}(x, 0) = -p(x), \quad |x| \leqslant c; \qquad u_y(x, 0) = 0, \quad |x| > c$$

corresponding to the opening up of each crack under the action of a symmetrical pressure $p(x)$ are satisfied if $\psi(\xi)$ is the solution of the pair of dual integral equations:

$$\frac{d}{dx} \mathscr{F}_s[\{1 + H(\xi b)\} \psi(\xi); x] = p(x), \qquad 0 \leqslant x \leqslant c,$$

$$\mathscr{F}_c[\psi(\xi); x] = 0, \qquad x > c.$$

The second of these equations is again satisfied by the integral representation (6.5), and it is easily shown that the first equation is satisfied if the function $g(t)$ is chosen in this case to be the solution of the Fredholm integral equation

$$g(t) + \int_0^c sK(s, t)\, g(s)\, ds = \hat{p}(t), \tag{6.25}$$

whose kernel is defined by the equation

$$K(s, t) = \frac{2}{\pi} \int_0^\infty \xi H(\xi b)\, J_0(\xi t)\, J_0(\xi s)\, d\xi \tag{6.26}$$

and in which the right-hand side is defined by the equation

$$\hat{p}(t) = \left(\frac{2}{\pi}\right)^{\frac{1}{2}} \int_0^t \frac{p(x)\, dx}{(t^2 - x^2)^{\frac{1}{2}}}. \tag{6.27}$$

It is also easily shown that the energy of the crack is again given by the equation (6.22).

The integral equation (6.25) has been derived by a different method by England and Green [6], who also derive an approximate solution of it in the case when $p(x) = p_0$, a constant, and $b \gg c$. For this solution the corresponding expression for $W$ is

$$W = \frac{\pi(1-\nu^2)p_0^2 c^2}{E} \left[1 - \frac{1}{2}\left(\frac{\pi c}{b}\right)^2 + \frac{1}{3}\left(\frac{\pi c}{b}\right)^4 + O(c^6/b^6)\right] \tag{6.28}$$

which represents the crack energy.

*Griffith crack in an infinite strip.* Next consider the solution of the equations of elastic equilibrium for the semi-infinite strip $|x| \leqslant a$, $y \geqslant 0$ subject to the boundary conditions

$$\sigma_{yy}(x, 0) = -p(x), \qquad |x| \leqslant c, \tag{6.29}$$

$$u_y(x, 0) = 0, \qquad c < |x| < a, \tag{6.30}$$

$$\sigma_{xy}(x, 0) = 0, \qquad |x| < a, \tag{6.31}$$

$$\sigma_{xx}(\pm a, y) = \sigma_{xy}(\pm a, y) = 0, \qquad |y| \geqslant 0, \tag{6.32}$$

This boundary value problem, which was solved by Sneddon and Srivastav [7], corresponds to the problem of a Griffith crack situated centrally in an infinite strip and oriented in a direction normal to the edges of the strip (cf. Figure 6.3). The crack is again assumed to be opened up by the application of a symmetrical pressure $p(x)$.

To obtain the solution of this boundary value problem a displacement field is assumed in the form

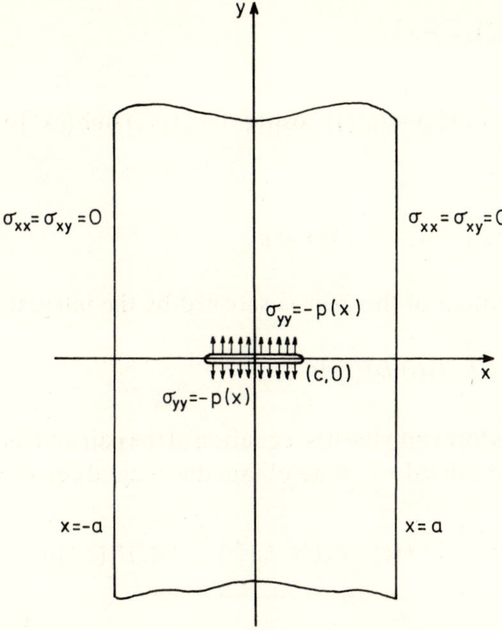


Figure 6.3

$$u_x(x,y) = -\tfrac{1}{2}\mathscr{F}_c[\xi^{-1}\{f(\xi)-(1-2\nu)g(\xi)\}\sinh(\xi x)+xg(\xi)\cosh(\xi x);\ \xi\to y]$$
$$-\tfrac{1}{2}\mathscr{F}_s[\zeta^{-1}\phi(\zeta)(1-2\nu-\zeta y)\mathrm{e}^{-\zeta y};\ \zeta\to x] \tag{6.33}$$

$$u_y(x,y) = -\tfrac{1}{2}\mathscr{F}_s[\xi^{-1}\{f(\xi)+2(1-\nu)g(\xi)\}\cosh(\xi x)+xg(\xi)\sinh(\xi x);\ \xi\to y]$$
$$+\tfrac{1}{2}\mathscr{F}_c[\zeta^{-1}\phi(\zeta)(2-2\nu+\zeta y)\mathrm{e}^{-\zeta y};\ \zeta\to x]\,, \tag{6.34}$$

in which the functions $f$, $g$ and $\phi$ are unknown. This form of solution ensures that the condition (6.31) is satisfied. The conditions (6.32) are satisfied if the functions $f$, $g$ and $\phi$ satisfy the equations

$$f(\xi)\cosh(\xi a)+a\xi g(\xi)\sinh(\xi a) = -\frac{\xi^2}{\pi^2}\int_0^\infty \frac{\zeta\phi(\zeta)\cos(\zeta a)}{(\xi^2+\zeta^2)^2}\,\mathrm{d}\zeta\,, \tag{6.35}$$

$$f(\xi)\sinh(\xi a)+g(\xi)\{\sinh(\xi a)+\xi a\cosh(\xi a)\} = \frac{4\xi}{\pi}\int_0^\infty \frac{\zeta^2\phi(\zeta)\sin(\zeta a)}{(\xi^2+\zeta^2)^2}\,\mathrm{d}\zeta\,. \tag{6.36}$$

The equations (6.29), (6.30) are equivalent to the pair of dual integral equations

$$\frac{d}{dx}\mathscr{F}_s[\zeta^{-1}\phi(\zeta);\zeta\to x] -$$

$$-\left(\frac{2}{\pi}\right)^{\frac{1}{2}}\int_0^\infty [\{f(\xi)+2g(\xi)\}\cosh(\xi x)+\xi x g(\xi)\sinh(\xi x)]\,d\xi = -p(x)/2\mu\,,$$

$$0\leqslant x\leqslant c\,,$$

$$\mathscr{F}_c[\zeta^{-1}\phi(\zeta);x]=0\,,\qquad c<x<a\,.$$

The second equation of this pair is satisfied by the integral representation

$$\phi(\zeta) = \left(\frac{\pi}{2}\right)^{\frac{1}{2}}\zeta\int_0^c G(t)\,J_0(\zeta t)\,dt\,. \tag{6.37}$$

Substituting this form into the first equation of the pair and then applying the operator $I^{-1}$ to both sides of it we obtain the integral equation

$$G(t) - \left(\frac{2}{\pi}\right)^{\frac{1}{2}} t\int_0^\infty [\{f(\xi)+2g(\xi)\}I_0(\xi t)+\xi t g(\xi)I_1(\xi t)]\,d\xi = t\hat{p}(t)\,,$$

$$0\leqslant t\leqslant c\,, \tag{6.38}$$

where the function $\hat{p}(t)$ is defined by the equation

$$\hat{p}(t) = \frac{1}{\pi\mu}\int_0^t \frac{p(x)\,dx}{(t^2-x^2)^{\frac{1}{2}}}\,,\qquad 0\leqslant t\leqslant c\,. \tag{6.39}$$

Eliminating the functions *f*, *g* and $\phi$ from the form equations (6.35) through (6.38) we get the integral equation

$$G(t) - \int_0^c \{H_1(u,t)+H_2(u,t)\}\,G(u)\,du = t\hat{p}(t)\,,\qquad 0\leqslant t\leqslant c\,, \tag{6.40}$$

where

$$H_1(u,t) = t\int_0^\infty \frac{x e^{-ax}}{\sinh(ax)}\,I_0(xu)\,I_0(xt)\,dx\,,\qquad 0\leqslant u,\ t\leqslant c\,,$$

$$H_2(u,t) = 2t\int_0^\infty \frac{x h(u,x)\,h(t,x)\,dx}{2ax+\sinh(2ax)}\,,\qquad 0\leqslant u,\ t\leqslant c\,,$$

with

$$h(u,x) = \{ax\coth(ax)-1\}\,I_0(ux)-ux\,I_1(ux)\,.$$

Simple calculations show that the energy of the crack is given in terms of the solution of equation (6.38) by the formula

$$W = 2\pi(1-\nu)\mu \int_0^c \hat{p}(t)\, G(t)\, \mathrm{d}t$$

and that

$$k = \lim_{x \to c+} [2(x-c)]^{\frac{1}{2}} \sigma_{yy}(x, 0) = \frac{2}{c^{\frac{1}{2}}} \mu G(c) .$$

The numerical work was carried out for the case $p(x) = p_0$, a constant. In this case $\hat{p}(t) = p_0/(2\mu)$. If $G(t)$ is replaced by $p_0 \chi(t)/(2\mu)$ it follows from the above equations that the crack energy $W$ and the stress intensity factor $k$ are given respectively by the equations

$$W = \frac{\pi(1-\nu^2)p_0^2}{E} \int_0^c \chi(t)\, \mathrm{d}t , \tag{6.41}$$

$$k = \frac{p_0}{c^{\frac{1}{2}}} \chi(c) , \tag{6.42}$$

where $\chi(t)$ is the solution of the Fredholm integral equation

$$\chi(t) - t \int_0^c \chi(u)[H_1(u, t) + H_2(u, t)]\, \mathrm{d}u = t , \qquad 0 \leqslant t \leqslant c . \tag{6.43}$$

It will be recalled from the definitions of $H_1$ and $H_2$ given above that the kernel of the integral equation (6.43) depends on the value of $a$.

To afford a direct comparison with the numerical values which are easily obtained in the case in which the edges of the strip are constrained, Sneddon and Srivastav derived the numerical solution of equation (6.43) for $a = \pi$ and a range of values of the crack length $c$. It turns out that even for the higher values of the ratio $c/a$, the function $\chi(t)$ differs only slightly from $t$. From the values of the function $\chi(t)$, the stress intensity factor $k$ and the crack energy $W$ were computed by means of (6.41) and (6.42). The results are shown in Tables I and II.

*Two coplanar Griffith cracks in an infinite elastic medium.* Lowengrub and Srivastava [8] have used the method of integral transforms to determine the stress field in the neighbourhood of two Griffith cracks defined by the relations $-b \leqslant x \leqslant -a$, $a \leqslant x \leqslant b$, $y = 0$ which are opened by constant pressure along the faces of the cracks.

TABLE I

*The variation with c/a of the stress intensity factor K*

| $c/a$ | $K/p_0$ | |
|---|---|---|
| | Constrained edge | Stress-free edge |
| 0.1 | 0.398 | 0.399 |
| 0.2 | 0.570 | 0.575 |
| 0.3 | 0.714 | 0.727 |
| 0.4 | 0.852 | 0.883 |
| 0.5 | 1.000 | 1.058 |
| 0.6 | 1.173 | 1.271 |
| 0.7 | 1.401 | 1.573 |
| 0.8 | 1.754 | 2.047 |
| 0.9 | 2.511 | 3.068 |

TABLE II

*The variation with c/a of the crack energy W*

| $c/a$ | $W/W_0$ | |
|---|---|---|
| | Constrained edge | Stress-free edge |
| 0.1 | 0.025 | 0.024 |
| 0.2 | 0.100 | 0.097 |
| 0.3 | 0.231 | 0.224 |
| 0.4 | 0.424 | 0.406 |
| 0.5 | 0.693 | 0.701 |
| 0.6 | 1.063 | 1.106 |
| 0.7 | 1.599 | 1.710 |
| 0.8 | 2.349 | 2.673 |
| 0.9 | 3.710 | 4.523 |

$W_0 = [\pi(1-\nu^2)p_0^2 a^2]/2E$

The problem is equivalent to that of determining the displacement field in the half-plane $y \geqslant 0$ when the boundary $y=0$ is subjected to the mixed boundary conditions

$$\sigma_{yy}(x, 0) = -p(x), \qquad a \leqslant |x| \leqslant b,$$
$$u_y(x, 0) = 0, \qquad |x| < a, \ |x| > b,$$
$$\sigma_{xy}(x, 0) = 0, \qquad -\infty < x < \infty,$$

where the function $p(x)$ is prescribed. By assuming a displacement field of the form given by equations (6.1) and (6.2) the above boundary conditions lead to the set of triple integral equations

$$\mathscr{F}_c[\psi(\xi); x] = 0, \qquad 0 < x < a,$$
$$\frac{d}{dx}\mathscr{F}_s[\psi(\xi); x] = \tfrac{1}{2}\pi p(x), \qquad a \leqslant x \leqslant b,$$
$$\mathscr{F}_c[\psi(\xi); x] = 0, \qquad x > b,$$

it being assumed that the prescribed pressure $p(x)$ is an even function of $x$. The solution of this set of equations is shown to be of the form

$$\psi(\xi) = \xi^{-1} \int_a^b h(t^2) \sin(\xi t) dt, \tag{6.44}$$

where

$$h(t^2) = -\frac{2}{\pi}\left(\frac{t^2+a^2}{b^2-t^2}\right)^{\frac{1}{2}} \int_a^b \left(\frac{b^2-y^2}{y^2-a^2}\right)^{\frac{1}{2}} \frac{yp(y)dy}{y^2-t^2} + C_1\{(t^2-a^2)(b^2-t^2)\}^{-\frac{1}{2}}, \tag{6.45}$$

the constant $C_1$ being defined in terms of the complete elliptic integral of the first kind $F = K\{(b^2-a^2)^{\frac{1}{2}}/b\}$ by the formula

$$C_1 = \frac{b}{\pi F}\int_a^b \left(\frac{b^2-y^2}{y^2-a^2}\right)^{\frac{1}{2}} \frac{yp(y)dy}{y^2-t^2} \int_a^b \left(\frac{t^2-a^2}{b^2-t^2}\right)^{\frac{1}{2}} \frac{dt}{y^2-t^2} \tag{6.46}$$

all integrals being interpreted as Cauchy principal values. Lowengrub and Srivastava show also that this solution can be put into the alternative form

$$h(t^2) = -\frac{2}{\pi}\left(\frac{b^2-t^2}{t^2-a^2}\right)^{\frac{1}{2}} \int_a^b \left(\frac{y^2-a^2}{b^2-y^2}\right)^{\frac{1}{2}} \frac{yp(y)dy}{y^2-t^2} + C_2\{(t^2-a^2)(b^2-t^2)\}^{-\frac{1}{2}}, \tag{6.47}$$

where the constant $C_2$ is defined by the formula

$$C_2 = \frac{b}{\pi F}\int_a^b \left(\frac{b^2-t^2}{t^2-a^2}\right)^{\frac{1}{2}} dt \int_a^b \left(\frac{y^2-a^2}{b^2-y^2}\right)^{\frac{1}{2}} \frac{yp(y)dy}{y^2-t^2}. \tag{6.48}$$

In terms of the constants $C_1, C_2$ and the constant $N$ defined by the equation

$$N=\frac{\pi}{2}(b^2-a^2)\int_a^b \frac{yp(y)\mathrm{d}y}{[(b^2-y^2)(y^2-a^2)]^{\frac{1}{2}}}, \tag{6.49}$$

the stress intensity factors $k_a$, $k_b$ defined by the equations

$$k_a=\lim_{x\to a-}\{2(a-x)\}^{\frac{1}{2}}\cdot\sigma_{yy}(x,0), \qquad k_b=\lim_{x\to b+}\{2(x-b)\}^{\frac{1}{2}}\cdot\sigma_{yy}(x,0)$$

are given by the formulas

$$k_a=\{a(b^2-a^2)\}^{-\frac{1}{2}}\cdot(N-C_2), \qquad k_b=\{b(b^2-a^2)\}^{-\frac{1}{2}}\cdot(N+C_1). \tag{6.50}$$

Further the crack energy is given by the formula

$$W=\frac{4(1-\nu^2)}{E}\int_a^b P(t)h(t^2)\mathrm{d}t,$$

in which the function $P(t)$ is defined by the equation

$$P(t)=\int_a^t p(x)\mathrm{d}x.$$

In the special case in which $p(x)=p_0$, a constant, it is found that

$$C_1=p_0\{a^2-(E/F)b^2\},$$
$$C_2=p_0b^2(1-E/F), \qquad h(t^2)=p_0\frac{t^2-(E/F)b^2}{[(t^2-a^2)(b^2-t^2)]^{\frac{1}{2}}},$$

where $E$ is the complete elliptic integral of the second kind, $E=E\{(b^2-a^2)^{\frac{1}{2}}/b\}$. These expressions lead to

$$k_a=\frac{p_0}{\{a(b^2-a^2)\}^{\frac{1}{2}}}\{(E/F)b^2-a^2\}, \quad k_b=\frac{p_0}{\{b(b^2-a^2)\}^{\frac{1}{2}}}(1-E/F), \tag{6.51}$$

for the stress intensity factors and the formula

$$W=\frac{\pi(1-\nu^2)p_0^2}{E}\{a^2+b^2-2(E/F)b^2\}. \tag{6.52}$$

*Two coplanar Griffith cracks in an elastic strip.* Lowengrub and Srivastava also considered the problem of determining the stress field in the vicinity of two coplanar Griffith cracks in an infinitely long elastic strip [9]. The problem they consider concerns a strip of material $-\infty<x<+\infty$, $-\delta\leqslant y\leqslant\delta$ with

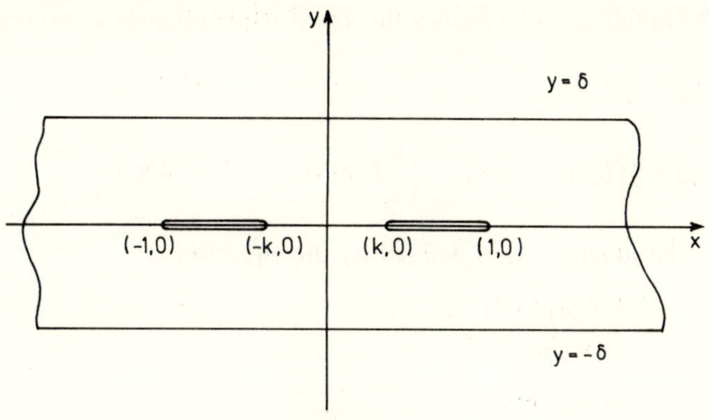


Figure 6.4

the cracks located in the interior of the material on the line $y=0$, $k \leqslant |x| \leqslant 1$, where the constant $k$ satisfies $0<k<1$ (cf. Figure 6.4). It is assumed that the cracks are opened up under the action of a completely symmetrical pressure $p(x)$ and that the edges of the strip are free from stress. The boundary conditions then become

$$\sigma_{yy}(x, \pm\delta) = \sigma_{xy}(x, \pm\delta) = 0\,, \tag{6.53}$$

$$\sigma_{yy}(x, 0) = -p(x) \qquad k \leqslant |x| \leqslant 1\,, \tag{6.54}$$

$$u_y(x, 0) = 0 \qquad |x| < k,\ |x| > 1\,, \tag{6.55}$$

$$\sigma_{xy}(x, 0) = 0 \qquad -\infty < x < \infty\,. \tag{6.56}$$

In the stresses and displacements in the half-strip $0 \leqslant y \leqslant \delta$ may be derived from the displacement field defined by the equations

$$(2\pi)^{\frac{1}{2}}\mu u_x(x, y) =$$

$$\mathscr{F}_s\left[\psi(\xi)\left\{\frac{\xi^2\delta^2\cosh(\xi y) + (1-2\nu)\sinh^2(\xi\delta)\cosh(\xi y) + \xi y\sinh^2(\xi\delta)\sinh(\xi y)}{2\xi\delta + \sinh 2\xi\delta} + \tfrac{1}{2}(1-2\nu)\sinh(\xi y) + \tfrac{1}{2}\xi y\cosh(\xi y)\right\};\quad \xi \to x\right]$$

$$(2\pi)^{\frac{1}{2}}\mu u_y(x, y) =$$

$$= \mathscr{F}_c\left[\psi(\xi)\left\{\frac{\xi^2\delta^2\sinh(\xi y) - 2(1-\nu)\sinh^2(\xi\delta)\sinh(\xi y) + \xi y\sinh^2(\xi\delta)\cosh(\xi y)}{2\xi\delta + \sinh 2\xi\delta} + (1-\nu)\cosh(\xi y) - \tfrac{1}{2}\xi y\sinh(\xi y)\right\};\quad \xi \to x\right]$$

where the function $\psi(\xi)$ satisfies the set of triple integral equations

$$\mathscr{F}_c[\psi(\xi); x] = 0, \qquad 0 \leqslant |x| < k, \ |x| > 1 ;$$
$$\frac{d}{dx}\mathscr{F}_s[1+M(\xi\delta)\psi(\xi); x] = \left(\frac{\pi}{2}\right)^{\frac{1}{2}} p(x), \qquad k \leqslant |x| \leqslant 1 \tag{6.57}$$

in which the function $M$ is defined by the equation

$$M(u) = \frac{e^{-2u} - 2u(u+1) - 1}{2u + \sinh 2u}.$$

Assuming the representation

$$\psi(\xi) = \xi^{-1}\int_k^1 h(t^2)\sin(\xi t)dt, \tag{6.58}$$

and using the method described in (*d*) above it is easily shown that $h(t^2)$ is the solution of the Fredholm integral equation of the second kind,

$$h(x^2) + \int_k^1 h(t^2)K(x^2, t)dt = F(x^2), \qquad k \leqslant x \leqslant 1, \tag{6.59}$$

satisfying the condition

$$\int_k^1 h(t^2)dt = 0, \tag{6.60}$$

where

$$K(x^2, t) = \frac{-4}{\pi^2}\left(\frac{x^2-k^2}{1-x^2}\right)^{\frac{1}{2}}\int_k^1\left(\frac{1-y^2}{y^2-k^2}\right)^{\frac{1}{2}}\frac{ym(y, t)}{y^2-x^2}dx \tag{6.61}$$

with

$$m(y, t) = \int_0^\infty M(\delta u)\cos(uy)\sin(ut)du \tag{6.62}$$

and

$$F(x^2) = -\frac{2}{\pi}\left(\frac{x^2-k^2}{1-x^2}\right)^{\frac{1}{2}}\int_k^1\left(\frac{1-y^2}{y^2-k^2}\right)^{\frac{1}{2}}\frac{yp(y)dy}{y^2-x^2} + C\{(x^2-k^2)(1-x^2)\}^{-\frac{1}{2}}, \tag{6.63}$$

$C$ being an arbitrary constant determined by the condition (6.60). Now consider the special case in which $p(x) = p_0$, a constant. The stress along the line of the cracks is given by the formula

$$\sigma_{yy}(x,0) = -\frac{2}{\pi}\int_k^1 \frac{th(t^2)\mathrm{d}t}{t^2-x^2} - \frac{2}{\pi}\int_k^1 h(t^2)m(x,t)\mathrm{d}t$$

from which it is a simple calculation to deduce that the stress intensity factors defined by the equations

$$K_k = \lim_{x\to k-} [2(k-x)^{\frac{1}{2}}]\cdot\sigma_{yy}(x,0)\,, \quad K_1 = \lim_{x\to 1+} [2(x-1)^{\frac{1}{2}}]\cdot\sigma_{yy}(x,0)$$

are determined by means of the formulas

$$K_k = \frac{p_0}{k'k^{\frac{1}{2}}}\Bigg[(E/F-k^2)\left\{1-\frac{I_0C_0}{2\delta^2}+\frac{I_0^2C_0^2}{4\delta^4}\right\}+ \\ + \frac{2I_1}{\delta^4}(3k^4+C_1k^2+C_2)+O(\delta^{-6})\Bigg] \tag{6.64}$$

$$K_1 = \frac{p_0}{k'}\left[(1-E/F)\left\{1-\frac{I_0C_0}{2\delta^2}+\frac{I_0^2C_0^2}{4\delta^4}\right\}+\frac{2I_1}{\delta^4}(3+C_1+C_2)+O(\delta^{-6})\right] \tag{6.65}$$

where $F=K(k')$, $E=E(k')$ are respectively the complete elliptic integrals of the first and second kinds with modulus $k'=(1-k^2)^{\frac{1}{2}}$ and the constants $C_0$, $C_1$ and $C_2$ are given by

$$C_0 = 1+k^2-2(E/F)\,, \\ C_1 = k'^4/(4C_0)-1-k^2\,, \quad C_2 = k^2+(E/F)C_1\,, \tag{6.66}$$

and $I_0$ and $I_1$ denote the integrals given by the equations

$$I_0 = \int_0^\infty uM(u)\mathrm{d}u\,, \quad I_1 = -\frac{1}{3!}\int_0^\infty u^3M(u)\mathrm{d}u\,. \tag{6.67}$$

The numerical values of $I_0$ and $I_1$ can be found from Ling's tables [10]; they are $I_0=-2.34974$, $I_1=1.66033$.

Similarly, the expression

$$u_y(x,0) = \frac{2(1-\nu^2)}{E}\int_x^1 h(t^2)\mathrm{d}t\,, \qquad k\leqslant x\leqslant 1$$

for the crack displacement is used to yield

$$W = \frac{2(1-\nu^2)p_0}{E}\int_k^1 th(t^2)\mathrm{d}t$$

for the crack energy. To the same order of approximation this gives

$$W = \frac{\pi p_0^2(1-\nu^2)}{E} C_0 \left[1 + \frac{\mu_1}{\delta^2} + \frac{\mu_2}{\delta^4} + O(\delta^{-6})\right] \tag{6.68}$$

where the constants $\mu_1$ and $\mu_2$ are given by the equations

$$\mu_1 = -\tfrac{1}{2}C_0 I_0, \quad \mu_2 = \tfrac{1}{4}C_0^2 I_0^2 - \tfrac{1}{4}(1+k^2)(2+k'^2/C_0^2)I_1 . \tag{6.69}$$

The case $k=0$ corresponds to the case in which the two cracks have merged to form a crack of length 2. Since in this case $E=1$, $E/F=0$ and $C_0=1$ it follows that $\mu_1=-\frac{1}{2}I_0$, $\mu_2=\frac{1}{4}I_0^2-\frac{3}{4}I_1$ in agreement with equation (6.23).

*Two parallel infinite rows of parallel cracks.* If equations (6.54) through (6.56) are retained but (6.53) is replaced by

$$\sigma_{xy}(x, \pm\delta)=0, \quad u_y(x, \pm\delta)=0, \qquad -\infty < x < \infty ,$$

the boundary value problem corresponding to the array of cracks shown in Figure 6.5 is obtained. In this case the displacement field is of the form

$$\mu u_x(x, y) = (2\pi)^{-\frac{1}{2}} \mathscr{F}_s \left[\left\{\frac{\xi\delta \cosh \xi y}{\sinh^2 \xi\delta} - \frac{(1-2\nu)\cosh \xi(\delta-y)}{\sinh \xi\delta} + \right.\right.$$
$$\left.\left. + \frac{\xi y \sinh \xi(\delta-y)}{\sinh \xi\delta}\right\}\psi(\xi);\ \xi\to x\right]$$

$$\mu u_y(x, y) = (2\pi)^{-\frac{1}{2}} \mathscr{F}_c \left[\left\{2(1-\nu)\frac{\sinh \xi(\delta-y)}{\sinh \xi\delta} - \frac{\xi\delta \sinh \xi y}{\sinh^2 \xi\delta} + \right.\right.$$
$$\left.\left. + \frac{\xi y \cosh \xi(\delta-y)}{\sinh \xi\delta}\right\}\psi(\xi);\ \xi\to x\right]$$

where $\psi(\xi)$ again satisfies the equations (6.57) but now the kernel $M(u)$ is defined by the equation

$$M(u) = \coth u - 1 + u \operatorname{cosech}^2 u .$$

The analysis proceeds in exactly the same way as before except that the numbers $I_0$ and $I_1$ defined by equations (6.67) with this new form for $M$ are found to have the values $I_0=\pi^2/4$, $I_1=\pi^2/144$. For instance, the crack energy is given by equation (6.68) with $\mu_1$ and $\mu_2$ defined by the equations

$$\mu_1 = -\tfrac{1}{4}\pi^2 C_0, \quad \mu_2 = \tfrac{1}{64}\pi^4 [C_0^2 + \tfrac{1}{9}(1+k^2)(2+k'^2/C_0^2)] .$$

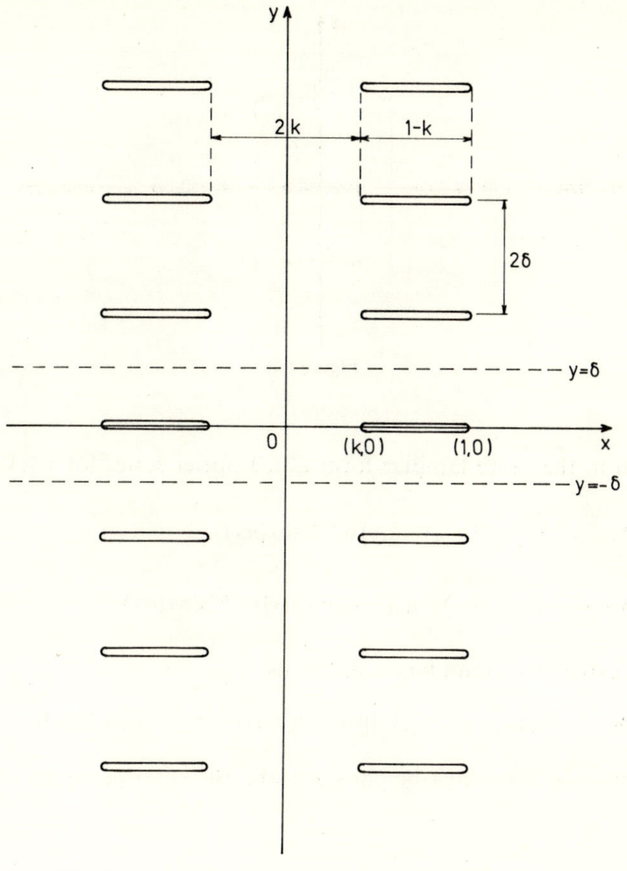


Figure 6.5

It is easily verified that these values reduce to those given by equation (6.28) in the case $k=0$.

*An infinite row of cracks.* The problem of determining the stress field in the vicinity of a row of identical cracks spaced periodically (cf. Figure 6.6) has been discussed by Sneddon and Srivastav [11] using the method of finite Fourier transforms to reduce the relevant mixed boundary value problem to the solution of a pair of dual *series* relations. The displacement field can

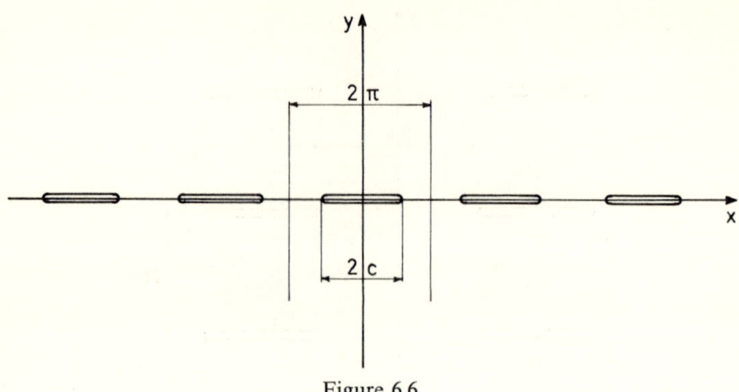


Figure 6.6

be written in the more familiar form of a Fourier series for $y \geqslant 0$:

$$u_x(x, y) = -\sum_{n=1}^{\infty} a_n(1-2\nu-ny)\mathrm{e}^{-ny}\sin(nx)$$
$$u_y(x, y) = (1-\nu)a_0 + \sum_{n=1}^{\infty} a_n(2-2\nu+ny)\mathrm{e}^{-ny}\cos(nx) \tag{6.70}$$

where to satisfy the boundary conditions

$$\sigma_{xy}(x,0)=0,\ |x|\leqslant\pi, \qquad \sigma_{yy}(x,0)=-p_0,\ |x|\leqslant c, \qquad u_y(x,0)=0,\ c<|x|<\pi,$$

the sequence $\{a_n\}_{n=0}^{\infty}$ must be chosen to be the solution of the dual series relations

$$\sum_{n=1}^{\infty} na_n\cos(nx) = \tfrac{1}{2}p_0, \qquad 0\leqslant x\leqslant c, \tag{6.71}$$

$$\tfrac{1}{2}a_0 + \sum_{n=1}^{\infty} a_n\cos(nx) = 0, \quad c<x<\pi. \tag{6.72}$$

To solve these equations a new function $g(t)$ is introduced by the assumption that the series on the left-hand side of equation (6.72) can be expressed in the range $0\leqslant x\leqslant c$ by the integral

$$\tfrac{1}{2}p_0\cos(\tfrac{1}{2}x)\int_x^c \frac{g(t)\mathrm{d}t}{(\cos x-\cos t)^{\frac{1}{2}}}.$$

Using the well-known formula for the calculation of the coefficients of a Fourier series we find that

$$a_0 = 2^{\frac{1}{2}} \cdot \int_0^c g(t)\mathrm{d}t\,,$$

$$a_n = 2^{-\frac{1}{2}} \cdot \int_0^c \{P_n(\cos t) + P_{n-1}(\cos t)\}\, g(t)\mathrm{d}t\,, \tag{6.73}$$

where $P_n$ denotes the Legendre polynomial of degree $n$. Substituting these expressions for the coefficients $a_n$ into the series on the left-hand side of equation (6.71), interchanging the order of summation and integration and summing the infinite series, we obtain an integral equation of Schlömilch type whose solution is easily found to be

$$g(t) = 2^{\frac{1}{2}} \cdot \tan(\tfrac{1}{2}t)\,.$$

It is a simple matter to show that for a spacing of $2\pi$ the stress intensity factor is given by the formula

$$k = \lim_{x \to c+} [2(x-c)]^{\frac{1}{2}} \sigma_{yy}(x, 0) = 2^{\frac{1}{2}} \cdot p_0 \tan^{\frac{1}{2}}(\tfrac{1}{2}c)\,.$$

Written in terms of conventional units with $2a$ as the spacing (instead of $2\pi$) this equation takes the form

$$k = p_0 \left[\frac{2a}{\pi} \tan \frac{\pi c}{2a}\right]^{\frac{1}{2}}. \tag{6.74}$$

The problem of determining the stress field in the vicinity of the irregularly spaced array of cracks shown in Figure 6.7 has been considered by Parihar

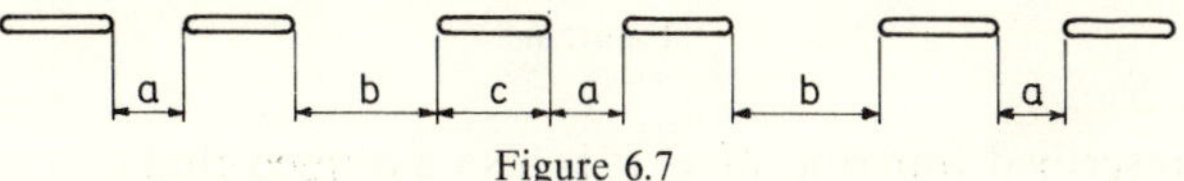


Figure 6.7

[12]. Starting from the displacement field (6.70) Parihar derives a set of triple series relations for the determination of the elements of the sequence $\{a_n\}_{n=1}^{\infty}$. His final result is complicated and the reader is referred to the original paper for details.

*Edge crack in an elastic half-plane.* An integral transform solution of the plane strain problem of determining the stress and displacement fields in an elastic half-plane containing an edge crack normal to the free surface when the faces of the crack are subjected to normal pressure has been given recently by Sneddon and Das [13].

Suppose that the elastic solid occupies the half-plane $x \geqslant 0$ and that we take the length of the crack as our unit of length so that it occupies the segment $0 \leqslant x \leqslant 1$, $y=0$. Because of the assumed symmetry with respect to the $x$-axis only the displacement field in the positive quadrant has to be considered. The boundary conditions are

$$\sigma_{yy}(x, 0) = -p_0 f(x), \qquad 0 \leqslant x \leqslant 1, \tag{6.75}$$

$$u_y(x, 0) = 0, \qquad x > 1, \tag{6.76}$$

$$\sigma_{xy}(x, 0) = 0, \qquad x \geqslant 0, \tag{6.77}$$

(cf. Figure 6.8) where $p_0$ is a constant with the dimensions of pressure and

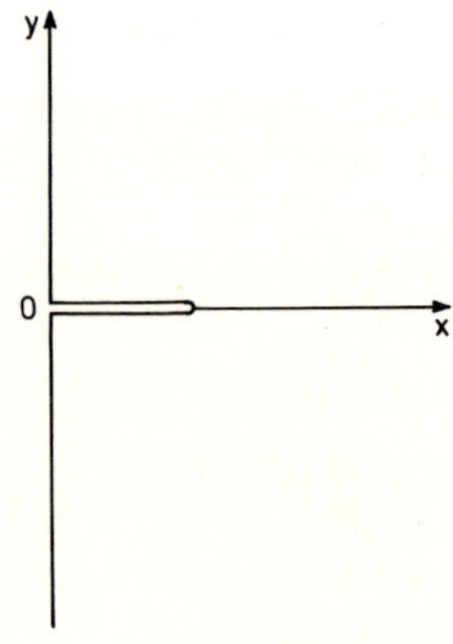


Figure 6.8

$f(x)$ is a prescribed function of $x$. It is also assumed that

$$\sigma_{xx}(0, y) = 0, \quad \sigma_{xy}(0, y) = 0, \qquad y \geqslant 0. \tag{6.78}$$

The displacement field may be represented by the equations

$$2\mu u_x(x, y) = -\mathscr{F}_s[\xi^{-1} A(\xi)(1-2\nu-\xi y)e^{-\xi y}; \xi \to x] + \mathscr{F}_c[\zeta^{-1} B(\zeta)(2-2\nu+\zeta x)e^{-\zeta x}; \zeta \to y], \tag{6.79}$$

$$2\mu u_y(x, y) = \mathscr{F}_c[\xi^{-1} A(\xi)(2-2\nu+\xi y)e^{-\xi y}; \xi \to x] - \mathscr{F}_s[\zeta^{-1} B(\zeta)(1-2\nu-\zeta x)e^{-\zeta x}; \zeta \to y]. \tag{6.80}$$

Whatever the forms of the functions $A$ and $B$, these displacement components are such that the conditions (6.77) and the second of (6.78) are satisfied. The boundary conditions (6.75) and (6.76) lead to the pair of dual integral equations

$$\mathscr{F}_s[\xi^{-1}A(\xi); x] + \left(\frac{2}{\pi}\right)^{\frac{1}{2}} x \int_0^\infty B(\zeta)e^{-\zeta x}d\zeta = p_0 F(x), \quad 0 \leqslant x \leqslant 1, \tag{6.81}$$

$$\mathscr{F}_c[\zeta^{-1}A(\zeta); x] = 0, \quad x > 1, \tag{6.82}$$

in which the function $F$ is defined in terms of the prescribed function $f$ by the equation

$$F(x) = \int_0^x f(t)dt.$$

Similarly, the first of the equations (6.78) is equivalent to the relation

$$\mathscr{F}_s[\zeta^{-1}B(\zeta); y] + \left(\frac{2}{\pi}\right)^{\frac{1}{2}} y \int_0^\infty A(\xi)e^{-\xi y}d\xi = 0, \quad y \geqslant 0. \tag{6.83}$$

Equation (6.82) is automatically satisfied by the integral representation

$$A(\xi) = \left(\frac{\pi}{2}\right)^{\frac{1}{2}} p_0 \xi \int_0^1 ta(t)J_0(\xi t)dt \tag{6.84}$$

of the unknown function $A(\xi)$; for the function $B(\zeta)$ a similar representation is

$$B(\zeta) = \left(\frac{\pi}{2}\right)^{\frac{1}{2}} p_0 \zeta \int_0^\infty tb(t)J_0(\zeta t)dt.$$

Substituting these expressions into equations (6.81) and (6.83) and operating on each with $I^{-1}$ we find that these equations are equivalent to

$$a(t) + \int_0^\infty K(t, u)b(u)du = \phi(t), \quad 0 \leqslant t \leqslant 1, \tag{6.85}$$

$$b(t) + \int_0^1 K(t, u)a(u)du = 0, \quad t \geqslant 0, \tag{6.86}$$

where the kernel $K$ is defined by the equation

$$K(t, u) = 4\pi^{-1}tu^2(u^2 + t^2)^{-2}, \tag{6.87}$$

and the function $\phi(t)$ by the equation

$$\phi(t) = \frac{2}{\pi}\int_0^t \frac{f(s)\,ds}{(t^2-s^2)^{\frac{1}{2}}}\,. \tag{6.88}$$

The stress intensity factor is given by the formula

$$k = \lim_{x\to 1+} [2(x-1)]^{\frac{1}{2}}\,\sigma_{yy}(x, 0) = p_0\,a(1) \tag{6.89}$$

and the crack energy by the formula

$$W = \frac{\pi(1-\nu^2)p_0^2}{E}\int_0^1 t a(t)\,\phi(t)\,dt\,. \tag{6.90}$$

Since these last two formulae involve only the function $a(t)$, $b(u)$ may be eliminated from the equations (6.85) and (6.86) to obtain the single integral equation

$$a(t) - \int_0^1 L(t, \tau)\,a(\tau)\,d\tau = \phi(t)\,, \qquad 0\leqslant t\leqslant 1\,, \tag{6.91}$$

whose kernel is given by the equation

$$L(t, \tau) = \frac{16t\tau^2}{\pi^2}\left[\frac{t^2+\tau^2}{(t^2-\tau^2)^3}\log\left(\frac{t}{\tau}\right) - \frac{1}{(t^2-\tau^2)^2}\right]. \tag{6.92}$$

This kernel appears to have a singularity at $t=\tau$ but in fact it is easily shown, by means of l'Hospital's rule, that $K(t, \tau)\to 4/(3\pi^2\tau)$ as $t\to\tau$.

It turns out that (as a result of numerical integration of the integral

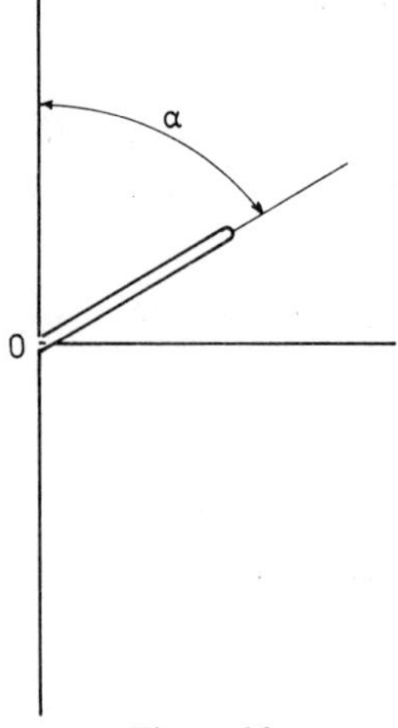


Figure 6.9

equation (6.91)) the stress intensity factor for a crack of length $c$ is

$$k = 0.793 p_0 (2c)^{\frac{1}{2}}$$

and the crack energy is

$$W = 0.988 W_0 ,$$

where $W_0 = 2(1-\nu^2) p_0^2 c^2 / E$.

The related problem of a line crack which makes an angle with the boundary of a half-plane (cf. Figure 6.9) has been considered by Krapkhov [14]. The method used is to express the displacement field in terms of inverse Mellin transforms and to solve the resulting dual integral equations by means of the Wiener–Hopf technique. The details are too complicated to be reproduced here; the reader is referred to the original paper for them.

## 6.4 External cracks

The problem of the determination of the displacement and stress fields in a plane weakened by two colinear external cracks, of the form shown in Figure 6.10, has been considered by Lowengrub [15]. To simplify the analysis the

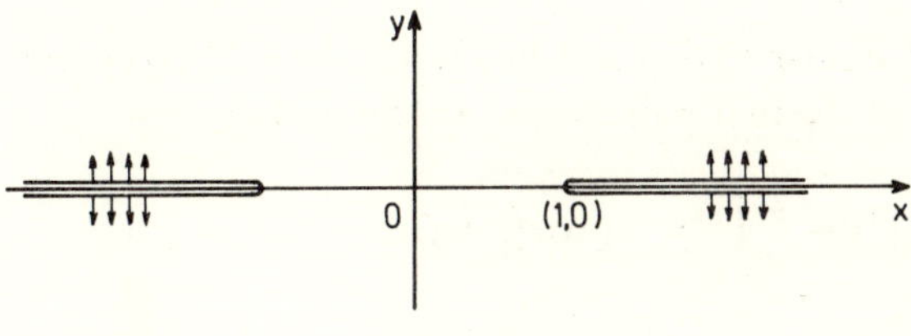


Figure 6.10

unit of length is taken to be one-half of the distance between the tips of the cracks. The boundary conditions for the corresponding mixed problem for the half-plane $y > 0$ are:

$$\sigma_{yy}(x, 0) = -p(x) , \quad |x| \geq 1 ,$$

$$u_y(x, 0) = 0 , \qquad |x| < 1 ,$$

$$\sigma_{xy}(x, 0) = 0 , \qquad -\infty < x < \infty ,$$

where the function $p(x)$ is prescribed for $|x| \geq 1$.

Consider a displacement field of the form

$$u_x(x,y)=\tfrac{1}{2}\int_0^\infty \xi^{-1}(1-2\nu-\xi y)e^{-\xi y}\{\psi_a(\xi)\cos(\xi x)-\psi_s(\xi)\sin(\xi x)\}\,d\xi\,, \tag{6.93}$$

$$u_y(x,y)=\tfrac{1}{2}\int_0^\infty \xi^{-1}(2-2\nu+\xi y)e^{-\xi y}\{\psi_s(\xi)\cos(\xi y)+\psi_a(\xi)\sin(\xi x)\}\,d\xi\,, \tag{6.94}$$

Now, if $p_s(x)=\frac{1}{2}\{p(x)+p(-x)\}$ and $p_a(x)=\frac{1}{2}\{p(x)-p(-x)\}$, then the unknown functions $\psi_s(\xi)$ and $\psi_a(\xi)$ respectively satisfy the pairs of dual integral equations:

$$\int_0^\infty \xi^{-1}\psi_s(\xi)\cos(\xi x)\,d\xi=0\,, \qquad 0\leqslant x<1\,,$$

$$\frac{d}{dx}\int_0^\infty \xi^{-1}\psi_s(\xi)\sin(\xi x)\,d\xi=p_s(x)\,, \qquad x\geqslant 1\,;$$

$$\int_0^\infty \xi^{-1}\psi_a(\xi)\sin(\xi x)\,d\xi=0\,, \qquad 0\leqslant x<1\,,$$

$$-\frac{d}{dx}\int_0^\infty \xi^{-1}\psi_a(\xi)\sin(\xi x)\,d\xi=p_a(x)\,, \qquad x\geqslant 1\,.$$

By a method similar to that adopted in section 6.2 Lowengrub showed that the solution of the first pair can be written in the form

$$\psi_s(\xi)=\psi_0(\xi)+\frac{2}{\pi}\int_1^\infty p_s(x)\cos(\xi x)\,dx\,, \tag{6.95a}$$

where

$$\psi_0(\xi)=-\frac{2}{\pi}[1-J_0(\xi)]\int_1^\infty p_s(t)\,dt$$
$$-\frac{2}{\pi}\int_0^1 J_0(\xi t)\,dt\int_1^\infty \frac{x p_s(x)\,dx}{(x^2-t^2)^{\frac{1}{2}}}\,, \tag{6.95b}$$

and that the solution of the second pair can be similarly written as

$$\psi_a(\xi)=\psi_1(\xi)+\frac{2}{\pi}\int_1^\infty p_a(t)\sin(\xi t)\,dt\,, \tag{6.96a}$$

$$\psi_1(\xi)=\frac{2}{\pi}\int_0^1 t^2 J_1(\xi t)\,dt\int_1^\infty \frac{p_a(x)\,dx}{(x^2-t^2)^{\frac{3}{2}}}\,. \tag{6.96b}$$

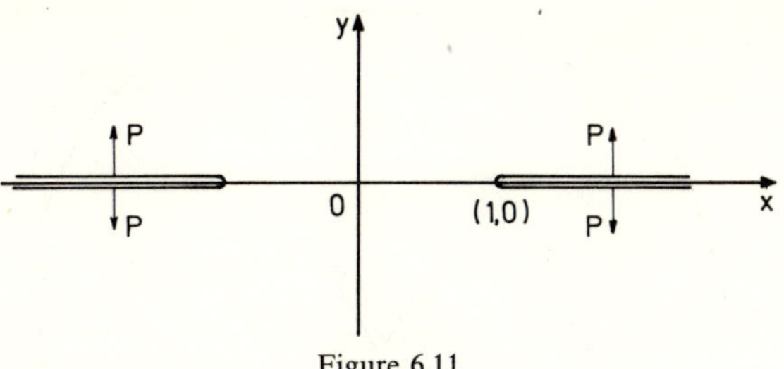


Figure 6.11

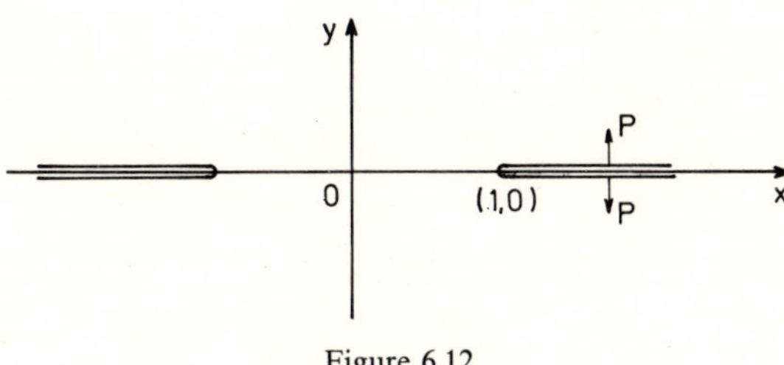


Figure 6.12

The case shown in Figure 6.11 corresponds to $p_s(x)=P\delta(c-x)$, $p_a(x)=0$ and it is easily deduced from the above solution that

$$k = \lim_{x\to 1-} [2(1-x)]^{\frac{1}{2}}\sigma_{yy}(x, 0) = \frac{2P}{\pi}\left[1 + \frac{c}{(c^2-1)^{\frac{1}{2}}}\right]. \tag{6.97}$$

Similarly the case shown in Figure 6.12 corresponds to $p_s(x)=0$, $p_a=P\delta(c-x)$ and subsequently the formula

$$k = \frac{P}{\pi}\left[1 + \frac{c}{(c^2-1)^{\frac{1}{2}}}\right]. \tag{6.98}$$

## 6.5 Star-shaped and cruciform shaped cracks

The problem of finding the stress field in the vicinity of a star-shaped crack of $2n$ arms (as shown in Figure 6.13) has been discussed by Srivastav and Narain [16]. It is assumed that the pressure is symmetrically distributed so that the problem, stated in terms of plane polar coordinates $(r, \theta)$, is equivalent to the mixed boundary value problem

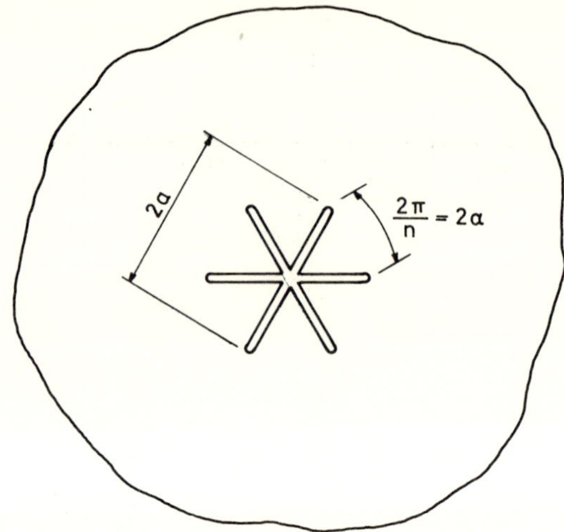


Figure 6.13

$$\sigma_{r\theta}(r, \pm\alpha) = 0\,, \quad u_\theta(r, \pm\alpha) = 0\,, \qquad r \geqslant 0\,,$$
$$\sigma_{\theta\theta}(r, 0) = -2\mu f(r)\,, \qquad 0 \leqslant r \leqslant 1$$
$$u_\theta(r, 0) = 0\,, \qquad r > 1\,,$$

for the infinite wedge $|\theta| \leqslant \alpha = \pi/n$. In terms of the inverse Mellin operator $\mathscr{M}^{-1}$ defined by the equation

$$\mathscr{M}^{-1}[\phi(s); x] = \frac{1}{2\pi i}\int_L \phi(s)x^{-s}\mathrm{d}s$$

(in the usual notation) and the differential operator $D = \partial/\partial\theta$, Srivastav and Narain represented the displacement field by the expressions

$$u_r(r, \theta) = +r^{-1}\mathscr{M}^{-1}[(s+1)^{-1}\{(1-\nu)D^2 - s(1+\nu s)\}\bar{\chi};\, s\to r]\,, \tag{6.99}$$

$$u_\theta(r, \theta) = -r^{-1}\mathscr{M}^{-1}[\{(s+1)(s+2)\}^{-1} \times \{(1-\nu)D^2 + (1+\nu)s^2 + (s+1)(s+2)\}D\bar{\chi};\, s\to r]\,, \tag{6.100}$$

where the function $\bar{\chi}(s, \theta)$ is given by the equation

$$\bar{\chi}(s, \theta) = = \psi(s)\left[\frac{(s+2)\cos\{(\theta-\alpha)s\}\sin\{(s+2)\alpha\} - s\sin(sx)\cos\{(s+2)(\theta-\alpha)\}}{s\cdot\sin(s\alpha)\sin\{(s+2)\alpha\}}\right]$$

The function $\psi(s)$ is determined by the dual integral equations

$$\mathscr{M}^{-1}[(s+1)\{(s+2)\cot(s\alpha)-s\cdot\cot(s+2)\alpha\}\,\psi(s);\,r]=r^2 f(r),\quad 0\leqslant r\leqslant 1\,,$$

$$\mathscr{M}^{-1}[\psi(s);\,r]=0 \qquad r>1\,.$$

If it is given the integral representation

$$\psi(s)=\frac{\Gamma(\frac{1}{2}s+\frac{1}{2})\,\Gamma(\frac{1}{2})}{\Gamma(\frac{1}{2}s+1)}\int_0^1 t^s g(t)\,\mathrm{d}t$$

it can be shown that the auxiliary function $g(t)$ is the solution of the Fredholm integral equation

$$g(t)+\int_0^1 g(u)\,K(u,t)\,\mathrm{d}u=h(t)\,, \tag{6.101}$$

whose kernel is defined by the equation

$$K(u,t)=\frac{t}{2}\mathscr{M}^{-1}[s\cdot\cot\{(s+2)\alpha\}-(s+2)\cot(s\alpha) + 2\cot(\tfrac{1}{2}s\pi)\}\tan(\tfrac{1}{2}s\pi);\,t/u] \tag{6.102}$$

and whose right-hand side is defined by the equation

$$h(t)=\frac{t}{\pi}\int_0^t\frac{f(x)\,\mathrm{d}x}{(t^2-x^2)^{\frac{1}{2}}}\,.$$

The integral equation (6.101) has been solved numerically by Srivastav and Narain; tabulated values of the function $g(t)$ for the cases in which the angle $\alpha$ takes the values 45°, 60°, 75° and 85° are given in [16]. From these values it is possible to calculate by a series of quadratures the stress and displacement fields near the tip of each arm of the crack.

A similar analysis has been given by Westmann [17]. He arrived also at the integral equation (6.201) but solved it by the Wiener–Hopf technique. Westmann derived the formula (in the case of constant pressure $p_0$)

$$k=\lim_{r\to 1+}[2(r-1)]^{\frac{1}{2}}\,\sigma_{\theta\theta}(r,0)=2^{\frac{1}{2}}\cdot p_0 K(n)$$

where $K(n)$ is a complicated function of $n$ which has the asymptotic behaviour

$$K(n)\sim(2/n)^{\frac{1}{2}}\,,\ \text{as}\ n\to\infty\,.$$

Figure 6.14 shows the variation of $K(n)$ with $n$ as determined by Westmann together with the variation of $(2/n)^{\frac{1}{2}}$ to show how good the asymptotic formula is.

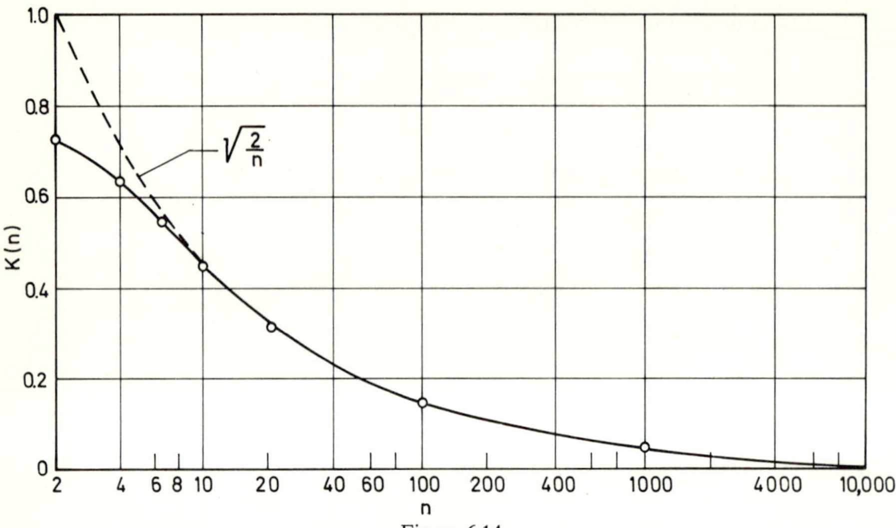

Figure 6.14

The related problem of a crack at the tip of a wedge has been discussed by Srivastav and Narain in [16]. The boundary conditions on the line $\theta=0$ are the same as before but on the boundaries $\theta=\pm\alpha$ of the wedge the stress-free conditions $\sigma_{\theta\theta}=\sigma_{rr}=0$ prevail. The displacement field is again given by equations (6.99) and (6.100) but now the appropriate form of the function $\bar{\chi}(s,\theta)$ is given by the equation

$$\begin{aligned}\bar{\chi}(s,\theta) = \psi(s)\Big[&\sin\{s(\theta-\alpha)\} - s(s+2)^{-1}\sin\{(s+2)(\theta-\alpha)\}\\ &+\frac{s\cdot\cos(s\alpha)-\cos\{(s+2)\alpha\}}{(s+2)\sin\{(s+2)\alpha\}-s\cdot\sin(s\alpha)}\\ &\times\{\cos\{s(\theta-\alpha)\}-\cos\{(s+2)(\theta-\alpha)\}\Big].\end{aligned}$$

If

$$\begin{aligned}\psi(s) = \frac{s+2}{s}&\left[\frac{s\cos\{(s+2)\alpha\}\sin(s\alpha)-(s+2)\cos(s\alpha)\sin\{(s+2)\alpha\}}{(s+2)\sin\{(s+2)\alpha\}-s\cdot\sin(s\alpha)}\right]^{-1}\\ &\times\frac{\Gamma(\frac{1}{2}s+\frac{1}{2})\Gamma(\frac{1}{2})}{\Gamma(\frac{1}{2}s+1)}\int_0^1 t^s g(t)\,\mathrm{d}t\end{aligned}$$

it is found that $g(t)$ satisfies a Fredholm equation of the type (6.101) with $h(t)$ defined in terms of $f$ as before but the kernel now being given by the formula

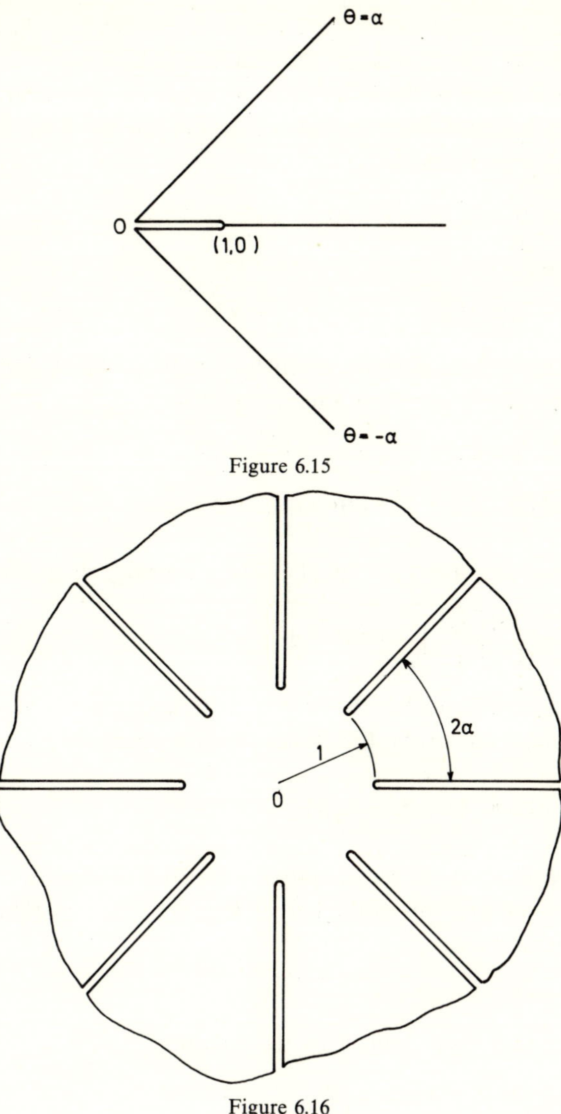


Figure 6.15



Figure 6.16

$K(u, t) =$

$$= \frac{t}{2} \mathcal{M}^{-1}\left[\left\{2 - \tan(\tfrac{1}{2}s\pi)\frac{2s(s+2)(1-\cos s\alpha) - 4\sin(s\alpha)\sin\{(s+2)\alpha\}}{s\cos\{(s+2)\alpha\}\sin(s\alpha) - (s+2)\sin\{(s+2)\alpha\}\cos(s\alpha)}\right\}; \frac{t}{u}\right]$$

This integral equation has been solved numerically by Srivastav and Narain but only in the case in which $\alpha=\pi/2$ which corresponds to the geometry of Figure 6.8 above. The results are tabulated in [16].

The corresponding problem for external cracks is shown in Figure 6.16. This problem has also been considered in [16] in the form of the mixed boundary value problem

$$\begin{aligned} &\sigma_{\theta\theta}(r,\pm\alpha)=0\,,\quad u_0(r,\pm\alpha)=0\,, && r\geqslant 0\,,\\ &u_\theta(r,0)=0\,, && 0\leqslant r<1\,,\\ &\sigma_{\theta\theta}(r,0)=-2\mu f(r)\,, && r\geqslant 1\,. \end{aligned}$$

The displacement field is again given by equations (6.99) and (6.100) but now $\psi(s)$ must satisfy the dual integral equations

$$\begin{aligned} &\mathscr{M}^{-1}[\psi(s);r]=0\,, && 0\leqslant r<1\,,\\ &\mathscr{M}^{-1}[k(s)\psi(s);r]=r\int_r^\infty f(u)\mathrm{d}u\,, && r\geqslant 1\,, \end{aligned}$$

where $k(s)=(s+2)\cot(s\alpha)-s\cdot\cot\{(s+2)\alpha\}$. To solve these dual equations

$$\psi(s)=\frac{\Gamma(\tfrac{1}{2}-\tfrac{1}{2}s)\Gamma(\tfrac{1}{2})}{2\Gamma(1-\tfrac{1}{2}s)}\int_1^\infty g(t)t^{s-1}\mathrm{d}t$$

which automatically satisfies the first of them and which when substituted into the second leads to the Fredholm integral equation

$$g(t)-\int_1^\infty g(u)K_1(u,t)\mathrm{d}u=h_1(t)$$

for the determination of the function $g(t)$. In this equation, the kernel $K_1(u,t)$ is defined in terms of the kernel $K(u,t)$ defined by equation (6.102) through the equation

$$K_1(t,u)=(ut)^{-1}K(u,t)\,,$$

and the function $h_1(t)$ is defined by the equation.

$$h_1(t)=-\frac{1}{\pi}\frac{\mathrm{d}}{\mathrm{d}t}\int_t^\infty\frac{x^2\mathrm{d}x}{(x^2-t^2)^{\frac{1}{2}}}\int_x^\infty f(y)\mathrm{d}y\,.$$

The case of a star-shaped crack when $n=2$ (i.e. when the crack is of cruciform shape) may be analysed by the method introduced at the end of section 6.3. In this case the geometry is the simple one shown in Figure 6.17. Assuming that the faces of the crack were subjected to a symmetric pressure $p_0f(x)$,

Rooke and Sneddon [18] took in the positive quadrant a displacement field of the type given by equations (6.79) and (6.80) but with $A \equiv B$. (The stress field in the remainder of the plane can easily be determined by a symmetry argument.) The boundary conditions

$$\begin{aligned}
\sigma_{xx}(0, y) &= -p_0 f(y), & 0 \leqslant y \leqslant 1, \\
u_x(0, y) &= 0, & y > 1, \\
\sigma_{xy}(0, y) &= 0, & y \geqslant 0, \\
\sigma_{yy}(x, 0) &= -p_0 f(x), & 0 \leqslant x \leqslant 1, \\
u_y(x, 0) &= 0, & x > 1, \\
\sigma_{xy}(x, 0) &= 0, & x \geqslant 0,
\end{aligned}$$

then lead to a pair of dual integral equations of the form (6.81), (6.82) but

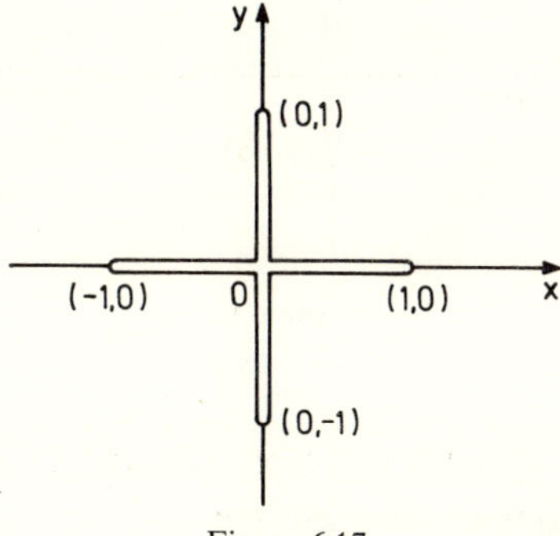


Figure 6.17

with $B$ replaced by $A$. It follows from equation (6.86) that if we represent $A(\xi)$ by a formula of the type (6.84) then the function $a(t)$ must satisfy the Fredholm integral equation

$$a(t) + \int_0^1 K(t, u)\, a(u) \mathrm{d}u = \phi(t), \qquad 0 \leqslant t \leqslant 1, \tag{6.103}$$

where $K(t, u)$ is given by equation (6.87) and $\phi(t)$ by equation (6.88). The stress intensity factor is given in terms of the solution of this equation by (6.89) and the crack energy by (6.90). In the case in which the applied internal pressure is constant, $p_0$, we have $\phi(t) = 1$ and

$$a(1) = 0.8636, \qquad \int_0^1 t a(t) \mathrm{d}t = 1.4914.$$

It follows therefore that the stress intensity factor is given by the expression

$$k = 0.8636 k_0 ,$$

and the crack energy by

$$W = 1.4914 W_0 ,$$

where $k_0 = p_0$ and $W_0 = 4\pi(1-\nu)p_0^2/E$.

The case of a crack with arms of unequal lengths $2a$ and $2b$ has been considered by Sneddon and Das [19]. In this case the cracks are defined by

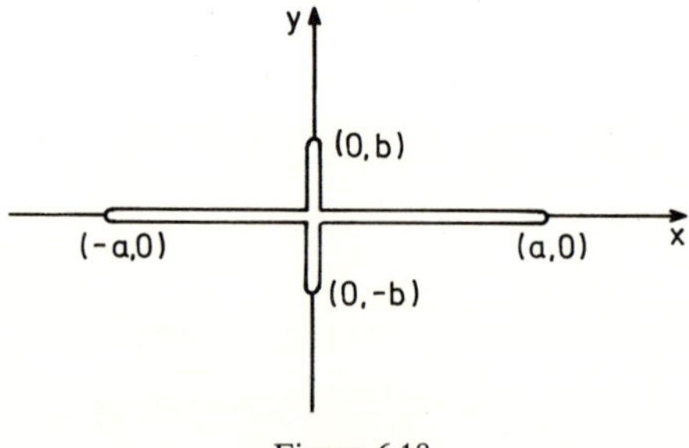


Figure 6.18

the relations $|x| \leqslant a$, $y=0$ and $|y| \leqslant b$, $x=0$ in the $xy$-plane (cf. Figure 6.18). The boundary conditions in this case are

$$\begin{aligned}
\sigma_{xx}(0, y) &= -p_0 g(y) , & 0 \leqslant |y| \leqslant b ,\\
u_x(0, y) &= 0 , & |y| > b\\
\sigma_{xy}(0, y) &= 0 , & -\infty < y < \infty\\
\sigma_{yy}(x, 0) &= -p_0 f(x) , & 0 \leqslant |x| \leqslant a ,\\
u_y(x, 0) &= 0 , & |x| > a ,\\
\sigma_{yy}(x, 0) &= 0 , & -\infty < x < \infty
\end{aligned}$$

where the functions $f$ and $g$ are prescribed and are for simplicity, assumed to be *even* functions.

## 6.6 Circular cracks

*Circular cracks in an infinite solid.* The problem of determining the distribution of stress in the neighbourhood of a circular—or "penny-shaped"—crack, specified by the relations $\rho \leqslant c$, $z=0$ in an infinite solid can be solved in a manner similar to that used in section 6.2 for a Griffith crack. In terms of cylindrical coordinate $(\rho, \phi, z)$, the problem is equivalent to finding the displacement field in the half-space $z \geqslant 0$ when its plane boundary is loaded in such a way that

$$\sigma_{zz}(\rho,0) = -p(\rho),\ 0 \leqslant \rho \leqslant c\,; \quad u_z(\rho,0)=0,\ \rho > c\,; \quad \sigma_{\rho z}(\rho,0)=0,\ \rho \geqslant 0\,, \tag{6.104}$$

where $\boldsymbol{u} = [u_\rho(\rho, z), 0, u_z(\rho, z)]$ denotes the displacement vector. By making use of the theory of Hankel transforms it can be shown that

$$\begin{aligned} u_\rho(\rho, z) &= -\mathscr{H}_1[\xi^{-1}(1-2\nu-\xi z)\psi(\xi c)\mathrm{e}^{-\xi z};\ \xi\to\rho] \\ u_z(\rho, z) &= \phantom{-}\mathscr{H}_0[\xi^{-1}(2-2\nu+\xi z)\psi(\xi c)\mathrm{e}^{-\xi z};\ \xi\to\rho] \end{aligned} \tag{6.105}$$

where the operator $\mathscr{H}_n$ is defined by the equation

$$\mathscr{H}_n[\bar{f}(\xi, z);\ \xi\to\rho] = \int_0^\infty \xi \bar{f}(\xi, z)\, J_n(\rho\xi)\mathrm{d}\xi\,. \tag{6.106}$$

By calculating $\sigma_{zz}(\rho, z)$ from the stress-strain relation and reformulating the conditions (6.104) we find that the unknown function $\psi(t)$ is the solution of the dual integral equations

$$\begin{aligned} \mathscr{H}_0[\psi(t);\ x] &= f(x)\,, \qquad 0 \leqslant x \leqslant 1\,; \\ \mathscr{H}_0[t^{-1}\psi(t);\ x] &= 0\,, \qquad\qquad x > 1\,. \end{aligned}$$

where the function $f(x)$ is defined in terms of $p$ by the equation

$$f(x) = (1+\nu)c^2 E^{-1} p(xc)\,, \qquad 0 \leqslant x \leqslant 1\,.$$

The second of the dual equations is automatically satisfied by taking $\psi(t)$ as the integral representation

$$\psi(t) = \int_0^1 g(x)\sin(xt)\mathrm{d}x\,. \tag{6.107}$$

Substituting in the first equation of the pair we find that the function $g(x)$ satisfies an integral equation of Abel type whose solution is easily found to be

$$g(x) = \frac{2}{\pi}\int_0^x \frac{sf(s)\mathrm{d}s}{(x^2-s^2)^{\frac{1}{2}}}\,. \tag{6.108}$$

(Cf. [4]).

It is readily deduced that the stress intensity factor

$$k = \lim_{\rho \to c+} [2(\rho - c)]^{\frac{1}{2}} \sigma_{zz}(\rho, 0)$$

is determined from the pair of formulae

$$k = \frac{2}{\pi} \frac{1}{c^{\frac{1}{2}}} G(c), \quad G(u) = \int_0^u \frac{\rho p(\rho) \mathrm{d}\rho}{(u^2 - \rho^2)^{\frac{1}{2}}}. \tag{6.109}$$

In the case in which $p(\rho) = p_0$, a constant, we find that $G(u) = p_0 u$ so that $k = (2/\pi) p_0 c^{\frac{1}{2}}$.

The energy of the formation of the crack can be expressed in terms of the function $G(u)$ by means of the formula

$$W = \frac{8(1 - \nu^2)}{E} \int_0^c \{G(u)\}^2 \mathrm{d}u. \tag{6.110}$$

In the case in which $p(\rho) = p_0$, this gives $W = 8(1 - \nu^2) p_0^2 c^3 / (3E)$.

The related problem of the opening up of circular cracks in an infinite solid as a result of thermal fields has been discussed by Olesiak and Sneddon [20] using integral transform techniques but no new matters of principle are involved and hence it will not be discussed here.

If, in addition to the applied pressure, there is a cohesive force of the type postulated by Barenblatt the formula for the stress intensity factor can be modified in an obvious way. Suppose that the cohesive force $c(s)$, with $s = c - \varepsilon$, acts over only a small zone near the rim of the crack then it is necessary to add to $G(c)$ the quantity $-(2c)^{-\frac{1}{2}} C$, where $C$ is the constant defined by the equation

$$C = \int_0^\varepsilon s^{\frac{1}{2}} \cdot c(s) \mathrm{d}s$$

and called by Barenblatt [21] the cohesive modulus of the material.

A model of the conditions near the rim of a circular crack when a constant pressure $p_0$ is applied to the faces of the crack has been analysed by Olesiak and Wnuk [22] by taking the elastic solution with

$$p(\rho) = \begin{cases} p_0, & 0 \leqslant \rho \leqslant c \cos \beta, \\ -p_1, & c \cos \beta < \rho \leqslant c, \end{cases} \tag{6.111}$$

where $\beta$ is very small. Barenblatt's condition is then replaced by the simple condition $G(c) = 0$, with $G(c)$ defined by the second equation of the pair (6.109) and the equations (6.111), and this yields the equation

$$p_1 = p_0(\operatorname{cosec} \beta - 1)$$

which connects $p_1$, $\beta$ and $p_0$. The crack energy is then given by the formula

$$W = \frac{8(1-\nu^2)p_0^2c^3}{3E}\{\cos(2\beta)+\tan^2(\tfrac{1}{2}\beta)(1+2\cos\beta)\}\,.$$

To calculate the value of the crack energy it is necessary to know the appropriate value of $\beta$, and the additional relation connecting the three quantities $p_1$, $\beta$, $p_0$ only one of which is prescribed ($p_0$). Olesiak and Wnuk assume that this is provided by making the hypothesis that the value of $p_1$ to be taken is that which leads to a certain combination of the stress components reaching a critical value which is characteristic of the material. The consequences of such a hypothesis are explored by these authors in the form of a series of numerical calculations.

*Circular cracks in cylinders.* The method of Hankel transforms outlined above has been extended in several papers in an attempt to estimate the effect of the geometry of the elastic solid upon the value of the stress intensity factor and that of the crack energy. We discuss briefly the problems concerning a penny-shaped crack in an infinite cylinder whose radius is chosen

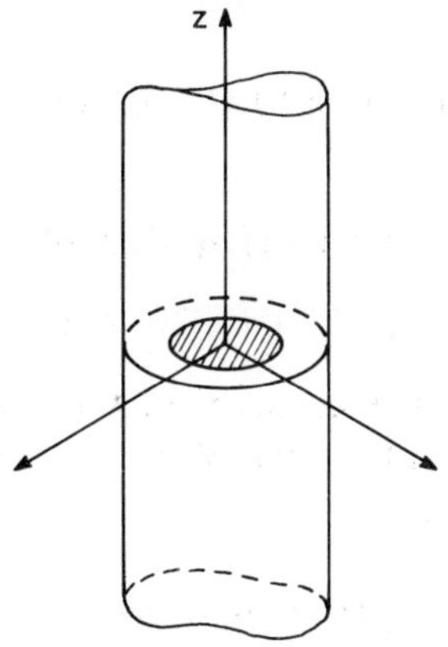


Figure 6.19

(in terms of the radius of the crack as unit of length) to be $a$. It is assumed that the crack lies in a plane normal to the axis of the cylinder and that its center lies on that axis (cf. Figure 6.19).

In the first problem [23] it is assumed that

$$u_\rho(a, z) = 0\,, \qquad \sigma_{\rho z}(a, z) = 0 \tag{6.112}$$

while in the second problem [24] it is assumed that

$$\sigma_{\rho\rho}(a, z) = \sigma_{\rho z}(a, z) = 0\,. \tag{6.113}$$

In both problems it is assumed that the crack faces are subjected to a pressure $2\mu f(\rho)$. The problem is therefore equivalent to a boundary value problem for the semi-infinite cylinder $0 \leqslant \rho \leqslant a$, $z \geqslant 0$, when the plane boundary is subjected to the mixed boundary conditions

$$\sigma_{zz}(\rho, 0) = -2\mu f(\rho)\,,\;\; 0 \leqslant \rho \leqslant 1; \qquad u_z(\rho, 0) = 0\,,\;\; 1 < \rho \leqslant a\,; \tag{6.114}$$

$$\sigma_{\rho z}(\rho, 0) = 0\,,\;\; 0 \leqslant \rho \leqslant a\,, \tag{6.115}$$

and the curved surface to either of the pairs of conditions (6.112) or (6.113).

The solution of the equations of equilibrium is taken in the form

$$u_\rho = \frac{\partial^2 U}{\partial \rho\, \partial z}, \qquad u_z = \frac{\partial^2 U}{\partial z^2} - 2(1-\nu)\Delta U\,, \tag{6.116}$$

where $\Delta$ denotes the axisymmetric Laplace operator

$$\Delta = \frac{\partial^2}{\partial \rho^2} + \frac{1}{\rho}\frac{\partial}{\partial \rho} + \frac{\partial^2}{\partial z^2}$$

and $U$ is a biharmonic function ($\Delta\Delta U = 0$) which is given by the integral representation

$$U(\rho, z) = \mathscr{F}_s[\xi^{-2}\{(A(\xi) + 4(1-\nu)B(\xi))I_0(\xi\rho) - \xi\rho B(\xi)I_1(\xi\rho)\}\,;\; \xi \to z] + \mathscr{H}_0[\xi^{-3}\psi(\xi)(2\nu + \xi z)e^{-\xi z}\,;\; \xi \to \rho]\,. \tag{6.117}$$

This representation has the property that it leads to $\sigma_{\rho z}(\rho, 0) = 0$. The remaining conditions (6.114) on the plane $z = 0$ are satisfied if $A, B$ and $\psi$ satisfy the dual equations

$$\mathscr{H}_0[\psi(\xi); \rho] + \frac{2}{\pi}\int_0^\infty [\{A(\xi) - 2\nu B(\xi)\}\, I_0(\xi\rho) - \xi\rho B(\xi)\, I_1(\xi\rho)]\, d\xi = f(\rho)\,, \qquad 0 \leqslant \rho \leqslant 1\,,$$

$$\mathscr{H}_0[\xi^{-1}\psi(\xi); \rho] = 0\,, \qquad 1 < \rho \leqslant a\,.$$

Again, let the function $\psi$ be represented by an integral of the type (6.107) and thereby ensure that the second equation of this pair is automatically satisfied.

When we substitute this expression into the first equation of the pair and then operate on both sides of the resulting equation by $I^{-1}$ we get the integral equation

$$g(t)+\frac{2}{\pi}\int_0^\infty [\{A(\xi)+(1-2\nu)B(\xi)\}\sinh(\xi t)-B(\xi)\xi t\cosh(\xi t)]\mathrm{d}\xi=h(t), \tag{6.118}$$

in which the function $h(t)$ is derived from the prescribed function $f(\rho)$ by means of the formula

$$h(t)=\frac{2}{\pi}\int_0^t \frac{\rho f(\rho)\mathrm{d}\rho}{(t^2-\rho^2)^{\frac{1}{2}}}. \tag{6.119}$$

So far the analysis is common to both problems. Writing down the integral equations corresponding to the boundary conditions (6.112) and eliminating the functions $A$ and $B$ from this pair of equations and equation (6.118) we find that the function $g(t)$ is the solution of the Fredholm equation

$$g(t)-\int_0^1 K(t,u)g(u)\mathrm{d}u=h(t) \tag{6.120}$$

whose kernel is defined by the equation

$$K(t,u)=H\left(\frac{u+t}{a}\right)-H\left(\frac{u-t}{a}\right),$$

with the function $H$ given by the expression

$$H(x)=\frac{2}{\pi^2 a}\int_0^\infty \frac{K_1(u)}{I_1(u)}[\cosh(ux)-1]\,du,$$

$K_1$ and $I_1$ denoting modified Bessel functions of the first order.

The stress intensity factor $k=\lim_{\rho\to 1+}[2(\rho-1)]^{\frac{1}{2}}\sigma_{zz}(\rho,0)$ is given in terms of this function $g$ by the simple formula

$$k=g(1), \tag{6.121}$$

and the crack energy by the formula

$$W=\frac{2\pi^2(1-\nu^2)}{E}\,\omega(a^{-1}),$$

where

$$\omega(a^{-1})=\int_0^1 h(t)g(t)\mathrm{d}t. \tag{6.122}$$

Of course, this last equation gives $W$ in terms of a system of units in which the unit of length is taken to be the radius of the circular crack. In terms of convential units in which the radius of the crack is $c$ the above formula is then replaced by

$$W = \frac{2\pi^2(1-\nu^2)c^3}{E}\,\omega(c/a)\,. \tag{6.123}$$

In particular, if a constant pressure $p_0$ is applied to the faces of the crack $h(t)=2p_0 t/\pi$ and this equation reduces to

$$W_1 = W_\infty\,\omega_1(c/a)\,, \tag{6.124}$$

where $W_\infty = 8(1-\nu^2)c^3 p_0^2/3E$ is the energy of the system in the case in which $a \gg c$,

$$\omega_1(c/a) = 3\int_0^1 t\phi(t)\,\mathrm{d}t \tag{6.125}$$

and $\phi(t)$ is the solution of the integral equation

$$\phi(t) - \int_0^1 K(t,u)\,\phi(u)\,\mathrm{d}u = t\,. \tag{6.126}$$

Since the function $g(t)=h(t)$ in the case in which $a \gg c$, the percentage increase of the stress intensity factor over its value in the case of an isolated crack is given by the formula

$$n = \frac{g(1)-h(1)}{h(1)} \times 100\,.$$

In the case of constant pressure $p_0$ this reduces to

$$n_1 = \{\phi(1)-1\} \times 100\,. \tag{6.127}$$

Two points in the analysis should be noticed: in the first instance the kernel of the integral equation (6.126) does not depend on the value of the Poisson's ratio of the material; secondly, to determine the stress intensity factor or the percentage increase $n$ only $g(1)$ has to be calculated.

The analysis in the second case based on the boundary conditions (6.113) goes through in exactly the same way, the only difference being that the kernel of the integral equation (6.126) is given by much more complicated formulas. In this case, too, the kernel does depend on the value of the Poisson's ratio. For details of the analysis an interested reader is referred to p. 204 *et seq.* of [25].

Numerical values of $n_1$ calculated from equation (6.127) are given in [23] and the corresponding factor in the case of the boundary conditions (6.113)—denoted by $n_2$—is tabulated in [24] on the assumption that the Poisson's ratio is 0.25. The variation of $n_1$ and $n_2$ with the ratio $r = c/a$ plotted from these values is shown graphically in Figure 6.20. From these curves it is observed that the difference between the values of $n_1$ and $n_2$ is not physically significant and that in either case the percentage increase of the value over that for an isolated crack is less than 10% if the radius of the crack is less than half of that of the cylinder.

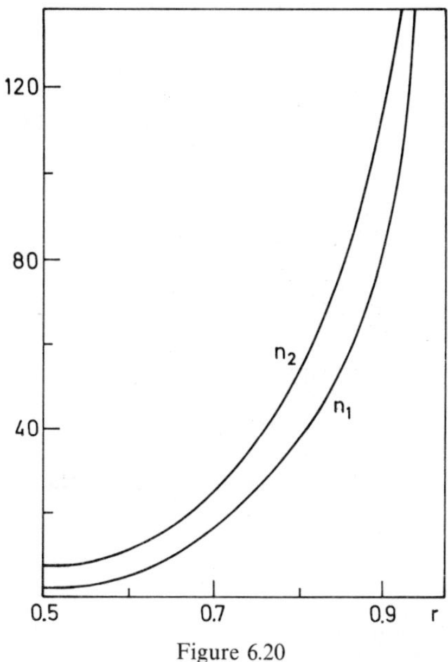


Figure 6.20

A comparison of the axisymmetric solution and that of the corresponding plane strain solution is shown in Figure 6.21, where $n_1$ has the above meaning and $n$ is the corresponding quantity calculated from formula (6.74):

$$n = 100\left\{\left(\frac{2a}{\pi c}\tan\frac{\pi c}{2a}\right)^{\frac{1}{2}} - 1\right\}.$$

In the axisymmetric case the independent variable $x$ is equal to the ratio of the area of the crack to the cross-sectional area of the cylinder while in the plane strain case $x$ is the ratio of the crack length to the distance between the centers of two neighbouring cracks. These graphs show that unless the radius of the crack is greater than one-half of the radius of the cylinder, there is little error incurred by taking the corresponding value of $n$ for either $n_1$ or $n_2$.

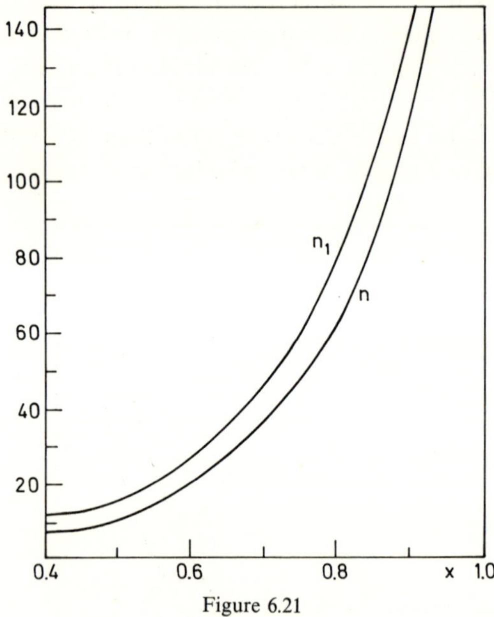


Figure 6.21

In other words a reasonable approximation for either $n_1$ or $n_2$ is given by

$$n = 100\left[\left(\frac{2a^2}{\pi c^2}\tan\frac{\pi c^2}{2a^2}\right)^{\frac{1}{2}} - 1\right],$$

which reduces when $c \ll a$ to

$$\frac{25\pi^2 c^4}{6a^4}\left[1 + O(c^4/a^4)\right].$$

*Circular cracks in thick plates.* Lowengrub [26] has considered two basic boundary value problems relative to the determination of the distribution of stress in the vicinity of a circular crack lying symmetrically in the central plane $z=0$ of a thick plate. If the radius of the crack is taken to be the unit length then the crack is determined by the relations $0 \leqslant \rho \leqslant 1$, $z=0$ and the plate is given by $-\delta \leqslant z \leqslant \delta$.

In the first problem it is assumed that the surfaces of the plate are given a uniform displacement $\varepsilon$ with the boundary conditions (for the elastic layer $0 \leqslant z \leqslant \delta$):

$$u_z(\rho, \delta) = \varepsilon\,, \quad \sigma_{\rho z}(\rho, \delta) = 0\,, \quad \rho \geqslant 0\,, \tag{6.128}$$

$$\sigma_{\rho z}(\rho, 0) = 0, \ \rho \geqslant 0; \quad \sigma_{zz}(\rho, 0) = 0, \ 0 \leqslant \rho \leqslant 1; \quad u_z(\rho, 0) = 0, \ \rho > 1\,. \tag{6.129}$$

It has been assumed that the faces of the crack are free from applied stresses. If the components of the displacement vector are taken to be $(u_\rho, 0, u_z + \varepsilon z/\delta)$ then $u_\rho$ and $u_z$ must be such as to satisfy the modified boundary conditions

$$u_z(\rho, \delta) = 0\,, \quad \sigma_{\rho z}(0, \delta) = 0\,, \qquad \rho \geqslant 0\,, \tag{6.130}$$

$$\sigma_{\rho z}(\rho, 0) = 0\,, \ \rho \geqslant 0; \quad \sigma_{zz}(\rho, 0) = -2\mu\beta^2\varepsilon/\delta\,, \ 0 \leqslant \rho \leqslant 1; \quad u_z(\rho, 0) = 0\,, \quad \rho > 1\,, \tag{6.131}$$

where $\beta^2 = 2(1-\nu)/(1-2\nu)$.

In the second problem it is assumed that the crack is opened up under internal pressure with the conditions

$$\sigma_{\rho z}(\rho, \delta) = \sigma_{zz}(\rho, \delta) = 0\,, \quad \rho \geqslant 0\,; \tag{6.132}$$

$$\sigma_{\rho z}(\rho, 0) = 0\,, \ \rho \geqslant 0; \quad \sigma_{zz}(\rho, 0) = -p(\rho)\,, \ 0 \leqslant \rho \leqslant 1; \quad u_z(\rho, 0) = 0\,, \quad \rho > 1\,. \tag{6.133}$$

If the displacement field is taken to be of the form

$$u_\rho = \left[ \tfrac{1}{2}\mathscr{H}_1 \left\{ \xi^{-1}\psi(\xi) \ \frac{\xi\delta \cosh \xi z}{\sinh^2 \xi\delta} - \frac{(1-2\nu)\cosh \xi(\delta - z)}{\sinh \xi\delta} + \frac{\xi z \sinh \xi(\delta - z)}{\sinh \xi\delta} \right\};\ \xi \to \rho \right]$$

$$u_z = -\tfrac{1}{2}\mathscr{H}_0 \left[ \xi^{-1}\psi(\xi) \left\{ \frac{\xi\delta \sinh \xi z}{\sinh^2 \xi\delta} - 2(1-\nu)\frac{\sinh \xi(\delta - z)}{\sinh \xi\delta} - \frac{\xi z \cosh \xi(\delta - z)}{\xi \sinh \xi\delta} \right\};\ \xi \to \rho \right]$$

then the boundary conditions of the first of these two problems are satisfied if the function $\psi(\xi)$ is taken to be the solution of the dual integral equations

$$\mathscr{H}_0[\{1 + H(\xi\delta)\}\psi(\xi);\, \rho] = \beta^2\varepsilon/\delta\,, \qquad 0 \leqslant \rho \leqslant 1\,,$$

$$\mathscr{H}_0[\xi^{-1}\psi(\xi);\, \rho] = 0\,, \qquad \rho > 1\,,$$

where the function $H$ is defined by the equation

$$H(x)=(x+1-e^{-2x})\operatorname{cosech}^2 x\,.$$

By using the method outlined for the cylinder, Lowengrub reduced the solution of these dual integral equations to the solution of a Fredholm integral equation of the second kind for an auxiliary function $g(t)$.

The same method can be used to derive the solution of the second of the above two problems.

The solution of the first of these two problems can be used to derive the solution of the problem of determining the stress field in an infinite solid containing an infinite set of circular cracks $0\leqslant\rho\leqslant c$, $z=\pm nh$, $(n=0, 1, 2, \ldots)$ when the faces of each crack are subjected to exactly the same pressure [27].

*External circular cracks.* The problem of determining the distribution of stress in an infinite elastic solid when pressure is applied to the faces of a flat "external" crack covering the outside of a circle has been discussed by Uflyand [28]. Uflyand expresses the Boussinesq–Papkovich solution of the equations of elastic equilibrium in toroidal coordinates and then uses the Mehler–Fock transform to solve the relevant boundary value problem. It should be observed that the boundary value problem is not a mixed problem in this case—the normal stress and the shearing stress are prescribed on the boundary—but the properties of the Mehler–Fock transform are by contrast much more complicated than those of the Hankel transform.

Lowengrub and Sneddon [29] use cylindrical coordinates $(\rho, \phi, z)$ and Muki's solution [30] of the elastic equations. To simplify the analysis the radius of the circle is taken to be the unit of length so that the crack faces are described by the relations $\rho\geqslant 1$, $z=0$, and it is assumed that:

(1) the face $\rho\geqslant 1$, $z=0+$ is loaded in exactly the same way as the face $\rho\geqslant 1$, $z=0-$;

(2) the applied pressure $p(\rho, \phi)$ is an even function of $\phi$.

The restriction (2) is not in any sense limiting; any problem in which it is not satisfied can be solved in exactly the same way except that the Fourier series representing $p(\rho, \phi)$ will then containing sine terms in addition to cosine terms. In the method used in [29] the condition (1) *cannot* be relaxed; it must be retained in order that the stated problem can be reduced to a mixed boundary value problem for a half-space. It should be emphasized that Uflyand's solution does not involve making this assumption of symmetry about the plane of the crack.

Because of the condition (1) the problem may be reduced to that of determining the displacement field in the half-space $z \geqslant 0$ when the plane $z=0$ is subjected to the boundary conditions

$$\begin{aligned} &u_z(\rho, \phi, 0) = 0\,, && 0 \leqslant \phi \leqslant 2\pi\,, \quad \rho \leqslant 1 \\ &\sigma_{zz}(\rho, \phi, 0) = -p(\rho, \phi)\,, && 0 \leqslant \phi \leqslant 2\pi\,, \quad \rho > 1 \\ &\sigma_{\rho z}(\rho, \phi, 0) = 0\,, && 0 \leqslant \phi \leqslant 2\pi\,, \quad \rho \geqslant 0\,. \end{aligned}$$

In these circumstances the displacement field may be given the Muki representation

$$\begin{aligned} u_\rho &= \sum_{m=0}^{\infty} \cos(m\phi)\,\mathscr{H}_{m-1}[\xi^{-1}\psi_m(\xi)(1-2\nu-\xi z)\mathrm{e}^{-\xi z};\ \xi\to\rho] \\ &\qquad - \sum_{m=0}^{\infty} m\cos(m\phi)\rho^{-1}\mathscr{H}_m[\xi^{-2}\psi_m(\xi)(1-2\nu-\xi z)\mathrm{e}^{-\xi z};\ \xi\to\rho] \\ u_\phi &= -2\rho^{-1}\sum_{m=0}^{\infty} m\sin(m\phi)\,\mathscr{H}_m[\xi^{-2}\psi_m(\xi)(1-2\nu-\xi z)\mathrm{e}^{-\xi z};\ \xi\to\rho] \\ u_z &= \sum_{m=0}^{\infty} \cos(m\phi)\,\mathscr{H}_m[\xi^{-1}\psi_m(\xi)(2-2\nu+\xi z)\mathrm{e}^{-\xi z};\ \xi\to\rho]\,, \end{aligned}$$

where the functions $\psi_m(\xi)$ are the solutions of the dual integral equations

$$\begin{aligned} &\mathscr{H}_m[\xi^{-1}\psi_m(\xi);\ \rho] = 0\,, && 0 \leqslant \rho < 1\,, \\ &\mathscr{H}_m[\psi_m(\xi);\ \rho] = g_m(\rho)\,, && \rho \geqslant 1\,, \end{aligned}$$

the functions $g_m(\rho)$ being defined by the expansion

$$p(\rho, \phi) = \frac{E}{1+\nu}\sum_{m=0}^{\infty} g_m(\rho)\cos(m\phi)\,.$$

The solution of these equations was derived by Lowengrub and Sneddon in the form

$$\psi_m(\xi) = \left(\frac{2\xi}{\pi}\right)^{\frac{1}{2}} \int_1^{\infty} t^{m+\frac{1}{2}} G_m(t)\, J_{m-\frac{1}{2}}(\xi t)\,\mathrm{d}t$$

where the functions $G_m(t)$ are defined by the equation

$$G_m(t) = \int_1^{\infty} \frac{\rho^{1-m} g_m(\rho)\,\mathrm{d}\rho}{(\rho^2 - t^2)^{\frac{1}{2}}} \tag{6.134}$$

from which it is easily deduced that the stress intensity factor

$$k = -\lim_{\rho \to 1-} [2(1-\rho)]^{\frac{1}{2}} \sigma_{zz}(\rho, \phi, 0)$$

is given by the formula

$$k = \frac{4E}{\pi^2(1+\nu)} \sum_{m=0}^{\infty} G_m(1) \cos(m\phi). \tag{6.135}$$

The case of loading by two equal and opposite point forces (cf. Figure 6.22) has been considered in detail in [29].

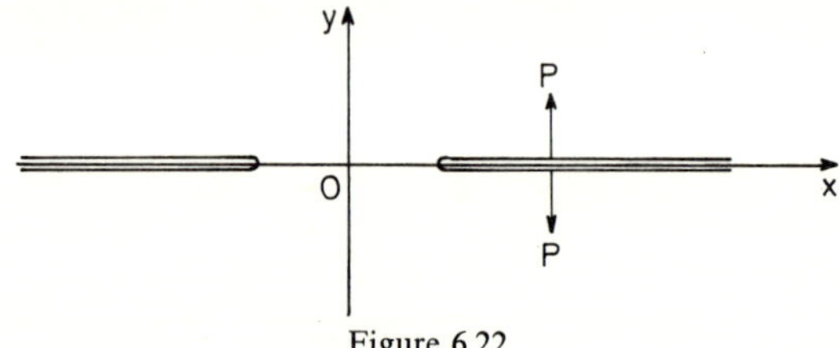


Figure 6.22

For an axisymmetric loading function $p(\rho)$, the displacement component $u_\phi(\rho, z)$ vanishes identically and the two remaining components of the displacement vector are given by the equations

$$u_\rho(\rho, z) = -\mathscr{H}_1[\xi^{-1}\psi(\xi)(1-2\nu-\xi z)e^{-\xi z};\ \xi \to \rho],$$

$$u_z(\rho, z) = \mathscr{H}_0[\xi^{-1}\psi(\xi)(2-2\nu+\xi z)e^{-\xi z};\ \xi \to \rho],$$

with the function $\psi(\xi)$ defined by the equation

$$\psi(\xi) = \int_1^\infty g(t) \cos(\xi t) dt,$$

in which

$$g(t) = \frac{2(1+\nu)}{\pi E} \int_t^\infty \frac{\rho p(\rho) d\rho}{(\rho^2 - t^2)^{\frac{1}{2}}}. \tag{6.136}$$

It is then easily shown that the stress intensity defined above is given by the formula

$$k = g(1). \tag{6.137}$$

Special cases of these formulas are considered in some detail in [29]. These correspond to the loadings:

(a) $p(\rho)=p_0 H(a-\rho)$, $k=(p_0/\pi)/(a^2-1)^{\frac{1}{2}}$, $(a=1.2, 1.6, 2.0)$

(b) $p(\rho)=p_0\rho^{-m}$, $(m>1)$, $k=\Gamma(\frac{1}{2}m-\frac{1}{2})p_0\{\Gamma(\frac{1}{2}m)\pi^{\frac{1}{2}}\}^{-1}$,

(c) $p(\rho)=p_0\rho^{-2}$, $k=p_0$,

(d) $p(\rho)=P\delta(\rho-c)/(2\pi 0)$, $k=\pi^{-2}(c^2-1)^{-\frac{1}{2}}P$, $(c>1)$.

## 6.7 Cracks in stress fields with body forces

The problem of determining the stress intensity factor in an elastic body in which there are body forces acting arises in the consideration of practical problems involving rivets and stiffeners. In a model describing such a situation the rivet reactions are often represented by body forces.

The problem of determining the displacement field in an infinite solid (in the plane strain case) is discussed in [31] in the situation in which there are symmetric body forces—i.e. it is assumed that the body force at $(x, y)$ in the $x$-direction, denoted by $X(x, y)$, is odd in $x$ and even in $y$, while $Y(x, y)$, the component in the $y$-direction is even in $x$ and odd in $y$. It is supposed that the faces of the crack are free from tractions (Cf. Figure 6.23).

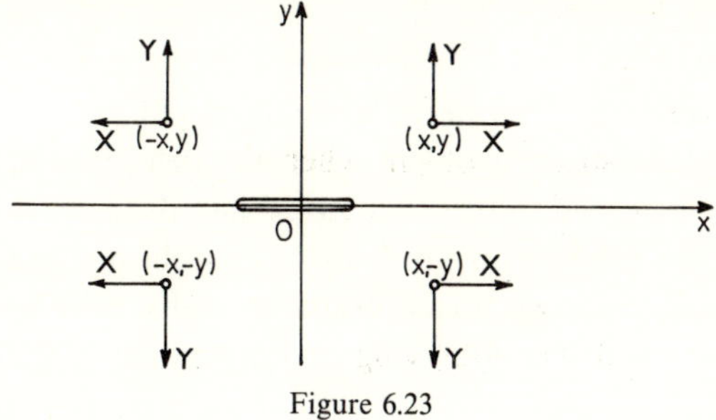

Figure 6.23

The solution is obtained as the sum of the solutions of the following two problems:

Problem 1 : *Find the displacement field* $\mathbf{u}^{(1)}$ *satisfying the equations of elastic equilibrium in the presence of the prescribed body forces, which tends to zero at infinity and satisfies the boundary conditions* $u_y^{(1)}(x,0)=0$, $\sigma_{xy}^{(1)}(x,0)=0$ *for all real values of* $x$.

For this solution it is a simple matter to calculate the normal stress on the $x$-axis, $\sigma_{yy}^{(1)}(x,0)$.

Problem 2: *Find the displacement field* $\boldsymbol{u}^{(2)}$ *in the neighbourhood of a Griffith crack when the faces of the crack are subjected to a pressure* $p(x) = \sigma_{yy}^{(1)}(x, 0)$.

*Problem* 2 has been discussed already in section 6.2; Sneddon and Tweed [31] have derived the solution of *Problem* 1 by applying the theory of Fourier transforms directly to the equations of elastic equilibrium with body forces present. Once the function $p(x)$ has been calculated from this solution we can use equation (6.11) to obtain the formula for the stress intensity factor at the tip of the crack. It turns out that

$$k = \frac{(1-2\nu)\sigma}{(1-\nu)\pi} c^{\frac{1}{2}} \int_0^\infty \int_0^\infty S(u, v) J_0(cu) \mathrm{d}u \mathrm{d}v\,, \tag{6.138}$$

where $\sigma$ denotes the density of the solid and the function $S$ is defined by

$$S(u, v) = \frac{2\nu u^3 - 2(1-\nu)uv^2}{(1-2\nu)(u^2+v^2)^2} \mathscr{F}_s[\mathscr{F}_c\{X(x, y); y\to v\}; x\to u] + \frac{2(2-\nu)u^2 v + 2(1-\nu)v^3}{(1-2\nu)(u^2+v^2)^2} \mathscr{F}_c[\mathscr{F}_s\{Y(x, y); y\to v\}; x\to u]\,. \tag{6.139}$$

Several special cases are considered in detail in [31].

The general plane strain problem when the distribution of body forces can no longer be assumed to be symmetrical has also been considered [32]. In this instance the problem is much more complicated but it can be simplified by reducing it to two mixed boundary value problems for the half-planes $y \geqslant 0$, and $y \leqslant 0$. The following decomposition of the body forces is made:

$$X(x, y) = \begin{cases} X^{(1)}(x, y)\,, & y > 0\,, \\ X^{(2)}(x, y)\,, & y < 0\,, \end{cases} \qquad Y(x, y) = \begin{cases} Y^{(1)}(x, y)\,, & y > 0\,, \\ Y^{(2)}(x, y)\,, & y < 0\,, \end{cases}$$

with the functions $X^{(1)}$, $Y^{(1)}$ uniquely defined in the upper half-plane and the functions $X^{(2)}$, $Y^{(2)}$ uniquely determined in the lower half-plane. In a similar manner the displacement vector $\boldsymbol{u}(x, y)$ can be resolved and determined according to the equations

$$\boldsymbol{u}(x, y) = \begin{cases} \boldsymbol{u}^{(1)}(x, y) + \boldsymbol{u}^{(3)}(x, y)\,, & y > 0\,, \\ \boldsymbol{u}^{(2)}(x, y) + \boldsymbol{u}^{(4)}(x, y)\,, & y < 0\,, \end{cases}$$

where

(1) $\boldsymbol{u}^{(1)}$ is the solution in the upper half-plane of the equations of equilibrium with the inclusion of body forces $(X^{(1)}, Y^{(1)})$, satisfying the boundary conditions $u_y^{(1)}(x, 0)=0$, $\sigma_{xy}^{(1)}(x, 0)=0$;

(2) $\boldsymbol{u}^{(2)}$ is the solution in the lower half-plane of the equations of equilibrium with the inclusion of body forces $(X^{(2)}, Y^{(2)})$, satisfying the boundary conditions $u_y^{(2)}(x, 0)=0$, $\sigma_{xy}^{(2)}(x, 0)=0$;

(3) $\boldsymbol{u}^{(3)}$ is the solution in the upper half-plane of the equations of elastic equilibrium in the absence of body forces;

(4) $\boldsymbol{u}^{(4)}$ is the solution of the equilibrium equations (without body forces) in the lower half-plane.

The solutions (3) and (4) each contain two arbitrary functions which can be determined by satisfying the prescribed boundary conditions on the crack surface

$$\sigma_{yy}^{(1)}(x, 0)+\sigma_{yy}^{(3)}(x, 0) = \sigma_{yy}^{(2)}(x, 0)+\sigma_{yy}^{(4)}(x, 0) = f(x)\,H(|x|-c)\,,$$

$$\sigma_{xy}^{(3)}(x, 0) = \sigma_{xy}^{(4)}(x, 0) = g(x)\,H(|x|-c)\,,$$

where $H$ denotes the Heaviside unit function and the functions $f$ and $g$ are not known, and the continuity conditions

$$u_x^{(1)}(x, 0)+u_x^{(3)}(x, 0) = u_x^{(2)}(x, 0)+u_x^{(4)}(x, 0)\,, \quad |x|>c\,,$$

$$u_y^{(3)}(x, 0) = u_y^{(4)}(x, 0)\,, \quad |x|>c\,.$$

When solutions of the type (6.1) and (6.2) are used for (3) and (4) it can be easily established that the arbitrary functions entering into these solutions must satisfy a set of relations which can be resolved into a set of integral equations whose solutions can be obtained by a direct application of the Fourier inversion theorem and a set of pairs of dual integral equations each pair of which can be solved by the method outlined in section 6.2 above. The analysis is fairly complicated but the final results turn out to be simple and capable of yielding expressions for the stress intensity factor in special cases.

An equation determining the stress intensity factor at the rim of a circular crack in an infinite solid in which there is an axisymmetric distribution of body forces acting in a direction perpendicular to the original plane of the crack has been derived by Sneddon and Tweed [33]. If the body force is always in the $z$-direction and has magnitude $Z(\rho, z)$, the corresponding formula for the stress intensity factor at the rim of a crack of radius $c$ is

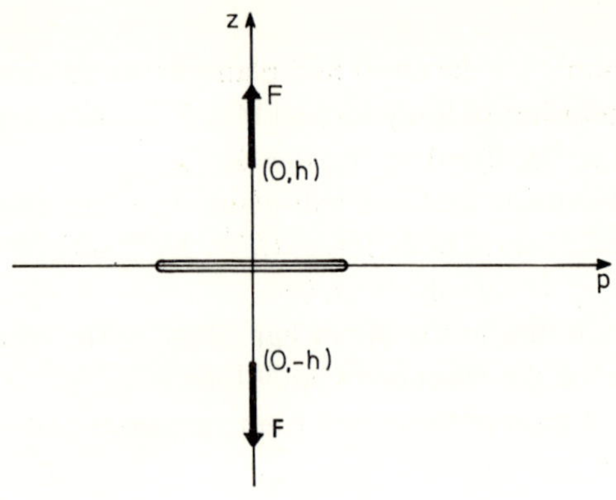


Figure 6.24

$$k = \frac{\sigma}{\pi}\frac{2}{c^{\frac{1}{2}}}\int_0^\infty\int_0^\infty Z(\rho, z)P(\rho, z, c)\mathrm{d}\rho\,\mathrm{d}z\,, \tag{6.140}$$

where $\sigma$ is the density and

$$P(\rho, z, c) = \int_0^\infty \{1+tz/2(1-\nu)\}\,\mathrm{e}^{-tz}J_0(t\rho)\sin(tc)\mathrm{d}t\,. \tag{6.141}$$

As an example of the use of this formula Sneddon and Tweed consider the effect on the crack of two point forces of magnitude $F$ acting at the points $(0, 0, \pm h)$ in the positive and negative $z$-directions respectively (Cf. Figure 6.24). In this case it is found that

$$k = \frac{F}{\pi^2}c^{\frac{1}{2}}\cdot\frac{c^2+\kappa h^2}{(c^2+h^2)^2}\,, \quad \kappa = \frac{2-\nu}{1-\nu}\,. \tag{6.142}$$

The variation of $k$ with $c$ and the corresponding variation of $u_z(o, 0)$, the normal component of the surface displacement, are shown graphically in [33].

## 6.8 Cracks under shear

In the above pages considerations were given only to those crack problems in which the deformation is produced by the application of pressure to the faces of the crack. The same methods may, of course, be applied in cases where an elastic solid containing one or more cracks is under shear. Examples of this type of analysis are to be found in [34], [35] and [36]. In each case the procedure is the same, consisting of the four stages listed at the end of section 6.2. The only difference is that, instead of getting a pair of dual integral equations, we get a system of simultaneous dual integral equations.

## References

[1] Sneddon, I. N., *Proc. Roy. Soc. London*, A187, p. 229 (1946).
[2] Sneddon, I. N. and Elliott, H. A., *Quart. Appl. Math.*, 4, p. 262 (1946).
[3] Lowengrub, M., *Proc. Edinburgh Math. Soc.*, 15, p. 131 (1966).
[4] Sneddon, I. N., *Proc. Cambridge Philos. Soc.*, 61, p. 609 (1965).
[5] Lowengrub, M., *Int. J. Engng. Sci.*, 4, p. 289 (1966).
[6] England, A. H. and Green, A. H., *Proc. Cambridge Philos. Soc.*, 59, p. 489 (1963).
[7] Sneddon, I. N. and Srivastav, R. P., *Int. J. Engng. Sci.*, 9, p. 479 (1971).
[8] Lowengrub, M. and Srivastava, K. N., *Int. J. Engng. Sci.*, 6, p. 359 (1968).
[9] Lowengrub, M. and Srivastava, K. N., *Int. J. Engng. Sci.*, 6, p. 425 (1968).
[10] Ling, C. B., *MTAC*, 11–12, p. 160 (1957–8).
[11] Sneddon, I. N. and Srivastav, R. P., *Proc. Roy. Soc. Edinburgh*, A67, p. 39 (1965).
[12] Parihar, K. S., *Proc. Roy. Soc. Edinburgh*, A69 (1971).
[13] Sneddon, I. N. and Das, S. C., *Int. J. Engng. Sci.*, 9, p. 25 (1971).
[14] Khrapkov, A. A., *Int. J. Fracture Mech.*, 7, p. 373 (1971).
[15] Lowengrub, M., *Int. J. Engng. Sci.*, 4, p. 69 (1966).
[16] Westmann, R. A., *J. Math. Phys.*, 43, p. 191 (1964).
[17] Srivastav, R. P. and Narain, R., *Proc. Cambridge Phil. Soc.*, 61, p. 945 (1965).
[18] Rooke, D. P. and Sneddon, I. N., *Int. J. Engng. Sci.*, 7, p. 1079 (1969).
[19] Sneddon, I. N. and Das, S. C., *Trends in Elasticity and Thermoelasticity: Witold Nowacki Anniversary Volume*, pp. 235–247. Wolters–Noordhoff Publ. Co. (1971).
[20] Sneddon, I. N. and Olesiak, Z., *Arch. Rat. Mech. and Anal.*, 4, p. 238 (1960).
[21] Barenblatt, G. I., *Prikh. Math. i Mekh.*, 23, p. 434 (1959).
[22] Olesiak, Z. and Wnuk, M., *Bull. de l'Acad. Polon. des Sci.* (*Ser. des Sci. Tech.*), 13, p. 445 (1965).
[23] Sneddon, I. N. and Tait, R. J., *Intl. J. Engng. Sci.*, 1, p. 391 (1963).
[24] Sneddon, I. N. and Welch, J. T., *Intl. J. Engng. Sci.*, 1, p. 411 (1963).
[25] Sneddon, I. N. and Lowengrub, M., *Crack Problems in the Classical Theory of Elasticity*, (John Wiley & Sons, Inc., New York, 1969).
✓[26] Lowengrub, M., *Quart. J. Appl. Math.*, 19, p. 119 (1961).
[27] Collins, W. D., *Proc. Edinburgh Math. Soc.*, (2), 13, p. 317 (1962).
[28] Uflyand, Ya. S., *Prikh. Math. i Mekh.*, 23, p. 101 (1959).
[29] Lowengrub, M. and Sneddon, I. N., *Int. J. Engng. Sci.*, 3, p. 451 (1965).
[30] Muki, R., *Progress in Solid Mechanics*, 1, p. 320 (1960).
[31] Sneddon, I. N. and Tweed, J., *Int. Jnl. Fracture Mech.*, 3, p. 317 (1967).
[32] Sneddon, I. N. and Tweed, J., *Proc. Roy. Soc. Edinburgh*, A69, p. 85 (1971).
[33] Sneddon, I. N. and Tweed, J., *Int. Jnl. Fracture Mech.*, 3, p. 291 (1967).
[34] Westmann, R. A., *J. Appl. Mech.*, 32, (1965).
[35] Keer, L. M., *J. Mech. Phys. Solids*, 14, p. 2 (1966).
[36] Keer, L. M. and Fu, W. S., *Int. J. Engng. Sci.*, 7, p. 361 (1969).

*F. Erdogan, G. D. Gupta and T. S. Cook*

# 7 Numerical solution of singular integral equations

## 7.1 Introduction

In this Chapter the numerical methods for the solution of two groups of singular integral equations will be described. These equations arise from the formulation of the mixed boundary value problems in applied physics and engineering. In particular, they play an important role in the solution of a great variety of contact and crack problems in solid mechanics. In the first group of integral equations the kernels have a simple Cauchy-type singularity. After separating the dominant parts, these equations may be expressed as

$$A(x)\psi(x) + \frac{1}{\pi}\int_a^b B(t)\psi(t)\frac{\mathrm{d}t}{t-x} + \int_a^b K(x,t)\psi(t)\mathrm{d}t = f(x)\,, \qquad (a<x<b) \tag{7.1}$$

where $\psi=(\psi_i)$, $(i=1, \ldots, N)$ is the unknown (vector) function, the square matrices $A=(a_{ij})$, $B=(b_{ij})$, $(i, j=1, \ldots, N)$ are known with $A \mp B$ nonsingular in $a<x<b$, the elements $k_{ij}(x, t)$ of the square matrix $K$ are Fredholm kernels*, and the vector $f=(f_i)$, $(i=1, \ldots, N)$ consists of the known functions satisfying a Hölder condition in the closed interval $[a, b]$.

The system of singular integral equations (7.1) may be obtained, for example, by considering the plane or the axisymmetric elastostatic problem for layered materials containing $N/2$ non-coplanar cracks located parallel to, and/or at the interfaces. Here, the unknown functions $\psi_i$ refer to the derivative of the crack opening displacements (or the dislocation densities if the crack is embedded in a homogeneous medium), and $f_i$ are the crack

* That is, $k_{ij}$ is square integrable in the square domain $a<(x, t)<b$. In practice these kernels are generally bounded and continuous in $a<(x, t)<b$. However, in axisymmetric problems $k_{ij}$ invariably contains a logarithmic singularity at $x=t$ (see Erdogan [1], Arin and Erdogan [2]).

surface tractions. If there are no interface cracks, the matrix $A$ is a null matrix and the resulting system of singular integral equations is of the first kind, otherwise at least some of the integral equations in (7.1) will be of the second kind (see Erdogan and Gupta [3, 4]). As will be seen from the examples given later in this Chapter, this will affect the nature of the singularity at the crack tips.

Equation (7.1) may also arise from the formulation of the elasticity problems for the parallel layers compressed by stamps with arbitrary profiles. In these problems if the contact between the stamp and the layer is frictionless, the corresponding component of the matrix $A$ is zero and the related integral equation is of the first kind. In the case of constant coefficient of friction or perfect adhesion between the stamp and the layer, the related equation is a (pair of) singular integral equation(s) of the second kind. It should be noted that, in general, the lengths of the cracks, or sizes of the stamps are not equal. However, since in the crack and stamp problems mentioned above the dominant parts of the integral equations are usually uncoupled, (i.e., the matrices $A$ and $B$ are in general diagonal), the range for $x$ can be normalized by simple changes in variable, and hence, the assumption of constant $a$ and $b$ does not affect the generality of the problem.

The distinguishing feature of the first group of singular integral equations given by (7.1) is that the dominant part of the system can easily be reduced to a Riemann–Hilbert problem for a set of sectionally holomorphic functions from which the matrix of the fundamental functions may be obtained (see Muskhelishvili [5]). A particular technique of reducing the problem to an eigenvalue problem and obtaining the fundamental matrix in a straightforward manner is described by Erdogan [6].

The second group of singular integral equations which will be considered in this Chapter is that in which the dominant part of the system has a generalized Cauchy kernel. A general form of such a system of singular integral equations may be expressed as

$$\begin{aligned} A\psi(x) &+ \frac{1}{\pi}\int_a^b B\psi(t)\frac{\mathrm{d}t}{t-x} + \frac{1}{\pi}\int_a^b \sum_0^K C_k\psi(t) \\ &\times (x-a)^k \frac{\mathrm{d}^k}{\mathrm{d}x^k}(t-z_1)^{-1}\mathrm{d}t + \frac{1}{\pi}\int_a^b \sum_0^J D_j\psi(t) \\ &\times (b-x)^j \frac{\mathrm{d}^j}{\mathrm{d}x^j}(t-z_2)^{-1}\mathrm{d}t + \int_a^b K(x,t)\psi(t)\mathrm{d}t = f(x), \end{aligned}$$

$$(a<x<b) \quad (7.2)$$

where $A$, $B$, $C_k$ and $D_j$ are $(N \times N)$ matrices which are generally constant, the matrix $K(x, t)$ consists of Fredholm kernels $k_{ij}(x, t)$, $(i, j=1, \ldots, N)$, and $f=(f_i(x))$, $(i=1, \ldots, N)$ is again the input vector which satisfies a Hölder condition in $a \leqslant x \leqslant b$.

In equation (7.2) the variables $z_1$ and $z_2$ are given by

$$z_1 = a+(x-a)e^{i\theta_1}, \quad z_2 = b+(b-x)e^{i\theta_2} \tag{7.3}$$

where $\theta_1$ and $\theta_2$ are known constants with $0<\theta_1<2\pi$ and $-\pi<\theta_2<\pi$. From equations (7.3) it is seen that as $x$ varies between $a$ and $b$, $z_1$ and $z_2$ will move on the lines described by (7.3), for $a<x<b$ the kernels in the third and fourth terms on the left-hand side of equation (7.2) are bounded, as $x \to a$, $z_1 \to a$, and as $x \to b$, $z_2 \to b$. Hence, as $x$ and $t$ go to $a$ simultaneously, the kernels in the third term, and as $x$ and $t$ go to $b$ simultaneously, the kernels in the fourth term of equation (7.2) will become unbounded. These kernels are thus singular in a special sense, and (even though at only the end points) since the corresponding singularities are still of the form $x^{-1}$, the kernels are called a generalized Cauchy kernel. As will be shown in this Chapter, the nature of the singularity of the unknown functions $\psi_j$ is strongly influenced by the presence of these additional kernels.

A typical application of equation (7.2) which will be discussed in this Chapter is the problem of a crack terminating at the interface of two bonded dissimilar half planes. If there is a symmetry with respect to the plane of the crack, there is only one unknown function which corresponds to the dislocation density. In this problem, since one end of the crack is embedded into a homogeneous medium, $A$ and $D_j$ are zero. In the plane strain problem $K=2$, and in the anti-plane shear problem $K=0$. In both cases $\theta_1=\pi$. A slightly modified form of equation (7.2) is found in the formulation of the torsion problem of a shaft bonded to a disk of finite dimensions (see Erdogan and Gupta [7]). A system of singular integral equations with $A=0$, $B=0$, $D_j=0$, $\theta_{11}=\pi/2$, and $\theta_{12}=3\pi/2$, and $K=1$ (i.e., the dominant part of the system of integral equations having the kernels of the form

$$\frac{t}{t^2+x^2} \quad \text{and} \quad x\frac{d}{dx}\left(\frac{x}{t^2+x^2}\right), \qquad 0<(x, t)<\infty)$$

is found in the formulation of the elasticity problem for a semi-infinite strip $(0<y<c,\ x>0)$ clamped along the boundaries $y=0$, $y=c$. A special case of equation (7.2) has recently been treated by Bueckner [8] and Bierman [9].

One of the characteristic features of the system of singular integral equations given by (7.2) is that the dominant part of the system (i.e., the equations without the terms containing the Fredholm kernels, $k_{ij}$) cannot readily be reduced to a Riemann–Hilbert problem. Hence, to obtain the matrix of the fundamental functions (which characterize the nature of the singularity of the unknown functions $\psi_i$), a more direct application of the function theoretic methods to the integral equations will be needed.

Sections 7.2–7.5 of this Chapter are devoted to the discussion of the methods of solution of the singular integral equations of the first kind with a simple Cauchy kernel (equation (7.1) with $A \equiv 0$). The solution of the singular integral equations of the second kind, as given by (7.1) is discussed in sections 7.6–7.8. Sections 7.9–7.11 deal with the determination of the fundamental function and the development of a method of solution for the singular integral equations with a generalized Cauchy kernel as shown by equation (7.2). One or more numerical examples for each class of equations are included to demonstrate the application of the techniques developed.

The integral equations with a simple Cauchy kernel given by (7.1) may, in principle, be regularized (i.e., may be reduced to a system of Fredholm-type integral equations with weakly singular kernels), and hence, may be solved numerically by conventional methods. The regularization may be accomplished either by using the method of Muskhelishvili and Vekua (see Muskhelishvili [5]) or by following the more direct method due to Carlemann and Vekua (see Pogorzelski [10], Gakhov [11]). In the former method the singular integral equation is operated upon by another singular operator of the same form (usually the adjoint operator). In the method of Carlemann and Vekua the terms containing the Fredholm kernels $k_{ij}$ are assumed to be part of the input functions and the method of solution of the dominant equations is followed, reducing the system of singular integral equations to that of Fredholm-type integral equations. In practice the difficulty arises in evaluating the Fredholm kernels in the regularized system (even in the simple cases where the original Fredholm kernels, $k_{ij}(x, t)$ are known in closed form, which, in crack problems, is seldom the case). Therefore, the primary emphasis of this Chapter will be on developing reliable numerical methods which preserve the correct nature of the singularity of the unknown functions, and which are applicable to the most general case of singular integral equations with complicated Fredholm kernels.

## 7.2 Singular integral equations of the first kind

In equation (7.1), let $A$ be a null matrix and $B$ be non-singular. Thus, by multiplying (7.1) by $B^{-1}$, the dominant part of the system may easily be uncoupled. Therefore, any method developed for the solution of a single equation may easily be extended to apply to the most general case. For this reason, and in order to simplify the manipulations in the derivations, in sections 7.2–7.5 the numerical methods will be discussed for a single integral equation only. Also, since the techniques are heavily based on the properties of the related Chebyshev and Jacobi polynomials, without any loss in generality, it will be assumed that the interval $(a, b)$ in which the functions $\psi_i$ and $f_i$ are defined is normalized to be $(-1, 1)$. Hence, the integral equation to be considered is the following:

$$\frac{1}{\pi}\int_{-1}^{1}\frac{\phi(t)\mathrm{d}t}{t-x}+\int_{-1}^{1}k(x,t)\phi(t)\mathrm{d}t=f(x)\,,\qquad(-1<x<1)\tag{7.4}$$

where $\phi(x)$ is unknown, and $f$ and $k$ are known functions which are $H$-continuous in the closed interval $[-1, 1]$.

*The fundamental function.* Since $\phi(t)$ and $k(x, t)$ are $H$-continuous functions, the second term in equation (7.4) is a bounded function of $x$. Hence, the singular behavior of $\phi$ may be obtained by studying the dominant part of (7.4) only, namely

$$\frac{1}{\pi}\int_{-1}^{1}\frac{\phi(t)}{t-x}\,\mathrm{d}t=F(x)\,,\qquad(-1<x<1)\tag{7.5}$$

where $F$ contains the input function $f$ and the term coming from the part of the integral equation with the Fredholm kernel.

Following Muskhelishvili [5], let

$$\Phi(z)=\frac{1}{2\pi\mathrm{i}}\int_{-1}^{1}\frac{\phi(t)\mathrm{d}t}{t-z}\,.\tag{7.6}$$

The boundary values of the sectionally holomorphic function, $\Phi(z)$ are related by the following Plemelj formulas:

$$\begin{aligned}\Phi^{+}(x)-\Phi^{-}(x)&=\phi(x)\,,\\ \Phi^{+}(x)+\Phi^{-}(x)&=\frac{1}{\pi\mathrm{i}}\int_{-1}^{1}\frac{\phi(t)\mathrm{d}t}{t-x}\,.\end{aligned}\tag{7.7}$$

From equations (7.5) and (7.7) we obtain the following Riemann–Hilbert problem to determine $\Phi(z)$:

$$\Phi^+(x)+\Phi^-(x)=-\mathrm{i}F(x), \qquad (-1<x<1). \tag{7.8}$$

The solution of equation (7.8) vanishing at infinity may easily be expressed as [5]

$$\Phi(z)=-\frac{X(z)}{2\pi}\int_{-1}^{1}\frac{F(t)\mathrm{d}t}{(t-z)X^+(t)}+CX(z), \tag{7.9}$$

where $C$ is a constant and $X(z)$ is the fundamental solution of the problem satisfying the following homogeneous boundary conditions:

$$\begin{aligned} &X^+(x)+X^-(x)=0, \qquad (|x|<1),\\ &X^+(x)-X^-(x)=0, \qquad (|x|>1). \end{aligned} \tag{7.10}$$

Ignoring an arbitrary multiplicative holomorphic function, the most general solution of equation (7.10) may be expressed as

$$X(z)=(z-1)^{\frac{1}{2}+N}(z+1)^{-\frac{1}{2}+M} \tag{7.11}$$

where $X(z)$ will be taken as the branch for which $z^{-N-M}X(z)\to 1$ as $z\to\infty$. It is clear that equation (7.10) is satisfied for any $N$ and $M$ provided they are integers.

On the other hand, from equations (7.7) and (7.9) the solution of the integral equation (7.5) is found to be

$$\begin{aligned} \phi(x)=\Phi^+(x)-\Phi^-(x)&=2CX^+(x)\\ &-X^+(x)\frac{1}{\pi}\int_{-1}^{1}\frac{F(t)\mathrm{d}t}{(t-x)X^+(t)}, \qquad (-1<x<1). \end{aligned} \tag{7.12}$$

It is seen that the singular nature of the unknown function $\phi(x)$ is characterized by that of the fundamental solution $X(z)$. In studying the boundary value problems such as (7.8), the following restriction is usually imposed on the integers $N$ and $M$:

$$-1<\tfrac{1}{2}+N<1, \qquad -1<-\tfrac{1}{2}+M<1. \tag{7.13}$$

This implies that at a given end $-1$ or 1 the unknown function $\phi$ is either bounded (having a value which is, in this case, necessarily zero) or has an integrable singularity. In a very large majority of physical problems, this of course is the case, that is, $\phi$ is generally either a "potential-type" quantity (such as displacement, temperature, electrostatic potential, velocity potential), or a "flux-type" quantity (such as stress, dislocation density, heat flux, charge density, velocity). However, there may be situations in which at the end points the solution may be required to have a stronger zero (e.g., stress function or lateral plate displacement) or a stronger singularity (e.g., the transverse shear in a fourth order plate theory). It is then clear that in selecting the arbitrary integers $N$ and $M$ in (7.13), that is in determining the index of the problem which is defined by

$$\kappa = -(N+M) \tag{7.14}$$

the physical nature of the singularity of the function $\phi$ at the end points $-1$ and $+1$ has to be considered.

The real function

$$w(x) = \mathrm{i}(-1)^N X^+(x) = (1-x)^{\frac{1}{2}+N}(1+x)^{-\frac{1}{2}+M}\,, \qquad (-1<x<1) \tag{7.15}$$

obtained from equation (7.11) is known as the "fundamental function" of the singular integral equation (7.5).

*The case of* $\kappa=1$. If the function $\phi$ has integrable singularities at both ends, then $N=-1$, $M=0$ in equations (7.13). Hence, equations (7.12), (7.14), and (7.15) lead to

$$\kappa = 1\,,\quad X(z) = (z^2-1)^{-\frac{1}{2}}\,,\quad w(x) = (1-x^2)^{-\frac{1}{2}}\,. \tag{7.16}$$

In this case, an additional condition is needed to determine the arbitrary constant $C$ which appears in equations (7.9) and (7.12), and thus obtain a unique solution. This condition is provided by the physics of the problem if it is properly formulated. It is usually either an equilibrium condition (if, for example, $\phi$ is a stress component), or a single-valuedness or compatibility condition (if $\phi$ is the tangential derivative of a "potential"). In either case, in addition to equation (7.5), $\phi$ must satisfy an additional condition which is invariably of the following form

$$\int_{-1}^{1} \phi(x)\mathrm{d}x = A \tag{7.17}$$

where $A$ is a known constant which may be zero.

*The case of* $\kappa=0$. If $\phi(x)$ is bounded at one end and has an integrable singularity at the other, one has

$$\begin{aligned} &N=0\,,\ M=0\,,\ \kappa=0\,,\\ &X(z)=(z-1)^{\frac{1}{2}}(z+1)^{-\frac{1}{2}}\,,\ w(x)=(1-x)^{\frac{1}{2}}(1+x)^{-\frac{1}{2}}\,, \end{aligned} \tag{7.18}$$

or

$$\begin{aligned} &N=-1\,,\ M=1\,,\ \kappa=0\,,\\ &X(z)=(z-1)^{-\frac{1}{2}}(z+1)^{\frac{1}{2}}\,,\ w(x)=(1-x)^{-\frac{1}{2}}(1+x)^{\frac{1}{2}}\,. \end{aligned} \tag{7.19}$$

In this case, the condition at infinity requires that in equation (7.9) the constant $C$ be zero, and with this, equation (7.12) gives the unique solution.

*The case of* $\kappa=-1$. If the function $\phi$ is bounded at both ends, then

$$\begin{aligned} &N=0\,,\ M=1\,,\ \kappa=-1\,,\\ &X(z)=(z^2-1)^{\frac{1}{2}}\,,\ w(x)=(1-x^2)^{\frac{1}{2}}\,. \end{aligned} \tag{7.20}$$

In this case, the condition at infinity requires that, in addition to $C=0$, the solution must satisfy the following consistency condition [5]:

$$\int_{-1}^{1}\frac{F(t)\mathrm{d}t}{w(t)}=0\,. \tag{7.21}$$

Finally, the general solution given by equation (7.12) indicates that the unknown function $\phi(x)$ is of the following form

$$\phi(x)=w(x)g(x)\,,\qquad(-1<x<1) \tag{7.22}$$

where $g(x)$ is a bounded continuous function in the closed interval $-1\leqslant x\leqslant 1$. The next two sections will be devoted to the development of numerical methods for the determination of $g(x)$.

## 7.3 Solution by Gaussian integration formulas

Consider the following Fredholm integral equation:

$$\int_a^b w(t)k(x,t)h(t)\mathrm{d}t=f(x)\,,\qquad(a<x<b)\,. \tag{7.23}$$

If equation (7.23) has no "closed form" solution, an effective numerical solution may be obtained by using a quadrature formula of the Gaussian type to evaluate the integral on the left-hand side for appropriately selected values of $x_i$, $(i=1,\ldots,n)$, thereby reducing the problem to a system of linear algebraic equations in the unknowns $h(t_j)$, $(j=1,\ldots,n)$. Thus, (7.23) may be written as

$$\sum_{j=1}^{n} W_j k(x_i, t_j) h(t_j) + R_n(x_i) = f(x_i), \qquad (a < x_i < b, \ i = 1, \ldots, n), \quad (7.24)$$

where $W_j$, $(j=1, \ldots, n)$ are the weights and $R_n$ is the remainder. In practice, by selecting $n$ sufficiently large, $R_n$ can be made as small as required, and hence, may be neglected. The locations $t_j$, $(j=1, \ldots, n)$ correspond to the zeros of the orthogonal polynomials related to the particular Gaussian quadrature*.

It is seen that if an equivalent Gaussian integration formula for singular integrals were to be available, the singular integral equation (7.4), or more specifically, by substituting from (7.22), the integral equation

$$\frac{1}{\pi}\int_{-1}^{1} \frac{g(t)}{t-x} w(t)\mathrm{d}t + \int_{-1}^{1} k(x, t) g(t) w(t)\mathrm{d}t = f(x), \quad (-1 < x < 1) \quad (7.25)$$

could be solved by a method similar to (7.24), that is, (7.25) could be reduced to a system of linear algebraic equations in $g(t_j)$, $(j=1, \ldots, n)$. In this section it will be shown that appropriate Gaussian integration formulas for singular integrals can be developed and equation (7.25) can be reduced to the following algebraic system:

$$\frac{1}{\pi}\sum_{j=1}^{n} W_j g(t_j)\left[\frac{1}{t_j - x_i} + \pi k(x_i, t_j)\right] = f(x_i), \qquad (i = 1, \ldots, n-\kappa). \qquad (7.26)$$

Here, the related orthogonal polynomials are selected in such a way that the fundamental function $w(x)$ is the corresponding weight function. Since, by equation (7.15) $w(x)$ is of the general form

$$w(x) = (1-x)^{\alpha}(1+x)^{\beta}, \qquad (-1 < (\alpha, \beta) < 1, \ |x| < 1), \qquad (7.27)$$

Jacobi polynomials $P_k^{(\alpha,\beta)}(x)$ are the related orthogonal polynomials**, and it will be shown that $t_j$ and $x_i$ are determined from

$$\begin{aligned}
&P_n^{(\alpha,\beta)}(t_j) = 0, \\
&P_{n-\kappa}^{(\alpha+\kappa,\beta+\kappa)}(x_i) = 0, \quad \text{for} \quad \kappa = \mp 1, \\
&P_n^{(\alpha+1,\beta-1)}(x_i) = 0, \quad \text{for} \quad \kappa = 0, \ -1 < \alpha < 0, \ 0 < \beta < 1, \\
&P_n^{(\alpha-1,\beta+1)}(x_i) = 0, \quad \text{for} \quad \kappa = 0, \ 0 < \alpha < 1, \ -1 < \beta < 0.
\end{aligned} \qquad (7.28)$$

* For a list of Gaussian quadrature formulas see Abramowitz and Stegun [12]. See also Stroud and Secrest [13].

** which reduce to Chebyshev polynomials of first ($T_n$) and second ($U_n$) kind for $\kappa=1$ and $\kappa=-1$, respectively.

In equations (7.26) the constants $W_j$ are, again, the appropriate weights corresponding to the particular quadrature formula. The development of the Gauss–Jacobi integration formulas is, in part, based on the following relations (see Szegö [14])

$$\frac{1}{\pi}\int_{-1}^{1} P_n^{(\alpha,\beta)}(t)\,w(t)\frac{dt}{t-x} = -\frac{\Gamma(\alpha)\Gamma(1-\alpha)}{\pi}\,2^{-\kappa}P_{n-\kappa}^{(-\alpha,-\beta)}(x), \quad (-1<x<1), \tag{7.29}$$

for $w$ and $(\alpha, \beta)$ as given by equation (7.27) (where $\kappa = -(\alpha+\beta)$ is $-1$, 0, or 1, otherwise $\alpha$ and $\beta$ may be arbitrary), and

$$\frac{d^m}{dx^m}\left[P_n^{(\alpha,\beta)}(x)\right] = \frac{1}{2^m}\frac{\Gamma(n+m+\alpha+\beta+1)}{\Gamma(n+\alpha+\beta+1)}P_{n-m}^{(\alpha+m,\beta+m)}(x) \tag{7.30}$$

which is valid for $-1<(\alpha, \beta)<1$.

*Gauss–Jacobi integration formulas for singular integrals,* $\kappa=0$. Consider the following integral

$$S(x) = \frac{1}{\pi}\int_{-1}^{1} g(t)\,w(t)\frac{dt}{t-x}, \qquad (-1<x<1) \tag{7.31}$$

where $g$ is bounded in the closed interval $-1 \leqslant t \leqslant 1$, and $w$ is given by

$$w(t) = (1-t)^{-\frac{1}{2}}(1+t)^{\frac{1}{2}}. \tag{7.32}$$

Noting that $w(t)$ is the weight function of the Jacobi polynomials $P_j^{(-\frac{1}{2},\frac{1}{2})}(t)$, it is assumed that the function $g$ can be approximated with a sufficient degree of accuracy by the following truncated series:

$$g(t) \simeq \sum_{0}^{p} B_j P_j^{(-\frac{1}{2},\frac{1}{2})}(t), \qquad (-1<t<1). \tag{7.33}$$

From equations (7.29), (7.31) and (7.33) it follows that

$$S(x) = \sum_{0}^{p} B_j P_j^{(\frac{1}{2},-\frac{1}{2})}(x). \tag{7.34}$$

Now consider the following simple fraction expansion:

$$-\frac{P_n^{(\frac{1}{2},-\frac{1}{2})}(x)P_j^{(-\frac{1}{2},\frac{1}{2})}(x)+P_n^{(-\frac{1}{2},\frac{1}{2})}(x)P_j^{(\frac{1}{2},-\frac{1}{2})}(x)}{P_n^{(-\frac{1}{2},\frac{1}{2})}(x)} = \sum_{1}^{n}\frac{a_i}{t_i-x} \tag{7.35}$$

where

$$j<n\,,\quad P_n^{(-\frac{1}{2},\frac{1}{2})}(t_i)=0\,,\qquad (i=1,\ldots,n)\,. \tag{7.36}$$

By using the relations [12]

$$\begin{aligned} P_n^{(-\frac{1}{2},\frac{1}{2})}(2z^2-1) &= \frac{\Gamma(n+\frac{1}{2})}{n!\,\pi^{\frac{1}{2}}}\,\frac{1}{z}\,T_{2n+1}(z)\,,\\ P_n^{(\frac{1}{2},-\frac{1}{2})}(2z^2-1) &= \frac{\Gamma(n+\frac{1}{2})}{n!\,\pi^{\frac{1}{2}}}\,U_{2n}(z)\,, \end{aligned} \tag{7.37}$$

and the recursion formula (see Erdelyi [15])

$$U_{k-i-1}(x)=T_i(x)\,U_{k-1}(x)-T_k(x)\,U_{i-1}(x) \tag{7.38}$$

the left-hand side of equation (7.35) can be modified and (7.35) may be expressed as

$$-\frac{\Gamma(j+\frac{1}{2})}{j!\,\pi^{\frac{1}{2}}}\,\frac{U_{2n-2j-1}(y)}{T_{2n+1}(y)}=\sum_1^n\frac{a_i}{t_i-x}\,,\qquad (x=2y^2-1)\,. \tag{7.39}$$

From equation (7.39), it is seen that in the rational function on the left-hand side of (7.35) the degree of the numerator is less than the degree of the denominator; hence the partial fraction expansion indicated in equation (7.35) is possible. By using equations (7.36), from (7.35) the coefficients $a_i$ may be obtained as

$$a_i=\frac{P_n^{(\frac{1}{2},-\frac{1}{2})}(t_i)\,P_j^{(-\frac{1}{2},\frac{1}{2})}(t_i)}{P_n^{\prime(-\frac{1}{2},\frac{1}{2})}(t_i)}\,. \tag{7.40}$$

On the other hand, using the recursion relations [14] for the Jacobi polynomials, it can be shown that, if $t_i$ are given by $P_n^{(\alpha,\beta)}(t_i)=0$, the derivative (7.30) at $x=t_i$ may be expressed as

$$\begin{aligned} P_n^{\prime(\alpha,\beta)}(t_i) &= \frac{n+\beta}{1+t_i}\,P_n^{(\alpha+1,\beta-1)}(t_i)\\ &= -\frac{n+\alpha}{1-t_i}\,P_n^{(\alpha-1,\beta+1)}(t_i)\,. \end{aligned} \tag{7.41}$$

Thus, from equations (7.40) and (7.41) we obtain

$$a_i=\frac{2(1+t_i)}{2n+1}\,P_j^{(-\frac{1}{2},\frac{1}{2})}(t_i)\,. \tag{7.42}$$

Now if a discrete set of points $x_k$ $(k=1, \ldots, n)$ is selected such that

$$P_n^{(\frac{1}{2}, -\frac{1}{2})}(x_k) = 0\,, \qquad (k = 1, \ldots, n) \tag{7.43}$$

equations (7.35) and (7.42) lead to

$$\sum_{i=1}^{n} \frac{2(1+t_i)}{2n+1} P_j^{(-\frac{1}{2}, \frac{1}{2})}(t_i) \frac{1}{t_i - x_k} = P_j^{(\frac{1}{2}, -\frac{1}{2})}(x_k)\,. \tag{7.44}$$

Substituting from equations (7.44), (7.34) becomes

$$S(x) = \sum_{i=1}^{n} \sum_{j=0}^{p} B_j P_j^{(-\frac{1}{2}, \frac{1}{2})}(t_i) \frac{2(1+t_i)}{2n+1} \frac{1}{t_i - x_k}\,, \tag{7.45}$$

or using equations (7.32) and (7.33) it is found that

$$\frac{1}{\pi} \int_{-1}^{1} \left(\frac{1+t}{1-t}\right)^{\frac{1}{2}} \frac{g(t)}{t - x_k}\, \mathrm{d}t = \sum_{i=1}^{n} \frac{2(1+t_i)}{2n+1} \frac{g(t_i)}{t_i - x_k} \tag{7.46}$$

where

$$\begin{aligned} &P_n^{(-\frac{1}{2}, \frac{1}{2})}(t_i) = 0\,, \ t_i = \cos\left(\frac{2i-1}{2n+1}\pi\right), \qquad (i = 1, \ldots, n)\,, \\ &P_n^{(\frac{1}{2}, -\frac{1}{2})}(x_k) = 0\,, \ x_k = \cos\left(\frac{2k\pi}{2n+1}\right), \qquad (k = 1, \ldots, n)\,. \end{aligned} \tag{7.47}$$

If one considers the Gauss–Jacobi integration formula for a bounded continuous function $f(t)$ given by [12]

$$\frac{1}{\pi} \int_{-1}^{1} f(t) \left(\frac{1+t}{1-t}\right)^{\frac{1}{2}} \mathrm{d}t \simeq \sum_{1}^{n} \frac{2(t_i+1)}{2n+1} f(t_i) \tag{7.48}$$

where $t_i$ is given by equations (7.47), it is seen that formally equation (7.46) may be considered as a Gauss–Jacobi integration formula for singular integrals which is valid only at certain discrete points $x_k$, $(k=1, \ldots, n)$. Note that if there is no approximation in equation (7.33), and if $n > p$, then the formula (7.46) gives the exact value of the integral at $x = x_k$.

Using now equations (7.46) and (7.48), the integral equation (7.25) may be reduced to the following system of linear algebraic equations in $g(t_i)$

$$\sum_{i=1}^{n} \frac{2(1+t_i)}{2n+1} g(t_i) \left[\frac{1}{t_i - x_k} + \pi k(x_k, t_i)\right] = f(x_k)\,, \qquad (k = 1, \ldots, n) \tag{7.49}$$

where $t_i$ and $x_k$ are given by (7.47).

In the second case of $\kappa=0$, i.e., for $\alpha=\frac{1}{2}$, $\beta=-\frac{1}{2}$, instead of equation (7.32) the weight function is given by

$$w(t)=(1-t)^{\frac{1}{2}}(1+t)^{-\frac{1}{2}}. \tag{7.50}$$

In this case, instead of equation (7.39), if the following partial fraction expansion is considered:

$$-\frac{\Gamma(j+\frac{1}{2})}{j!\,\pi^{\frac{1}{2}}}\frac{U_{2n-2j-1}(y)}{yU_{2n}(y)}=\sum_{1}^{n}\frac{a_i}{t_i-x},$$
$$x=2y^2-1\,,\quad P_n^{(\frac{1}{2},-\frac{1}{2})}(t_i)=0\,, \tag{7.51}$$

then it may be easily shown that

$$\frac{1}{\pi}\int_{-1}^{1}\left(\frac{1-t}{1+t}\right)^{\frac{1}{2}}\frac{g(t)}{t-x_k}\,\mathrm{d}t=\sum_{i=1}^{n}\frac{2(1-t_i)}{2n+1}\frac{g(t_i)}{t_i-x_k} \tag{7.52}$$

where $t_i$ and $x_k$ are now given by

$$P_n^{(\frac{1}{2},-\frac{1}{2})}(t_i)=0\,,\quad t_i=\cos\left(\frac{2i\pi}{2n+1}\right),\qquad (i=1,\dots,n)\,,$$
$$P_n^{(-\frac{1}{2},\frac{1}{2})}(x_k)=0\,,\quad x_k=\cos\left(\frac{2k-1}{2n+1}\pi\right),\qquad (k=1,\dots,n)\,. \tag{7.53}$$

Thus, equations (7.25) may be reduced to

$$\sum_{i=1}^{n}\frac{2(1-t_i)}{2n+1}g(t_i)\left[\frac{1}{t_i-x_k}+\pi k(x_k,t_i)\right]=f(x_k)\,,\qquad (k=1,\dots,n) \tag{7.54}$$

with $t_i$ and $x_k$ as given by (7.53).

*Gauss–Chebyshev integration formula for* $\kappa=1$. If the solution of the integral equation (7.4), or (7.25) has integrable singularities at the end points, as shown in section 7.2, $\kappa=1$, $\alpha=-\frac{1}{2}=\beta$ and $w=(1-t^2)^{-\frac{1}{2}}$. In this case the related orthogonal polynomials $P_n^{(\alpha,\beta)}(t)$ reduce to $T_n(t)$, the Chebyshev polynomials of the first kind. For this problem, following a procedure similar to that just presented, and substituting from (7.22), the integral equation (7.4) and the additional condition (7.17) may be expressed as

$$\sum_{i=1}^{n}\frac{1}{n}g(t_i)\left[\frac{1}{t_i-x_k}+\pi k(x_k,t_i)\right]=f(x_k)\,, \tag{7.55}$$

$$T_n(t_i) = 0\,, \quad t_i = \cos\left(\pi\,\frac{2i-1}{2n}\right), \qquad (i = 1, \ldots, n)\,,$$

$$U_{n-1}(x_k) = 0\,, \quad x_k = \cos\frac{\pi k}{n}\,, \qquad (k = 1, \ldots, n-1)\,, \tag{7.56}$$

$$\sum_1^n \frac{\pi}{n}\, g(t_i) = A\,. \tag{7.57}$$

Equations (7.55) and (7.57) provide $n$ linear algebraic equations to determine $g(t_i)$, $(i = 1, \ldots, n)$. The details of the derivation of equations (7.55) to (7.57) are given by Erdogan and Gupta [16].

*Gauss–Chebyshev integration formula for* $\kappa = -1$. If the function $\phi(t)$ is bounded at both ends, in section 7.2 it was shown that $\kappa = -1$, $\alpha = \frac{1}{2} = \beta$, and $w(t) = (1-t^2)^{\frac{1}{2}}$. Here the related Jacobi polynomials, $P_n^{(\frac{1}{2},\frac{1}{2})}(t)$ reduce to the Chebyshev polynomials of the second kind, $U_n(t)$, and one can again develop a Gauss–Chebyshev integration formula for the singular integral involving the particular weight function (see [16] for details). Thus, with equation (7.22), (7.4) may be expressed as

$$\sum_{i=1}^{n} \frac{1-t_i^2}{n+1}\, g(t_i) \left[\frac{1}{t_i - x_k} + \pi k(x_k, t_i)\right] = f(x_k)\,, \tag{7.58}$$

$$U_n(t_i) = 0\,, \qquad t_i = \cos\left(\frac{i\pi}{n+1}\right), \qquad (i = 1, \ldots, n)\,,$$

$$T_{n+1}(x_k) = 0\,, \quad x_k = \cos\left(\frac{\pi}{2}\,\frac{2k-1}{n+1}\right), \qquad (k = 1, \ldots, n+1)\,. \tag{7.59}$$

By assuming that

$$g(t) \simeq \sum_0^p B_j U_j(t) \tag{7.60}$$

and using the relations

$$\frac{1}{\pi}\int_{-1}^{1} U_j(t)(1-t^2)^{\frac{1}{2}}\,\frac{\mathrm{d}t}{t-x} = -T_{j+1}(x)\,, \qquad (-1 < x < 1)\,, \tag{7.61}$$

$$\frac{1}{\pi}\int_{-1}^{1} T_i(t)\,T_k(t)(1-t^2)^{-\frac{1}{2}}\mathrm{d}t = \begin{cases} 0\,, & (i \neq k) \\ 1\,, & (i = k = 0) \\ \frac{1}{2}\,, & (i = k > 0)\,. \end{cases} \tag{7.62}$$

From equation (7.4) it is found that

$$\int_{-1}^{1}\left[f(x)-\int_{-1}^{1}k(x,t)\phi(t)\mathrm{d}t\right]\frac{\mathrm{d}x}{(1-x^2)^{\frac{1}{2}}}$$
$$=-\sum_{0}^{p}B_j\int_{-1}^{1}T_{j+1}(x)\frac{\mathrm{d}x}{(1-x^2)^{\frac{1}{2}}}=0\,. \tag{7.63}$$

Thus, using the present technique to solve the integral equation, the consistency condition (7.21) is seen to be automatically satisfied.

Equation (7.58) is satisfied for $n+1$ values of $x_k$. Since there are only $n$ unknowns, $g(t_i)$, in (7.58) an equation corresponding to one of the $x_k$'s must be ignored. In practice, the most harmless point to neglect would be the one closest to $x=0$, and usually $n$ is selected to be an even number and the equation corresponding to $x_{1+n/2}=0$ is ignored.

## 7.4 Solution by Jacobi polynomials

Another method to solve the singular integral equation (7.4) numerically would be to reduce it to an infinite system of linear algebraic equations. In (7.4) let

$$-1<(\alpha,\beta)<1\,,\quad \alpha+\beta=-\kappa=(0\,,\ \mp 1)\,,$$
$$\phi(t)=w(t)g(t)\,,\quad w(t)=(1-t)^{\alpha}(1+t)^{\beta}\,, \tag{7.64}$$
$$g(t)=\sum_{0}^{\infty}B_jP_j^{(\alpha,\beta)}(t)\,.$$

Substituting from equations (7.64) into (7.4) and using (7.29) we obtain

$$\sum_{0}^{\infty}B_j\left[-\frac{2^{-\kappa}}{\sin\pi\alpha}P_{j-\kappa}^{(-\alpha,-\beta)}(x)+p_j(x)\right]=f(x)\,,\qquad(-1<x<1) \tag{7.65}$$

where

$$p_j(x)=\int_{-1}^{1}k(x,t)P_j^{(\alpha,\beta)}(t)w(t)\mathrm{d}t\,. \tag{7.66}$$

The functional equation (7.65) may easily be reduced to an infinite system of algebraic equations in the unknown coefficients $B_j$ by multiplying both sides of (7.65) by $w(-\alpha,-\beta,t)P_k^{(-\alpha,-\beta)}(t)$, ($k=0, 1, 2, \ldots$) and integrating in $(-1, 1)$. First, note the following orthogonality relations [12]:

$$\int_{-1}^{1} P_n^{(\alpha,\beta)}(t)\, P_k^{(\alpha,\beta)}(t)\, w(t)\,\mathrm{d}t = \begin{cases} 0\,, & n \neq k \\ \theta_k^{(\alpha,\beta)}\,, & n = k \end{cases} \qquad (k = 0, 1, \ldots) \tag{7.67}$$

where

$$\theta_k^{(\alpha,\beta)} = \frac{2^{\alpha+\beta+1}}{2k+\alpha+\beta+1} \frac{\Gamma(k+\alpha+1)\,\Gamma(k+\beta+1)}{k!\,\Gamma(k+\alpha+\beta+1)}\,, \qquad (k = 0, 1, \ldots)\,. \tag{7.68}$$

Equation (7.68) is valid in all cases except for $k=0$, $\kappa=1$ for which, noting that $P_0^{(\alpha,\beta)}(t)=1$, one has (Gradshteyn and Ryzhik [17])

$$\theta_0^{(\alpha,\beta)} = \int_{-1}^{1} w(t)\,\mathrm{d}t = \frac{2^{\alpha+\beta+1}\,\Gamma(\alpha+1)\,\Gamma(\beta+1)}{\Gamma(\alpha+\beta+2)}\,. \tag{7.69}$$

Thus, using equations (7.67) and defining the constants

$$\begin{aligned} b_{kj} &= \int_{-1}^{1} P_k^{(-\alpha,-\beta)}(x)\, p_j(x)(1-x)^{-\alpha}(1+x)^{-\beta}\,\mathrm{d}x\,, \\ c_k &= \int_{-1}^{1} f(x)\, P_k^{(-\alpha,-\beta)}(x)(1-x)^{-\alpha}(1+x)^{-\beta}\,\mathrm{d}x \end{aligned} \tag{7.70}$$

from (7.65) it is found that

$$-\frac{2^{-\kappa}}{\sin \pi\alpha}\,\theta_k^{(-\alpha,-\beta)}\, B_{k+\kappa} + \sum_{j=0}^{\infty} b_{kj} B_j = c_k\,, \qquad (k = 0, 1, \ldots) \tag{7.71}$$

where in the case of $\kappa=-1$, $B_{-1}=0$.

The infinite system may be solved by the method of reduction (i.e., by truncating the series at the $n$ th term and in equations (7.71) considering only the first $n$ equations) provided the system is completely or quasi regular (Kantorovich and Krylov [18]). If $\kappa=0$ then the $n\times n$ equations obtained from (7.71) will give a unique solution. For $\kappa=1$ equations (7.71) contain one more unknown than the number of equations, and the equilibrium or compatibility condition (7.17) provides the necessary additional equation. Substituting from equations (7.64) into (7.17), and using the orthogonality relations (7.67) to (7.69), it is easily seen that

$$B_0 = A/\theta_0^{(\alpha,\beta)} \tag{7.72}$$

and the remaining constants $B_1, \ldots, B_n$ are determined from (7.71) by letting $k=0, 1, \ldots, n-1$.

For $\kappa=-1$ the assumed solution must satisfy the consistency condition

(7.21). Since $B_{-1}=0$, $P_0^{(-\alpha,-\beta)}(x)=1$, from equations (7.64) to (7.66), and (7.70) it is seen that the first equation in (7.71) obtained by letting $k=0$ is equivalent to the condition (7.21). This means that by using this method, the consistency condition is automatically satisfied.

The special case of the technique described in this section for $\alpha=-\frac{1}{2}=\beta$ and $\alpha=\frac{1}{2}=\beta$ is considered by Erdogan [6]. In this case the related orthogonal polynomials are $T_n(x)$, $U_n(x)$, the Chebyshev polynomials of first and second kind, respectively. As a result, the manipulations are somewhat simpler and the logarithmic part of $k(x,t)$ (if there is one) can easily be treated separately. The results for these special cases $\alpha=-\frac{1}{2}=\beta$ and $\alpha=\frac{1}{2}=\beta$ may be obtained from those given in this section by simply substituting

$$
\begin{aligned}
P_n^{(-\frac{1}{2},-\frac{1}{2})}(x) &= \frac{\Gamma(n+\frac{1}{2})}{n!\,\pi^{\frac{1}{2}}}\,T_n(x)\,,\\
P_n^{(\frac{1}{2},\frac{1}{2})}(x) &= \frac{2\Gamma(n+\frac{3}{2})}{(n+1)!\,\pi^{\frac{1}{2}}}\,U_n(x)\,.
\end{aligned}
\tag{7.73}
$$

It should be noted that the main part of the numerical work in this technique involves the evaluation of the constants $b_{kj}$ and $c_k$ defined by equations (7.70). However, since the related integrals are of the Gauss–Jacobi type, these constants may be evaluated rather accurately without an extensive computational effort. The quadrature formula which should be used for this purpose is the following* [13]:

$$
\begin{aligned}
&\int_{-1}^{1} F(x)(1-x)^{-\alpha}(1+x)^{-\beta}\,\mathrm{d}x \simeq \sum_{k=1}^{N} A_k F(x_k)\,,\\
&P_N^{(-\alpha,-\beta)}(x_k)=0\,,\qquad (k=1,\ldots,N)\\
&A_k = -\frac{2N-\alpha-\beta+2}{(N+1)!\,(N-\alpha-\beta+1)}\,\frac{\Gamma(N-\alpha+1)\,\Gamma(N-\beta+1)}{\Gamma(N-\alpha-\beta+1)}\times\\
&\qquad\times\frac{2^{-\alpha-\beta}}{P_N'^{(-\alpha,-\beta)}(x_k)\,P_{N+1}^{(-\alpha,-\beta)}(x_k)}\,.
\end{aligned}
\tag{7.74}
$$

## 7.5 Examples

In this section the solution of some crack problems will be given in order to demonstrate the application of the methods described in sections 7.3 and 7.4.

* For the special cases $(\alpha,\beta)=(\mp\frac{1}{2},\mp\frac{1}{2})$ see the quadrature formulas given in section 7.3.

*A finite crack perpendicular to the interface in two bonded half planes.* Consider the plane elasticity problem described in Figure 7.1. Assume that the composite plane is symmetrically loaded and, through superposition, the perturbation problem is separated in such a way that in Figure 7.1 the normal

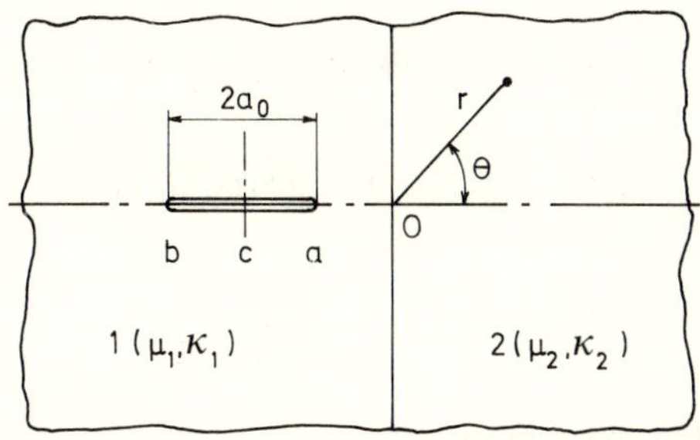


Figure 7.1. A finite crack perpendicular to the bi-material interface.

component of the crack surface traction is the only external load. Thus, the problem has a plane of symmetry, and it is sufficient to consider only one half ($r \geqslant 0$, $0 < \theta < \pi$) of the medium. To derive the integral equation, first consider the problem under the following boundary conditions:

$$\sigma_{2r\theta} = 0\,, \qquad u_{2\theta} = 0\,, \qquad (\theta = 0\,,\ 0 < r < \infty)\,, \tag{7.75}$$

$$\sigma_{1\theta\theta} = \sigma_{2\theta\theta}\,, \quad \sigma_{1r\theta} = \sigma_{2r\theta}\,, \qquad (\theta = \pi/2\,,\ 0 < r < \infty)\,, \tag{7.76}$$

$$u_{1r} = u_{2r}\,, \qquad u_{1\theta} = u_{2\theta}\,, \qquad (\theta = \pi/2\,,\ 0 < r < \infty)\,, \tag{7.77}$$

$$\sigma_{1r\theta} = 0\,, \qquad \frac{\partial u_{1\theta}}{\partial r} = -\tfrac{1}{2}\phi\delta(r - r_0)\,, \qquad (\theta = \pi\,,\ 0 < r < \infty) \tag{7.78}$$

where $\phi$ is a constant. This corresponds to two bonded half planes containing an edge dislocation* at $r = r_0$, $\theta = \pi$.

Using the basic two-dimensional wedge solution (see, for example, Hein and Erdogan [19]), all the desired field quantities may be expressed in terms of Mellin inversion integrals. In particular, by evaluating the related inversion integral for $\sigma_{1\theta\theta}(r, \pi)$ by using the method of residues and summing the

* More appropriately, equations (7.78) may be expressed as

$$\sigma_{1r\theta} = 0\,, \quad \frac{\partial}{\partial r} u_{1\theta}(r, \pi + 0) - \frac{\partial}{\partial r} u_{1\theta}(r, \pi - 0) = \phi\delta(r - r_0)\,.$$

resulting infinite series, after some lengthy manipulations it is found that★ (see Cook [20] for details)

$$\sigma_{1\theta\theta}(r, r_0, \pi) = \phi \frac{2\mu_1}{\pi(1+\kappa_1)}\left\{\frac{1}{r_0 - r} + \frac{1}{2(1+m\kappa_1)(m+\kappa_2)} \times \right.$$

$$\times \left[\frac{1}{r_0+r}\{(1+m\kappa_1)(m+\kappa_2) - m(1+\kappa_1)(1+m\kappa_1)\right.$$

$$- 3(1-m)(m+\kappa_2)\} + 12(1-m)(m+\kappa_2)\frac{r}{(r_0+r)^2}$$

$$\left.\left. - 8(1-m)(m+\kappa_2)\frac{r^2}{(r_0+r)^3}\right]\right\} \tag{7.79}$$

where $\mu_1$, $\mu_2$ are the shear moduli, $\kappa_i = (3-\nu_i)/(1+\nu_i)$ for generalized plane stress, $\kappa_i = 3-4\nu_i$ for plane strain, $\nu_i$, $(i=1, 2)$ being the Poisson's ratio, and $m = \mu_2/\mu_1$.

Now, if the medium contains a crack along $\theta = \pi$, $a < r < b$, clearly $\phi = \phi(r_0)$ is the unknown function satisfying

$$\phi(r_0) = 0\,, \qquad (0 < r_0 < a\,,\ b < r_0 < \infty)\,, \tag{7.80}$$

$$\int_0^\infty \sigma_{1\theta\theta}(r, r_0, \pi)\mathrm{d}r_0 = p(r)\,, \qquad (a < r < b) \tag{7.81}$$

where $\sigma_{1\theta\theta}(r, r_0, \pi)$ is the "Green's Function" given by equation (7.79) with $\phi = \phi(r_0)$, and

$$p(r) = \sigma_{1\theta\theta}(r, \pi)\,, \qquad (a < r < b) \tag{7.82}$$

is the crack surface traction. Using equations (7.79) and (7.80), the integral equation (7.81) may be expressed as

$$\frac{1}{\pi}\int_a^b \frac{\phi(r_0)}{r_0 - r}\mathrm{d}r_0 + \frac{1}{\pi}\int_a^b H(r, r_0)\phi(r_0)\mathrm{d}r_0 = \frac{1+\kappa_1}{2\mu_1}p(r)\,, \qquad (a < r < b) \tag{7.83}$$

★ In the inversion, for $r > r_0$ and $r < r_0$ the contour is closed respectively to the right, and to the left of the line $s = c$ going through the strip of regularity which gives different expressions for $\sigma_{1\theta\theta}$ in the two intervals. However, after summing the infinite series, the two expressions become identical.

$$H(r, r_0) = \frac{1}{2(1+m\kappa_1)(m+\kappa_2)} \left\{ \frac{1}{r_0+r} [(1+m\kappa_1)(m+\kappa_2) - m(1+\kappa_1)(1+m\kappa_1) - 3(1-m)(m+\kappa_2)] + 12(1-m)(m+\kappa_2) \frac{r}{(r_0+r)^2} - 8(1-m)(m+\kappa_2) \frac{r^2}{(r_0+r)^3} \right\}. \tag{7.84}$$

For $a>0$, the kernel $H(r, r_0)$ is a bounded function, and the singular integral equation (7.83) is identical to (7.4) the solution of which is discussed in the previous sections. Since the Green's functions similar to (7.79) can easily be evaluated for all the desired field quantities, these quantities may all be expressed as definite integrals similar to (7.83) and may be calculated once the density function $\phi(r_0)$ is determined. Note that for the homogeneous medium $m=1$ which gives $H(r, r_0)=0$ in equation (7.84), and (7.83) reduces to the singular integral equation for a plane containing a crack along $a<r<b$. Similarly, with a change in variable $r=c+r', r_0=c+r_0', c=(b+a)/2$, it is easy to show that for a fixed crack length, $b-a$, as $c\to\infty$, $H(r, r_0)\to 0$, and the problem again reduces to that of a homogeneous plane.

To normalize the interval $(a, b)$ let (see Figure 7.1)

$$x=(r-c)/a_0, \quad t=(r_0-c)/a_0, \quad a_0=(b-a)/2,$$

$$\frac{1+\kappa_1}{2\mu_1} p(r)=f(x), \quad \phi(r_0)=F(t), \tag{7.85}$$

$$\frac{1}{\pi} H(r, r_0) = \frac{1}{a_0} k(x, t).$$

Thus, equation (7.83) becomes

$$\frac{1}{\pi}\int_{-1}^{1} \frac{F(t)\mathrm{d}t}{t-x} + \int_{-1}^{1} k(x, t) F(t)\mathrm{d}t = f(x), \qquad (|x|<1). \tag{7.86}$$

The density function $\phi(r)=\partial u_{1\theta}/\partial r$ or $F(t)$ is a "flux-type" quantity. Therefore it has integrable singularities at the end points, and from section 7.2 it follows that

$$\kappa=1, \quad \alpha=-\tfrac{1}{2}=\beta, \quad w=(1-t^2)^{-\frac{1}{2}}. \tag{7.87}$$

The solution will then be determinate within an arbitrary constant $C$ (see equation (7.12)). Noting that in equation (7.83) the $r$-derivative of $u_{1\theta}$ rather than $u_{1\theta}$ itself is assumed to vanish outside the interval $(a, b)$, theoretically the constant $C$ must be determined from (or, the numerical solution of (7.86) must be obtained under) the following compatibility condition

$$\int_{-1}^{1} F(t)\mathrm{d}t = 0\,. \tag{7.88}$$

If we let

$$F(t) = g(t)(1-t^2)^{-\frac{1}{2}} \tag{7.89}$$

from the analysis given in section 7.2, or from the study of the singular behavior of stresses and displacements around the crack tips, the stress intensity factors may be obtained as

$$\begin{aligned} k(a) &= \lim_{r\to a}\,[2(a-r)]^{\frac{1}{2}}\sigma_{1\theta\theta}(r,\pi) = \frac{2\mu_1}{1+\kappa_1}\lim_{r\to a}\,[2(r-a)]^{\frac{1}{2}}\phi(r) \\ &= \frac{2\mu_1}{1+\kappa_1}\,a_0^{\frac{1}{2}}\,g(-1)\,, \\ k(b) &= \lim_{r\to b}\,[2(r-b)]^{\frac{1}{2}}\sigma_{1\theta\theta}(r,\pi) = -\frac{2\mu_1}{1+\kappa_1}\lim_{r\to b}\,[2(b-r)]^{\frac{1}{2}}\phi(r) \\ &= -\frac{2\mu_1}{1+\kappa_1}\,a_0^{\frac{1}{2}}\,g(1)\,. \end{aligned} \tag{7.90}$$

The numerical solution of the singular integral equation (7.86) is obtained by using the Gauss–Chebyshev integration formula (7.55) given in section 7.3. The results are shown in Tables I, II, and III. Table I shows the stress

TABLE I

*Stress intensity factors for the crack perpendicular to the boundary of a half plane*

| $c/a_0$ | $k(b)/p_0(a_0)^{\frac{1}{2}}$ | $k(a)/p_0(a_0)^{\frac{1}{2}}$ |
|---|---|---|
| 1.00 | (1.590) | $(\infty)$ |
| 1.01 | 1.330 | 3.720 |
| 1.05 | 1.254 | 2.159 |
| 1.10 | 1.211 | 1.759 |
| 1.15 | 1.183 | 1.575 |
| 1.20 | 1.163 | 1.464 |
| 1.25 | 1.146 | 1.388 |
| 1.50 | 1.097 | 1.204 |
| 2.00 | 1.054 | 1.091 |
| 5.00 | 1.009 | 1.011 |
| 10.00 | 1.002 | 1.003 |
| $\infty$ | 1.000 | 1.000 |

TABLE II

*Stress intensity factors for two materials*

Epoxy-aluminum, $\mu_2/\mu_1 = 23.08$

| | *Plane stress* | | | | *Plane strain* | | | |
|---|---|---|---|---|---|---|---|---|
| | $p=-p_0$ | | $p=-p_0$ | | $p=-p_0 x$ | | $p=-p_0 x^2$ | |
| $c/a_0$ | $\frac{k(b)}{p_0(a_0)^{\frac{1}{2}}}$ | $\frac{k(a)}{p_0(a_0)^{\frac{1}{2}}}$ | $\frac{k(b)}{p_0(a_0)^{\frac{1}{2}}}$ | $\frac{k(a)}{p_0(a_0)^{\frac{1}{2}}}$ | $\frac{k(b)}{p_0(a_0)^{\frac{1}{2}}}$ | $\frac{k(a)}{p_0(a_0)^{\frac{1}{2}}}$ | $\frac{k(b)}{p_0(a_0)^{\frac{1}{2}}}$ | $\frac{k(a)}{p_0(a_0)^{\frac{1}{2}}}$ |
| 1.00 | 0.8789 | 4.1760 | 0.8827 | 2.6237 | 0.5110 | −1.6867 | 0.4699 | 1.6660 |
| 1.10 | 0.8957 | 0.6485 | 0.8985 | 0.6674 | 0.5084 | −0.4058 | 0.4743 | 0.3916 |
| 1.15 | 0.9027 | 0.7046 | 0.9051 | 0.7178 | 0.5075 | −0.4300 | 0.4760 | 0.4133 |
| 1.25 | 0.9148 | 0.7764 | 0.9165 | 0.7838 | 0.5062 | −0.4564 | 0.4790 | 0.4380 |
| 2.00 | 0.9614 | 0.9344 | 0.9616 | 0.9349 | 0.5019 | −0.4942 | 0.4904 | 0.4833 |
| 5.00 | 0.9929 | 0.9912 | 0.9929 | 0.9912 | 0.5002 | −0.4997 | 0.4982 | 0.4978 |
| 10.00 | 0.9981 | 0.9979 | 0.9981 | 0.9979 | 0.5000 | −0.4999 | 0.4995 | 0.4995 |
| Aluminum-epoxy, $\mu_2/\mu_1 = 0.043$ | | | | | | | | |
| 1.00 | 1.3525 | 0.0744 | 1.3552 | 0.0700 | 0.4243 | −0.0303 | 0.6090 | 0.0314 |
| 1.10 | 1.1806 | 1.6451 | 1.1817 | 1.6487 | 0.4799 | −0.6692 | 0.5471 | 0.7032 |
| 1.15 | 1.1581 | 1.4942 | 1.1591 | 1.4971 | 0.4842 | −0.6161 | 0.5407 | 0.6489 |
| 1.25 | 1.1275 | 1.3369 | 1.1283 | 1.3389 | 0.4892 | −0.5661 | 0.5325 | 0.5960 |
| 2.00 | 1.0477 | 1.0808 | 1.0480 | 1.0813 | 0.4976 | −0.5073 | 0.5120 | 0.5208 |
| 5.00 | 1.0082 | 1.0100 | 1.0082 | 1.0106 | 0.4998 | −0.5003 | 0.5021 | 0.5025 |
| 10.00 | 1.0021 | 1.0023 | 1.0021 | 1.0023 | 0.5000 | −0.5000 | 0.5005 | 0.5006 |

intensity factors for the special case of half plane, i.e., $\mu_2=0$, and the crack surface traction $\sigma_{1\theta\theta}(r, \pi) = -p_0 = \text{constant}$. Here the value of $k(b)$, corresponding to $c=a_0$ (the edge crack) is obtained from Koiter's paper [21].

Tables II and III show the results for two materials. In Table II the plane strain results are obtained for a linear and a quadratic, as well as a constant traction on the crack surface. In these tables $k(b)$ and $k(a)$ correspond to the stress intensity factors at the end points $b$ and $a$, respectively.

With the exception of the values given for $c=a_0$, the results of the tables are self-explanatory. The case of $c=a_0$ corresponding to a crack which terminates at the interface will be discussed later in this Chapter. The details of the numerical analysis and the study of convergence of the results

TABLE III

*Stress intensity factors for two materials*

Epoxy-boron, $\mu_2/\mu_1 = 138.46$, $p = -p_0$

| $c/a_0$ | Plane stress | | Plane strain | |
|---|---|---|---|---|
| | $\frac{k(b)}{p_0(a_0)^{\frac{1}{2}}}$ | $\frac{k(a)}{p_0(a_0)^{\frac{1}{2}}}$ | $\frac{k(b)}{p_0(a_0)^{\frac{1}{2}}}$ | $\frac{k(a)}{p_0(a_0)^{\frac{1}{2}}}$ |
| 1.00 | 0.8706 | 4.9218 | 0.8751 | 3.0723 |
| 1.10 | 0.8879 | 0.6227 | 0.8915 | 0.6447 |
| 1.15 | 0.8953 | 0.6822 | 0.8984 | 0.6980 |
| 1.25 | 0.9080 | 0.7587 | 0.9104 | 0.7680 |
| 2.00 | 0.9581 | 0.9288 | 0.9586 | 0.9297 |
| 5.00 | 0.9922 | 0.9905 | 0.9923 | 0.9905 |
| 10.00 | 0.9980 | 0.9977 | 0.9980 | 0.9977 |

Boron-epoxy, $\mu_2/\mu_1 = 0.0072$, $p = -p_0$

| $c/a_0$ | Plane stress $k(b)$ | Plane stress $k(a)$ | Plane strain $k(b)$ | Plane strain $k(a)$ |
|---|---|---|---|---|
| 1.00 | 1.5087 | 0.0079 | 1.5102 | 0.0078 |
| 1.10 | 1.2053 | 1.7383 | 1.2055 | 1.7383 |
| 1.15 | 1.1786 | 1.5605 | 1.1788 | 1.5608 |
| 1.25 | 1.1430 | 1.3784 | 1.1432 | 1.3787 |
| 2.00 | 1.0528 | 1.0894 | 1.0529 | 1.0895 |
| 5.00 | 1.0090 | 1.0110 | 1.0090 | 1.0110 |
| 10.00 | 1.0023 | 1.0026 | 1.0023 | 1.0026 |

may be found in [20]★. For this case (i.e., $a > 0$) the convergence is extremely good. Some results showing the convergence of the results for the more important case of $a = 0$ will be given later in this Chapter.

In the problems using this method, the most stable values for $g(\mp 1)$ were obtained from a quadratic extrapolation employing $g(t_2)$, $g(t_3)$, $g(t_4)$, and $g(t_{n-3})$, $g(t_{n-2})$, $g(t_{n-1})$. However, the difference between these results and those obtained from the quadratic extrapolation using the last three points, or the linear extrapolation using the last or the next-to-last two points were insignificant.

★ For convergence analysis and the comparison of the results obtained by using various methods, see also [16].

*A half plane containing a crack parallel to the boundary.* As a simple example for the application of a coupled system of singular integral equations of the first kind the problem described in Figure 7.2, namely an elastic half plane containing a crack parallel to its free boundary will be considered.

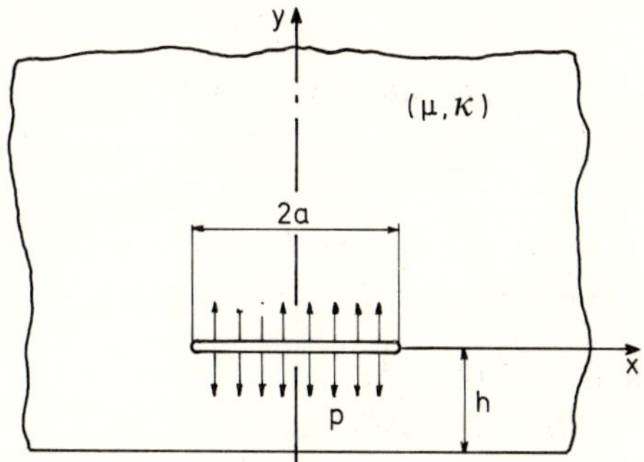


Figure 7.2. Crack parallel to a free boundary.

Assume that a uniform pressure applied to the crack surfaces is the only external load. Thus, by defining

$$\phi_1(x) = \frac{\partial}{\partial x}[u(x, +0)-u(x, -0)],$$

$$\phi_2(x) = \frac{\partial}{\partial x}[v(x, +0)-v(x, -0)], \tag{7.91}$$

$$\sigma_{yy}(x, 0) = -p_0, \quad \sigma_{xy}(x, 0) = 0, \qquad (-1<x<1)$$

and using Fourier transforms, the problem may be formulated as (see [3] for the analysis of the general problem of layered materials)

$$\frac{1+\kappa}{2\mu}\sigma_{iy}(x, 0) = \frac{1}{\pi}\int_{-1}^{1}\frac{\phi_i(t)}{t-x}dt + \frac{1}{\pi}\int_{-1}^{1}\sum_{1}^{2}k_{ij}(x, t)\phi_j(t)dt$$
$$= f_i(x), \quad (i=1, 2;\ -1<x<1;\ \sigma_{1y}=\sigma_{xy},\ \sigma_{2y}=\sigma_{yy}) \tag{7.92}$$

where the crack length is normalized to be 2, $h$ is the distance of the crack from the boundary, and

$$f_1(x)=0, \quad f_2(x) = -p_0\frac{1+\kappa}{2\mu},$$

$$k_{11}(x,t) = -\frac{t-x}{(t-x)^2+4h^2} + \frac{8h^2(t-x)}{[(t-x)^2+4h^2]^2} - \frac{4h^2(t-x)[12h^2-(t-x)^2]}{[(t-x)^2+4h^2]^3},$$

$$k_{12}(x,t) = k_{21}(x,t) = -\frac{8h^3[4h^2-3(t-x)^2]}{[(t-x)^2+4h^2]^3},$$

$$k_{22}(x,t) = -\frac{t-x}{(t-x)^2+4h^2} - \frac{8h^2(t-x)}{[(t-x)^2+4h^2]^2} - \frac{4h^2(t-x)[12h^2-(t-x)^2]}{[(t-x)^2+4h^2]^3}. \tag{7.93}$$

The functions $\phi_i$, again, have integrable singularities at the end points $-1$ and 1. Hence, the index for both equations in (7.92) is $\kappa=1$. The two arbitrary constants resulting from the general solution of equations (7.92) must be determined from (or, the numerical solution of (7.92) must be obtained under) the following compatibility conditions:

$$\int_{-1}^{1} \phi_1(t)\mathrm{d}t = 0, \qquad \int_{-1}^{1} \phi_2(t)\mathrm{d}t = 0. \tag{7.94}$$

To solve the system of singular integral equations (7.92), use will be made of the method described in section 7.4. The fundamental function for both equations in (7.92) is $w(t)=(1-t^2)^{-\frac{1}{2}}$. Hence, the related orthogonal polynomials are the Chebyshev polynomials of the first kind, $T_n(t)$. From symmetry, it is seen that $\phi_1(x)=\phi_1(-x)$ and $\phi_2(x)=-\phi_2(-x)$. Then, the unknown functions may be expressed as follows:

$$\phi_1(x) = (1-t^2)^{-\frac{1}{2}} \sum_{0}^{\infty} A_n T_{2n}(x),$$

$$\phi_2(x) = (1-t^2)^{-\frac{1}{2}} \sum_{1}^{\infty} B_n T_{2n-1}(x). \tag{7.95}$$

With equations (7.95), from the orthogonality conditions (7.62) it follows that the second condition in (7.94) is satisfied and the first gives $A_0=0$. Substituting from (7.95) into (7.92) and using the following relations

$$\frac{1}{\pi}\int_{-1}^{1} T_n(t)(1-t^2)^{-\frac{1}{2}}\frac{\mathrm{d}t}{t-x} = \begin{cases} U_{n-1}(x), & (|x|<1), \\ G_n(x) = \dfrac{[(x^2-1)^{\frac{1}{2}}-x]^n}{(-1)^{n+1}(x^2-1)^{\frac{1}{2}}}, & (x>1) \end{cases} \tag{7.96}$$

it is found that

$$\begin{aligned} &\sum_1^{\infty}[A_n U_{2n-1}(x)+A_n H_n^{11}(x)+B_n H_n^{12}(x)] = f_1(x), \\ &\sum_1^{\infty}[B_n U_{2n-2}(x)+A_n H_n^{21}(x)+B_n H_n^{22}(x)] = f_2(x), \end{aligned} \qquad (-1<x<1) \tag{7.97}$$

where

$$\begin{aligned} H_n^{11}(x) &= \frac{1}{\pi}\int_{-1}^{1} k_{11}(x,t)\,T_{2n}(t)(1-t^2)^{-\frac{1}{2}}\mathrm{d}t, \\ H_n^{12}(x) &= \frac{1}{\pi}\int_{-1}^{1} k_{12}(x,t)\,T_{2n-1}(t)(1-t^2)^{-\frac{1}{2}}\mathrm{d}t, \\ H_n^{21}(x) &= \frac{1}{\pi}\int_{-1}^{1} k_{21}(x,t)\,T_{2n}(t)(1-t^2)^{-\frac{1}{2}}\mathrm{d}t, \\ H_n^{22}(x) &= \frac{1}{\pi}\int_{-1}^{1} k_{22}(x,t)\,T_{2n-1}(t)(1-t^2)^{-\frac{1}{2}}\mathrm{d}t. \end{aligned} \tag{7.98}$$

To solve the functional equations, (7.97) a weighted residual method will be used (or simply both sides will be expanded into series of Chebyshev polynomials and the coefficients will be compared) to reduce them to a system of algebraic equations. Thus, multiplying the first equation by $U_{2k-1}(x)(1-x^2)^{\frac{1}{2}}$, the second by $U_{2k-2}(x)(1-x^2)^{\frac{1}{2}}$, truncating the series at the $N$th term and integrating in $(-1, 1)$ we obtain

$$\begin{aligned} &\frac{\pi}{2}A_k + \sum_1^N (a_{kn}A_n + b_{kn}B_n) = F_{1k}, \\ &\frac{\pi}{2}B_k + \sum_1^N (c_{kn}A_n + d_{kn}B_n) = F_{2k}, \qquad (k=1, \ldots, N) \end{aligned} \tag{7.99}$$

where

$$
\begin{aligned}
a_{kn} &= \int_{-1}^{1} U_{2k-1}(x)\, H_n^{11}(x)(1-x^2)^{\frac{1}{2}}\,\mathrm{d}x\,,\\
b_{kn} &= \int_{-1}^{1} U_{2k-1}(x)\, H_n^{12}(x)(1-x^2)^{\frac{1}{2}}\,\mathrm{d}x\,,\\
c_{kn} &= \int_{-1}^{1} U_{2k-2}(x)\, H_n^{21}(x)(1-x^2)^{\frac{1}{2}}\,\mathrm{d}x\,,\\
d_{kn} &= \int_{-1}^{1} U_{2k-2}(x)\, H_n^{22}(x)(1-x^2)^{\frac{1}{2}}\,\mathrm{d}x\,,\\
F_{1k} &= 0\,, \qquad (k=1, \ldots, N)\,,\\
F_{21} &= -\frac{\pi}{2}\, p_0 \frac{1+\kappa}{2\mu}\,, \quad F_{2k}=0\,, \qquad (k=2, \ldots, N)\,.
\end{aligned} \tag{7.100}
$$

After obtaining the constants $A_k$ and $B_k$ from equations (7.99), any desired field quantity may be calculated by evaluating the related definite integral. For example, noting that equations (7.92) give the stresses on $y=0$ for $|x|>1$ as well as $|x|<1$, using (7.95), the second relation in (7.96), and (7.98), it is found that

$$
\begin{aligned}
\frac{1+\kappa}{2\mu}\,\sigma_{xy}(x,0) &= \sum_1^{\infty} [A_n G_{2n}(x) + A_n H_n^{11}(x) + B_n H_n^{12}(x)]\,,\\
& \qquad\qquad (x>1)\,.\\
\frac{1+\kappa}{2\mu}\,\sigma_{yy}(x,0) &= \sum_1^{\infty} [B_n G_{2n-1}(x) + A_n H_n^{21}(x) + B_n H_n^{22}(x)]\,,
\end{aligned} \tag{7.101}
$$

Since $H_n^{ij}(x)$, $(i, j=1, 2)$ are bounded functions, from equations (7.101) and (7.96) the stress intensity factors may be obtained as

$$
\begin{aligned}
k_1 &= \lim_{x\to 1}\, (x^2-1)^{\frac{1}{2}}\, \sigma_{yy}(x,0) = -\frac{2\mu}{1+\kappa} \sum_1^{\infty} B_n\,,\\
k_2 &= \lim_{x\to 1}\, (x^2-1)^{\frac{1}{2}}\, \sigma_{xy}(x,0) = -\frac{2\mu}{1+\kappa} \sum_1^{\infty} A_n\,.
\end{aligned} \tag{7.102}
$$

These expressions can also be obtained from equations (2.95) by noting that $T_n(1)=1$, $(n=0, 1, \ldots)$ and using the following relations between the stress intensity factors and the displacement derivatives (see equations (7.90)):

TABLE IV

*Stress intensity factors for the crack parallel to the boundary*

| $h$ | 0.2 | | | |
|---|---|---|---|---|
| $N$ | 2 | 4 | 6 | 8 |
| $k_1$ | 4.6298 | 4.6981 | 4.7621 | 4.7600 |
| $k_2$ | 2.0294 | 1.6925 | 1.7173 | 1.7181 |

| $h$ | 0.4 | | | |
|---|---|---|---|---|
| $N$ | 2 | 4 | 6 | 8 |
| $k_1$ | 2.5437 | 2.5949 | 2.5957 | 2.5956 |
| $k_2$ | 0.7649 | 0.7364 | 0.7376 | 0.7375 |

| $h$ | 0.6 | | | 0.8 | | |
|---|---|---|---|---|---|---|
| $N$ | 2 | 4 | 6 | 2 | 4 | 6 |
| $k_1$ | 1.9226 | 1.9616 | 1.9608 | 1.6386 | 1.6610 | 1.6609 |
| $k_2$ | 0.4265 | 0.4298 | 0.4296 | 0.2678 | 0.2716 | 0.2715 |

| $h$ | 1.0 | | | 1.2 | | |
|---|---|---|---|---|---|---|
| $N$ | 2 | 4 | 6 | 2 | 4 | 6 |
| $k_1$ | 1.4741 | 1.4860 | 1.4860 | 1.3660 | 1.3722 | 1.3722 |
| $k_2$ | 0.1775 | 0.1796 | 0.1796 | 0.1223 | 0.1234 | 0.1234 |

| $h$ | 1.5 | | 2.0 | | 3.0 | | $\infty$ |
|---|---|---|---|---|---|---|---|
| $N$ | 2 | 4 | 2 | 4 | 2 | 4 | 1 |
| $k_1$ | 1.2603 | 1.2628 | 1.1615 | 1.1621 | 1.0778 | 1.0778 | 1.0 |
| $k_2$ | 0.0743 | 0.0746 | 0.0366 | 0.0367 | 0.0123 | 0.0123 | 0 |

$$
\begin{aligned}
k_1 &= -\frac{2\mu}{1+\kappa}\lim_{x\to 1}(1-x^2)^{\frac{1}{2}}\phi_2(x)\,,\\
k_2 &= -\frac{2\mu}{1+\kappa}\lim_{x\to 1}(1-x^2)^{\frac{1}{2}}\phi_1(x)\,.
\end{aligned}
\tag{7.103}
$$

For $p_0 = 1$ and $a = 1$ some of the calculated results are shown in Table IV⋆.

⋆ Alternatively, $h$, $k_1$, and $k_2$ shown in the table may be interpreted as $h/a$, $k_1/(p_0 a^{\frac{1}{2}})$, and $k_2/(p_0 a^{\frac{1}{2}})$, respectively.

For $h=\infty$, $k_{ij}(x, t)=0$ and from equations (7.95), (7.99) and (7.100) the exact solution of (7.92) is obtained as

$$\phi_1(x)=0\,, \quad \phi_2(x)=-p_0\frac{1+\kappa}{2\mu}x(1-x^2)^{-\frac{1}{2}} \tag{7.104}$$

which gives

$$k_1=p_0\,, \quad k_2=0\,.$$

This is shown in the last column of Table IV. The table shows the rapid convergence of the results even for relatively small values of $N$.

*A simple example for $\kappa=0$.* As an application for $\kappa=0$ consider the punch

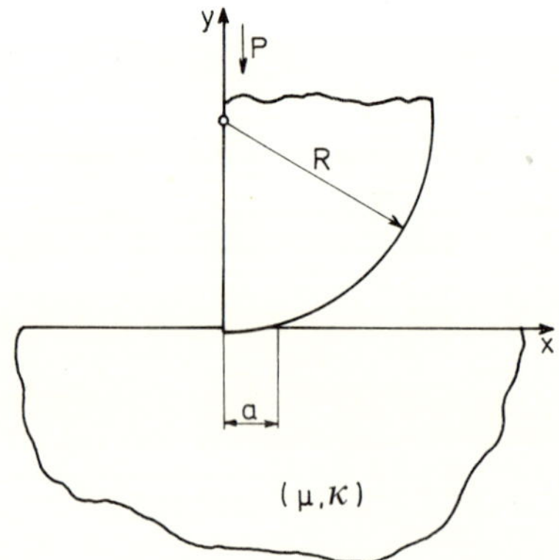


Figure 7.3. Rigid punch on an elastic half plane ($\kappa=0$).

problem shown in Figure 7.3, where a rigid half cylinder of radius $R$ is pressed on an elastic half plane. Neglecting the friction and defining the pressure in the contact area as $p(x)=-\sigma_{yy}(x, 0)$, the integral equation of the problem may be written as

$$\frac{1}{\pi}\int_0^a p(t)\frac{\mathrm{d}t}{t-x} = -\frac{4\mu}{1+\kappa}\frac{\partial}{\partial x}v(x, 0) = -\frac{4\mu}{1+\kappa}\frac{x}{R}, \quad (0<x<a) \tag{7.105}$$

where the unknown constant $a$ is determined from

$$\int_0^a p(t)\mathrm{d}t = P\,. \tag{7.106}$$

Here the unknown function $p(t)$ has an integrable singularity at $t=0$ and is bounded at $t=a$. Hence, the index and the fundamental function of the integral equation, and the unknown function may be expressed as

$$\kappa=0\,,\quad w(t)=(a-t)^{\frac{1}{2}}t^{-\frac{1}{2}}\,,\quad p(t)=w(t)g(t)\,,\qquad (0<t<a) \tag{7.107}$$

where $g(t)$ is a bounded function.

Defining

$$r=\frac{2t}{a}-1\,,\quad s=\frac{2x}{a}-1\,,\quad g(t)=G(r) \tag{7.108}$$

one obtains

$$\begin{aligned}&\frac{1}{\pi}\int_{-1}^{1}G(r)\left(\frac{1-r}{1+r}\right)^{\frac{1}{2}}\frac{\mathrm{d}r}{r-s}=-\frac{4\mu}{1+\kappa}\frac{a}{2R}(1+s)\,,\qquad (|s|<1)\,,\\ &\frac{a}{2}\int_{-1}^{1}G(r)\left(\frac{1-r}{1+r}\right)^{\frac{1}{2}}\mathrm{d}r=P\,.\end{aligned} \tag{7.109}$$

If one further defines

$$G(r)=\sum_{0}^{\infty}c_nP_n^{(\frac{1}{2},-\frac{1}{2})}(r) \tag{7.110}$$

and substitutes into equations (7.109), using (7.29) and observing that

$$P_0^{(-\frac{1}{2},\frac{1}{2})}(s)=1\,,\quad P_1^{(-\frac{1}{2},\frac{1}{2})}(s)=s-\tfrac{1}{2}\,,$$

the result can be written as

$$-\sum_{0}^{\infty}c_nP_n^{(-\frac{1}{2},\frac{1}{2})}(s)=\frac{2\mu}{1+\kappa}\frac{a}{R}\left[\tfrac{3}{2}P_0^{(-\frac{1}{2},\frac{1}{2})}(s)+P_1^{(-\frac{1}{2},\frac{1}{2})}(s)\right],\qquad(-1<s<1)\,. \tag{7.111}$$

From (7.111) it is seen that

$$c_0=-\frac{3\mu}{1+\kappa}\frac{a}{R}\,,\quad c_1=-\frac{2\mu}{1+\kappa}\frac{a}{R}\,,\quad c_n=0\,,\qquad(n=2,3,\ldots)\,,$$

$$G(r)=\frac{\mu}{1+\kappa}\frac{a}{R}\left[3P_0^{(\frac{1}{2},-\frac{1}{2})}(r)+2P_1^{(\frac{1}{2},-\frac{1}{2})}(r)\right]=\frac{2\mu}{1+\kappa}\frac{a}{R}(2+r)\,, \tag{7.112}$$

or from equations (7.107), (7.108), and (7.111) the pressure may be expressed as

$$p(t)=\frac{2\mu}{1+\kappa}\frac{a}{R}\left(1+\frac{2t}{a}\right)\left(\frac{a-t}{t}\right)^{\frac{1}{2}}. \tag{7.113}$$

Substituting from equation (7.112) into (7.109) and using the orthogonality condition (7.67) a relation between $a$ and $P$ is obtained as follows:

$$P = \frac{3\pi\mu}{1+\kappa}\,\frac{a^2}{2R}\,. \tag{7.114}$$

Here it should be emphasized that since $G(r)$ can be expressed as a truncated series of Jacobi polynomials in an exact form (see equation (7.112)), by applying the method of Gauss–Jacobi integration formulas described in section 7.3 (i.e., the formula (7.52)) to this problem, the exact value of $G(r)$ at each location $r_i$, $(i=1, \ldots, n)$ would be obtained for any $n$ provided $n>1$. This is implicit in the assumption (7.33) and the subsequent analysis.

## 7.6 Singular integral equations of the second kind

In the general system of the singular integral equations (7.1), diagonalizing the matrices $A$ and $B$ simultaneously, the dominant part of the system can be uncoupled. Thus, any numerical technique developed for the solution of a single equation may easily be generalized to apply to a system of equations. It will then be sufficient to consider the solution of the following singular integral equation:

$$a\phi(x) + \frac{b}{\pi}\int_{-1}^{1} \phi(t)\,\frac{dt}{t-x} + \int_{-1}^{1} k(x,t)\,\phi(t)\,dt = f(x)\,, \qquad (-1<x<1) \tag{7.115}$$

where the interval is again normalized to be $(-1, 1)$ without any loss in generality, and it will be assumed that $a$ and $b$ are constant and the known functions $f$ and $k$ are $H$-continuous. In equation (7.115) the functions $\phi$, $k$, $f$, and the constants $a$, $b$ may be real or complex.

*The fundamental function.* The fundamental function of the integral equation (7.115) may again be obtained by considering only the dominant part of (7.115), or by writing (7.115) as

$$a\phi(x) + \frac{b}{\pi}\int_{-1}^{1} \phi(t)\,\frac{dt}{t-x} = F(x)\,, \qquad (|x|<1) \tag{7.116}$$

where $F$ contains the input function $f$ and the bounded term containing the Fredholm kernel $k(x, t)$. Defining the sectionally holomorphic function

$$\Phi(z) = \frac{1}{2\pi i}\int_{-1}^{1} \frac{\phi(t)}{t-z}\,dt \tag{7.117}$$

and using the Plemelj formulas, (7.116) becomes

$$(a+\mathrm{i}b)\,\Phi^{+}(x)-(a-\mathrm{i}b)\,\Phi^{-}(x)=F(x)\,,\qquad(-1<x<1)\,. \tag{7.118}$$

The fundamental solution of equation (7.118) may be obtained [5] as

$$\begin{aligned}
X(z) &= (z-1)^{\alpha}(z+1)^{\beta}\\
\alpha &= \frac{1}{2\pi\mathrm{i}}\log\left(\frac{a-\mathrm{i}b}{a+\mathrm{i}b}\right)+N\\
\beta &= -\frac{1}{2\pi\mathrm{i}}\log\left(\frac{a-\mathrm{i}b}{a+\mathrm{i}b}\right)+M\\
&-1<\mathrm{Re}(\alpha)<1\,,\quad -1<\mathrm{Re}(\beta)<1\\
\kappa &= -(\alpha+\beta)=-(N+M)
\end{aligned} \tag{7.119}$$

where $X(z)$ is the branch for which $z^{\kappa}X(z)\to 1$ as $z\to\infty$, $N$ and $M$ are integer, and again it is assumed that $\phi$ is either a "potential" or a "flux", hence the restriction on the real parts of $\alpha$ and $\beta$. In this problem too the index $\kappa$ is $-1$, 0, or $+1$, and is determined by the physics of the problem. The following function of $x$ obtained from equations (7.119) is known as the fundamental function of the integral equation:

$$w(x)=(-1)^{-\alpha}X^{+}(x)=(1-x)^{\alpha}(1+x)^{\beta}\,. \tag{7.120}$$

Similar to the singular integral equation of the first kind, in this problem, too,

(1) for $\kappa=1$ the solution contains one arbitrary constant which is determined from an equilibrium or a compatibility condition of the form

$$\int_{-1}^{1}\phi(t)\,\mathrm{d}t=A\,, \tag{7.121}$$

(2) for $\kappa=0$ the inversion of equation (7.115) does not contain any arbitrary constants, and

(3) for $\kappa=-1$, there is no arbitrary constant to be determined; however, the solution must satisfy the following consistency condition:

$$\int_{-1}^{1}\left[f(x)-\int_{-1}^{1}k(x,t)\,\phi(t)\,\mathrm{d}t\right]\frac{\mathrm{d}t}{w(t)}=0\,. \tag{7.122}$$

**Solution by Jacobi polynomials.** Once the fundamental function, $w(t)$ of the integral equation is determined, the solution of (7.115) may be expressed as

$$\phi(t) = g(t)\, w(t)\,, \qquad (-1 < t < 1) \tag{7.123}$$

where $g(t)$ is a bounded continuous function and can always be represented by an infinite series. Observing that $w(t)$ defined by (7.120) is the weight function of $P_n^{(\alpha,\beta)}(t)$, $(n=0, 1, \ldots)$, one may write

$$\phi(t) = \sum_0^\infty c_n w(t)\, P_n^{(\alpha,\beta)}(t) \tag{7.124}$$

where $c_n$, $(n=0, 1, \ldots)$ are undetermined constants.

Before substituting equation (7.124) into (7.115) note that for $\kappa = (-1, 0, 1)$ (see Tricomi [22], Szegö [14])

$$\frac{1}{\pi}\int_{-1}^{1} w(t)\, P_n^{(\alpha,\beta)}(t)\, \frac{\mathrm{d}t}{t-x} = \cot \pi\alpha w(x)\, P_n^{(\alpha,\beta)}(x)$$

$$- \frac{2^{\alpha+\beta}\,\Gamma(\alpha)\,\Gamma(n+\beta+1)}{\pi\Gamma(n+\alpha+\beta+1)}\, F\left(n+1\,,\; -n-\alpha-\beta\,;\; 1-\alpha\,;\; \frac{1-x}{2}\right),$$

$$\Bigl(-1<x<1,\ \mathrm{Re}(\alpha) > -1,\ \mathrm{Re}(\beta) > -1,\ \alpha+\beta = -\kappa,\ \mathrm{Re}(\alpha) \neq (0, 1, \ldots)\,,$$

$$\cot \pi\alpha = \cot \pi\left(\frac{1}{2\pi \mathrm{i}} \log\left(\frac{a-\mathrm{i}b}{a+\mathrm{i}b}\right) + N\right) = -\frac{a}{b}\Bigr), \tag{7.125}$$

and

$$P_{n-\kappa}^{(-\alpha,-\beta)}(x) = \frac{\Gamma(n-\kappa-\alpha+1)}{\Gamma(n-\kappa+1)\,\Gamma(1-\alpha)}\, F\left(n+1,\, -n+\kappa;\, 1-\alpha;\, \frac{1-x}{2}\right). \tag{7.126}$$

Combining equations (7.125) and (7.126) yield

$$\frac{1}{\pi}\int_{-1}^{1} w(t)\, P_n^{(\alpha,\beta)}(t)\, \frac{\mathrm{d}t}{t-x} = -\frac{a}{b}\, w(x)\, P_n^{(\alpha,\beta)}(x)$$

$$- 2^{-\kappa}\, \frac{\Gamma(\alpha)\,\Gamma(1-\alpha)}{\pi}\, P_{n-\kappa}^{(-\alpha,-\beta)}(x)\,, \qquad (|x| < 1)\,. \tag{7.127}$$

If equation (7.124) is substituted into (7.115) and use is made of (7.127), there results

$$\sum_0^\infty c_n \left[ -\frac{2^{-\kappa} b}{\sin \pi\alpha} P_{n-\kappa}^{(-\alpha, -\beta)}(x) + h_n(x) \right] = f(x)\,, \qquad (7.128)$$

$$h_n(x) = \int_{-1}^{1} w(t)\, P_n^{(\alpha,\beta)}(t)\, k(x, t)\,\mathrm{d}t\,, \qquad (-1 < x < 1)\,. \qquad (7.129)$$

The functional equation (7.128) can be reduced to an infinite system of algebraic equations in $c_n$ by expanding both sides into series of Jacobi polynomials $P_k^{(-\alpha,-\beta)}(x)$ $(k=0, 1, \ldots)$, and comparing the respective coefficients. Thus, using the orthogonality relations (7.67) to (7.69) and truncating the series give

$$-\frac{2^{-\kappa} b}{\sin \pi\alpha}\, \theta_k(-\alpha, -\beta)\, c_{k+\kappa} + \sum_{k=0}^{N} d_{nk} c_n = F_k\,, \qquad (k = 0, 1, \ldots, N) \quad (7.130)$$

where

$$d_{nk} = \int_{-1}^{1} P_k^{(-\alpha,-\beta)}(x)\, w(-\alpha, -\beta, x)\, h_n(x)\,\mathrm{d}x\,, \qquad (7.131)$$

$$F_k = \int_{-1}^{1} P_k^{(-\alpha,-\beta)}(x)\, w(-\alpha, -\beta, x) f(x)\,\mathrm{d}x\,,$$

$$w(-\alpha, -\beta, x) = (1-x)^{-\alpha}(1+x)^{-\beta} = w^{-1}(x)\,. \qquad (7.132)$$

(1) $\kappa = -1$. In this case note that the first term in the series (7.128) is a constant times $c_0 P_1^{(-\alpha,-\beta)}(x)$. Hence, in solving (7.130) it can be formally assumed that $c_{-1}=0$. Also, from (7.128) to (7.132) it is seen that, since $P_0^{(-\alpha,-\beta)}(x)=1$, the first equation obtained from (7.130) for $k=0$ is seen to be equivalent to the consistency condition (7.122). Thus, (7.130) provides $N+1$ linear algebraic equations for the unknown constants $c_0, \ldots, c_N$.

(2) $\kappa=0$. In this case there are no additional arbitrary constants or conditions, and (7.130) gives the unique solution.

(3) $\kappa=1$. In this case the $N+1$ equations given by (7.130) contains $N+2$ unknown constants, $c_0, \ldots, c_{N+1}$. The additional equation for a unique solution is provided by the equilibrium or compatibility condition given by

(7.121), from which, by substituting from (7.124) and using the orthogonality condition (7.130), one finds

$$c_0\theta_0(\alpha, \beta) = A\,. \tag{7.133}$$

The method described in this section is valid for the most general singular integral equation given by (7.115), and may easily be reduced to the special case described in section 7.4 by substituting $a=0$, $b=1$. A Gauss–Jacobi type integration method similar to that described in section 7.3 may also be developed for the solution of (7.115). However, in this case because of the existence of the free term $a\phi(x)$ in the integral equation (which has zeros or singularities at the end points), such a method may not be as effective as that described in this section. The application of the method will be demonstrated by considering the following two examples:

**Example 1:** *Layered materials with an interface crack.* Consider the plane strain problem for two semi-infinite elastic media bonded through an elastic layer of thickness $h$. Let the composite medium contain a crack of length $2a$ between the layer and the material 3 as shown in Figure 7.4. Assume that the crack length is normalized to be 2 and the symmetric tractions on the crack surface

$$\begin{aligned}&\sigma_{yy}(x, 0) = p_1(x) = p_1(-x)\,,\\&\sigma_{xy}(x, 0) = p_2(x) = -p_2(-x)\,, \qquad (|x|<1)\end{aligned} \tag{7.134}$$

are the only external loads applied to the medium. Using the Fourier transforms, the integral equations of the problem may be obtained as (see [3, 4] for details)

$$\begin{aligned}&\gamma f_1(x) + \frac{1}{\pi}\int_{-1}^{1} f_2(t)\frac{\mathrm{d}t}{t-x} + \frac{c}{\pi}\int_{-1}^{1}\sum_{1}^{2} k_{1j}(x,t)f_j(t)\mathrm{d}t\\&\qquad = -\frac{1+\kappa_3}{\mu_3}\,cp_1(x)\,, \qquad (-1<x<1)\,,\\&-\gamma f_2(x) + \frac{1}{\pi}\int_{-1}^{1} f_1(t)\frac{\mathrm{d}t}{t-x} + \frac{c}{\pi}\int_{-1}^{1}\sum_{1}^{2} k_{2j}(x,t)f_j(t)\mathrm{d}t\\&\qquad = -\frac{1+\kappa_3}{\mu_3}\,cp_2(x)\,, \qquad (-1<x<1)\end{aligned} \tag{7.135}$$

where

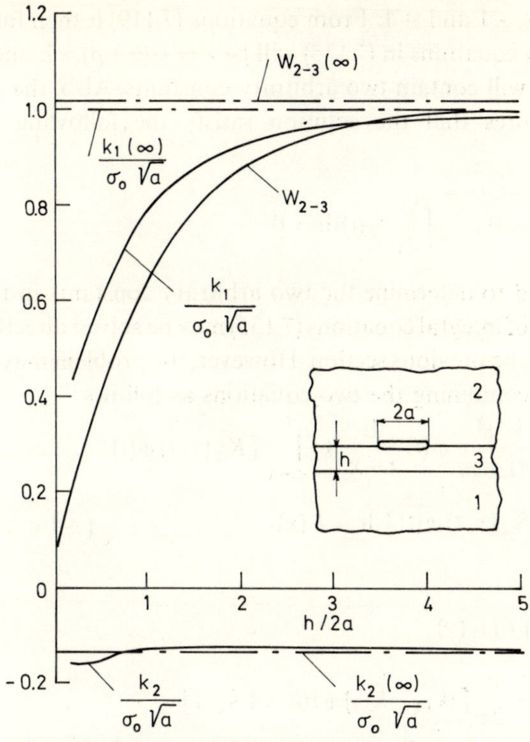


Figure 7.4. Stress intensity factors and the strain energy release rate vs. $h/2a$. Materials: (1) Aluminum, (2) Aluminum, (3) Epoxy.

$$f_1(x) = \frac{\partial}{\partial x}(u_2^+ - u_3^-), \quad f_2(x) = \frac{\partial}{\partial x}(v_2^+ - v_3^-)$$

$$\gamma = \frac{(\mu_3+\kappa_3\mu_2)-(\mu_2+\kappa_2\mu_3)}{(\mu_2+\kappa_2\mu_3)+(\mu_3+\kappa_3\mu_2)}$$

$$c = C_1/(1+2C_1-C_1C_2) \tag{7.136}$$

$$C_1 = (\mu_3+\mu_2\kappa_3)/(\mu_2-\mu_3)$$

$$C_2 = (\kappa_2\mu_3-\kappa_3\mu_2)/(\mu_2+\kappa_2\mu_3)$$

and the bounded functions $k_{ij}(x, t), (i, j=1, 2)$ are given in [4]. The expressions for $k_{ij}$ are rather lengthy and will not be repeated here.

From the definitions (7.136) it is seen that the unknown functions, $f_1$ and $f_2$ are "flux-type" quantities, and hence, will have integrable singularities at the end points $-1$ and $+1$. From equations (7.119) it then follows that the index for both equations in (7.135) will be $\kappa = -(\alpha+\beta) = 1$, and the solution of the system will contain two arbitrary constants. Also, the physics of the problem requires that the solution satisfy the following compatibility conditions:

$$\int_{-1}^{1} f_1(t)\mathrm{d}t = 0\,, \qquad \int_{-1}^{1} f_2(t)\mathrm{d}t = 0 \tag{7.137}$$

which are used to determine the two arbitrary constants in the solution.

The system of integral equations (7.135) may be solved directly by applying the method in the previous section. However, the problem may be somewhat simplified by combining the two equations as follows:

$$\begin{aligned} -\gamma\phi(x) + \frac{1}{\pi \mathrm{i}}\int_{-1}^{1} \phi(t)\frac{\mathrm{d}t}{t-x} + \int_{-1}^{1} [K_1(x,t)\phi(t) \\ + K_2(x,t)\overline{\phi(t)}]\,\mathrm{d}t = F(x)\,, \qquad (-1<x<1) \end{aligned} \tag{7.138}$$

where

$$\begin{aligned} \phi(x) &= f_2(x) + \mathrm{i} f_1(x)\,, \\ K_1(x,t) &= -\frac{c}{2\pi}[(k_{11}-k_{22}) + \mathrm{i}(k_{12}+k_{21})]\,, \\ K_2(x,t) &= \frac{c}{2\pi}[(k_{11}+k_{22}) - \mathrm{i}(k_{12}-k_{21})]\,, \\ F(x) &= -\frac{1+\kappa_3}{\mu_3}c(p_2 - \mathrm{i}p_1)\,. \end{aligned} \tag{7.139}$$

By letting

$$a = -\gamma\,,\quad b = -\mathrm{i}\,,\quad -1 < \mathrm{Re}(\alpha) < 0\,,\quad -1 < \mathrm{Re}(\beta) < 0$$

from equations (7.119) and (7.120) the fundamental function of (7.138) is obtained to be

$$\begin{aligned} & w(t) = (1-t)^{\alpha}(1+t)^{\beta}\,, \\ & \alpha = -\tfrac{1}{2} - \mathrm{i}\omega\,,\quad \beta = -\tfrac{1}{2} + \mathrm{i}\omega\,,\quad \omega = \frac{1}{2\pi}\log\left(\frac{1+\gamma}{1-\gamma}\right). \end{aligned} \tag{7.140}$$

If the unknown function is expressed as

$$\phi(t) = \sum_{0}^{N} c_n w(t) P_n^{(\alpha,\beta)}(t), \tag{7.141}$$

the equations (7.137) and (7.133) give $c_0=0$. The remaining constants $c_1, \ldots, c_N$ are obtained from (7.130). In the special case when $h\to\infty$, $k_{ij}\to 0$ and the singular integral equation reduces to its dominant part. For the case of uniform surface tractions, one has

$$p_1(x) = -\sigma_0, \quad p_2(x) = 0. \tag{7.142}$$

From equations (7.131) and (7.132) one finds

$$d_{nk} = 0, \qquad (n, k = 0, 1, \ldots)$$

$$F_0 = -\frac{1+\kappa_3}{\mu_3} c\sigma_0 \mathrm{i}, \quad F_k = 0, \qquad (k = 1, 2, \ldots)$$

and (7.130) gives

$$c_1 = -2\sigma_0 c(1-\gamma^2)^{-\frac{1}{2}}(1+\kappa_3)/\mu_3, \quad c_n = 0, \qquad (n > 1)$$

or

$$\phi(x) = \frac{2c\sigma_0(1+\kappa_3)}{\mu_3(1-\gamma^2)^{\frac{1}{2}}} w(x) P_1^{(\alpha,\beta)}(x). \tag{7.143}$$

In general, defining the stress intensity factors $k_1$, $k_2$ by

$$k_1 + \mathrm{i}k_2 = \lim_{x\to 1} (x-1)^{-\alpha}(x+1)^{-\beta}(\sigma_{yy} + \mathrm{i}\sigma_{xy}) \tag{7.144}$$

it was found that (see Erdogan and Gupta [4], Erdogan and Ozbek [23])

$$k_1 + \mathrm{i}k_2 = \lim_{x\to 1} \frac{\mu_3}{c(1+\kappa_3)} (1-\gamma^2)^{\frac{1}{2}} w^{-1}(x)\phi(x)$$

$$= \frac{\mu_3}{c(1+\kappa_3)} (1-\gamma^2)^{\frac{1}{2}} \sum_{1}^{\infty} c_n P_n^{(\alpha,\beta)}(1). \tag{7.145}$$

In the special case considered, noting that

$$P_1^{(\alpha,\beta)}(x) = \tfrac{1}{2}[\alpha-\beta+(\alpha+\beta+2)x] = \frac{x}{2} - \mathrm{i}\omega,$$

one finds from equations (7.143) and (7.145) that*

$$k_1 = \sigma_0, \quad k_2 = -2\omega\sigma_0. \tag{7.146}$$

* Here, $k_1$, $k_2$ should be multiplied by $a_0^{\frac{1}{2}}$ if $a_0 \neq 1$, $2a_0$ being the crack length.

TABLE V

*Stress intensity factors for the interface crack*

| $h/2a=0.2$ | | | | | |
|---|---|---|---|---|---|
| $N$ | 4 | 6 | 8 | 10 | 12 |
| $k_1$ | 0.4245 | 0.4357 | 0.4455 | 0.4449 | 0.4448 |
| $k_2$ | −0.1669 | −0.1648 | −0.1617 | −0.1620 | −0.1621 |

| $h/2a=0.4$ | | | | | |
|---|---|---|---|---|---|
| $N$ | 4 | 6 | 8 | 10 | 12 |
| $k_1$ | 0.5576 | 0.5542 | 0.5524 | 0.5525 | 0.5525 |
| $k_2$ | −0.1611 | −0.1603 | −0.1598 | −0.1598 | −0.1598 |

| $h/2a=0.6$ | | | | | |
|---|---|---|---|---|---|
| $N$ | 4 | 6 | 8 | 10 | 12 |
| $k_1$ | 0.6513 | 0.6497 | 0.6484 | 0.6484 | 0.6484 |
| $k_2$ | −0.1478 | −0.1476 | −0.1475 | −0.1476 | −0.1476 |

| | $h/2a=0.8$ | | | $h/2a=1.0$ | | |
|---|---|---|---|---|---|---|
| $N$ | 2 | 4 | 6 | 2 | 4 | 6 |
| $k_1$ | 0.7356 | 0.7311 | 0.7300 | 0.7795 | 0.7931 | 0.7927 |
| $k_2$ | −0.1383 | −0.1379 | −0.1378 | −0.1313 | −0.1326 | −0.1326 |

| | $h/2a=2.0$ | | | $h/2a=\infty$ |
|---|---|---|---|---|
| $N$ | 2 | 4 | 6 | 1 |
| $k_1$ | 0.9289 | 0.9302 | 0.9302 | 1 |
| $k_2$ | −0.1295 | −0.1297 | −0.1297 | −0.1342 |

For the external loads (7.142) and the material combination Aluminum ($E=10^7$ psi, $\nu=0.3$)—Epoxy ($E=4.5\times10^5$ psi, $\nu=0.35$)—Aluminum, the results obtained from (7.130) and (7.145) are shown in Table V and Figure 7.4. The table indicates that the convergence of the calculated results is quite rapid*. Other results for different material combinations and configurations are shown in Figures 7.5 to 7.8. The quantity, $W_{2-3}$, shown in these figures is the measure of the strain energy release rate, $(\partial U/da)$ for the crack lying

* The values of $k_1$ and $k_2$ shown in the table are calculated for $\sigma_0=1$, $a_0=1$; if $a_0\neq1$ they should be multiplied by $\sigma_0 a_0^{\frac{1}{2}}$.

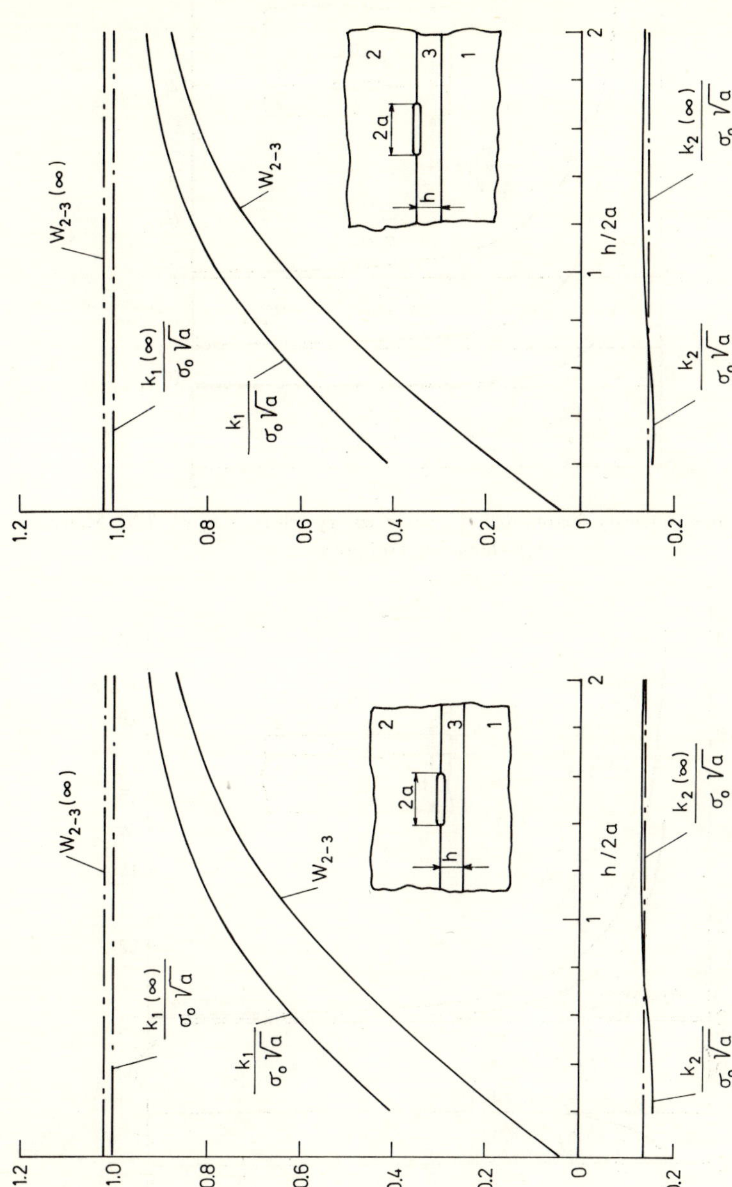


Figure 7.5. Stress intensity factors and the strain energy release rate vs. $h/2a$. Materials: (1) Steel, (2) Aluminum, (3) Epoxy.

Figure 7.6. Stress intensity factors and the strain energy release rate vs. $h/2a$. Materials: (1) Aluminum, (2) Steel, (3) Epoxy.

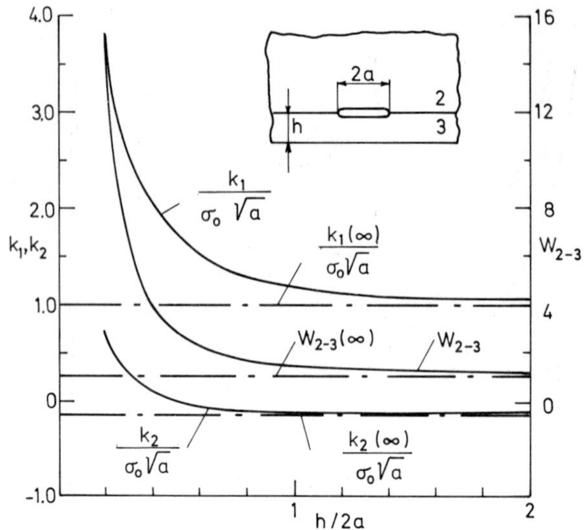


Figure 7.7. Stress intensity factors and the strain energy release rate vs. $h/2a$. Materials: (2) Aluminum, (3) Epoxy.

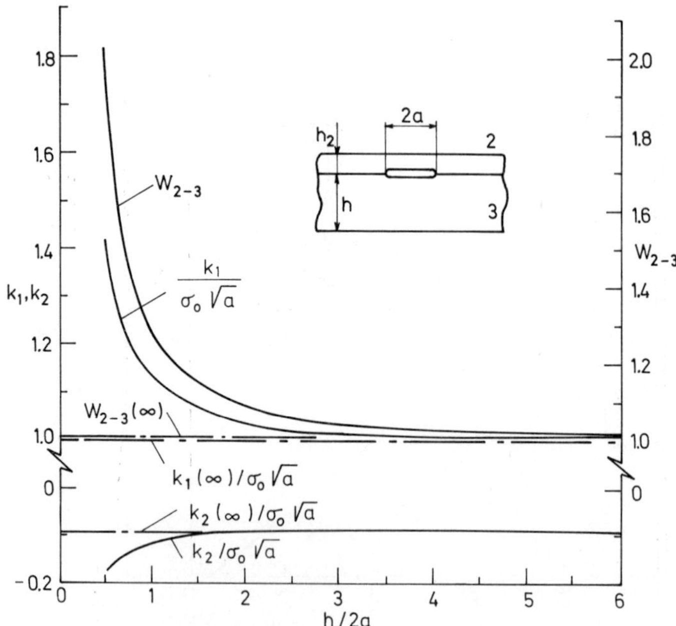


Figure 7.8. Stress intensity factors and the strain energy release rate vs. $h_2/2a$ for $h/h_2 = 3$. Materials: (2) Steel, (3) Aluminum.

between materials 2 and 3 and is calculated from (see Malyshev and Salganik [24])

$$W_{2-3} = -\frac{2\mu_3}{c\sigma_0^2 a\pi(1+\kappa_3)}\left(\frac{\partial U}{\partial a}\right)_{2-3} = \frac{k_1^2+k_2^2}{\sigma_0^2 a}. \tag{7.147}$$

For a homogeneous medium, $c=-\frac{1}{2}$ and equation (7.147) becomes

$$\left(\frac{\partial U}{\partial a}\right)_3 = \frac{\pi(1+\kappa_3)}{4\mu_3}(k_1^2+k_2^2). \tag{7.148}$$

Note that for $h=0$ one has

$$\lim_{h\to 0}\left(\frac{\partial U}{\partial a}\right)_{2-3} = \left(\frac{\partial U}{\partial a}\right)_{2-1}$$

or

$$\lim_{h\to 0} W_{2-3}(h) = \frac{\mu_3}{\mu_1}\frac{1+\kappa_1}{1+\kappa_3}\frac{c_{2-1}}{c_{2-3}} W_{2-1}(\infty) \tag{7.149}$$

which gives the points on the figures corresponding to $h=0$.

**Example 2:** *A punch with constant friction on an elastic half plane.* Consider the plane elasticity problem for a punch pressed on a half-plane at an angle.

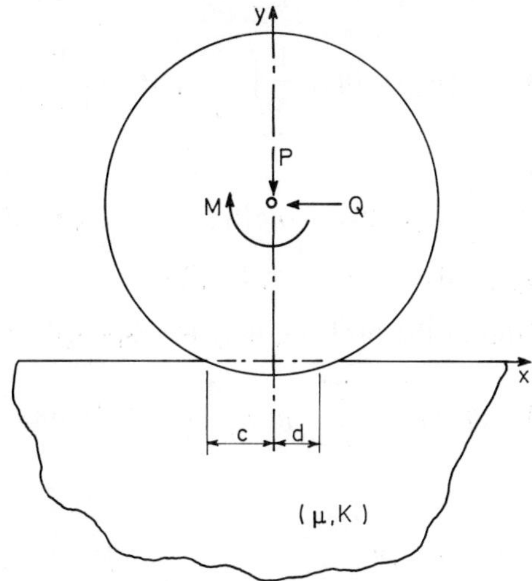


Figure 7.9. Rigid cylindrical punch on an elastic half plane with constant coefficient of friction.

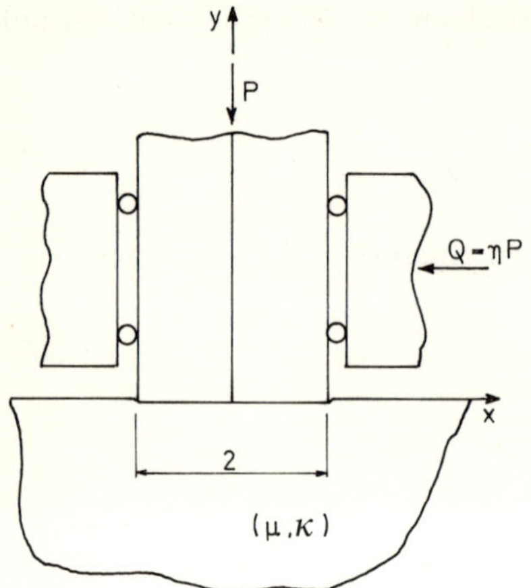


Figure 7.10. Flat punch on an elastic half plane with constant coefficient of friction.

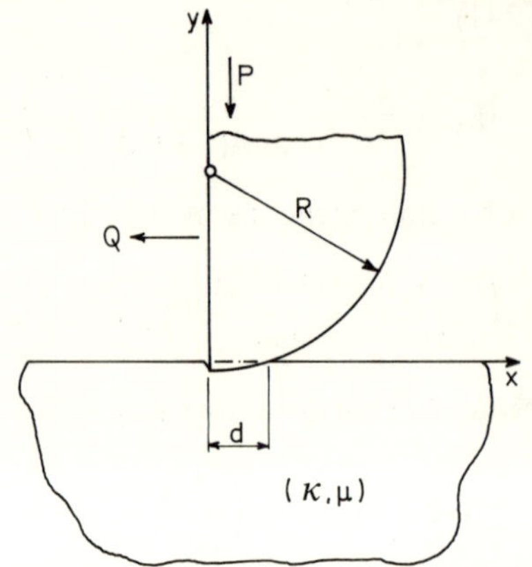


Figure 7.11. Rigid punch on an elastic half plane with constant coefficient of friction ($\kappa=0$).

Let the coefficient of friction, $\eta$ be constant (Figures 7.9 to 7.11). From the following equations for a half plane $y<0$ subjected to tractions $\sigma_{xy}$ and $\sigma_{yy}$ at $y=0$ (Figure 7.9)

$$\begin{aligned}\frac{4\mu}{1+\kappa}\frac{\partial}{\partial x}u(x,0)&=\gamma\sigma_{yy}(x,0)+\frac{1}{\pi}\int_{-\infty}^{\infty}\sigma_{xy}(t,0)\frac{dt}{t-x},\\ \frac{4\mu}{1+\kappa}\frac{\partial}{\partial x}v(x,0)&=-\gamma\sigma_{xy}(x,0)+\frac{1}{\pi}\int_{-\infty}^{\infty}\sigma_{yy}(t,0)\frac{dt}{t-x},\end{aligned}\tag{7.150}$$

by letting

$$\begin{aligned}&\sigma_{yy}(t,0)=-p(t)\,,\quad \sigma_{xy}(t,0)=-\eta p(t)\,,\quad (-c<t<d)\,,\\ &\sigma_{yy}(t,0)=0=\sigma_{xy}(t,0)\,,\quad (t<-c,\ t>d)\end{aligned}\tag{7.151}$$

the integral equation of the problem may be obtained as

$$\frac{4\mu}{1+\kappa}\frac{\partial}{\partial x}v(x,0)=ap(x)-\frac{1}{\pi}\int_{-c}^{d}p(t)\frac{dt}{t-x},\qquad (-c<x<d)\tag{7.152}$$

where

$$\gamma=(\kappa-1)/(\kappa+1)\,,\quad a=\gamma\eta\,.\tag{7.153}$$

(a) $\kappa=-1$. First consider the case of a spinning cylinder of radius $R$

(Figure 7.9) for which the input function is

$$f(x) = \frac{4\mu}{1+\kappa}\frac{\partial v}{\partial x} = \frac{4\mu}{1+\kappa}\frac{x}{R}, \qquad (-c<x<d). \tag{7.154}$$

To normalize the interval $(-c, d)$ the following changes in the variables $r$ and $s$ are made:

$$r=(2t-d+c)/(d+c), \quad s=(2x-d+c)/(d+c). \tag{7.155}$$

Let $p(t)=\phi(r)$. With equation (7.154), (7.152) becomes

$$a\phi(s) - \frac{1}{\pi}\int_{-1}^{1} \phi(r)\frac{dr}{r-s} = As+B, \qquad (-1<s<1) \tag{7.156}$$

where

$$A = \frac{2\mu}{1+\kappa}\frac{d+c}{R}, \quad B = \frac{2\mu}{1+\kappa}\frac{d-c}{R}. \tag{7.157}$$

Here, because of the smooth contact, the real parts of $\alpha$ and $\beta$ must be positive*. Thus from equations (7.119) and (7.120) the fundamental function of (7.157) is found to be

$$\begin{aligned} &w(r)=(1-r)^{\alpha}(1+r)^{\beta}, \\ &\tan \pi\alpha = 1/a, \quad \beta=1-\alpha, \quad \kappa=-1. \end{aligned} \tag{7.158}$$

Assuming a solution to (7.156) of the form (7.124), (7.128) becomes

$$\sum_{0}^{\infty} \frac{2}{\sin \pi\alpha} c_n P_{n+1}^{(-\alpha,-\beta)}(s) = As+B$$

or

$$\begin{aligned} &\frac{2}{\sin \pi\alpha}\left[c_0 P_1^{(-\alpha,-\beta)}(s)+c_1 P_2^{(-\alpha,-\beta)}(s)+\ldots\right] \\ &\quad =\left[B-(1-2\alpha)A\right]P_0^{(-\alpha,-\beta)}(s)+2AP_1^{(-\alpha,-\beta)}(s). \end{aligned} \tag{7.159}$$

Comparing the coefficients of the Jacobi polynomials on both sides of equation (7.159) gives

$$B-(1-2\alpha)A=0, \tag{7.160}$$

* Physically the cylinder spins about a shaft fixed to the cylinder and the load $P$ is applied to the cylinder through the shaft in $-y$-direction. The shaft is also under a torque $M$ (see Figure 7.9). In order to hold the cylinder in equilibrium a force $Q=\eta P$ must also be applied to the shaft in $-x$-direction which is the $x$-component of the bearing reaction.

$$c_0 = A \sin \pi\alpha, \quad c_n = 0, \qquad (n = 1, 2, \ldots). \tag{7.161}$$

In this problem $c$ and $d$, which determine the contact area, are unknown. Equation (7.160) with (7.157) gives one expression relating $c$ and $d$, namely

$$d - c - (1 - 2\alpha)(d + c) = 0. \tag{7.162}$$

A second equation for $c$ and $d$ may be obtained by considering the following equilibrium condition:

$$\int_{-c}^{d} p(t)\,dt = \frac{d+c}{2} \int_{-1}^{1} \phi(r)\,dr = P. \tag{7.163}$$

Using now (7.161), the solution of the integral equation is obtained as

$$\phi(r) = A \sin \pi\alpha (1 - r)^{\alpha} (1 + r)^{\beta} \tag{7.164}$$

or

$$p(t) = \frac{4\mu}{1+\kappa} \frac{\sin \pi\alpha}{R} (d - t)^{\alpha} (c + t)^{\beta}. \tag{7.165}$$

Substituting from equations (7.164) and (7.157), (7.163) becomes

$$\begin{aligned} &\frac{\mu}{1+\kappa} \frac{\sin \pi\alpha}{R} \theta_0^{(\alpha,\beta)} (d + c)^2 = P, \\ &\theta_0^{(\alpha,\beta)} = \frac{2\pi\alpha(1-\alpha)}{\sin \pi\alpha}. \end{aligned} \tag{7.166}$$

Finally, the external torque $M$ which has to be applied to the cylinder is obtained from the following moment equilibrium*:

$$M = \eta P R + \int_{-c}^{d} p(t)\, t\, dt \tag{7.167}$$

which, by using equations (7.165), (7.164), and the orthogonality condition, becomes

$$M = \eta P R + \pi\alpha(1-\alpha) \frac{\mu}{1+\kappa} \frac{(d+c)^2}{R} \left[ d - c + \frac{1-2\alpha}{3}(d + c) \right]. \tag{7.168}$$

(b) $\kappa = 1$. As the next example consider the rigid flat punch of width 2 which is constrained against rotation and is under the loads $P$ and $Q = \eta P$ as shown in Figure 7.10. In this problem the integral equation (7.152) becomes

* Note that $P$ and $M$ are loads per unit thickness of the cylinder.

$$ap(x) - \frac{1}{\pi}\int_{-1}^{1} p(t)\frac{dt}{t-x} = 0, \qquad (-1<x<1) \tag{7.169}$$

subject to

$$\int_{-1}^{1} p(t)dt = P. \tag{7.170}$$

Here, $p$ has integrable singularities at $\mp 1$, $\kappa=1$, and from equations (7.119) and (7.120) it is found that

$$\begin{aligned} &w(t) = (1-t)^{\alpha}(1+t)^{\beta}, \\ &\alpha = -(1-\theta/\pi), \ \beta = -\theta/\pi, \quad \tan\theta = 1/a. \end{aligned} \tag{7.171}$$

In this case since $f=0$, from (7.128) it follows that $c_n=0$, $(n=1, 2, \ldots)$, and equations (7.124) and (7.133) give the solution as

$$p(t) = \frac{Pw(t)}{\theta_0(\alpha, \beta)} \tag{7.172}$$

(c) $\kappa=0$. Consider the contact problem shown in Figure 7.11. The integral equation of the problem is given by (7.52) to (7.54), with $0<x<d$. With the following change in variables

$$\begin{aligned} &r = (2t-d)/d, \quad s = (2x-d)/d, \\ &p(t) = \phi(r), \end{aligned} \tag{7.173}$$

equation (7.52) becomes

$$a\phi(s) - \frac{1}{\pi}\int_{-1}^{1} \phi(r)\frac{dr}{r-s} = \frac{4\mu}{1+\kappa}\frac{d}{R}(s+1), \qquad (-1<s<1). \tag{7.174}$$

$\phi$ has an integrable singularity at $-1$ and is bounded at $+1$; hence from (7.119) and (7.120) it follows that

$$\begin{aligned} &w(r) = (1-r)^{\alpha}(1+r)^{\beta}, \\ &\kappa = 0, \ \alpha = \theta/\pi, \ \beta = -\theta/\pi, \ \tan\theta = 1/a. \end{aligned} \tag{7.175}$$

Using

$$s+1 = P_1^{(-\alpha,-\beta)}(s) + (1+\alpha)P_0^{(-\alpha,-\beta)}(s)$$

from equation (7.128) it is found that

$$c_0 = \frac{4\mu}{1+\kappa}\frac{d}{R}(1+\alpha)\sin\pi\alpha\,,\qquad c_1 = \frac{4\mu}{1+\kappa}\frac{d}{R}\sin\pi\alpha\,,$$

$$c_n = 0\,,\quad (n>1)\,,\qquad \phi(s) = [c_0 P_0^{(\alpha,\beta)}(s) + c_1 P_1^{(\alpha,\beta)}(s)]\,w(s)\,, \tag{7.176}$$

or the pressure may be expressed as

$$p(t) = \left(\frac{d-t}{t}\right)^{\alpha}\left[c_0 + c_1\left(\alpha + \frac{2t-d}{d}\right)\right]. \tag{7.177}$$

The constant $d$ is obtained from

$$P = \int_0^d p(x)\mathrm{d}x = \frac{d}{2}\int_{-1}^{1}\phi(s)\mathrm{d}s$$

which gives

$$c_0\theta_0^{(\alpha,\beta)}\frac{d}{2} = P\,. \tag{7.178}$$

In Figure 7.11 if the direction of the tangential load $Q$ is reversed, in $a=\gamma\eta$ the sign of the coefficient of friction, $\eta$ must be reversed. In this case the foregoing analysis remains valid with the following change:

$$\alpha = \tfrac{1}{2} + \frac{\theta}{\pi}\,,\quad \tan\theta = 1/\gamma\eta\,,\quad \beta = -\alpha$$

where the coefficient of friction, $\eta$ is a positive constant.

## 7.7 Singular integral equations with generalized Cauchy kernels

In some physical problems, in addition to the part having a Cauchy singularity $(t-x)^{-1}$, the kernel of the integral equation may contain terms which may be unbounded only if $t$ and $x$ simultaneously approach one or both end points of the interval. For example, in the first problem considered in section 7.5, if $a=0$, that is, if one end of the crack terminates at the interface, then it is clear that $H(r, r_0)$ is no longer a Fredholm kernel, since for $r=r_0$, $r_0\to 0$, $H$ becomes unbounded at $r_0=0$ as $r_0^{-1}$. In this section it will be shown that such additional terms in the kernel affect the nature of singularity of the unknown function at the related end point. The general problem was considered by Erdogan [25]. Here, the discussion will be restricted to one singular integral equation having a kernel which becomes unbounded at one end only. Thus, the dominant part of the integral equation to be considered is

$$\frac{1}{\pi}\int_a^b \frac{\phi(t)}{t-x}\,dt + \frac{1}{\pi}\int_a^b \phi(t)\sum_0^n c_k(x-a)^k \frac{d^k}{dx^k}(t-z_1)^{-1}\,dt = f(x)\,,$$

$$(a<x<b)\,, \qquad (7.179)$$

$$z_1 = a+(x-a)\,e^{i\theta_1}$$

where $c_k$, $(k=0, 1, \ldots, n)$ and $\theta_1$, $(0<\theta_1<2\pi)$ are known constants. As $x$ varies on $L$, $z_1$ varies on $L_1$ (see Figure 7.12) such that as $x \to a$, $z_1 \to a$.

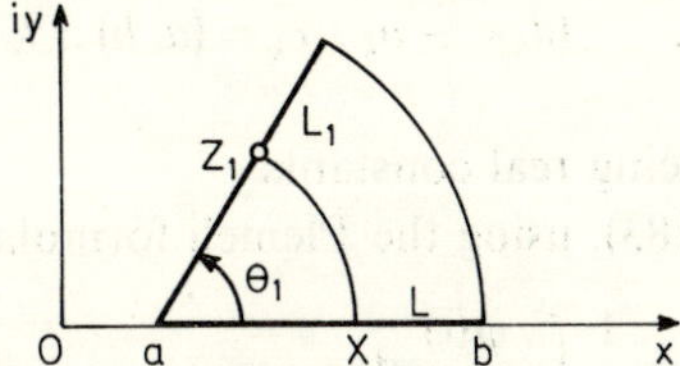


Figure 7.12. The branch cut $L$ and the line $L_1$ in the complex plane.

*The fundamental function.* The integral equation (7.179) cannot be reduced to a simple Riemann–Hilbert problem in order to obtain the fundamental function of the solution. For this, the function-theoretic method outlined by Muskhelishvili [5] has to be applied directly to the integral equation. Following Muskhelishvili, the most general class of solution of equation (7.179) may be expressed as

$$\begin{aligned}\phi(t) &= g(t)\,w(t) = g(t)(b-t)^{\alpha}(t-a)^{\beta}\\ &= g(t)\,e^{-\pi i\alpha}(t-b)^{\alpha}(t-a)^{\beta}\end{aligned} \qquad (7.180)$$

where $g(t)$ is $H$-continuous in $a \leqslant t \leqslant b$, $(t-b)^{\alpha}(t-a)^{\beta}$ is any definite branch which varies continuously on $a<t<b$, and

$$\alpha = a_1 + ib_1\,,\ \ \beta = a_2 + ib_2\,,\ \ 0 > a_k > -1\,, \qquad (k=1, 2)\,. \qquad (7.181)$$

Now consider the following sectionally holomorphic function

$$\Phi(z) = \frac{1}{\pi}\int_a^b \frac{\phi(t)}{t-z}\,dt = \frac{e^{-\pi i\alpha}}{\pi}\int_a^b (t-b)^{\alpha}(t-a)^{\beta} g(t)\,\frac{dt}{t-z}\,. \qquad (7.182)$$

Following the technique outlined by Muskhelishvili [5, Chapter 4], the singular behavior of $\Phi(z)$ near the end points may be examined and $\Phi(z)$ may be expressed as

$$\Phi(z) = -g(a)(b-a)^{\alpha} \frac{e^{-\pi i\beta}}{\sin \pi\beta} (z-a)^{\beta}$$

$$+g(b)(b-a)^{\beta} \frac{1}{\sin \pi\alpha} (z-b)^{\alpha} + \Phi_0(z) \tag{7.183}$$

where the function $\Phi_0$ is bounded everywhere except possibly at the ends near which

$$|\Phi_0(z)| < \frac{C_k}{|z-c_k|^{d_k}}, \qquad (d_k < -a_k\,,\ c_k = (a, b)\,,\ k = 1, 2)\,, \tag{7.184}$$

$C_k$ and $d_k$, $(k=1, 2)$ being real constants.

From equation (7.183), using the Plemelj formula

$$\tfrac{1}{2}[\Phi^+(x) + \Phi^-(x)] = \frac{1}{\pi}\int_a^b \frac{\phi(t)}{t-x}\,dt \tag{7.185}$$

we obtain

$$\frac{1}{\pi}\int_a^b \frac{\phi(t)}{t-x}\,dt = -g(a)(b-a)^{\alpha} \cot \pi\beta (x-a)^{\beta}$$

$$+g(b)(b-a)^{\beta} \cot \pi\alpha (b-x)^{\alpha} + F_0(x)\,, \qquad (a<x<b) \tag{7.186}$$

where the behavior of $F_0$ near the ends is similar to that of $\Phi_0$ as given by (7.184).

From equation (7.182) it is clear that $L$ is the line of discontinuity and $\Phi(z)$ is holomorphic everywhere on $L_1$ except at the end point $z_1 = a$. Thus,

$$\frac{1}{\pi}\int_a^b \frac{\phi(t)}{t-z_1}\,dt = \Phi(z_1)\,, \qquad (z_1 \in L_1)\,, \tag{7.187}$$

or substituting $z_1 - a = (x-a)e^{i\theta_1}$, from equation (7.183) we obtain

$$\Phi(z_1) = -g(a)(b-a)^{\alpha} \frac{e^{-\pi i\beta}}{\sin \pi\beta} e^{i\beta\theta_1}(x-a)^{\beta} + F_1(x)\,, \qquad (a<x<b)\,. \tag{7.188}$$

The behavior of $F_1(x)$ near the end $x=a$ is similar to that of $\Phi_0$ given by (7.184), otherwise $F_1$ is a bounded function.

Since $\Phi$ is holomorphic at $z=z_1$, using (7.188) the second group of terms in equation (7.179) may be expressed as

$$\frac{1}{\pi}\int_a^b \phi(t)(x-a)^k \frac{\mathrm{d}^k}{\mathrm{d}x^k}(t-z_1)\mathrm{d}t = (x-a)^k \frac{\mathrm{d}^k}{\mathrm{d}x^k}\Phi(z_1)$$

$$= -g(a)(b-a)^\alpha \frac{\mathrm{e}^{-\pi i\beta}}{\sin \pi\beta}\mathrm{e}^{i\beta\theta_1}\beta(\beta-1)\dots(\beta-k+1)(x-a)^\beta$$

$$+(x-a)^k \frac{\mathrm{d}^k}{\mathrm{d}x^k}F_1(x), \qquad (a<x<b). \tag{7.189}$$

Substituting now from equations (7.186), (7.188), and (7.189) into (7.179) gives

$$-g(a)(b-a)^\alpha \cot \pi\beta\,(x-a)^\beta + g(b)(b-a)^\beta \cot \pi\alpha\,(b-x)^\alpha + F_0(x)$$

$$-c_0 g(a)(b-a)^\alpha \frac{\mathrm{e}^{-\pi i\beta}}{\sin \pi\beta}\mathrm{e}^{i\beta\theta_1}(x-a)^\beta + F_1(x)$$

$$+\sum_1^n c_k\left[-g(a)(b-a)^\alpha \frac{\mathrm{e}^{-\pi i\beta}}{\sin \pi\beta}\mathrm{e}^{i\beta\theta_1}\beta(\beta-1)\dots(\beta-k+1)(x-a)^\beta\right.$$

$$\left.+(x-a)^k \frac{\mathrm{d}^k}{\mathrm{d}x^k}F_1(x)\right] = f(x), \qquad (a<x<b). \tag{7.190}$$

Observing that $g(a)$ and $g(b)$ are nonzero and

$$-1<\mathrm{Re}(\alpha)<0, \quad -1<\mathrm{Re}(\beta)<0,$$

$$\lim_{x\to a}(x-a)^{-\beta}[f(x),\ F_0(x),\ (x-a)^k F_1^{(k)}(x)] = 0, \qquad (k=0,1,\dots,n),$$

$$\lim_{x\to b}(b-x)^{-\alpha}[f(x),\ F_0(x),\ (x-a)^k F_1^{(k)}(x)] = 0, \qquad (k=0,\dots,n),$$

from equation (7.190), by multiplying both sides first by $(x-a)^{-\beta}$ and letting $x\to a$, and then by $(b-x)^{-\alpha}$ and letting $x\to b$, the following characteristic equations for the unknown constants $\alpha$ and $\beta$ are obtained:

$$\cot \pi\alpha = 0, \quad \alpha = -\tfrac{1}{2} \tag{7.191}$$

$$\cos \pi\beta + \mathrm{e}^{i\beta(\theta_1-\pi)}\left[c_0 + \sum_1^n c_k\beta(\beta-1)\dots(\beta-k+1)\right] = 0. \tag{7.192}$$

With $\alpha$ and $\beta$ determined from equations (7.191) and (7.192), the fundamental function of the singular integral equation with a generalized Cauchy kernel may be expressed as

$$w(t) = (b-t)^\alpha (t-a)^\beta. \tag{7.193}$$

From the foregoing derivation, it is obvious that, instead of the simple case of $z_1 = a+(x-a)^{i\theta_1}$ varying on $L_1$, if the kernel contains terms involving $z_{1j}=a+(x-a)^{i\theta_{1j}}$, $0<\theta_{1j}<2\pi$, $z_{1j}\in L_{1j}$, and $z_{2p}=b+(b-x)e^{i\theta_{2p}}$, $-\pi<\theta_{2p}<\pi$, $z_{2p}\in L_{2p}$, one could easily extend this derivation and obtain a pair of characteristic equations to determine $\alpha$ and $\beta$ (see [25]). Also, the case of a coupled system of singular integral equations with generalized Cauchy kernels may be treated in the same manner. Such a case, for example, arises from the analysis of two bonded half planes containing a finite crack which makes an arbitrary angle with and terminates at the interface. Instead of having an integrable singularity, if the function $\phi(t)$ is bounded at an end, say $x=a$, then the foregoing analysis remains the same, except that the corresponding exponent $\beta$ in the fundamental function must now be replaced by $1-\beta$.

**Solution by Gauss–Jacobi integration formula.** Consider now the following singular integral equation with a generalized Cauchy kernel:

$$\frac{1}{\pi}\int_{-1}^{1}\frac{\phi(t)\mathrm{d}t}{t-x}+\frac{1}{\pi}\int_{-1}^{1}K(x,t)\phi(t)\mathrm{d}t$$

$$+\int_{-1}^{1}k(x,t)\phi(t)\mathrm{d}t=f(x)\,,\qquad(-1<x<1)\tag{7.194}$$

where, without any loss in generality, the interval $(a, b)$ has been normalized to be $(-1, 1)$, $k(x, t)$ is a Fredholm kernel, $f(x)$ is $H$-continuous in $(-1\leqslant x\leqslant 1)$ and is known, and

$$K(x,t)=\sum_{0}^{n}c_k(x+1)^k\frac{\mathrm{d}^k}{\mathrm{d}x^k}(t-z_1)^{-1}$$

$$+\sum_{0}^{m}b_j(1-x)^j\frac{\mathrm{d}^j}{\mathrm{d}x^j}(t-z_2)^{-1}\,,\tag{7.195}$$

$$z_1=-1+(x+1)e^{i\theta_1}\,,\quad z_2=1+(1-x)e^{i\theta_2}\,,$$

$$-1<(x,t)<1\,,\quad 0<\theta_1<2\pi\,,\quad -\pi<\theta_2<\pi\,.$$

If the function $\phi$ has integrable singularities at the end points, in addition to equation (7.194) it must satisfy an equilibrium or a compatibility condition of the form

$$\int_{-1}^{1} \phi(t)\mathrm{d}t = A \tag{7.196}$$

where $A$ is a known constant.

To obtain a numerical solution of equation (7.194), first following the derivation given in section 7.9, the fundamental function $w(t)=(1-t)^{\alpha}\cdot(1+t)^{\beta}$ of the integral equation is determined. Then noting that the singular behavior of the unknown function $\phi(t)$ near the end points $-1$ and $+1$ is completely characterized by that of $w(t)$, it is assumed that

$$\phi(t) = w(t)g(t), \qquad (-1<t<1) \tag{7.197}$$

where $g(t)$ is a bounded function in the closed interval $(-1\leqslant t\leqslant 1)$. Note that $w(t)$ is the weight function of the Jacobi polynomials $P_n^{(\alpha,\beta)}(t)$. Thus, equation (7.194) may be solved by using a numerical method based on a Gauss–Jacobi integration formula similar to that outlined in section 7.3. In this case, the basic quadrature formula is (see [13])

$$\int_{-1}^{1} F(t,x)(1-t)^{\alpha}(1+t)^{\beta}\mathrm{d}t \simeq \sum_{1}^{N} W_k F(t_k, x), \qquad (-1<[\mathrm{Re}(\alpha),\ \mathrm{Re}(\beta)]<1) \tag{7.198}$$

where $t_k$ are the roots of

$$P_N^{(\alpha,\beta)}(t_k) = 0, \qquad (k=1, \ldots, N), \tag{7.199}$$

and the weights are given by

$$W_k = -\frac{2N+\alpha+\beta+2}{(N+1)!\,(N+\alpha+\beta+1)}\,\frac{\Gamma(N+\alpha+1)\,\Gamma(N+\beta+1)}{\Gamma(N+\alpha+\beta+1)}\times$$

$$\times\frac{2^{\alpha+\beta}}{P_N'^{(\alpha,\beta)}(t_k)\,P_{N+1}^{(\alpha,\beta)}(t_k)}. \tag{7.200}$$

For $-1<[\mathrm{Re}(\alpha), \mathrm{Re}(\beta)]<0$, substituting from (7.197) into equations (7.194) and (7.196), and formally using (7.198) we obtain (see section 7.3)

$$\frac{1}{\pi}\sum_{k=1}^{N} g(t_k)\,W_k\left[\frac{1}{t_k-x_j} + K(x_j, t_k)+\pi k(x_j, t_k)\right] = f(x_j), \qquad (j=1, \ldots, N-1), \tag{7.201}$$

$$\sum_{1}^{N} g(t_k)\,W_k = A \tag{7.202}$$

where $t_k$ and $x_j$ are given by

$$P_N^{(\alpha,\beta)}(t_k)=0\,, \qquad (k=1,\,\ldots,\,N)\,,$$
$$P_{N-1}^{(\alpha+1,\beta+1)}(x_j)=0\,, \qquad (j=1,\,\ldots,\,N-1) \tag{7.203}$$

and the weights $W_k$ are given by (7.200). Equations (7.201) and (7.202) provide $N$ linear algebraic equations to determine $g(t_1),\,\ldots,\,g(t_N)$.

For other combinations of $\alpha$ and $\beta$, the equation (7.201) and the first equation of (7.203) remain valid. However, the points $x_j$ must be selected as follows:

$$0<\mathrm{Re}(\alpha)<1\,, \quad 0<\mathrm{Re}(\beta)<1: \quad P_{N+1}^{(\alpha-1,\beta-1)}(x_j)=0\,,$$
$$0<\mathrm{Re}(\alpha)<1\,, \quad -1<\mathrm{Re}(\beta)<0: \quad P_N^{(\alpha-1,\beta+1)}(x_j)=0\,,$$
$$-1<\mathrm{Re}(\alpha)<0\,, \quad 0<\mathrm{Re}(\beta)<1: \quad P_N^{(\alpha+1,\beta-1)}(x_j)=0\,.$$

There is no additional condition in any of these three cases. In the case where the function is bounded at both ends, (i.e., for $0<[\mathrm{Re}(\alpha),\,\mathrm{Re}(\beta)]<1$), equations (7.201) may be expressed at $N+1$ points $x_j$, $(j=1,\,\ldots,\,N+1)$. Since there are only $N$ unknowns $g(t_k)$, $(k=1,\,\ldots,\,N)$, in this case the equation corresponding to one of the $x_j$'s (usually the point closest to $x=0$) must be ignored.

**Example:** *Two bonded half planes with a crack terminating at the interface.* The elasticity problem in Figure 7.1 for two bonded half planes containing a finite crack located perpendicular to the interface will now be reconsidered. In that problem it was assumed that the "symmetric" surface traction $\sigma_{1\theta\theta}=-p(r_0)$ acting on the crack is the only external load. The case of a crack fully embedded in one of the homogeneous half planes, (i.e., $b>r>a>0$) was considered as the first example in section 7.5. In that case, aside from a term having a Cauchy singularity, the kernel was continuous and bounded in the closed domain $a\leqslant(r,\,r_0)\leqslant b$.

If one end of the crack terminates at the interface i.e., by letting $a\to 0$, it is clear that the integral equation (7.83) or (7.86), and the compatibility condition (7.88) are still valid. However, the kernel $H(r,\,r_0)$ given by (7.84) is no longer a Fredholm kernel, as it becomes unbounded for $r\to 0$, $r_0\to 0$ with $r=r_0$. Thus, letting $a=0$ and with the following change in variable

$$t = (r_0 - a_0)/a_0\,, \quad x = (r - a_0)/a_0\,, \qquad 2a_0 = b\,,$$
$$\frac{1+\kappa_1}{2\mu_1}\, p(r) = f(x)\,, \quad \phi(r_0) = F(t)\,, \tag{7.204}$$

from equations (7.83), (7.84), and (7.88) one obtains

$$\frac{1}{\pi}\int_{-1}^{1} F(t)\frac{dt}{t-x} + \frac{1}{\pi}\int_{-1}^{1} K(x,t)F(t)dt = f(x)\,, \qquad (-1<x<1)\,, \tag{7.205}$$

$$\int_{-1}^{1} F(t)dt = 0 \tag{7.206}$$

where

$$K(x,t) = \frac{c_0}{t+x+2} + c_1(1+x)\frac{d}{dx}(t+x+2)^{-1}$$
$$+ c_2(1+x)^2\frac{d^2}{dx^2}(t+x+2)^{-1}\,,$$
$$c_0 = \tfrac{1}{2}\left[1 - \frac{m(1+\kappa_1)}{m+\kappa_2} - \frac{3(1-m)}{1+m\kappa_1}\right], \tag{7.207}$$
$$c_1 = -\frac{6(1-m)}{1+m\kappa_1}\,, \quad c_2 = -\frac{2(1-m)}{1+m\kappa_1}\,, \quad m = \mu_2/\mu_1\,.$$

It is seen that $K(x, t)$ is of the form (7.195) with $\theta_1 = \pi$, $L_1 = (-3, -1)$.

Assuming now the solution of equation (7.205) to be

$$F(t) = w(t)\,g(t) = (1-t)^{\alpha}(1+t)^{\beta}g(t)\,, \qquad (|t|<1) \tag{7.208}$$

and using $c_0$, $c_1$, $c_2$ as given by (7.207), after some manipulations, the characteristic equations (7.191) and (7.192) are obtained as

$$\cot \pi\alpha = 0\,, \quad \alpha = -\tfrac{1}{2}\,, \tag{7.209}$$

$$2d_1 \cos \pi(\beta+1) - d_2(\beta+1)^2 - d_3 = 0 \tag{7.210}$$

where

$$d_1 = (m+\kappa_2)(1+m\kappa_1)$$
$$d_2 = -4(m+\kappa_2)(1-m) \tag{7.211}$$
$$d_3 = (1-m)(m+\kappa_2) + (1+m\kappa_1)(m+\kappa_2) - m(1+\kappa_1)(1+m\kappa_1)\,.$$

Since the end $t=1$ is embedded into a homogeneous medium, (7.209) gives

TABLE VI

*The first root of the characteristic equation,* (7.210)

| *Material pair* | $m=\mu_2/\mu_1$ | $-\beta$ | |
|---|---|---|---|
| | | *Plane strain* | *Plane stress* |
| Aluminum-rigid | $\infty$ | 0.2888 | 0.2417 |
| Epoxy-rigid | $\infty$ | 0.3203 | 0.2616 |
| Epoxy-boron | 138.46 | 0.3234 | 0.2666 |
| Epoxy-aluminum | 23.08 | 0.3381 | 0.2890 |
| Aluminum-steel | 3.00 | 0.4007 | 0.3892 |
| Aluminum-aluminum | 1.00 | 0.5000 | 0.5000 |
| Steel-aluminum | 0.333 | 0.6205 | 0.6210 |
| Aluminum-epoxy | 0.043 | 0.8248 | 0.8242 |
| Boron-epoxy | 0.0072 | 0.9258 | 0.9251 |
| Rigid-aluminum | 0 | 1.0000 | 1.0000 |

the expected $\alpha=-\frac{1}{2}$ singularity. It can be shown that (7.210) is identical to the characteristic equation obtained from the analysis of bonded wedges through the use of Mellin transforms (see, for example, [19, 20]), or from the eigenfunction expansion (see Zak and Williams [26]). The examination of equations (7.210) and (7.211) shows that the first acceptable root of (7.210) is always real for all possible material combinations. Tables VI and VII show the roots of equation (7.210) for various material combinations.

After determining the constants $\alpha$ and $\beta$, equations (7.201) and (7.202) give the solution of the integral equation. The values of the function $g(t)$ at the end points are related to the stress intensity factors. This relationship can be established by obtaining the asymptotic expressions for $\sigma_{1\theta\theta}(r,\pi)$ and $\sigma_{2\theta\theta}(r,0)$ near the points $r=b$ and $r=0$, respectively, by using the respective Green's functions and the density function $\phi(r_0)$ (see [20]). The stress intensity factors can also be obtained directly from the following general expression relating the asymptotic values of displacement derivative $\partial u_{1\theta}/\partial r$ and the cleavage stress $\sigma_{2\theta\theta}$ (see [20]):

$$\sigma_{2\theta\theta}(r,0)=-2\mu^*\frac{\partial}{\partial r}u_{1\theta}(r,\pi-0)+O(r^\lambda)\,,\qquad(\lambda>0)\,,$$

$$\mu^*=\mu_1 m\,\frac{(3+2\beta)(1+m\kappa_1)-(1+2\beta)(m+\kappa_2)}{(m+\kappa_2)(1+m\kappa_1)\sin\pi(1+\beta)}\,. \tag{7.212}$$

TABLE VII

*The first root of the characteristic equation,* (7.210)

| $\nu_1=\nu_2=0.30$ | | | $\nu_1=0.35,\ \nu_2=0.30$ | | |
|---|---|---|---|---|---|
| | *Plane strain* | *Plane stress* | | *Plane strain* | *Plane stress* |
| $\mu_2/\mu_1$ | $-\beta$ | $-\beta$ | $\mu_2/\mu_1$ | $-\beta$ | $-\beta$ |
| 1000 | 0.2893 | 0.2424 | 1000 | 0.3207 | 0.2623 |
| 100 | 0.2939 | 0.2490 | 100 | 0.3246 | 0.2684 |
| 44.44 | 0.2999 | 0.2578 | 46.15 | 0.3295 | 0.2759 |
| 22.22 | 0.3013 | 0.2724 | 23.07 | 0.3381 | 0.2890 |
| 10 | 0.3328 | 0.3033 | 10.0 | 0.3583 | 0.3186 |
| 1.02 | 0.4980 | 0.4979 | 1.02 | 0.5094 | 0.5038 |
| 1.0 | 0.5000 | 0.5000 | 1.00 | 0.5114 | 0.5059 |
| 0.98 | 0.5021 | 0.5022 | 0.98 | 0.5134 | 0.5080 |
| 0.10 | 0.7536 | 0.7513 | 0.10 | 0.7608 | 0.7548 |
| 0.045 | 0.8258 | 0.8230 | 0.043 | 0.8344 | 0.8289 |
| 0.01 | 0.9148 | 0.9130 | 0.01 | 0.9178 | 0.9146 |
| 0.001 | 0.9728 | 0.9722 | 0.001 | 0.9738 | 0.9727 |

Equation (7.212) is valid for small values of $r$, i.e., for the stress and the displacement derivative around the end point $r=0$. For the end point $r=b$ equation (7.212) is modified as

$$\sigma_{1\theta\theta}(r,0)|_{r>b}=2\mu^*\frac{\partial}{\partial r}u_{1\theta}(r,\pi-0)|_{r<b}+O[(r-b)^\lambda]. \tag{7.213}$$

From the definition of $\phi$ given by (7.78) it follows that (see equation (7.90))

$$\begin{aligned}&\sigma_{2\theta\theta}(r,0)=\mu^*\phi(r)+O(r^\lambda), \qquad (r>0),\\ &\sigma_{1\theta\theta}(r,0)|_{r>b}=-\frac{2\mu_1}{1+\kappa_1}\phi(r)|_{r<b}+O[(r-b)^\lambda].\end{aligned} \tag{7.214}$$

Now defining the stress intensity factors as

$$\begin{aligned}k(b)&=\lim_{r\to b}[2(r-b)]^{\frac{1}{2}}\sigma_{1\theta\theta}(r,\pi),\\ k(a)&=\lim_{r\to 0}2^{\frac{1}{2}}\cdot r^{-\beta}\sigma_{2\theta\theta}(r,0),\end{aligned} \tag{7.215}$$

from equations (7.214) it is found that

$$k(b) = -\frac{2\mu_1}{1+\kappa_1} \lim_{r\to b} [2(b-r)]^{\frac{1}{2}} \phi(r),$$
$$k(a) = \lim_{r\to 0} \mu^* 2^{\frac{1}{2}} \cdot r^{-\beta} \phi(r). \tag{7.216}$$

To solve equation (7.205), define

$$\phi(r) = h(r) r^{\beta} (b-r)^{\alpha} = F(x) = g(x)(1-x)^{\alpha}(1+x)^{\beta}. \tag{7.217}$$

From equations (7.204) and (7.217) it is seen that $h(r) = g(x) a_0^{-(\alpha+\beta)}$. Substituting now $\alpha = -\frac{1}{2}$ and using (7.216) and (7.217) result in

$$k(b) = -\frac{2\mu_1}{1+\kappa_1} 2^{\frac{1}{2}+\beta} a_0^{\frac{1}{2}} g(1),$$
$$k(a) = \mu^* a_0^{-\beta} g(-1), \tag{7.218}$$

where the bielastic constant $\mu^*$ is defined by equation (7.212).

The numerical solution of (7.205) and (7.206) is obtained by using equations (7.201) and (7.202) (where $A=0$). The resulting stress intensity factors for various material combinations and loading conditions are shown in Tables II and III as the first line corresponding to $c=a_0$. In this problem, too, the most stable results for $g(\mp 1)$ are obtained by ignoring the last point (i.e., $g(t_1)$ and $g(t_N)$) and using a quadratic extrapolation formula based on the next three values of $g(t_k)$, (i.e., $k=2, 3, 4$ and $k=N-3, N-2, N-1$).

In order to have some idea about the convergence of the results, in equations (7.201) the number of points in the Gauss–Jacobi quadrature formula,

TABLE VIII

*The effect of the number of points in the Gauss–Jacobi quadrature formula on the stress intensity factors*

| $p(x) = -p_0$ | $\mu_2/\mu_1 = 23.08$ | |
|---|---|---|
| $N$ | $\frac{k(b)}{p_0(a_0)^{\frac{1}{2}}}$ | $\frac{k(a)}{p_0(a_0)^{\frac{1}{2}}}$ |
| 20 | 0.883063837 | 2.63008398 |
| 40 | 0.882810532 | 2.62481953 |
| 48 | 0.882759563 | 2.62447473 |
| 60 | 0.882716454 | 2.62417608 |
| 98 | 0.882649609 | 2.62365727 |

$N$ was varied between 20 and 98. The effect of this variation is shown in Table VIII. The speed of convergence appears to be the same at both ends, and is quite satisfactory★.

## References

[1] Erdogan, F., *J. Appl. Mech., Trans. ASME*, 32, p. 829 (1965).
[2] Arin, K. and Erdogan, F., *Int. J. Engrg. Sci.*, 9, p. 213 (1971).
[3] Erdogan, F. and Gupta, G. D., *Int. J. Solids, Structures*, 7, p. 39 (1971a).
[4] Erdogan, F. and Gupta, G. D., *Int. J. Solids, Structures*, 7, p. 1089 (1971b).
[5] Muskhelishvili, N. I., *Singular Integral Equations*, P. Noordhoff, Groningen, The Netherlands (1953).
[6] Erdogan, F., *SIAM J. Appl. Math.*, 17, p. 1041 (1969).
[7] Erdogan, F. and Gupta, G. D., *Int. J. Solids, Structures*, 8, p. 93 (1972a).
[8] Bueckner, H. F., *J. Math. Anal. Appl.*, 14, p. 392 (1966).
[9] Bierman, G. J., *SIAM J. Appl. Math.*, 20, p. 99 (1971).
[10] Pogorzelski, W., *Integral Equations and their Applications*, Pergamon Press, New York (1966).
[11] Gakhov, F. D., *Boundary Value Problems*, Pergamon Press, New York (1966).
[12] Abramowitz, M. and Stegun, I. A., *Handbook of Mathematical Functions*, National Bureau of Standards, Appl. Math. Series 55 (1964).
[13] Stroud, A. H. and Secrest, D., *Gaussian Quadrature Formulas*. Prentice-Hall, New York (1966).
[14] Szegö, G., *Orthogonal Polynomials*, Colloquium Publications, 23, Am. Math. Soc. (1939).
[15] Erdelyi, A., *Higher Transcendental Functions*, Vol. 2, McGraw-Hill, New York (1953).
[16] Erdogan, F. and Gupta G. D., *Quart. Appl. Math.*, 30, p. 525 (1972b).
[17] Gradshteyn, I. S. and Ryzhik, I. M., *Table of Integrals, Series, and Products*, Academic Press, New York (1965).
[18] Kantorovich, L. V. and Krylov, V. I., *Approximate Methods of Higher Analysis*. Interscience, New York (1958).
[19] Hein, V. L. and Erdogan, F., *Int. J. Fracture Mechanics*, 7, p. 317 (1971).
[20] Cook, T. S., Ph. D. Dissertation, Lehigh University (1971).
[21] Koiter, W. T., *J. Appl. Mech., Trans. ASME*, 32, p. 237 (1965).
[22] Tricomi, F. G., *Integral Equations*, Interscience, New York (1957).
[23] Erdogan, F. and Ozbek, T., *J. Appl. Mech., Trans. ASME*, 35, p. 865 (1969).
[24] Malyshev, B. M. and Salganik, R. L., *Int. J. Fracture Mechanics*, 1, p. 114 (1965).
[25] Erdogan, F., *Treatise on Continuum Physics*, edited by A. C. Eringen, Academic Press, New York (1972).
[26] Zak, A. R. and Williams, M. L., *J. Appl. Mech., Trans. ASME*, 30, p. 142 (1963).

★ For a more general discussion of the question of convergence of this and the other methods considered in this Chapter, and for a technique of estimating the values such as $g(\mp 1)$ as $N \to \infty$ see Erdogan and Gupta [16].

For another example in which the kernel $K(x, t)$ in equation (7.205) becomes unbounded at both ends, see Erdogan and Gupta [7].

*P. D. Hilton and G. C. Sih*

# 8 *Applications of the finite element method to the calculations of stress intensity factors*

## 8.1 Introduction

According to linear elastic fracture mechanics theory in two-dimensions, a crack will begin to propagate when the amplitude of the stress field in the immediate vicinity of the crack tip, the stress intensity factor, reaches a critical value. This is equivalent to the original concept of Griffith based on energy considerations. Thus, the prediction of strength for a specimen requires the knowledge of the stress intensity factor in terms of the specimen geometry and loading conditions. While analytic solutions for crack problems are available for many configurations [1], they are limited to idealized geometries. Hence it is important to develop a numerical approach which is capable of determining accurate values for stress intensity factors for a wide range of specimen geometries and loading conditions. The problem is complicated by the fact that the quantity of primary interest is the amplitude of a singular solution; and the accuracy of most standard numerical techniques without special treatment of this singularity deteriorates quickly in the neighborhood of the crack tip.

One approach for the numerical determination of stress intensity factors which has received considerable attention because of its ability to treat very general geometric and loading conditions is the finite element technique. Many articles and a number of books have been written about various aspects and applications of the finite element method. These will not be reviewed here; rather, the reader who is unfamiliar with this technique, is referred to [2] as a starting point for learning the finite element method.

## 8.2 Some aspects of finite element method

A brief description of some aspects of the finite element displacement method

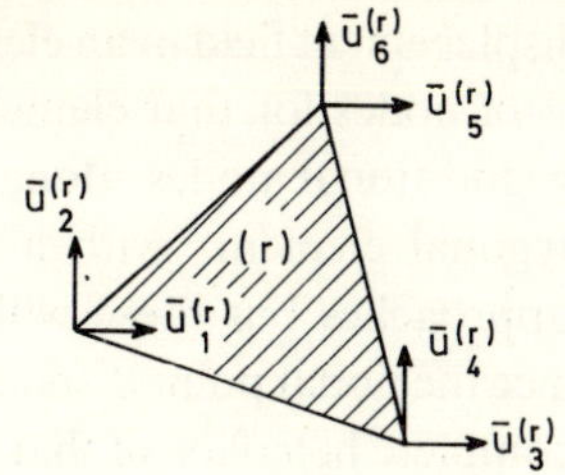


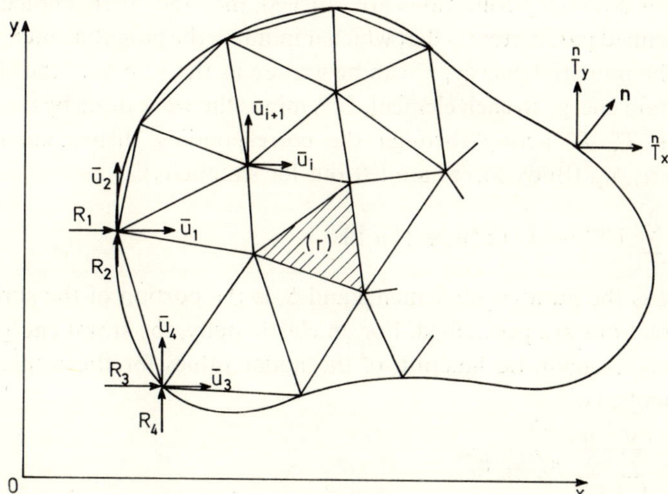


Figure 8.1. A typical finite element idealization of a two-dimensional solid.

for two-dimensional plane problems which will be used to discuss fracture calculations follows. The two-dimensional body is divided into a number of sub-regions called elements, Figure 8.1. Triangular elements are frequently employed because they lead to a relatively simple mathematical formulation while giving some flexibility for the choice of subdivisions and thus the ability to solve a rather large set of problems. A functional representation of the displacement components in each element known as the displacement shape function is then chosen which contains a number of undetermined parameters. The most elementary choice for the displacement shape function associated with triangular elements gives a linear displacement field in the element and the coefficients can be determined as a function of the displacement component values at the corners of the triangle, the nodal points. Higher order polynomial expressions for the displacement field in an element are generally associated with a larger number of nodes for that element [2]. This is accomplished by either including additional nodes along the edges of the element or by considering polygonal elements with a larger number of corners. There are alternative approaches but these will not be discussed here. It suffices to observe that once the nodal point displacements have been determined for an element, the complete behavior of that element is known.

If the displacement shape functions are chosen such that continuity of displacement components across element boundaries is insured and displacement boundary conditions are satisfied, then the "best" choice for the undetermined parameters is that which minimizes the potential energy of the body. The potential energy, $V$, can be written as the sum over the elements of the strain energy in each element, $U^{(r)}$, minus the work done by the surface tractions $T_x^n$, $T_y^n$ acting through the corresponding displacement components $u_x$, $u_y$. (Body forces are left out for simplicity).

$$V = \sum_{r=1}^{N} U^{(r)} - \int_{S'} (T_x^n u_x + T_y^n u_y) \mathrm{d}s \tag{8.1}$$

where $N$ is the number of elements and $S'$ is the portion of the surface on which tractions are prescribed. For an elastic body, the strain energy of an element is a quadratic function of the nodal values for the displacement components, i.e.,

$$U^{(r)} = \sum_{i=1}^{M} \sum_{j=1}^{M} k_{ij}^{(r)} \bar{u}_i^{(r)} \bar{u}_j^{(r)} \tag{8.2}$$

where $\bar{u}_j^{(r)}$ is the $j$th nodal displacement component of the $r$th element, $M$ is the number of nodal displacement components associated with that element, and $k_{ij}^{(r)}$ is the $(i, j)$ component of the element stiffness matrix, $[\mathbf{k}^{(r)}]$. The traction integral can be evaluated in a manner consistent with the finite element approach by expressing the displacement components on the surface $S'$ in terms of the nodal displacement components as prescribed by the displacement shape functions for those elements which are in contact with $S'$. Integration over $S'$ gives the result in the form

$$\int_{S'} (T_x^n u_x + T_y^n u_y) \mathrm{d}s = \left[ \sum_{r=1}^{N'} \sum_{j=1}^{M} P_j^{(r)} \bar{u}_j^{(r)} \right]_{S'} \tag{8.3}$$

where $N'$ is the number of elements in contact with the portion of the surface on which tractions are applied, $S'$, and $P_j^{(r)}$ may be regarded as the contribution from element $r$ to the equivalent force at node $j$.

The potential energy is now expressible in terms of the nodal displacement components as

$$V = \sum_{r=1}^{N} \left( \sum_{i=1}^{M} \sum_{j=1}^{M} k_{ij}^{(r)} \bar{u}_i^{(r)} \bar{u}_j^{(r)} - \sum_{j=1}^{M} P_j^{(r)} \bar{u}_j^{(r)} \right). \tag{8.4}$$

Here, $P_j^{(r)}$ is understood to be zero for elements not in contact with $S'$. A global numbering system is introduced for the nodal displacement components, i.e., each component is given a number, independent of the elements with which it is associated, to identify it in the grid pattern. The order of summation in Equation 8.4 is inverted to give

$$K_{ij} = \sum_{r=1}^{N} k_{ij}^{(r)}$$

where $(i, j)$, which now refer to the global numbering system, vary from 1 to $n$, the total number of nodal displacement components (or degrees of freedom) in the finite element grid and $K_{ij}$ are the elements of the master stiffness matrix $[\mathbf{K}]$. Thus the potential energy of the body, as approximated by the finite element procedure, is expressed in terms of the nodal displacement components as

$$V = \sum_{i=1}^{n} \sum_{j=1}^{n} K_{ij} \bar{u}_i \bar{u}_j - \sum_{i=1}^{n} R_i \bar{u}_i \tag{8.5}$$

and the equivalent nodal force, $R_i$, associated with the $i$th degree of freedom, is given by

$$R_i = \sum_{r=1}^{N} P_i^{(r)}.$$

Displacement boundary conditions are satisfied by constraining appropriate nodal displacement components. For definiteness, let us say that the numbers of the constrained components are $c_1, c_2, \ldots, c_m$ where $m$ is the total number of such constraints. Then minimization of the potential energy with respect to the undetermined parameters (unconstrained nodal displacement components) leads to a set of $(n-m)$ linear algebraic equations for these displacement component values, i.e.,

$$\frac{\partial V}{\partial \bar{u}_i} = 0, \qquad i = 1, 2, \ldots, n \text{ but not } c_1, c_2, \ldots, c_m. \tag{8.6}$$

Substitution into the expression for the potential energy, Equation (8.5), gives

$$\sum_{i=1}^{n} K_{il} \bar{u}_i = R_l. \tag{8.7}$$

If the known values $u_{c_1}, u_{c_2}, \ldots, u_{c_m}$ are substituted into this system of equations, the result is a set of symmetric equations of the form

$$\sum_{i=1}^{n-m} K'_{ij}\delta_i = B_j \tag{8.8}$$

where $[\mathbf{K}']$ is known as the reduced master stiffness matrix and

$$\begin{aligned} \delta_i &= \bar{u}_i\,, \quad B_i = R_i - \sum_{l=c_1,c_2,\ldots}^{c_m} K_{i,l}u_l \quad \text{for} \quad i<c_1 \\ \delta_i &= \bar{u}_{i+1}\,, \quad B_i = R_{i+1} - \sum_{l=c_1,c_2,\ldots}^{c_m} K_{i+1,l}u_l \quad \text{for} \quad c_1 \leqslant i \leqslant c_2, \text{etc.} \end{aligned} \tag{8.9}$$

Solution of these equations completes the procedure; stresses and strains within an element can then be determined directly from the corresponding displacement shape functions.

The finite element formulation for axisymmetrical problems is similar to that for the plane theory of elasticity discussed above, except that, in this case, the elements are toroidal in shape, Figure 8.2. The cross section of the toroidal elements, which results from intersecting an element with a plane that contains the axis of symmetry, has the same shape as the corresponding two-dimensional element, i.e., triangular, quadrilateral, etc. The stresses, strains, and displacement components within each axisymmetric element are expressible in terms of the nodal values of the radial and axial displacement components via the displacement shape function. Thus the strain energy of an element will again be quadratic in these nodal displacement

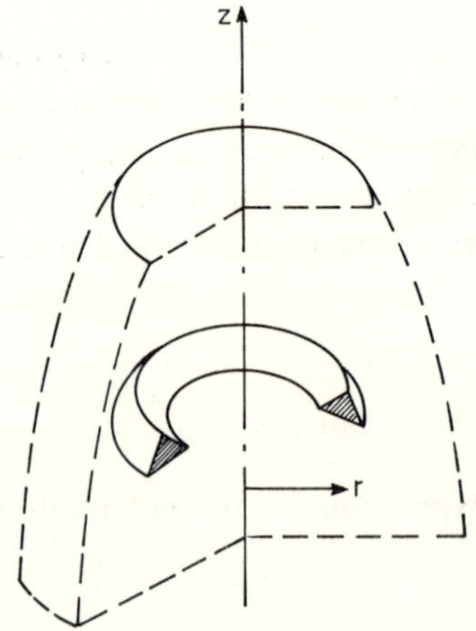


Figure 8.2. A typical finite element for an axi-symmetric problem.

components as expressed by Equation (8.2), but, the numerical values of the coefficients for the axisymmetric element will, in general, differ from those for the corresponding two-dimensional one for two reasons:

(1) There is a non-zero contribution to the strain energy density associated with the circumferential stress and strain components which does not exist for the planar elements.

(2) Integration to obtain the strain energy of an element from the strain energy density distribution is performed over the toroid as opposed to the corresponding cylinder for the planar case.

With these changes, the formulation described for the two-dimensional problems is applicable to the axisymmetric case. A set of linear algebraic equations for the nodal values of the unconstrained radial and axial displacement components, of the form given by Equation (8.8), governs the finite element calculation.

## 8.3 Basic equations of linear fracture mechanics

The character of the stress, strain, and displacement fields in the vicinity of the crack tip is known for the linear two-dimensional theory of elasticity [1] under the conditions of generalized plane stress, plane strain, and anti-plane shear. The local behavior of the in-plane solution also holds for the axisymmetric case. For all but the axisymmetric case, it suffices to consider a through crack directed along the $x$-axis (the out-of-plane direction is denoted by $z$). For in-plane loading situations, the stress and displacement components asymptotically close to the crack tip are

$$\begin{aligned}
\sigma_{xx} &= \frac{k_1}{(2\rho)^{\frac{1}{2}}}\cos(\theta/2)[1-\sin(\theta/2)\sin(3\theta/2)] \\
\sigma_{yy} &= \frac{k_1}{(2\rho)^{\frac{1}{2}}}\cos(\theta/2)[1+\sin(\theta/2)\sin(3\theta/2)] \\
\sigma_{xy} &= \frac{k_1}{(2\rho)^{\frac{1}{2}}}\sin(\theta/2)\cos(\theta/2)\cos(3\theta/2) \qquad (8.10) \\
u_x &= \frac{k_1(2\rho)^{\frac{1}{2}}}{8\mu}[(2\kappa-1)\cos(\theta/2)-\cos(3\theta/2)] \\
u_y &= \frac{k_1(2\rho)^{\frac{1}{2}}}{8\mu}[(2\kappa+1)\sin(\theta/2)-\sin(3\theta/2)]
\end{aligned}$$

for the symmetric case associated with the opening mode of crack extension; while those for the skew-symmetric case of in-plane sliding of one crack surface over the other are

$$\begin{aligned}
\sigma_{xx} &= -\frac{k_2}{(2\rho)^{\frac{1}{2}}}\sin(\theta/2)[2+\cos(\theta/2)\cos(3\theta/2)] \\
\sigma_{yy} &= \frac{k_2}{(2\rho)^{\frac{1}{2}}}\sin(\theta/2)\cos(\theta/2)\cos(3\theta/2) \\
\sigma_{xy} &= \frac{k_2}{(2\rho)^{\frac{1}{2}}}\cos(\theta/2)[1-\sin(\theta/2)\sin(3\theta/2)] \\
u_x &= \frac{k_2(2\rho)^{\frac{1}{2}}}{8\mu}[(2\kappa+3)\sin(\theta/2)+\sin(3\theta/2)] \\
u_y &= -\frac{k_2(2\rho)^{\frac{1}{2}}}{8\mu}[(2\kappa-3)\cos(\theta/2)+\cos(3\theta/2)]
\end{aligned} \tag{8.11}$$

where $\kappa = 3-4\nu$ for plane strain and $(3-\nu)/(1+\nu)$ for generalized plane stress; $(\rho, \theta)$ are polar coordinates centered at the crack tip with $\theta$ measured from the line of expected crack extension. The corresponding expression for the only non-zero displacement component associated with a crack in a body governed by the theory of anti-plane shear is

$$\sigma_{xz} = -\frac{k_3}{(2\rho)^{\frac{1}{2}}}\sin(\theta/2) \quad \sigma_{yz} = \frac{k_3}{(2\rho)^{\frac{1}{2}}}\cos(\theta/2) \quad u_z = \frac{k_3(2\rho)^{\frac{1}{2}}}{\mu}\sin(\theta/2)\,. \tag{8.12}$$

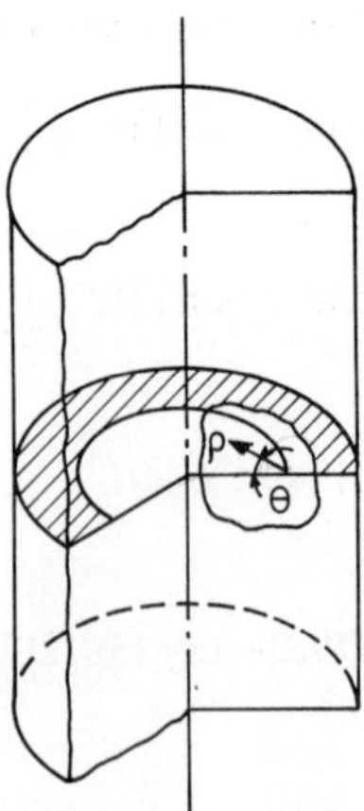


Figure 8.3. A circumferential crack in a solid of revolution.

For axisymmetric problems concerned with a body of revolution containing a penny-shaped crack centered on the axis, the radial and axial displacement components near the crack edge are identical to the plane strain results, Equation (8.10), with $u_x$ and $u_y$ interpreted as the radial and axial components of displacement, $u_r$ and $u_z$ respectively, and $\theta$ measured from the direction of increasing radius. If instead a circumferential crack emanating from the surface of the body is to be considered, then the radial component $u_r$ of displacement becomes the negative of $u_y$, Equation (8.10), and $\theta$ is measured from the direction of decreasing radius, Figure 8.3.

## 8.4 Conventional finite element method applied to crack problems

Early studies on fracture mechanics problems involving the finite element method have been carried out by Swedlow [3], Tuba [4], and Kobayashi [5]. They attempted a straight forward application of the technique with no special attention given to the stress singularity. The specimen was conceptually divided into finite elements with a relatively high element concentration near the crack tip (Figure 8.4 shows a typical grid pattern). The boundary conditions were specified and the resulting equations were solved. The character of the finite element method is such that the numerical values of the displacement components at the element corners are solved for directly

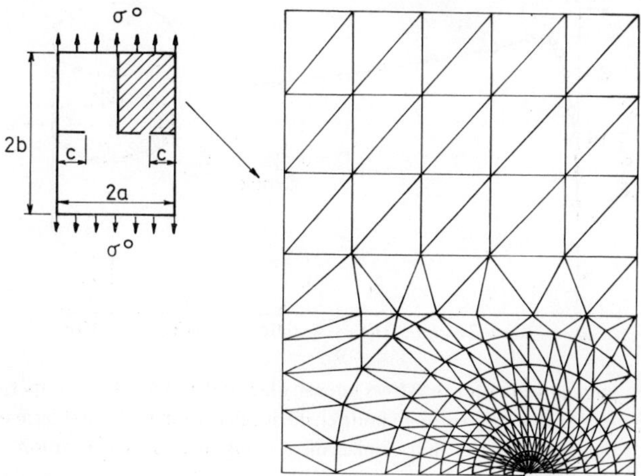

Figure 8.4. A typical finite element grid pattern for a crack problem.

from which the displacement, strain, and stress fields within each element are determined. The stress intensity factor is then found by comparing a displacement component at a node close to the crack tip to the asymptotic solution for the displacement component at the position of that node, Equations (8.10)–(8.12).

Computational experiments indicate the need to include a very large number of degrees of freedom in order to obtain reasonable accuracy [3, 4, 5]. Further study of this technique [6, 7, 8] brings up serious questions about its convergence, reliability, and convenience. The numerical value of the stress intensity factor, determined in the manner described above, varies over a considerable range, depending upon which node is chosen for its calculation. This problem is illustrated by results obtained for the case of a solid cylinder with a circular notch subjected to tensile loading*. Values for the stress intensity factor are obtained by comparing the asymptotic solution for a displacement component at a nodal position with the finite element result for that displacement component at that node. In Figure 8.5, the stress intensity factor, normalized by the product of the net section stress

$$\sigma_{\text{net}} = \sigma^0 \left(\frac{D}{D-2c}\right)^2$$

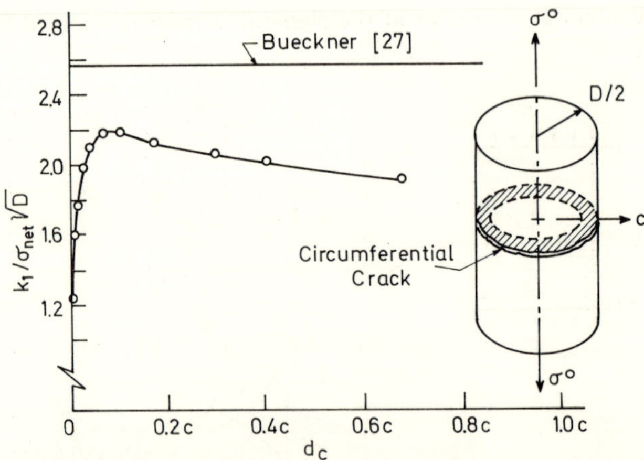


Figure 8.5. Numerical values for the stress intensity factor determined by comparing the nodal point displacement from a conventional finite element calculation to the first term of the asymptotic expansion for that displacement component at that position.

* Calculations performed by J. Oglesby, Naval Ship Research and Development Center.

and the square root of the cylinder diameter, is plotted against the distance, $d_c$, from the crack tip to the node employed for that calculation of the stress intensity factor. Bueckner's result [27] for the stress intensity factor associated with this problem, which is believed to be within one percent of the exact value, is indicated for comparison.

The results based on nodes very near the crack tip and those based on nodes relatively far from it are highly inaccurate. The large errors in numerical values of the stress intensity factor which were obtained via nodes very near the crack tip are directly attributable to the inaccuracy of the finite element method in that region. Specifically, the polynomial form assumed for the displacement fields within the elements does not properly describe the square root behavior of the displacement components near the crack tip. Approximations of this kind obviously lead to finite stress and strain fields at the crack which is not a satisfactory representation of the asymptotic solution, Equations (8.10)–(8.12), of infinite stress and strain. The poor results observed for calculations based on nodes farther from the crack tip are, at least in part, attributed to the inaccuracy of the one-term asymptotic expansion in this region.

Somewhat more reliable results can be obtained by taking an average of the predicted values for the stress intensity factor from the nodes on a circular ring centered at the crack tip. The averaging process is carried out for a number of different ring radii and the results are plotted against distance from the crack tip in the manner shown in Figure 8.5. An extrapolation scheme is then applied.

For a given finite element calculation, significant improvements on the values of the stress intensity factor can be obtained by retaining high order terms in the asymptotic expansion of the displacement components about the crack tip. For example, in the case of a crack in an axisymmetric body subjected to Mode I loading these expansions become [9]

$$
\begin{aligned}
u_r &= \frac{k_1 (2\rho)^{\frac{1}{2}}}{8\mu} \left[(5-8\nu)\cos(\theta/2) - \cos(3\theta/2)\right] \\
&\quad + \frac{\alpha\rho}{2\mu}(1-\nu)\cos\theta + \dots \qquad (8.13) \\
u_z &= \frac{k_1 (2\rho)^{\frac{1}{2}}}{\mu} \left[(7-8\nu)\sin(\theta/2) - \sin(3\theta/2)\right] - \frac{\alpha\rho\nu}{2\mu}\sin\theta + \dots
\end{aligned}
$$

where $\alpha$ is the coefficient of the second term in the asymptotic expansion. Equating these expressions to the nodal point, displacement component values obtained via the finite element procedure at a number of nodes equal to the number of terms retained in Equations (8.13) results in a set of simultaneous equations for $k_1$, $\alpha$, etc. Solution gives the stress intensity factor.

Inclusion of terms of order beyond $\rho^{\frac{1}{2}}$ in Equations (8.13) increases the accuracy of the asymptotic expansions at moderate distances from the crack tip where the finite element results are more accurate. Thus the resulting values for the stress intensity factor are expected to be better than those obtained with the one-term expansion and the best numerical results will be associated nodes which are a bit further from the crack tip than was the case before.

To illustrate the effect of including higher order terms, such as $\rho$, $\rho^{\frac{3}{2}}$ etc., stress intensity factors are recalculated employing the first two terms of the asymptotic expansion (i.e., containing $\rho^{\frac{1}{2}}$ and $\rho$) for the solid cylinder with a circumferential crack, Equations (8.13). Identical grid patterns were used for both calculations. A comparison of the results is shown in Figure 8.6 in the form of a plot of the calculated value for the stress intensity factor against the distance from the crack tip to the nodes used for the calculation.

Note that improvement in the numerical values of the stress intensity factor is observed for nodes further from the crack tip and that the peak value obtained is considerably more accurate than that obtained using the

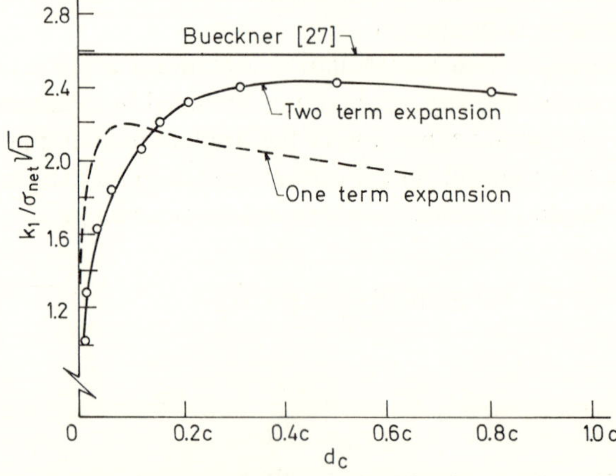


Figure 8.6. A comparison of numerical values obtained for the stress intensity factor via conventional finite element calculations in conjunction with one and two term asymptotic expansions.

one-term expansion. This result supports the explanation given above for the large errors which were observed when using a one-term expansion in conjunction with nodes that were not very near the crack tip to determine the stress intensity factor.

It is appropriate to briefly review the convergence statements for the finite element method to determine whether they are applicable to the calculations described above. Namely, the finite element displacement method with compatible elements (those which enforce continuity of the displacement components across element boundaries) can be shown to have the properties of the Ritz technique. Therefore the convergence theorems for the Ritz method [10] are applicable to such finite element procedures and, in particular, to the crack solution calculations described. The consequence of these theorems is that convergence can be assured only when the approximate displacement field and its first derivatives can be made arbitrarily close to the corresponding true fields everywhere in the region by increasing the number of elements. This condition is violated by any of the standard elements in the immediate vicinity of the crack tip where the displacement derivatives of the exact solution become unbounded. Thus there is no assurance that the finite element solution is accurate near the crack tip and that errors in this region do not propagate into the entire solution.

## 8.5 Finite element method with embedded singularity

As was observed, the direct application of the finite element method to crack problems has a number of drawbacks including the need for an excessively large number of degrees of freedom, the difficulties in determining the stress intensity factor from the finite element results, and the lack of assurance of convergence. In an attempt to eliminate some of these undesirable features, an alternative approach has been developed by Wilson [6] for elastic problems and by Hilton and Hutchinson [11] for elastic-plastic crack problem calculations. It directly incorporates both the finite element method and the analytical crack tip expansions. The philosophy behind this approach is based on the mathematical properties of the two numerical techniques employed, i.e., the asymptotic expansion becomes increasingly more accurate as one approaches the singularity, while the finite element approximation can be made to be very accurate everywhere except near the crack tip. Thus it is natural to attempt to combine these two numerical tools so that each is used in the region it is most accurate and not employed where its accuracy becomes questionable. This thought process has led to a special finite element technique for fracture problems which embeds the asymptotic expansion into the finite element grid at the crack tip.

*Cracked bodies subjected to anti-plane loading.* The embedded singularity, finite element approach to crack problems will first be illustrated for Mode III fracture, i.e., crack growth under out-of-plane or longitudinal shear loading, because the mathematical description of the anti-plane theory is somewhat simpler than that for plane problems. Briefly, the displacement field for longitudinal shear problems is assumed to be of the form

$$u_x = u_y = 0\ ;\quad u_z = u_z(x, y)\,. \tag{8.14}$$

Substitution into Hooke's Law and the strain-displacement relations gives the stress components as

$$\begin{aligned}&\sigma_x=\sigma_y=\sigma_z\sigma_{xy}=0\\&\sigma_{xz}=\mu\gamma_{xz}=\mu(\partial u_z/\partial x)\\&\sigma_{yz}=\mu\gamma_{yz}=\mu(\partial u_z/\partial y)\end{aligned} \tag{8.15}$$

where the shear modulus of elasticity is denoted by $\mu=E/[2(1+\nu)]$. This formulation describes the state of a cylindrical body of infinite height subjected to loading along the generators of the cylinder. Note that the entire theory is expressible in terms of one unknown which is dependent on the two in-plane coordinates, i.e., $u_z=u_z(x, y)$.

In particular, consider the problem of a rectangular solid with in-plane dimensions $2a$ by $2b$, containing a crack of length $2c$, which is subjected to anti-plane shear tractions as illustrated in Figure 8.7. The variational

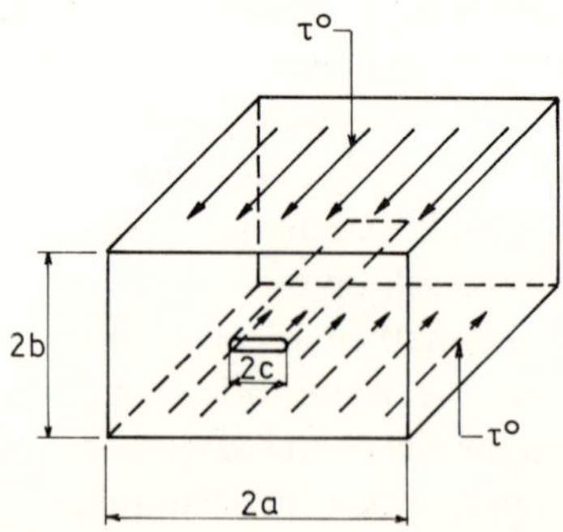


Figure 8.7. A cracked body subjected to anti-plane shearing tractions.

principle of minimum potential energy forms the bases of this analysis. Symmetry conditions permit consideration of the first quadrant alone with the following additional requirements:

$$\begin{aligned} &\sigma_{xz}(0, y) = 0 \\ &\sigma_{yz}(x, 0) = 0\,, \quad x < c \\ &u_z(x, 0) = 0\,, \quad c < x < a\,. \end{aligned} \tag{8.16}$$

For purposes of analysis the first quadrant of the solid is conceptually divided into two regions by a circular arc $\Gamma_1$ of radius $R_1$ centered at the crack tip, Figure 8.8.

The first term of the asymptotic expansion about the crack tip, expressible in polar coordinates centered at the tip as

$$u_z(\rho, \theta) = \frac{k_3(2\rho)^{\frac{1}{2}}}{\mu} \sin(\theta/2) \tag{8.17}$$

is employed to describe the full solution on and within $\Gamma_1$; thus it is necessary

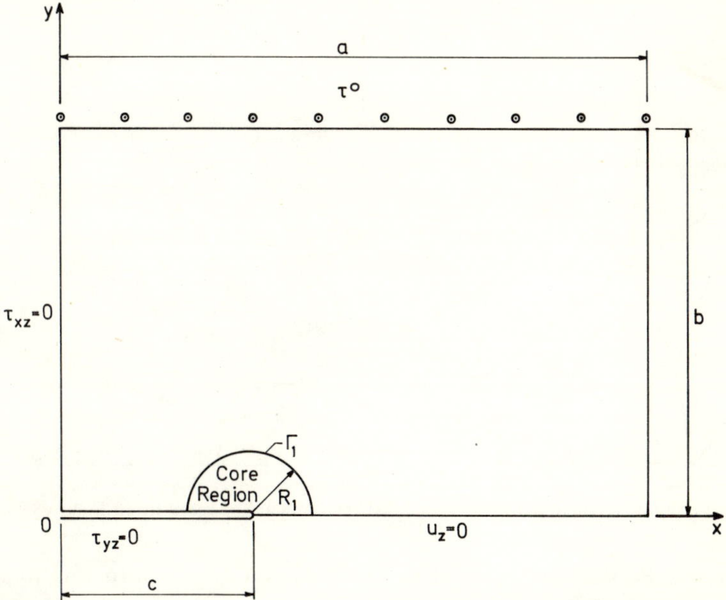


Figure 8.8. Division of the specimen geometry into the core region associated with the asymptotic expansion and the outer region where finite elements are employed.

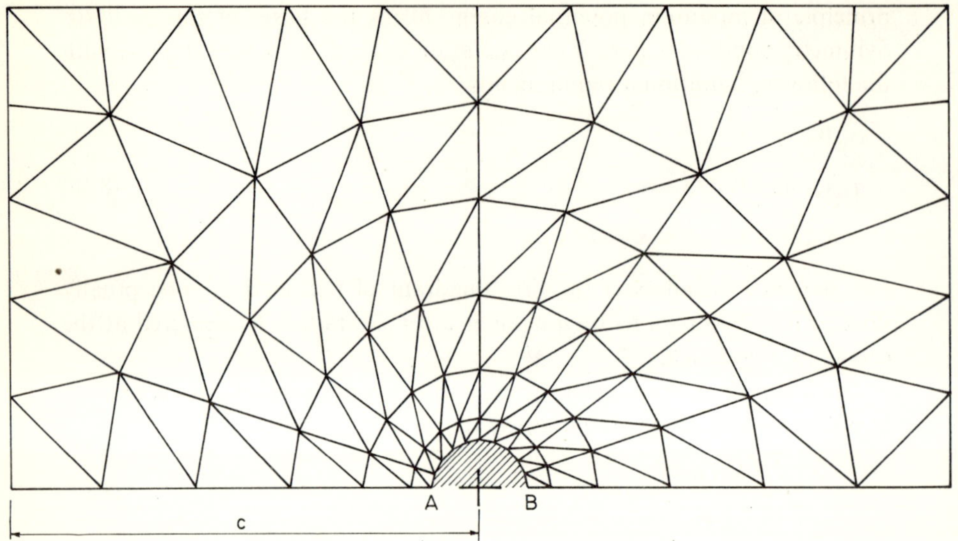


Figure 8.9. A typical embedded singularity, finite element grid pattern. (See also Figure 8.10).

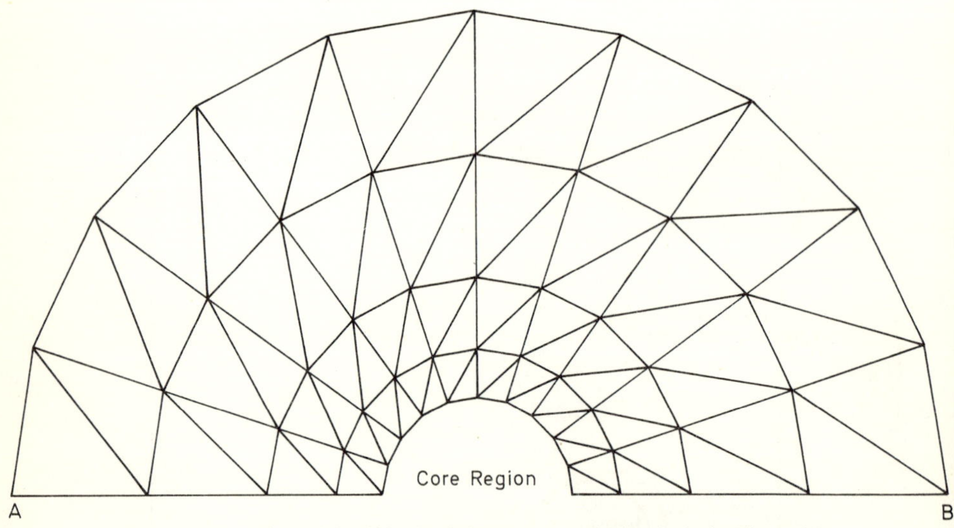


Figure 8.10. Enlarged detail in shaded region near the crack tip in Figure 8.9.

to choose $R_1$ sufficiently small so that this is a good approximation. The region outside $\Gamma_1$ is divided into finite elements as illustrated in Figures 8.9 and 8.10. For clarity, the fine mesh in the vicinity of the crack tip is shown separately at an enlarged scale, Figure 8.10. Discussion of the number, type, and distribution of these elements is deferred until the formulation is completed.

The purpose here is to construct the potential energy for the first quadrant of the solid as the sum of the contributions from the core region (that enclosed by $\Gamma_1$) and those from the finite elements. With this in mind, the strain energy per unit thickness associated with the core region $U_c$ is calculated as

$$\begin{aligned} U_c &= \frac{1}{2}\int_{\text{Core region}} \sigma_{ij}\varepsilon_{ij}\,\mathrm{d}s \\ &= \frac{1}{2}\int_0^{R_1}\int_0^{\pi} (\sigma_{\rho z}\gamma_{\rho z}+\sigma_{\theta z}\gamma_{\theta z})\,\rho\,\mathrm{d}\theta\,\mathrm{d}\rho \\ &= \frac{1}{2\mu}\int_0^{R_1}\int_0^{\pi}\left[\left(\frac{\partial u_z}{\partial\rho}\right)^2+\left(\frac{1}{\rho}\frac{\partial u_z}{\partial\theta}\right)^2\right]\rho\,\mathrm{d}\theta\,\mathrm{d}\rho \end{aligned} \tag{8.18}$$

where $\gamma_{ij}=2\varepsilon_{ij}$, $i\neq j$. Substitution of the first term of the asymptotic expansion for $u_z(\rho,\theta)$, Equation (8.11), into the above equation and performance of the indicated integrations gives

$$U_c = \frac{k_3^2}{4\mu}\pi R_1\,. \tag{8.19}$$

The finite element method is used to describe the solution everywhere outside the core region. The strain energy per unit thickness in each element can be calculated in the standard manner as described by Equation (8.2) to give

$$U^{(r)} = \sum_{i=1}^{M}\sum_{j=1}^{M} k_{ij}^{(r)}\bar{u}_{z_i}^{(r)}\bar{u}_{z_j}^{(r)}\,. \tag{8.20}$$

The potential energy per unit dimension in the $z$-direction of the first quadrant of the solid as obtained by summing the contributions from the core region and the finite elements is

$$V=\frac{k_3^2}{4\mu}\pi R_1+\sum_{r=1}^{N}\left[\sum_{i=1}^{M}\sum_{j=1}^{M} k_{ij}^{(r)}\bar{u}_{z_i}^{(r)}\bar{u}_{z_j}^{(r)}\right]-\int_0^a \tau_0 u_z(x,b)\,\mathrm{d}x\,. \tag{8.21}$$

The traction integral can be evaluated explicitly in terms of the displacements at the nodes along $y=b$, $x$ on $(0,a)$ by using the appropriate displacement shape functions to describe $u_z(x,b)$ between adjacent nodes.

The nodes of $\Gamma_1$ are treated separately. On these nodes, the displacement is required to match that given by the asymptotic expansion. Using the

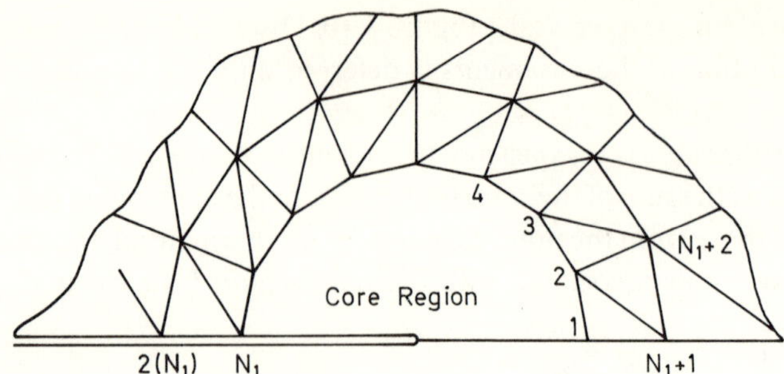


Figure 8.11. A global numbering system for nodes in the vicinity of the core region.

numbering system shown in Figure 8.11, the displacement of the $i$th node on the first ring is

$$\bar{u}_i = \bar{u}_{z_i} = u_z(R_1, \theta_i) = k_3 \frac{(2R_1)^{\frac{1}{2}}}{\mu} \sin(\theta_i/2) . \tag{8.22}$$

Thus the displacement field along the first ring is prescribed except for one unknown parameter, the stress intensity factor. In this way, the asymptotic expansion is connected to the finite element grid. The question of compatibility along $\Gamma_1$ arises but discussion is deferred, to be included in a more complete study of convergence. Minimization of the potential energy expression leads to the following set of equations which will govern this calculation:

$$\frac{\partial V}{\partial k_3} = 0 , \quad \frac{\partial V}{\partial \bar{u}_i} = 0 \tag{8.23}$$

where $i$ goes over the unconstrained nodes (not those on $\Gamma_1$). The potential energy of the quarter plane will have the form

$$V = Ak_3^2 + \sum_i B_i k_3 \bar{u}_i + \sum_i \sum_j C_{ij} \bar{u}_i \bar{u}_j + \sum_i D_i \bar{u}_i \tag{8.24}$$

where the coefficient of the first term, $A$, includes contributions from the core region and the first ring of nodes. The second set of terms with coefficients $B_i$ represents the coupling of the first and second ring of nodes through the first ring of elements. The last two sets of terms are of the standard finite element form, being associated with the nodes beyond the first ring and the traction boundary conditions, respectively.

It is not difficult to alter a general anti-plane shear, finite element program so as to embed the asymptotic expansion discussed above at the singularity

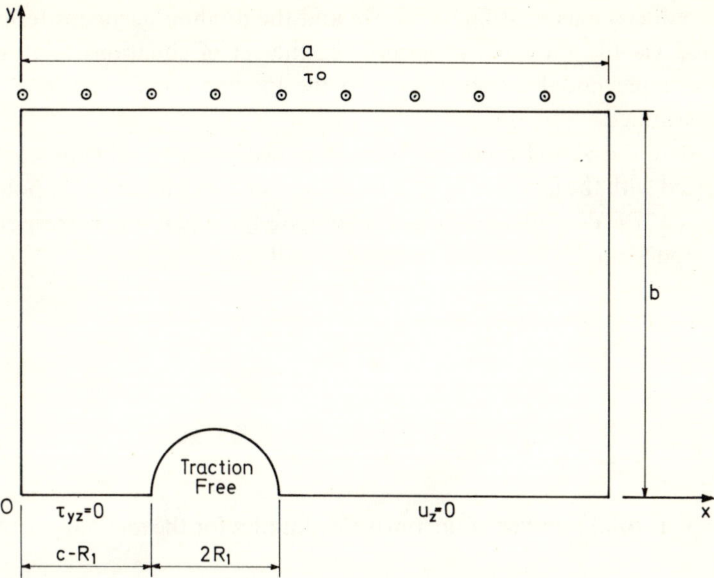


Figure 8.12. The auxiliary problem for embedded singularity, finite element calculations.

and thereby make the program directly applicable to crack problems [6]. To do this, consider the auxiliary problem for which the core region is replaced by a traction-free hole, Figure 8.12.

This problem is amenable to solution by a standard finite element program. It is convenient to use a grid pattern which contains a number of rings of elements about the void in conjunction with the numbering system shown in Figure 8.13.

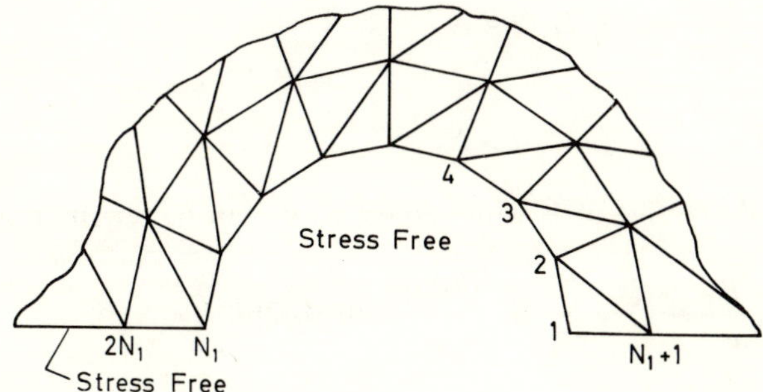


Figure 8.13. The global numbering system in the vicinity of the hole for the auxiliary problem.

The reduced master stiffness matrix and the nonhomogeneous terms are obtained via the standard program, i.e., the set of equations relating the unconstrained nodal displacements to the loading conditions ($K'_{ij}\bar{u}_j = B_j$) are determined.

To solve the actual crack problem it is necessary to add the equation associated with the fact that $k_3$ is to be chosen so as to minimize the potential energy, i.e., $\partial V/\partial k_3 = 0$, and then to eliminate the equations corresponding to the nodes on $\Gamma_1$. With this in mind, recall

$$V = U_c + \sum_{r=1}^{N} U^{(r)} - \tau_0 \int_0^a u_z(x, b)\,\mathrm{d}x \tag{8.25}$$

and

$$\sum_{r=1}^{N} U^{(r)} = \sum_{i=1}^{N_g} \sum_{j=1}^{N_g} K'_{ij}\bar{u}_i\bar{u}_j$$

where $N_g$ is total number of unconstrained nodes for the auxiliary problem. Thus

$$\frac{\partial V}{\partial k_3} = \frac{k_3 \pi R_1}{2\mu} + \sum_{i=1}^{N_g} \sum_{j=1}^{N_g} K'_{ij}\bar{u}_i \frac{\partial \bar{u}_j}{\partial k_3}\,. \tag{8.26}$$

The nodal displacements on the first ring are given by

$$\bar{u}_j = u_z(R_1, \theta_j) = k_3 \frac{(2R_1)^{\frac{1}{2}}}{\mu} \sin(\theta_j/2)\,, \qquad j = 1, N_1 \tag{8.27}$$

where $N_1$ is the number of nodal points on the boundary of the core region. Thus

$$\frac{\partial \bar{u}_j}{\partial k_3} = \begin{cases} \dfrac{(2R_1)^{\frac{1}{2}}}{\mu} \sin(\theta_j/2) & j \leqslant N_1 \\ 0 & j > N_1 \end{cases} \tag{8.28}$$

The partial derivative of $V$ with respect to $k_3$ can be rewritten as

$$\frac{\partial V}{\partial k_3} = \frac{k_3 \pi R_1}{2\mu} + \sum_{i=1}^{N_g} \sum_{j=1}^{N_1} K'_{ij} \frac{(2R_1)^{\frac{1}{2}}}{\mu} \sin(\theta_j/2)\bar{u}_i \tag{8.29}$$

which can be further separated into

$$\frac{\partial V}{\partial k_3} = k_3 \left[ \frac{\pi R_1}{2\mu} + \sum_{i=1}^{N_1} \sum_{j=1}^{N_1} K'_{ij} \frac{2R_1}{\mu^2} \sin(\theta_i/2) \sin(\theta_j/2) \right]$$

$$+ \sum_{i=N_1+1}^{2(N_1)} \sum_{j=1}^{N_1} K'_{ij} \frac{(2R_1)^{\frac{1}{2}}}{\mu} \sin(\theta_j/2) \bar{u}_i \,. \tag{8.30}$$

The goal is to determine the master stiffness matrix for the crack problem which will be denoted as $[\mathbf{K}^*]$. The elements of $[\mathbf{K}^*]$ can now be found directly from the auxiliary problem as

$$K^*_{11} = \frac{\pi R_1}{2\mu} + \sum_{i=1}^{N_1} \sum_{j=1}^{N_1} K'_{ij} \frac{2R_1}{\mu^2} \sin(\theta_i/2) \sin(\theta_j/2)$$

and

$$K^*_{1i} = \sum_{j=1}^{N_1} K'_{i+N_1-1,j} \frac{(2R_1)^{\frac{1}{2}}}{\mu} \sin(\theta_j/2) \,, \qquad i=2\,,\ N_1+1\,.$$

The remaining terms in the master stiffness matrix are unchanged, so that

$$K^*_{ij} = K'_{i+N_1-1,j+N_1-1}\,; \qquad i\,,\, j=2\,,\ N_g-N_1+1\,. \tag{8.31}$$

The nonhomogeneous terms are shifted correspondingly by the transformation

$$B^*_i = B_{i+N_1}\,; \qquad i=1\,,\ N_g-N_1+1\,. \tag{8.32}$$

The standard finite element program is readily augmented, as shown above, to give the governing equations of the embedded singularity-finite element technique for crack problems.

The calculations described above were carried out* for the cracked rectangular body shown in Figure 8.7. Figures 8.9 and 8.10 are typical of the grid patterns employed. A sequence of results is reported for the particular body dimensions $a/c=b/c=2.0$ to demonstrate the sensitivity of this approach to the various approximations involved and to show the kind of accuracies attainable via the embedded singularity, finite element method. In particular, the influence of the size of the core region and the number of nodes along the asymptotic expansion-finite element grid pattern interface, $\Gamma_1$, is studied. Numerical values of the stress intensity factor are given in Table I for various numbers of nodal points along $\Gamma_1$ and corresponding numbers of degrees of freedom at a constant core radius of $R_1=0.01\,c$. In all cases, the grid patterns are chosen so that the shapes of the triangular

* Calculations were performed by Marlin Kipp, Research Assistant, Lehigh University.

TABLE I

*Variations of $k_3$ with the number of nodes*

| Number of nodes along $\Gamma_R$ | Number of degrees of freedom | $k_3(\text{cal})/(\tau_0^0 c^{\frac{1}{2}})$ |
|---|---|---|
| 11 | 227 | 1.15581 |
| 13 | 272 | 1.15972 |
| 16 | 386 | 1.16552 |
| 19 | 550 | 1.16902 |
| 21 | 639 | 1.17051 |

TABLE II

*Variations of $k_3$ with $R_1/c$*

| $R_1/c$ | Number of degrees of freedom | $k_3(\text{cal})/(\tau_0^0 c^{\frac{1}{2}})$ |
|---|---|---|
| 0.008 | 550 | 1.16846 |
| 0.01 | 550 | 1.16902 |
| 0.012 | 550 | 1.16940 |
| 0.02 | 550 | 1.17044 |

elements do not change drastically from one calculation to the next and should not appreciably affect the numerical results. The calculated values for the stress intensity factor are seen to increase as the grid pattern is refined and the maximum change in its value over this set of calculations is less 1.5% of the lowest result.

Next a test of the core size is made; i.e., calculations are performed for 19 nodes along $\Gamma_1$ at various values of the core radius $R_1$. Results are shown in Table II.

The stress intensity factor is found to be very insensitive to the actual value chosen for $R_1$ within the range considered and thus the core radii considered are assumed to be sufficiently small so that the one-term expansion is a good approximation to the true solution everywhere in the core region. Note, however, that as the core radius is decreased, presumably increasing the accuracy of the approximations, the calculated values for the stress intensity factor actually decrease. This may occur because the one-term expansion underestimates the strain energy in the core region which may force correspondingly large values for the stress intensity factors in the numerical calculation. In any case, it should be realized that the embedded singularity, finite element results may not always bound the true solution from either above or below. These results do, however, appear to be convergent to the true solution as will be discussed later.

With this background, the embedded singularity, finite element procedure is used to determine the variation of the stress intensity factor with the geometric parameters of the cracked, rectangular body subjected to anti-plane shearing tractions, Figure 8.7. Graphs of the stress intensity factor against $c/a$ for $b/c = 1.0, 2.0$ are shown in Figure 8.14.

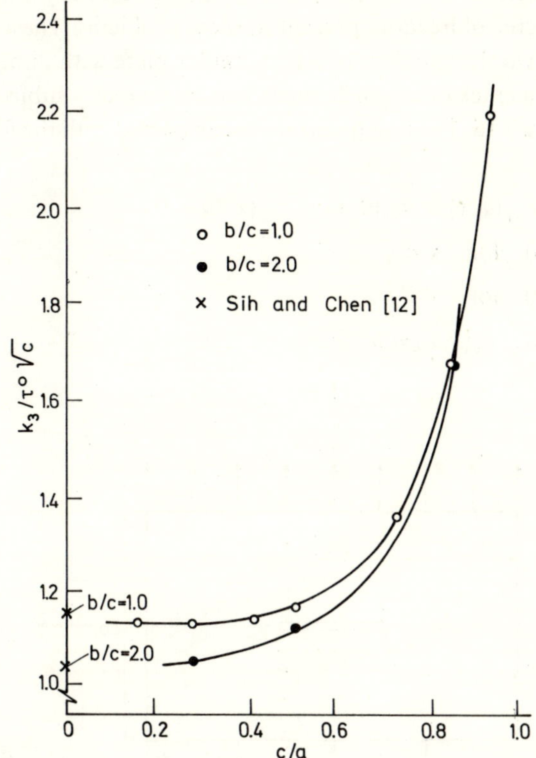


Figure 8.14. The stress intensity factor against the specimen width for a body with an interior crack.

The limiting values of these results as $c/a$ approaches zero can be compared to those obtained via a different procedure by Sih and Chen [12] for the corresponding infinite strip problem as indicated in Figure 8.14. Note that in both the cases of $b/c = 1.0, 2.0$ the values calculated here for the stress intensity factor are below those reported by Sih and Chen.

Computation times for the calculations reported were about 20 seconds per run on a CDC 6400 computer.

*Cracked bodies subjected to in-plane loading.* The procedure for obtaining stress intensity factors for anti-plane shear by using the finite element technique with an embedded asymptotic expansion is also applicable to problems with crack opening Modes I or II for generalized plane stress, plane strain, and axisymmetric problems. Minor alterations are necessary to accommodate the second degree of freedom present in these problems. These changes will be illustrated via the problem of a rectangular plate with dimensions $2a$ by $2b$ containing a crack of length $2c$ on its line of symmetry subjected to tensile loading, Figure 8.15. The first quadrant is considered with the following conditions:

$$\begin{aligned} &\sigma_{xy}(x, b) = \sigma_{xy}(a, y) = \sigma_{xy}(0, y) = \sigma_{xy}(x, 0) = 0 \\ &\sigma_y(x, 0) = 0 \quad \text{for} \quad x < c \\ &u_y(x, 0) = 0 \quad \text{for} \quad x > c \\ &u_x(0, y) = 0\,, \quad \sigma_y(x, b) = \sigma^0\,. \end{aligned} \tag{8.33}$$

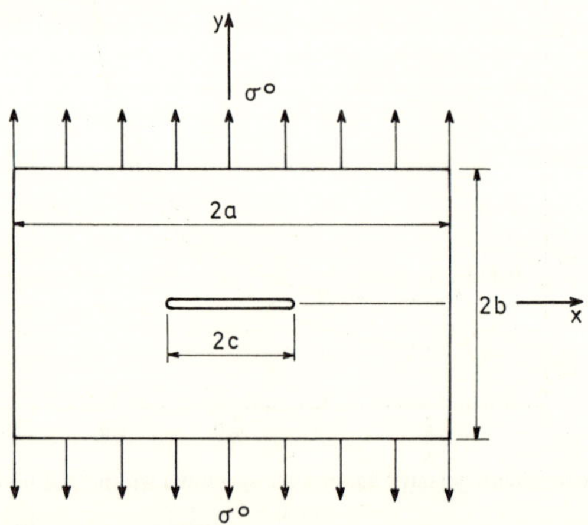


Figure 8.15. An illustrative problem for embedded singularity finite element calculations associated with the theories of plane stress and plane strain.

Note that as a result of the symmetry conditions, displacement of the crack tip in the $y$-direction is prevented; however, the tip may displace in the $x$-

direction. This is accounted for by writing the near-tip displacement components as

$$\begin{aligned} u_x &= \delta + u_x(\rho, \theta) \\ u_y &= u_y(\rho, \theta) \end{aligned} \qquad (8.34)$$

and choosing $\delta$ so as to minimize the potential energy. The asymptotic behavior of the displacement components $u_x(\rho, \theta)$ and $u_y(\rho, \theta)$ relative to the crack tip is given in Equations (8.10) for generalized plane stress, plane strain and axisymmetric problems. Spatial derivatives of the displacement components are independent of $\delta$, so $\delta$ can have no effect on the strain, stress, or strain energy in the core region. The potential energy of these two-degree of freedom problems is functionally dependent on $k_1$, $\delta$ and $\bar{u}_j$, i.e.,

$$V = V(k_1, \delta, \bar{u}_j)$$

where

$$\begin{aligned} \bar{u}_j &= (u_x)_{(j+1)/2}\,, \quad j \text{ odd} \\ \bar{u}_j &= (u_y)_{(j/2)}\,, \quad j \text{ even} \end{aligned}$$

and $j$ varies over the unconstrained degrees of freedom. The governing minimization equations are

$$\frac{\partial V}{\partial k_1} = 0\,, \quad \frac{\partial V}{\partial \delta} = 0\,, \quad \frac{\partial V}{\partial \bar{u}_j} = 0\,. \qquad (8.35)$$

Standard two-dimensional finite element programs can be augmented to include the embedded singularity for plane crack problems. To accomplish this, consider the auxiliary problem for which the core region is replaced by a hole with a traction-free boundary; all other boundary conditions for the actual and auxiliary problems are identical. A grid pattern which includes a number of rings of elements about the void with the numbering system shown in Figure 8.13 is again used for convenience. Let $[\mathbf{K}']$ be the reduced master stiffness matrix with elements $K'_{ij}$ and $B_j$ the set of corresponding nonhomogeneous terms for the auxiliary problem, i.e.,

$$\frac{\partial V}{\partial \bar{u}_i} = \sum_{j=1}^{N_g} K'_{ij}\bar{u}_j - B_i = 0\,, \qquad i = 1\,,\, N_g \qquad (8.36)$$

where $N_g$ is the number of unconstrained degrees of freedom.

For the actual problem, all the nodes adjacent to the core region are constrained to the extent that their behavior is completely determined by the two parameters $k_1$ and $\delta$. Thus the number of degrees of freedom for the actual problem is $N_g-2(N_1)+2$ where $N_1$ is the number of nodes along the ring surrounding the core. The governing equations for the crack problem can be written as (the superscript * is used to indicate the actual problem as contrasted to the auxiliary one).

$$\frac{\partial V^*}{\partial k_1} = K^*_{11}k_1 + K^*_{12}\delta + \sum_{j=3}^{N_g-2N_1+2} K^*_{1j}\bar{u}_{j+2N_1-2} = 0$$

$$\frac{\partial V^*}{\partial \delta} = K^*_{21}k_1 + K^*_{22}\delta + \sum_{j=3}^{N_g-2N_1+2} K^*_{2j}\bar{u}_{j+2N_1-2} = 0$$

$$\frac{\partial V^*}{\partial \bar{u}_{i+2N_1-2}} = K^*_{i1}k_1 + K^*_{i2}\delta + \sum_{j=3}^{N_g-2N_1+2} K^*_{ij}\bar{u}_{j+2N_1-2} - B^*_i = 0 \tag{8.37}$$

where $i=3, N_g-2(N_1)+2$. The elements of the master stiffness matrix for the crack problem can be determined from the corresponding quantities associated with the auxiliary problem as

$$K^*_{11} = \frac{1}{k_1}\frac{\partial U_c}{\partial k_1} + \frac{1}{k_1^2}\sum_{i=1}^{2N_1}\sum_{j=1}^{2N_1} K'_{ij}\tilde{u}_i\tilde{u}_j$$

$$K^*_{12} = \frac{1}{k_1}\sum_{i=1,3,5}^{2N_1-1}\sum_{j=1}^{2N_1} K'_{ij}\tilde{u}_j$$

$$K^*_{22} = \sum_{i=1,3,5}^{2N_1-1}\sum_{j=1,3,5}^{2N_1-1} K'_{ij}$$

$$\left.\begin{aligned} K^*_{1j} &= \frac{1}{k_1}\sum_{i=1}^{2N_1} K'_{i,j+2N_1-2}\tilde{u}_i \\ K^*_{2j} &= \sum_{i=1,3,5}^{2N_1-1} K'_{1,j+2N_1-2} \end{aligned}\right\} \quad j=3,4,\ldots,2N_1+2$$

$$\left.\begin{aligned} K^*_{ij} &= K'_{i+2N_1-2,\,j+2N_i-2} \\ B^*_i &= B_{i+2N_1-2} \end{aligned}\right\} \quad i,j=3,4,\ldots,N_g-2N_1+2 \tag{8.38}$$

where

$$U_c = \frac{1}{2}\int_{\text{Core region}} \sigma_{ij}\varepsilon_{ij}\,\mathrm{d}A \tag{8.39}$$

$$\text{and } \tilde{u}_i = \begin{cases} u_x(R_1, \theta_{(i+1)/2}), & i \text{ odd} \\ u_y(R_1, \theta_{i/2}), & i \text{ even} \end{cases}$$

$u_x(R_1, \theta)$ and $u_y(R_1, \theta)$ are given by their one term expansions in Equations 8.10 for generalized plane stress, plane strain, and axisymmetric problems, and $\sigma_{ij}$, $\varepsilon_{ij}$ can be calculated from them.

Consider the axisymmetric problems of either a penny-shaped crack centered on the axis of a body of revolution or a circumferential crack in an axisymmetric body. There is a close similarity between the axisymmetric and planar problems which enables almost identical embedded singularity, finite element formulations for the two cases. For example, the problem of a solid circular cylinder, containing a penny-shaped crack, subjected to tensile loading on its end surfaces corresponds to the plane strain problem considered previously if the $(x, y)$ coordinate system in Figure 8.15 is interpreted as a measure of position in the radial and axial directions, respectively. The variations of the displacement components with angle about the crack tip, $u_r(\rho, \theta)$ and $u_t(\rho, \theta)$ employed in Equation (8.34), are given by Equation (8.10) with $\kappa = 3 - 4\nu$ (note the comments pertaining to axisymmetric problems in Section 8.3). For numerical calculations, the only distinction that need be made between the penny-shaped crack problem and the plane strain formulation for the through crack problem is that the strain energy in the core region will differ as a result of the different shapes of that region for the two cases (for the axisymmetric problem the core region is a toroid; while, it is cylindrical for the plane strain problem). Clearly, the standard finite element stiffness matrices for the two cases will be different as discussed in Section 8.2. Thus a standard axisymmetric, finite element program can be altered to give the embedded singularity, finite element procedure for the penny-shaped crack just as the planar program was changed to include the special element for the two-dimensional crack problem.

There is a similar correspondence between the axisymmetric problem of a body of revolution containing an exterior circumferential crack and the double notched, plane strain specimen problem.

Results of a number of embedded singularity, finite element calculations for plane strain and axisymmetric crack problems are presented in the Appendix. They tend, in a limited way, to illustrate both the accuracy and versatility afforded by the use of the embedded singularity, finite element approach for solving crack problems.

## 8.6 Finite element procedures for infinite regions

Solutions to infinite region problems are of interest in that they provide a means of comparison with available analytic solutions so as to determine the accuracy of the numerical technique. To accomplish this by the finite element method, calculations are usually carried out for large finite bodies whose boundaries are sufficiently far away from the crack so that the results are a good approximation to the infinite region solution. Such an approach clearly requires a large number of elements and correspondingly long computer times.

Alternatively, it is possible to deal directly with boundary value problems over infinite regions. The example of an infinite plate with a finite crack in a uniform far anti-plane shear field, Figure 8.16, will illustrate the approach. The potential energy of this problem is infinite, however, the problem can be expressed as the superposition of two problems, Figure 8.16. The solution to problem (1) is trivial ($u_z = (\tau^\infty/\mu)y$) and the potential energy of problem (2) is finite. Thus a solution to problem (2) is sought.

The first quadrant is considered with the boundary conditions shown in Figure 8.17. The potential energy is given by

$$V = U - \tau^\infty \int_0^c u_z(x, 0)\,\mathrm{d}x \tag{8.40}$$

where $U$ is the strain energy of the quarter plane. For the evaluation of $U$, the quarter plane is conceptually divided into three regions, Figure 8.17, to be associated with three different numerical techniques. A circular arc, $\Gamma_2$, centered at the crack midpoint and of radius $R_2$ ($R_2 > c$) separates the space exterior to the core region into two portions, denoted as $A_1$ and $A_2$ respectively. As before, the asymptotic expansion is employed in the core region and finite elements are used in region $A_1$. In considering region $A_2$, note that there are no singularities in the quarter plane outside the circular arc $\Gamma_2$; thus the anti-plane shear solution can be expressed in terms of one function which is analytic on $\Gamma_2$ and in $A_2$ (two analytic functions are required for plane stress and plane strain problems), i.e.,

$$\nabla^2 u_z = 0 \tag{8.41}$$

implies that $u_z$ is equal to the real part of an analytic function, $\phi(z)/\mu$ of the complex variable $z = x + \mathrm{i}y$. The stress components are given by [13]

$$\tau_{xz} - \mathrm{i}\tau_{yz} = \phi'(z)\,. \tag{8.42}$$

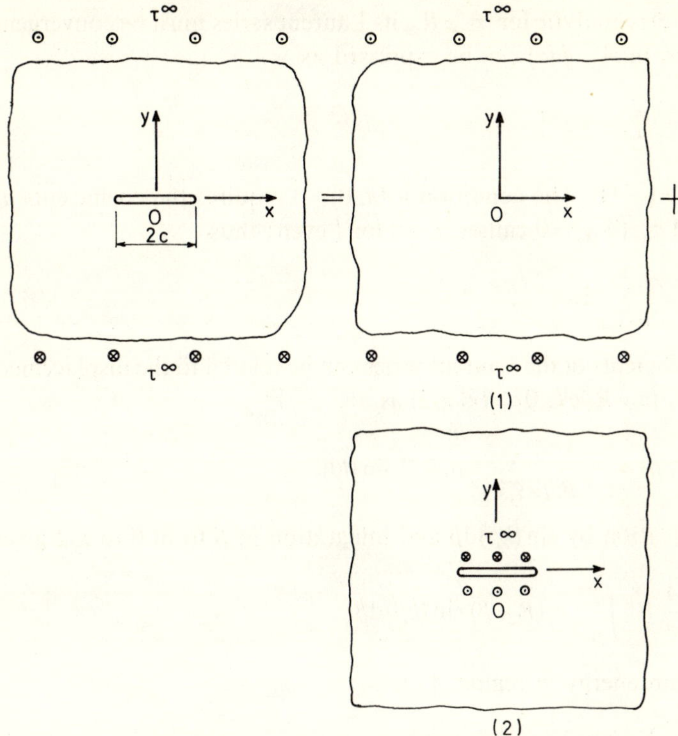


Figure 8.16. The superposition principle for an infinite body containing a line crack.

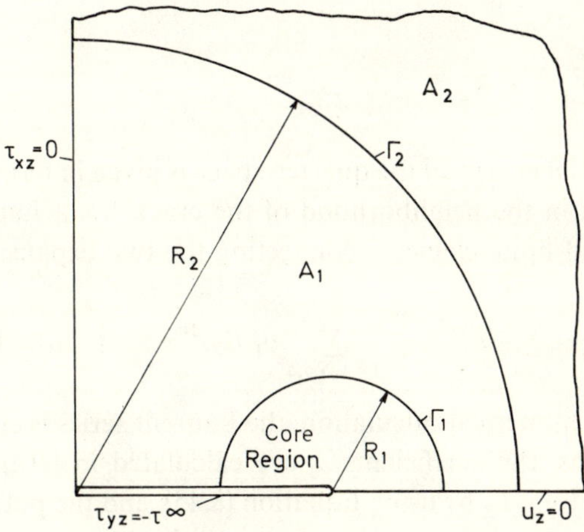


Figure 8.17. Division of the quarter-plane into three regions; each is associated with a different numerical procedure.

Since $\phi(z)$ is analytic for $|z| \geqslant R_2$, its Laurent series must be convergent there and thus, in $A_2$. $\phi(z)$ can be expressed as

$$\phi(z) = \sum_{l=0}^{\infty} \mathrm{i} a_l z^{-l} \tag{8.43}$$

where $\mathrm{i}=(-1)^{\frac{1}{2}}$. The condition $u_z(x, 0)=0$ requires the coefficients $a_l$ to be real and $\tau_{xz}(0, y)=0$ causes $a_l=0$ for $l$ even; thus

$$\phi(z) = \sum_{l=1,3}^{\infty} \mathrm{i} a_l z^{-l} .$$

The coefficients of the Laurent series can be related to the displacement field along $\Gamma_2$ $(z=R_2 \mathrm{e}^{\mathrm{i}\beta},\ 0 \leqslant \beta \leqslant \pi/2)$ as

$$u_z(R_2, \beta) = -\frac{1}{\mu} \sum_{l=1,3,\dots}^{\infty} a_l R_2^{-l} \sin(l\theta) . \tag{8.44}$$

Multiplication by $\sin(k\beta)\mathrm{d}\beta$ and integration in $\beta$ from 0 to $\pi/2$ gives

$$a_k = \frac{4}{\pi} R_2^k \int_0^{\pi/2} u_z(R_2, \beta) \sin(k\beta) \mathrm{d}\beta \tag{8.45}$$

The strain energy in region $A_2$ is

$$\begin{aligned} U_{A_2} &= \frac{1}{2} \int_{R_2}^{\infty} \rho \mathrm{d}\rho \int_0^{\pi/2} (\tau_{xz}\gamma_{xz} + \tau_{yz}\gamma_{yz}) \mathrm{d}\beta \\ &= \frac{\pi}{8} \sum_{l=1,3,\dots}^{\infty} a_l^2 R_2^{-2l} . \end{aligned} \tag{8.46}$$

Thus the potential energy of the quarter-space is given in terms of an asymptotic expansion in the neighborhood of the crack tip, a Laurent series far from the tip and finite elements connecting the two expansions as

$$V = \frac{k_3^2 \pi R_1}{4\mu} + \sum_{r=1}^{N} U^{(r)} + \frac{\pi}{8} \sum_{l=1,3,\dots}^{\infty} a_l^2 R_2^{-2l} - \tau^{\infty} \int_0^c u_z(x, 0) \mathrm{d}x . \tag{8.47}$$

To perform the numerical calculation, the Laurent series is cut-off at a finite number of terms, the coefficients $a_l$ are calculated in terms of the nodal displacements along $\Gamma_2$ by using Equation (8.45), and the potential energy is minimized with respect to the stress intensity factor and the displacements of the unconstrained nodes, i.e.,

$$\frac{\partial V}{\partial k_3} = \frac{\partial V}{\partial \bar{u}_j} = 0\,.$$

This calculation was performed under the following conditions:

546 elements

$R_1 = 0.02\,c$

$N_1 = 11$

11 nodes along $\Gamma_2$

11 terms retained in Laurent series

The numerical result was

$$k_3(\text{calculated}) = 0.994\,\tau^{\infty} c^{\frac{1}{2}} = 0.994\,k_3\,.$$

The percentage error is less than 1%.

## 8.7 Convergence and accuracy of the singularity embedded finite element approach

Convergence of the finite element displacement method for compatible elements follows from the theorem of minimum potential energy and the convergence theorems for the Ritz technique. The potential energy theorem states that of all displacements satisfying displacement boundary conditions, the displacement field which satisfies equilibrium renders the potential energy functional an absolute minimum. Further, it can be shown from the potential energy function that if a sequence of approximations for the displacement field and its first derivatives can be made arbitrarily close to the true fields everywhere in the body by including a sufficient number of parameters (nodal displacement components), this sequence of solutions will converge to the true solution.

For the embedded singularity, finite element scheme described above the approximate displacement field and its first derivatives can be made arbitrarily accurate everywhere, except in the immediate vicinity of the circular arc $\Gamma_1$, by decreasing the size of the finite elements and the radius $R_1$ of the core region. For infinite region problems it is also necessary to increase the number of terms retained in the Laurent series. Along $\Gamma_1$ the displacement field as given by the asymptotic expansions, Equations (8.10)–(8.12) differs from that which results from the finite element approximation (a piecewise linear function) except at the nodal points. Thus the displacement components are not, in general, continuous across $\Gamma_1$ and the compatibility condition is violated there. Further, the strain energy in the region between the circular arc $\Gamma_1$ and the sequence of line segments approximating this arc for the adjacent finite elements has been included twice in the potential energy expression. Both the displacement incompatibility and the error in the potential energy will vanish as the number of finite element nodes along $\Gamma_R$ approaches infinity. Thus, the embedded singularity, finite element approach is expected to converge in the limit to the exact solution. The numerical results presented indicate the high degree of accuracy attainable with this technique.

The parameters which affect the accuracy of the embedded singularity, finite element method are the core radius, $R_1$, the number of nodes on the first ring, $N_1$, the number and distribution of finite elements, the number of nodes in $\Gamma_2$, and the number of terms included in the Laurent series. In particular, $R_1$ must be chosen sufficiently small in comparison with all other geometric parameters so that the one term asymptotic expansion is an accurate description of the solution in the core region. For the calculations discussed, this requirement was met by choosing $R_1$ small enough so that the solution was insensitive to small changes in $R_1$.

An improved version of the embedded singularity, finite element is discussed by Wilson [7]. Higher order terms are retained in the asymptotic expansion used to describe the solution within the core regions; thereby permitting the choice of a larger radius for the core and the use of correspondingly fewer finite elements. It is not yet clear how many terms should be included in the asymptotic expansion in order to obtain the highest degree of accuracy for a given amount of computation time.

## 8.8 An alternative finite element approach to crack problems

In a recent paper, Tracey [14] has introduced an alternative technique for applying finite elements to crack problems. This approach is based on the development of "special" triangular elements which enable the representation of the near tip displacement field with approximately the same degree of accuracy as the usual finite element representation away from the crack tip.

Tracey employs quadrilateral isoparametric elements [2] which can be made triangular for use at the crack tip by requiring that two nodes coincide. The elements are mapped to a square in the auxiliary coordinates $(\xi, \eta)$,

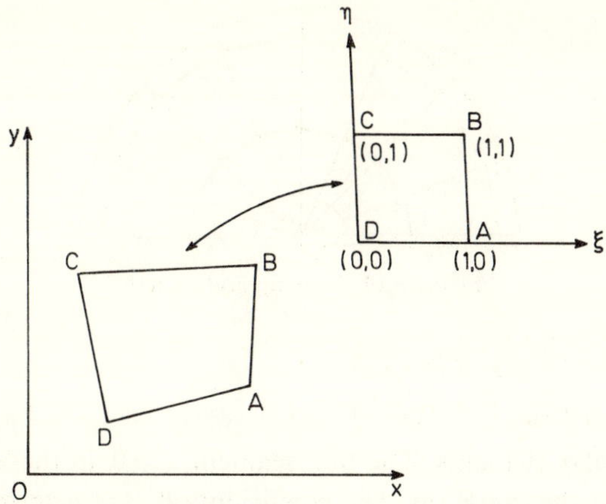


Figure 8.18. Mapping for isoparametric quadralateral elements.

Figure 8.18, by the transformation

$$\begin{Bmatrix} x \\ y \end{Bmatrix} = \begin{Bmatrix} x_A \\ y_A \end{Bmatrix} \xi(1-\eta) + \begin{Bmatrix} x_B \\ y_B \end{Bmatrix} (\xi\eta) + \begin{Bmatrix} x_C \\ y_C \end{Bmatrix} (1-\xi)\eta + \begin{Bmatrix} x_D \\ y_D \end{Bmatrix} (1-\xi)(1-\eta) \tag{8.48}$$

Thus along each edge of the element one of the auxiliary coordinates is constant while the other varies linearly with respect to $x$ and $y$. The displacement field within each element is expressed in terms of the transform coordinates $(\xi, \eta)$ in such a way that, the displacement components along an edge of the element are uniquely determined by the displacements of the nodes adjacent to that edge, and thus, displacement components are continuous across element boundaries. It is customary to satisfy this condition of compatibility by requiring a linear variation of displacement components along element edges. The displacement components within an element are then of the form

$$\begin{aligned} u_x(\xi, \eta) &= \alpha_{11} + \alpha_{12}\xi + \alpha_{13}\eta + \alpha_{14}\xi\eta \\ u_y(\xi, \eta) &= \alpha_{21} + \alpha_{22}\xi + \alpha_{23}\eta + \alpha_{24}\xi\eta \end{aligned} \tag{8.49}$$

where the constants $\alpha_{ij}$ are expressible in terms of the nodal displacement components. Indeed, Tracey does use elements of this type everywhere except very near the crack tip. He employs the grid pattern shown in Figure 8.19 in the neighborhood of the crack tip and chooses displacement functions of the form

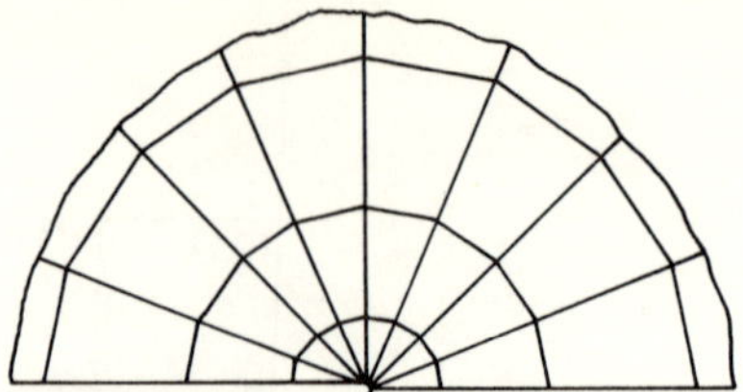

Figure 8.19. Near-tip grid pattern.

$$\begin{aligned} u_x(\xi, \eta) &= \beta_{11} + \beta_{12}\xi^{\frac{1}{2}} + \beta_{13}\xi^{\frac{1}{2}}\eta \\ u_y(\xi, \eta) &= \beta_{21} + \beta_{22}\xi^{\frac{1}{2}} + \beta_{23}\xi^{\frac{1}{2}}\eta \end{aligned} \tag{8.50}$$

for the triangular elements. The line segment, $\xi = 0$, in the auxiliary plane corresponds to the crack tip. The coefficients $\beta_{ij}$ are determined from the nodal values of the displacement components. Note that this displacement field satisfies the conditions described above resulting in compatible displacements both between adjacent triangles and from the triangular elements to the quadrilateral elements with displacement functions given by Equation (8.49).

Through this procedure, Tracey has obtained a compatible finite element grid which requires the near-tip displacement components to be proportional the square root of the distance from the crack tip. He has suggested that improved accuracy can be gained by using "special" quadrilaterals for a few

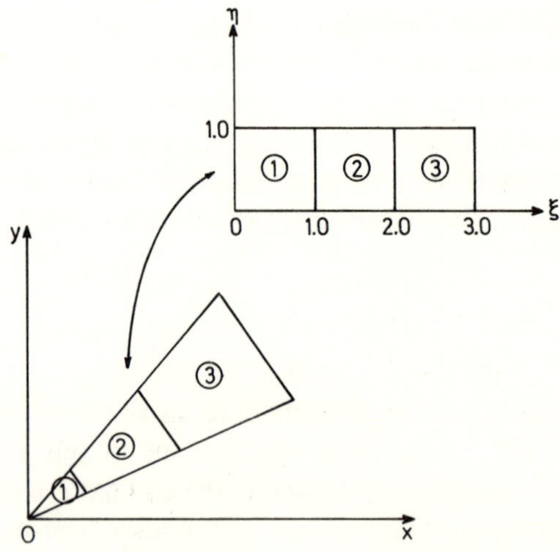


Figure 8.20. Mapping of special elements.

rings of elements surrounding the tip; i.e., these elements have the displacement function

$$u_x(\xi, \eta) = \gamma_{11} + \gamma_{12}\xi^{\frac{1}{2}} + \gamma_{13}\xi^{\frac{1}{2}}\eta + \gamma_{14}\eta$$
$$u_y(\xi, \eta) = \gamma_{21} + \gamma_{22}\xi^{\frac{1}{2}} + \gamma_{23}\xi^{\frac{1}{2}}\eta + \gamma_{24}\eta\,. \qquad (8.51)$$

In order to model a displacement variation which is proportional to the square root of the distance from the crack tip it is necessary to map these elements without radial translation as shown in Figure 8.20.

Calculations are reported for the doubly-notched plate specimen and for the solid cylinder with a circumferential edge crack, both subjected to uniform tensile loading. With the procedure described above, it is necessary to again compare the asymptotic expansions for displacement components with the finite element nodal results. Tracey averages the values obtained for $k_1$ from each node on the first ring using each displacement component. Sample results are given for the doubly cracked, plane strain specimen under uniform tensile loading, Figure 8.21. The length of the edge cracks is *c*. The stress intensity factor for this problem is usually expressed in the form

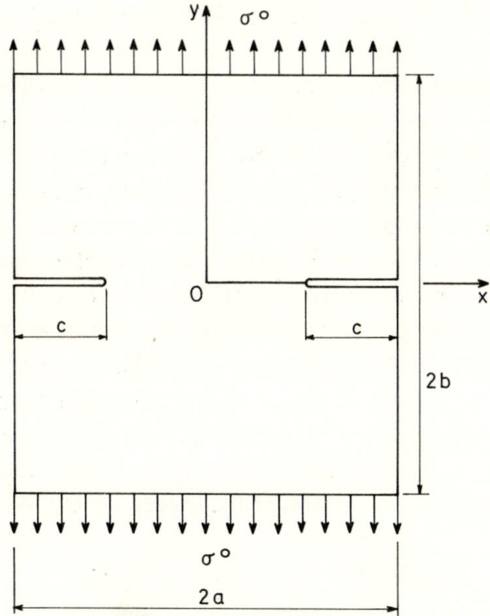


Figure 8.21. Double-notched tensile specimen.

TABLE III

*Correction factor*

| | *Number of degrees of freedom* | $C$ |
|---|---|---|
| Bowie | — | 1.02 |
| Tracey | 248 | 1.02±5% |
| Tracey | 548 | 1.06 |

$$k_1 = \sigma^0 (\pi c)^{\frac{1}{2}} \left[ \frac{2b}{\pi a} \tan\left(\frac{\pi a}{2b}\right) \right] C \tag{8.52}$$

where $C$ is the correction factor multiplying the solution for a periodic array of cracks in a body under a remote uniform tensile stress, $\sigma^0$. Tracey's averaged numerical values for $C$ are compared with those obtained by Bowie [15] (which are believed to be within 1% of the exact value) in Table III. The calculations by Tracey also give excellent results for the circumferential variation of the stress components about the crack tip.

In comparing the embedded singularity approach to crack problems with that suggested by Tracey it should be observed that the embedded singularity technique results in an approximate displacement field which is asymptotically exact at the crack tip but which lacks compatibility along the asymptotic expansion-finite element interface. On the other hand, Tracey's approach of incorporating special triangular elements about the crack tip satisfies compatibility everywhere but approximates the circumferential variation of the displacement components near the crack tip by a piecewise linear function. The special quadrilateral elements, which admit a square-root stress singularity, used by Tracey to surround the triangular crack tip elements are a clear improvement over the constant strain triangles employed by Wilson and Hilton in that region. This disadvantage of the embedded singularity approach is eliminated by the use of a larger core region which includes higher order terms in the crack tip expansion of the displacement components [7]. Finally, the direct calculation of stress intensity factors made possible by the embedded singularity technique appears preferable to the averaging scheme necessitated by Tracey's method.

## 8.9 Elastic-plastic crack analysis

There have been studies by a number of authors related to the extension of fracture mechanics to take into account the effects of nonlinear material behavior in the vicinity of crack tips [16, 17]. In particular, Hilton and Hutchinson [11] have carried out an analysis concerned with the amplitude of the plastic singularity at the crack tip, the plastic intensity factor. This analysis is based on a small strain deformation theory of plasticity.

The material behavior is modeled by the non-dimensional tensile stress-strain relations

$$\varepsilon = \begin{cases} \sigma\,; & \sigma \leqslant 1 \\ \sigma^{p}\,; & \sigma > 1 \end{cases} \tag{8.53}$$

where $p$ is the hardening coefficient. Here, the non-dimensionalized stress $(\sigma_{ij})$ and strain $(\varepsilon_{ij})$ components are related to the actual stresses and strains by

$$\sigma_{ij} = \bar{\sigma}_{ij}/\bar{\sigma}_{yd}\,; \quad \varepsilon_{ij} = \bar{\varepsilon}_{ij}/\bar{\varepsilon}_{yd} \tag{8.54}$$

where $\bar{\sigma}_{yd}$ is the yield stress in tension and $\bar{\varepsilon}_{yd} = \bar{\sigma}_{yd}/E$. The plastic deformation is assumed to be independent of the hydrostatic component of stress, $\sigma_{kk}$, and completely determined by the first invariant of the stress deviator, $s_{ij} = \sigma_{ij} - \frac{1}{3}\sigma_{kk}\delta_{ij}$. This invariant, the "effective stress" $\sigma_e$ is defined by $\sigma_e^2 = \frac{3}{2}s_{ij}s_{ij}$. For simple tension $\sigma_e = \sigma$ and the Mises yield condition is $\sigma_e = 1$. The generalized stress-strain relationship which reduces to Equation (8.53) for simple tension is

$$e_{ij} = \begin{cases} (1+\nu)s_{ij}\,; & \sigma_e \leqslant 1 \\ [(1+\nu) + (3/2)(\sigma_e^{p-1} - 1)]\,s_{ij}\,; & \sigma_e > 1 \end{cases} \tag{8.55}$$

where $e_{ij}$ is the strain deviator $e_{ij} = \varepsilon_{ij} - \frac{1}{3}\varepsilon_{kk}\delta_{ij}$ and $\varepsilon_e^2 = \frac{3}{2}e_{ij}e_{ij}$. It is assumed that no unloading will occur for the problems to be considered and solutions based on this formulation must be checked to assure that the assumption is not violated.

For the plasticity theory described, it has been shown that the dominant singular term describing the asymptotic behavior of the stress and strain components at the crack tip for Mode I, II, and III can be written in the form [18]

$$\begin{aligned} \sigma_{ij} &= k_\sigma \rho^{1/(p+1)} S_{ij}(\theta) \\ \varepsilon_{ij} &= k_\varepsilon \rho^{-p/(p+1)} E_{ij}(\theta) \end{aligned} \tag{8.56}$$

where $S_{ij}(\theta)$ and $E_{ij}(\theta)$ describe the angular distribution of the stresses $\sigma_{ij}$ and strains $\varepsilon_{ij}$. The plastic strain intensity factor $k_\varepsilon$ is related to the plastic stress intensity factor, $k_\sigma$ by $k_\varepsilon=(k_\sigma)^p$ and $(\rho, \theta)$ are polar coordinates centered at the crack tip. This expansion is independent of the plastic zone size.

Note that this singular solution is proportional, i.e., the relative magnitude of the stress components do not change with increasing applied load. Thus it is valid for flow theory as well as deformation theory; however, in general, the amplitudes $k_\sigma$ and $k_\varepsilon$ will depend on the theory employed. Additional information concerning the character of these solutions for each of the modes of crack opening will be presented in conjunction with separate descriptions for each case in the sections to follow.

Before attempting to study the elastic-plastic crack analysis in detail, it is worthwhile to study the general implications of the results when the plastic region at the crack tip is small in comparison with the characteristic geometric parameters of the specimen. When this occurs, it is known that the current theory of linear fracture mechanics is applicable to fracture initiation predicting even though the material ahead of the crack tip flows plastically. This situation has been referred to as "small scale yielding" by Rice [16] based on the solution of a special elastic-plastic crack problem in Mode III.

A necessary condition of small scale yielding is that there exists a region outside the crack tip plastic zone where the elastic-plastic solution can be characterized by the singular solution (equations (8.10)–(8.12)) of the corresponding elastic problem. Hence the elastic stress intensity factor is a measure of the amplitude of the elastic-plastic stress field. In fact, it can be shown that in this case the plastic stress or strain intensity factors can be expressed in terms of the elastic stress intensity factor normalized by the yield stress, i.e., $k_e=\bar{k}_e/\sigma_{yd}$, in the form [16, 18]

$$\begin{aligned} k_\sigma &= C_p (k_e)^{2/(p+1)} \\ k_\varepsilon &= (k_\sigma)^p \end{aligned} \tag{8.57}$$

where $C_p$ is a constant depending on the hardening coefficient $p$. (Refer to the forthcoming sections for the derivation of equations (8.57).) Mathematically, equations (8.57) represent the condition of small scale yielding; but they are only correct for vanishingly small plastic zone sizes. Such a strong restriction obviously has no practical value in deciding the limitations on the application of linear fracture mechanics.

In a given problem, it is proposed that an elastic-plastic analysis be carried out with no *a priori* restrictions on the plastic size. The embedded singularity, finite element method may be used to take into account the effect of plasticity [11]. In what follows, this method will be applied to solve the problem of an infinite body with a crack subjected to anti-plane shear and that of an infinite cracked plate in a far uniform tensile field. For these examples, good agreement was observed, at even moderately large plastic zone sizes, between the numerical results obtained for the plastic intensity factors from the large scale yielding, finite element calculations and the small scale yielding values of the plastic intensity factors as given by equations (8.57). Thus there appears to be a significant range over which the small scale yielding conditions, as described above, are violated; yet, the value obtained for the plastic stress intensity factor from equation (8.57) is still accurate. This result explains, at least in part, the success that has been observed in applying elastic fracture mechanics to situations in which plastic deformation is observed in the vicinity of the crack tip.

*Elastic-plastic solution procedure for anti-plane shear crack problems.* Consider first the problem of a body containing a crack which is subjected to out-of-plane shear loading. Neuber [19] has shown that the particular form of the dominant solution in the plastic zone near the crack tip, Equation (8.57), for this case is

$$u_z(\rho, \theta) = k_\varepsilon \left[\frac{h(\Theta)}{\rho}\right]^{-(p+1)^{-1}} \sin \Theta$$

$$\begin{bmatrix} \tau_{xz}(\rho, \theta) \\ \tau_{yz}(\rho, \theta) \end{bmatrix} = k_\sigma \left[\frac{h(\Theta)}{\rho}\right]^{(p+1)^{-1}} \begin{bmatrix} -\sin \Theta \\ \cos \Theta \end{bmatrix} \tag{8.58}$$

$$\begin{bmatrix} \gamma_{xz}(\rho, \theta) \\ \gamma_{yz}(\rho, \theta) \end{bmatrix} = k_\varepsilon \left[\frac{h(\Theta)}{\rho}\right]^{p(p+1)^{-1}} \begin{bmatrix} -\sin \Theta \\ \cos \Theta \end{bmatrix}$$

where

$$k_\varepsilon = (k_\sigma)^p ,$$

$$\Theta = \theta/2 + \tfrac{1}{2} \arcsin\left[\left(\frac{p-1}{p+1}\right) \sin \theta\right], \qquad h(\Theta) = \frac{\sin (2\Theta)}{2 \sin (\theta)} .$$

Rice [16] observed the exceptional implications of this particular solution for the small scale yielding problem. Specifically, if $k_\sigma = (k_3)^{2/(p+1)}$, then Equation (8.58) matches the elastic singular solution exactly on a circular arc of radius $(k_3/2)^2$ centered a distance $(k_3/2)^2[(p-1)/(p+1)]$ ahead of the crack tip. Thus, the first term of the asymptotic expansion is the complete solution throughout the plastic zone, which in turn, is bounded by the circular arc just described. As this deformation theory solution is proportional throughout the plastic region, Equation (8.58) is also the solution to the corresponding flow theory formulation for the case of monotonically increasing applied load.

Note that in the small scale yielding range, the amplitude of the plastic singularity, $k_\sigma$, is related to the stress intensity factor $k_\sigma$ by $[k_\sigma = (k_3)^{2/(p+1)}]$. When the plastic zone increases in size, the relationship of the plastic behavior to the elastic singular solution breaks down. The plastic stress intensity factor, while no longer related directly to $k_3$, is still the amplitude of the plastic singularity, and as such, is a measure of the load transfer to the crack tip for elastic-plastic problems.

The embedded singularity finite element procedure described for elastic crack problems will be modified so that it is applicable to the calculation of plastic intensity factors for situations where the small scale yielding conditions may be violated. Specifically, a finite element grid will be employed to connect the plastic singular solution, through the elastic-plastic region to the material boundaries. Consider again the problem of a rectangular body containing a crack subjected to out-of-plane shear loading, Figure 8.7, as an illustrative example. The potential energy of the first quadrant can be written in the form:

$$V = U_c + \sum_{r=1}^{N} U^{(r)} - \tau_0 \int_0^a u_z(x, b)\,dx \tag{8.59}$$

where the strain energy of the core region as calculated from Equation (8.58) is

$$U_c = (k_\varepsilon)^{(p+1)p^{-1}} S_p = (k_\varepsilon)^{(1-p)p^{-1}} \cdot (k_\varepsilon^2 S_p) \tag{8.60}$$

and $S_p$ is a constant which depends on the material hardening parameter $p$. The displacement field along the first ring of nodes, $\Gamma_1$, is

$$u_z(R_1, \theta) = k_\varepsilon \left[\frac{h(\Theta)}{R_1}\right]^{-(p+1)^{-1}} \sin\Theta \tag{8.61}$$

Since the first term of an asymptotic expansion within the plastic zone is used to describe the complete solution in the core region, it is necessary that the radius of the core, $R_1$, be small with respect to the plastic zone size as well as the geometric parameters of the specimen, e.g., the crack length. Thus for elastic-plastic calculations, $R_1$ is dependent on the loading conditions as well as the specimen geometry.

Consider the finite element grid next. The strain energy of element $(r)$ is

$$U^{(r)} = \int_{\text{Area}\,(r)} \mathrm{d}A \int_0^{\varepsilon_{ij}} \sigma_{ij}\mathrm{d}\varepsilon_{ij} = \int_{\text{Area}\,(r)} \mathrm{d}A \times \left[\int_0^{\gamma_{xz}} \tau_{xz}\mathrm{d}\gamma_{xz} + \int_0^{\gamma_{yz}} \tau_{yz}\mathrm{d}\gamma_{yz}\right]. \tag{8.62}$$

Substituting from the stress-strain relations for $\tau_{xz}$ and $\tau_{yz}$ into Equation (8.62) yields

$$U^{(r)} = \begin{cases} \displaystyle\int_{\text{Area}\,(r)} \mathrm{d}A \int_0^{\gamma} \frac{\gamma}{1+v}\,\mathrm{d}\gamma\,, & \gamma \leqslant \gamma_{yd} \\ \displaystyle\int_{\text{Area}\,(r)} \mathrm{d}A \int_0^{\gamma} \left(\frac{\gamma}{1+v}\right)^{(1-p)p^{-1}} \frac{\gamma}{1+v}\,\mathrm{d}\gamma\,, & \gamma > \gamma_{yd} \end{cases} \tag{8.63}$$

where $\gamma = (\gamma_{xz}^2 + \gamma_{yz}^2)^{\frac{1}{2}}$. Because this result will be substituted into a variational principle, it is only necessary to determine the variation of the element strain energy. This quantity can be expressed in the convenient form

$$\delta[U^{(r)}] = \begin{cases} \displaystyle\frac{1}{1+v}\int_{\text{Area}\,(r)} [\gamma\delta(\gamma)]\mathrm{d}A\,, & \gamma \leqslant \gamma_{yd} \\ \displaystyle\frac{1}{1+v}\int_{\text{Area}\,(r)} \left[\left(\frac{\gamma}{1+v}\right)^{(1-p)p^{-1}} \gamma\delta(\gamma)\right]\mathrm{d}A\,, & \gamma > \gamma_{yd}\,. \end{cases} \tag{8.64}$$

For constant strain triangular elements

$$\delta[U^{(r)}] = \begin{cases} \displaystyle\frac{\gamma}{1+v}\,\delta(\gamma)\,\text{Area}\,, & \gamma \leqslant \gamma_{yd} \\ \displaystyle\left(\frac{\gamma}{1+v}\right)^{(1-p)p^{-1}} \left(\frac{\gamma}{1+v}\right)\delta(\gamma)\,\text{Area}\,, & \gamma < \gamma_{yd}\,. \end{cases} \tag{8.65}$$

Thus, finally the variation of the strain energy for an element can be written in the form

$$\delta[U^{(r)}] = D^{(r)}(\gamma)\,\delta\left[\sum_{i=1}^{M}\sum_{j=1}^{M} k_{ij}\bar{u}_i\bar{u}_j\right] \tag{8.66}$$

where

$$D^{(r)}(\gamma) = \begin{cases} 1.0\,, & \gamma \leqslant \gamma_{yd} \\ \left(\dfrac{\gamma}{1+\nu}\right)^{(1-p)p^{-1}}, & \gamma > \gamma_{yd}\,. \end{cases}$$

The governing equations for this numerical calculation are

$$\frac{\partial V}{\partial k_\varepsilon} = 0\,, \quad \frac{\partial V}{\partial \bar{u}_i} = 0\,, \quad i = 1, n-m\,. \tag{8.67}$$

When the hardening coefficient $p$ is equal to 1.0 this formulation reduces exactly to the elastic problem described earlier. For $p \neq 1.0$ an iterative technique is employed; initial values for $k_\varepsilon^{p/(1-p)}$ and $D^{(r)}(\gamma)$, $r = 1, N$ are chosen, the resulting set of equations are solved and new values for $k_\varepsilon^{p/(1-p)}$ and $D^{(r)}(\gamma)$ are calculated. The calculation is repeated, using values obtained for $k_\varepsilon^{p/(1-p)}$ and $D^{(r)}(\gamma)$ from the previous iteration, until convergence is attained.

Such a calculation was performed for the problem of an infinite plate with a line crack in a remote anti-plane shear field [11] and an accuracy study was made by comparing the results with those reported by Rice [16] using an alternative technique. This calculation required the use of a modified potential energy functional which remains finite for all admissible displacement fields over the infinite region. With this aim in mind, the quarter plane is divided as shown in Figure 8.22 and $R_2$ is chosen sufficiently large so that the plastic zone is contained within $\Gamma_2$. In $A_2$, the displacements are written as the sum of $u_z^{(0)}$ and $u_z^{(1)}$, $u_z^{(0)}$ is chosen so that far from the crack,

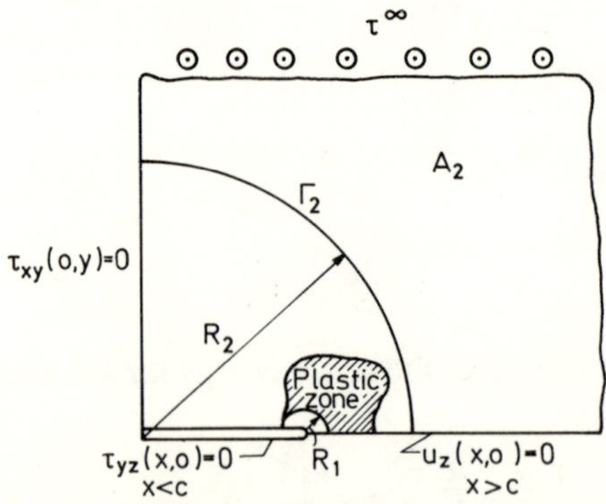


Figure 8.22. Division of quarter-plane into regions for numerical computation.

$u_z^{(1)}$ decays as the inverse of the distance from the crack. For this problem $u_z^{(0)}=\tau^\infty y$ and the corresponding stress components are $\tau_{xz}^{(0)}=0$, $\tau_{yz}^{(0)}=\tau^\infty$. The modified potential energy functional $V'$ is

$$V' = \int_{A_1} U(u_z)\mathrm{d}A + \int_{A_2} U(u_z^{(1)})\mathrm{d}A - \int_{\Gamma_2} \tau^\infty u_z^{(1)}\mathrm{d}x \tag{8.68}$$

where

$$U(u_z) = \int_0^{\varepsilon_{ij}} \sigma_{ij}\mathrm{d}\varepsilon_{ij}, \qquad U(u_z^{(1)}) = \tfrac{1}{2}\sigma_{ij}^{(1)}\varepsilon_{ij}^{(1)}.$$

It can be shown that the modified potential $V'$ for hardening materials is minimized by the exact solution among admissible displacement fields $u_z$ in $A_1$ and $u_z^{(1)}$ in the infinite domain $A_2$ which preserve continuity of displacements across $\Gamma_2$.

As in the elastic case, a Laurent series description of the displacement field is used in $A_2$; otherwise the procedure follows exactly as that employed for the finite plate, elastic-plastic calculation described above. Calculations were performed using 546 elements, 11 nodes on the first ring and a value of $R_1$ equal to 2% of the half crack length. Convergence was attained in six iterations and the calculated values for the plastic strain intensity factor were less than 2% below Rice's result for $p=10/3$ and up to 5% low at $p=10$ for large values of the applied load.

For the problem of an infinite body with finite line crack, the plastic strain intensity factor has the special form $k_\varepsilon=f(\tau^\infty)c^{p/(p+1)}$. Thus the ratio of $k_\varepsilon$ to the small scale yielding result, $k_\varepsilon^{(s)}=k_3^{2p/(p+1)}$, only depends on $\tau^\infty$. The numerical values obtained for the normalized plastic strain intensity factor,

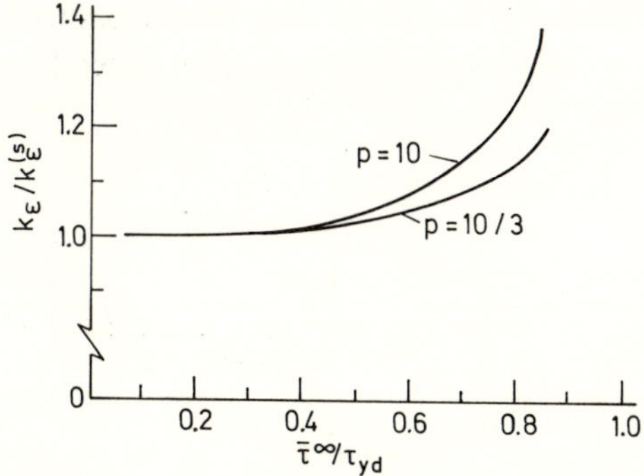


Figure 8.23. The plastic strain intensity factor as a function of the applied stress for anti-plane shear.

$\lambda(\tau^\infty)=k_\varepsilon/k_\varepsilon^{(s)}$, are plotted against the ratio of the applied stress to the yield stress in Figure 8.23 (reproduced from [11]). Observe that for the problem under consideration, the magnitude of the stress and strain fields in the plastic zone at the crack tip is accurately predicted by the elastic analysis via the small scale yielding argument for applied stresses below about half the yield stress. For large values of the applied stress, the elastic calculation is seen to underestimate the magnitude of the near tip field.

*In-plane loading of cracked bodies.* The approach employed to consider the elastic-plastic behavior of cracked bodies subjected to anti-plane shearing tractions is also applicable to the case of in-plane loading. Hutchinson [18, 20] and Rice [21] have shown that the asymptotic character of the solution within the fully plastic region near the crack tip is of the form given by equations (8.57). Hutchinson has numerically solved the nonlinear ordinary differential equations which describe the circumferential variation of the near tip fields under plane stress and plane strain conditions for a number of values of the hardening coefficient $p$.

The plastic stress (or strain) intensity factor for problems of plane stress or plane strain can be related directly to the $J$-integral [20] for deformation theory plasticity with no unloading, where

$$J = \int_\Gamma U\,\mathrm{d}y - \sigma_{ij} n_j u_{i,x}\,\mathrm{d}s \tag{8.69}$$

and

$$U = \int_0^{\varepsilon_{ij}} \sigma_{ij}\,\mathrm{d}\varepsilon_{ij}\,.$$

Under these conditions, $J$ is zero for any closed path which encloses no singularities for any solution to the equations of generalized plane stress or

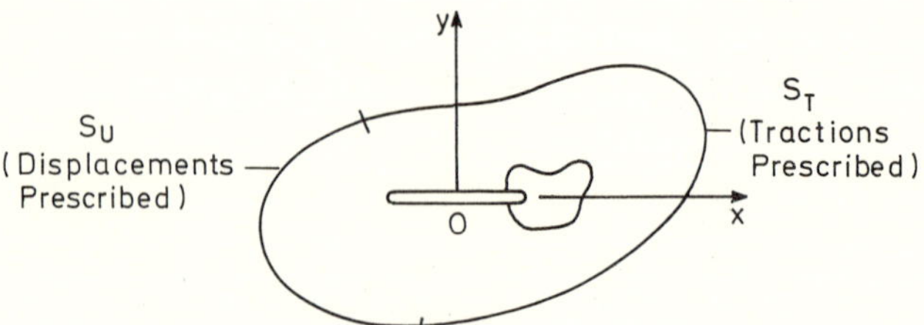


Figure 8.24. The $J$-integral has the same value on all contours $\Gamma$, from one crack edge to the other.

plane strain. Further $J$ has a constant value for any contour $\Gamma$ enclosing the crack tip, Figure 8.24. Thus $J$ can be evaluated via the plastic singular solution, equation (8.57) by choosing a contour sufficiently close to the crack tip. Therefore the plastic stress intensity factor is related to the $J$-integral by an expression of the form

$$k_\sigma = A_p(J)^{1/(p+1)} \tag{8.70}$$

where values of $A_p$ can be calculated for either generalized plane stress or plane strain conditions. As a consequence of this relationship, the $J$-integral may be considered as a measure of the intensity of the near tip stress and strain fields. For the special case of small scale yielding it is convenient to use the $J$-integral for the determination of plastic intensity factors as a function of the elastic stress intensity factor. To do this, simply evaluate $J$ from the elastic singular solution and substitute into equation (8.70). The result is

$$k_\sigma = C_p(k_1)^{2/(p+1)} \tag{8.71}$$

and values of $C_p$ are reproduced in Table IV from reference [18].

TABLE IV

*Values of* $C_p$ [18]

| $p$ | 3 | 5 | 9 | 13 |
|---|---|---|---|---|
| $C_p$ | 0.949 | 0.987 | 1.004 | 1.006 |

The embedded singularity finite element technique was employed to solve the problem of an infinite plate with a crack subjected to a far tensile field under generalized plane stress conditions. The procedure parallels that for the anti-plane, elastic-plastic calculation in much the same manner as the in-plane elastic procedure followed from the anti-plane elastic one. Thus, only results will be presented and the reader is referred to reference [11] for details. The input data used is:

$R_1 = 0.02c$

21 nodes on $\Gamma_1$

1,066 triangular elements.

The same general behavior was observed for the mode I calculation as has been reported earlier for the corresponding mode III problem. Graphs of

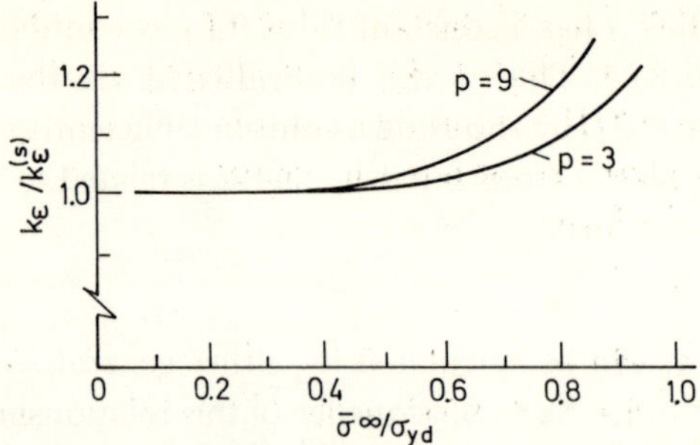


Figure 8.25. The plastic strain intensity factor as a function of the applied stress for tensile loading.

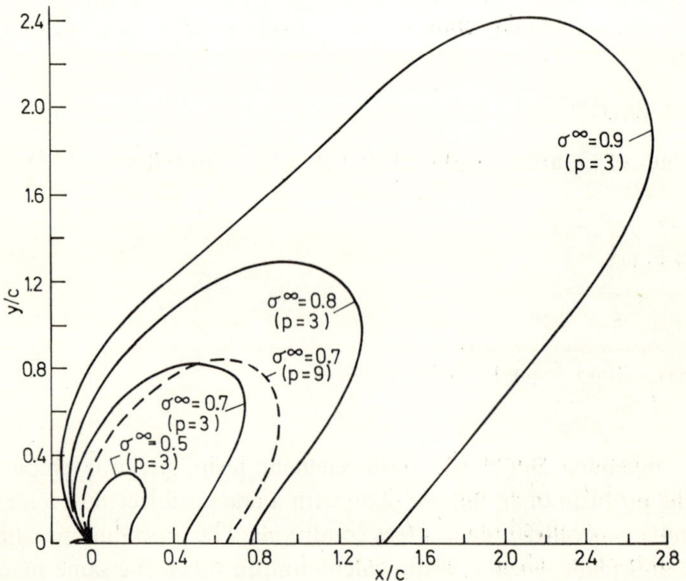


Figure 8.26. Elastic-plastic boundaries for a crack subjected to a remote field.

the plastic strain intensity factor normalized by the small scale yielding results $k_\varepsilon/k_\varepsilon^{(s)}$ against the applied stress over the yield stress $(\sigma^\infty/\sigma_{yd})$ are presented in Figure 8.25*. Further, these calculations yield the information necessary to determine the elastic-plastic boundaries and some representative curves are included in Figure 8.26.

* Reproduced from Reference [11].

Note, the results presented indicate that the numerical values obtained for the plastic strain intensity factor, with no restriction on the plastic zone size, are in close agreement with those given by the small scale yielding formulation for applied loads up to approximately fifty percent of the yield stress for the particular problem considered. In this loading range, the maximum extent of the plastic zone can be approximately equal to the quarter crack length, a clear violation of the usual definition for small scale yielding given in the beginning of this section. These results suggest that, for engineering purposes, the range of applicability of elastic fracture mechanics can be extended beyond the conditions of small scale yielding to include all situations where the numerical values of the plastic strain intensity factor do not differ appreciably from the small scale yielding results, equation (8.71).

In the case considered above, for which the specimen boundaries are sufficiently far from the crack tip so that they do not significantly influence the near tip solution, the results based on linear elasticity are accurate for applied loads up to about fifty percent of the yield stress. The range of validity of elastic fracture mechanics, in terms of applied stress to yield stress, is expected to be strongly influenced by possible interactions of the specimen boundaries with the plastic zone emanating from the crack tip. Thus, the solution for applied loads of fifty percent of yield is only known to be accurate for situations where the size of the plastic zone is small in comparison with the distance from any point in it to the nearest material boundary. The range of validity for elastic fracture mechanics and the influence of plasticity for finite specimens containing cracks is currently under consideration.

A brief summary of some of the research contributions which have been discussed in this chapter for elastic-plastic crack problems follows. For small strain, deformation theory plasticity associated with a power-hardening model of the material behavior, the form of the near tip solution to crack problems based on the two-dimensional theories of generalized plane stress, plane strain and anti-plane shear have been determined [16–21]. The amplitude of the singular solution, the plastic stress (or strain) intensity factor, can be related directly to the elastic stress intensity factor under the conditions of small scale yielding [18–21]. The embedded singularity, finite element method is applicable to the determination of the plastic intensity factors in a manner which is independent of the plastic zone size [11]. Calculations have been carried out for the classical problems of (1) an infinite body with a line crack in a remote, uniform, anti-plane shearing field and (2) an infinite plate with a line crack subjected to a uniform, far tensile field. In both cases, the calculated values of plastic intensity factors are in agreement with the corresponding semi-analytical, small scale yielding results for applied stresses below about fifty percent of the yield stress. For higher values of the applied stresses, the finite element calculations indicate that small scale yielding underestimates the magnitude of the near tip solution.

*Biaxial loading of a cracked plate.* Based on the information presented in the previous sections, it is possible to consider other elastic-plastic crack problems of interest. One such problem, which was recently studied by Hilton [22], is that of an infinite plate with a line crack subjected to biaxial tension. The purpose here is to study the experimental observation that the component of loading parallel to the crack affects the elastic-plastic crack solution [23]. It can be seen that the linear elastic solution, being free from any plastic flow around the crack tip, does not account for the interaction between the local plastic flow and parallel loading as shown in Figure 8.27.

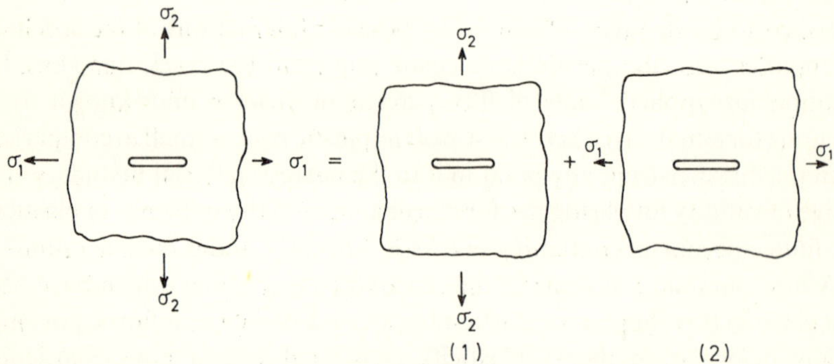


Figure 8.27. Superposition principle for the elastic solution of the biaxial loading problem.

The solution to problem (2), loading parallel to the crack edge, is non-singular, given by

$$\sigma_x = \sigma_1 , \quad \sigma_y = \tau_{xy} = 0$$

and thus can have no effect on the elastic singular solution for loading normal to the crack face.

Thus the interaction of the two loading components in the biaxial tension problem for a cracked plate is inherently nonlinear and as such an elastic-plastic analysis is expected to give results which differ qualitatively from the

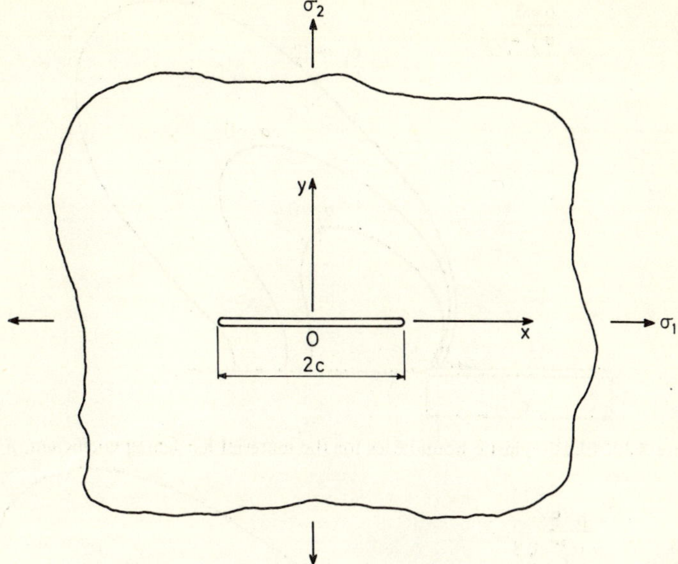


Figure 8.28. Cracked plate subjected to biaxial tension.

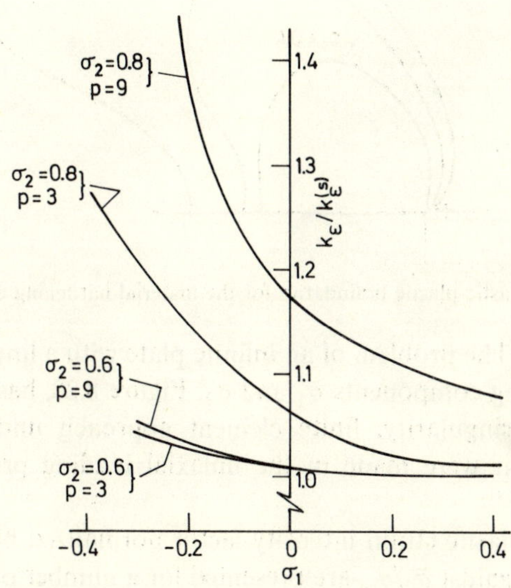


Figure 8.29. The plastic strain intensity factor as a function of the applied stress components.

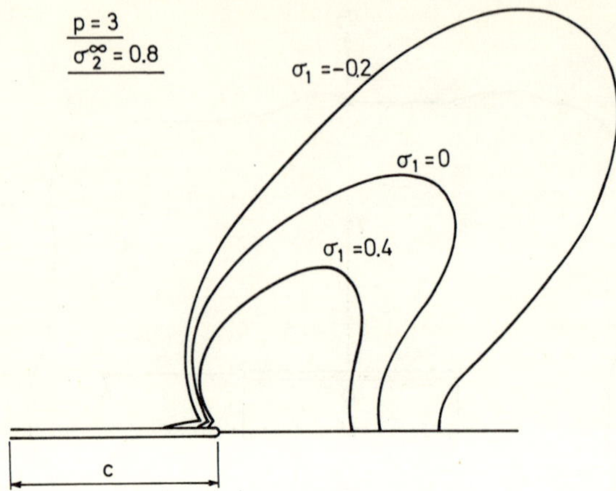

Figure 8.30. Elastic-plastic boundaries for the material hardening coefficient, $p=3$.

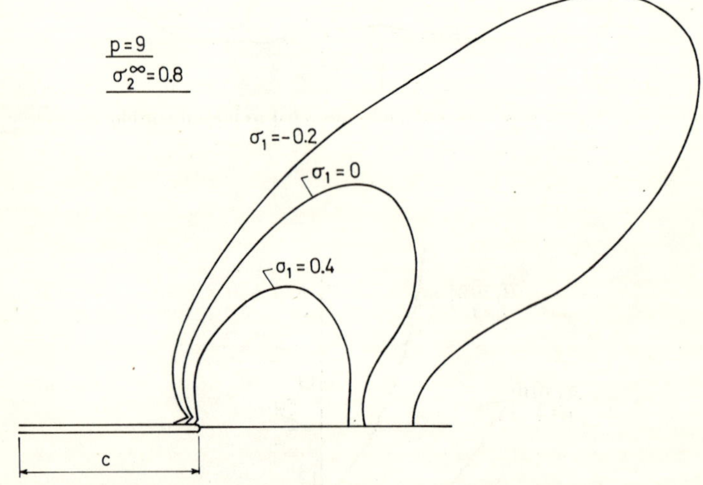

Figure 8.31. Elastic-plastic boundaries for the material hardening coefficient, $p=9$.

elastic solution. The problem of an infinite plate with a line crack subjected to biaxial loading components $\sigma_1$ and $\sigma_2$, Figure 8.28, has been solved via the embedded singularity, finite element approach under the identical assumptions that were made in the uniaxial loading problem described earlier.

Plots of the plastic strain intensity factor normalized by the small scale yielding result against $\bar{\sigma}_2/\sigma_{yd}$ are presented for a number of values of $\bar{\sigma}_1/\sigma_{yd}$ in Figure 8.29 for the hardening coefficients $p=3,9$.

The elastic-plastic boundaries corresponding to some of the loading conditions considered above are shown in Figures 8.30 and 8.31 for $p=3,9$ respectively.

These results indicate that tensile loading parallel to the crack direction has the effect of increasing the "strength" of the specimen, i.e., it increases the magnitude of the loading component applied normal to the crack for which the amplitude of the near tip solution reaches a specified value. In this sense, the load transfer to the crack tip is reduced by the addition of a tensile loading component parallel to the crack edge. Note that the corresponding plastic zone size is decreased as a result of this tensile load. It appears that for a given component of applied stress normal to the crack edge, further increases of the tensile load parallel to the crack edge will continue to decrease both the plastic strain intensity factor and the plastic zone size such that the small scale yielding conditions are approached in the limit. Compressive loading parallel to the crack is observed to have the opposite effect, i.e., it lowers the magnitude of the normal stress $\bar{\sigma}_2/\sigma_{yd}$ for which the plastic strain intensity factor reaches a prescribed value and increases the plastic zone size. These results are in qualitative agreement with the experimental observation of Kibler and Roberts [23].

The inclusion of nonlinearity in the stress-strain relations into the mathematical model results in some behavior predictions which were observed experimentally but could not be explained by linear fracture mechanics. However, a caution is required in the interpretation of these results; possible nonlinearities in the strain-displacement relations have not been included in this formulation and the range over which their effects are important is not yet known.

*Future research on elastic-plastic crack problems.* The effects of plasticity, as modeled in this chapter, on cracks in finite bodies is a subject of current research by the authors and it is hopefully anticipated that some numerical results will be available in the near future. For finite bodies with cracks, the deviation of the actual values for the plastic intensity factors from the small scale yielding results may be (1) observed at lower values of the applied tractions and (2) more pronounced at any given load as a consequence of the boundary influence. In particular, the authors are interested in comparing the situation where the plastic zone is completely contained within an elastic region to that where the plastic zone extends to a material boundary.

Elastic-plastic analysis of combined mode, crack problems with the aim of predicting fracture initiation is clearly important. The extension of the work presented here to include the simultaneous application of normal and shearing tractions to a cracked specimen is nontrivial because of the inherent nonlinearity of the model and the consequence that the superposition principle does not apply.

The range of validity of the small strain, plasticity approach should be examined. It is necessary that there exists a region near the crack tip but beyond the zone of finite strains where the small strain plastic singular solution is an accurate representation of the full solution. For example, the analysis presented here is not applicable to large ductile fracture phenomena associated with gross blunting of the crack tip. In general, corresponding finite strain calculations will be necessary to determine more precisely the range of applicability of small strain theory, plasticity and to predict crack growth outside that range.

The present analysis is based on deformation theory plasticity with no unloading. The results of such a theory are identical to those obtained from the corresponding flow theory when the deformation is proportional, i.e., the relative magnitudes of the stress components do not change with loading. Careful examination of the numerical results, associated with the calculations presented in this Chapter, indicates that the stress field at most points in the body did not deviate significantly from proportionality⋆; and further for the anti-plane shear problem, where comparison with flow theory is possible, the numerical results of deformation and incremental theory calculations were in close agreement. In general, it is believed that deformation theory calculations will give an accurate description of the solution for bodies containing stationary cracks, that are subjected to loading distributions whose magnitude is monotonically increased. The accuracy of the numerical solutions reported here for the anti-plane shear problem was observed to decrease slightly for applied stresses which approached the yield stress. On the other hand, for situations which involve substantial slow crack growth before ultimate failure, large deviations from proportionality in the stress field and significant unloading will occur, thus incremental theory analysis is clearly required. Fatigue crack propagation is an excellent example because 1) the loading is not monotonically increasing and 2) a large range of stable crack growth is observed.

⋆ Deformation theory is known to be inadequate when there are large departures from proportionality; however when this does not occur, results from deformation theory are considered to be as good as those from the corresponding flow theory [24].

## 8.10 Appendix: Elastic stress intensity factor solutions

A number of numerical solutions to plane strain and axi-symmetric problems are presented to demonstrate the versatility as well as reliability of the embedded singularity, finite element procedure [8]. Where applicable, these numerical results are compared to those obtained by alternative techniques.

**PLANE STRAIN**

*Single edge-cracked specimen.* Consider a single edge-cracked body with height $b$ and width $a$ subjected to a uniform tensile load $\sigma^0$, in the direction normal to the crack. The pertinent parameters for the finite element calculations are:

$b/a = 1.4$

Core radius $R_1/c = 0.02$

Number of nodes on $\Gamma_R = 21$

Total number of nodes $= 297$

Computation time $\simeq 2$ minutes on CDC 6400

where $\bar{k}_1 = k_1/(\sigma^0 c^{\frac{1}{2}})$. Comparative values for $\bar{k}_1$ in Table V were obtained by Gross and Brown [25] using a collocation procedure.

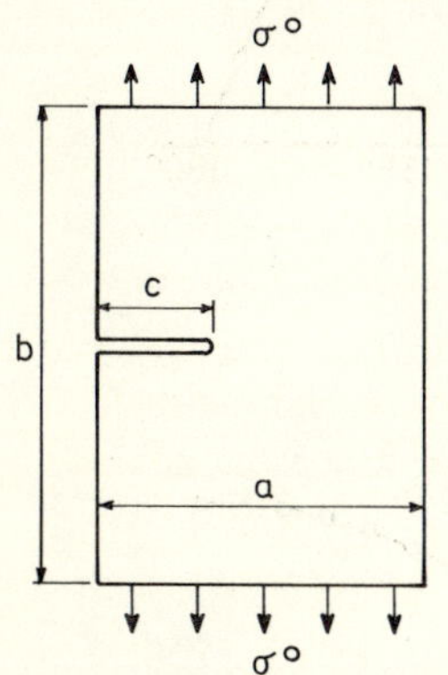


Figure 8.32. A single-notched specimen.

TABLE V

$k_1$ *for a single-edge crack system*

| $c/a$ | $\bar{k}_1$ (calculated) | $\bar{k}_1$ (comparative) |
|---|---|---|
| 0.3 | 1.59 | 1.66 |
| 0.5 | 2.70 | 2.82 |

*System of double-symmetric edge cracks.* Consider the case of two edge cracks each of length $c$ in a specimen as shown in Figure 8.21. For this problem, it is assumed that

$b/a = 1.4$

Core radius $R_1/c = 0.02$

Number of nodes on $\Gamma_R = 21$

Total number of nodes = 340

and

$$\bar{k}_1 = \frac{k_1}{\sigma^0 c^{\frac{1}{2}} \left[\frac{2a}{\pi c} \tan\left(\frac{\pi c}{2a}\right)\right]^{\frac{1}{2}}}.$$

Comparative values for $\bar{k}_1$ were obtained by Bowie [15].

A comparison of the theoretical and numerical values for the radial and circumferential components of displacement $(u_\rho, u_\theta)$ described in polar

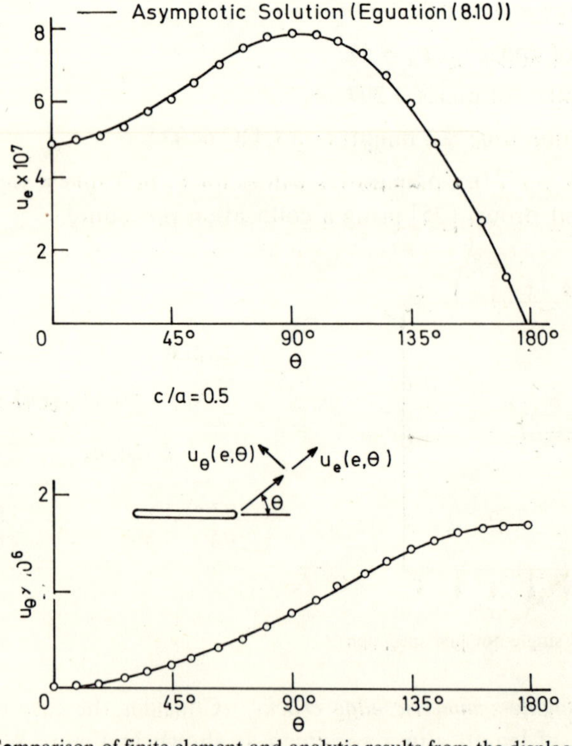


Figure 8.33. Comparison of finite element and analytic results from the displacement components near the crack tip.

TABLE VI

$k_1$ *for a double-edge crack system*

| $c/a$ | $\bar{k}_1$ (calculated) | $\bar{k}_1$ (comparative) |
|---|---|---|
| 0.3 | 1.13 | 1.13 |
| 0.5 | 1.15 | 1.15 |

coordinates $(\rho, \theta)$ centered at the crack tip, is shown in Figure 8.33 for points along the second ring of nodes about the crack tip.

*Central-cracked specimen subjected to tension.* For this crack problem, the following information

$b/a = 1.4$

Core radius $R_1/c = 0.2$

Number of nodes on $\Gamma_R = 21$

Total number of nodes $= 340$

is to be observed with $\bar{k}_1 = k_1/\sigma^0 c^{\frac{1}{2}}$. The two values of $\bar{k}_1$ used for comparison are respectively, 1.04, an approximate result derived by Paris and Sih [1] from the solution for a periodic array of cracks, and 1.07 obtained by Isida [26] via a mapping procedure.

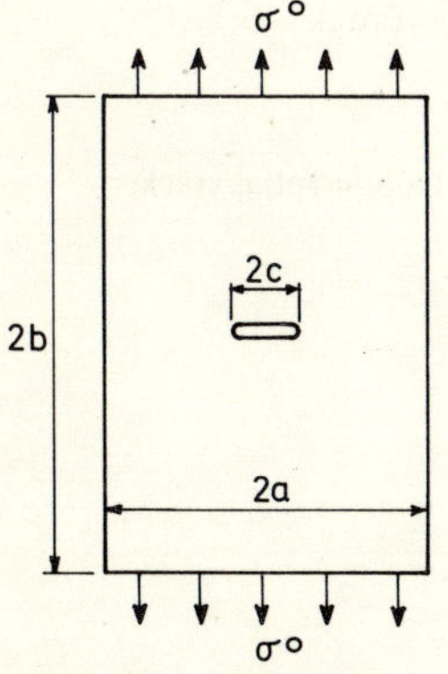


Figure 8.34. Specimen with an internal crack.

TABLE VII

$k_1$ *for a central crack*

| $c/a$ | $\bar{k}_1$ (calculated) | $\bar{k}_1$ (comparative) |
|---|---|---|
| 0.3 | 1.03 | 1.04 |
| | | 1.07 |

**AXI-SYMMETRIC CRACK PROBLEMS**

*Solid circular cylinder with a circumferential edge crack.* Referring to the crack geometry in Figure 8.35; the cylinder is subjected to a uniform tensile loading and

$h/R_0 = 1.4$

Core radius $R_1/c = 0.02$

Number of nodes on $\Gamma_R = 21$

Total number of nodes $= 297$

Here, the normalized stress intensity factor is

$$\bar{k}_1 = \frac{k_1}{\sigma_0 (2R_0)^{\frac{1}{2}}}$$

with $R_0$ being the radius of the cylinder and $\sigma_0$ the stress in the neck section. The comparative values for $\bar{k}_1$ were obtained from Bueckner's work [27].

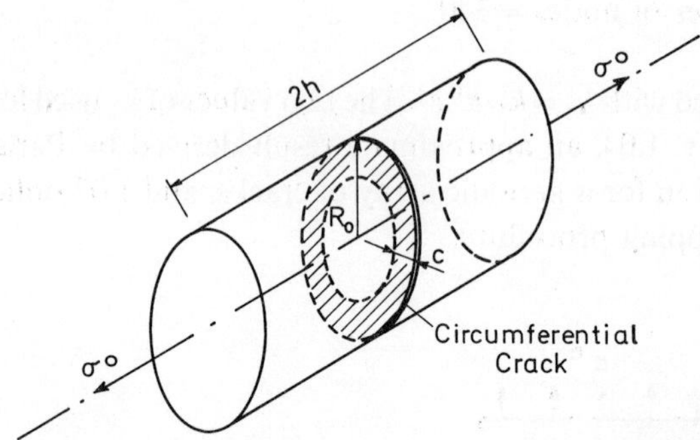


Figure 8.35. A solid cylinder with a circumferential crack.

TABLE VIII

*$k_1$ for an external crack in cylinder*

| $c/R_0$ | $\bar{k}_1$ (calculated) | $\bar{k}_1$ (comparative) |
|---|---|---|
| 0.4 | 0.250 | 0.255 |
| 0.5 | 0.235 | 0.240 |

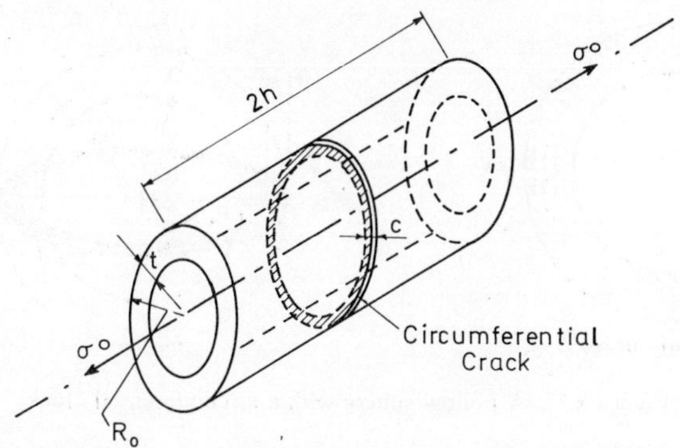


Figure 8.36. A hollow cylinder with a circumferential crack.

*Hollow circular cylinder with an external circumferential crack.* Let a hollow circular cylinder with external radius $R_0$ and thickness $t$ be subjected to a uniform tensile loading $\sigma^0$ as shown in Figure 8.36. More specifically, using the given information

$h/R_0 = 0.265$

$R_m/t = 10$

$c/t = 0.5$

Core radius $R_1/c = 0.02$

Number of nodes on $\Gamma_R = 21$

Total number of nodes $= 297$

the stress-intensity factor is found as

$k_1$ (calculated) $= 2.70\,\sigma^0 c^{\frac{1}{2}}$

where $c$ is the crack depth. For equivalent crack face loading (i.e., uniform tension applied to the crack surfaces), the calculated value of the stress intensity factor was $k_1 = 2.60\,\sigma^0 c^{\frac{1}{2}}$. The discrepancy observed appears to be due to the difficulty of approximating the continuous load by equivalent nodal forces in the vicinity of the crack edge. The solution for remote loading is believed to be more accurate; however, the crack face loading result is given for comparison with the next example.

*Hollow spherical shell with an external circumferential crack.* Referring to Figure 8.37, a hollow spherical shell with an external radius of $R_0$ and wall

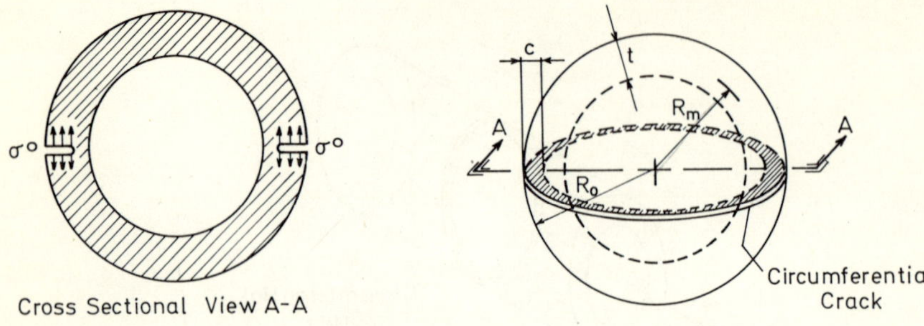


Figure 8.37. A hollow sphere with a circumferential crack.

thickness $t$ is pressurized uniformally. Using the numerical values

$R_m/t = 10$
$c/t = 0.5$
Core radius $R_1/c = 0.02$
Number of nodes on $\Gamma_R = 21$

it is found that

$$k_1 \text{ (calculated)} = 1.70\sigma^0 c^{\frac{1}{2}}$$

A comparison with the previous example indicates that the stress intensity factor for the spherical shell is substantially lower than that for the corresponding circular cylinder.

## References

[1] Sih, G. C. and Paris, P. C., Symposium of Fracture Toughness Testing and its Applications, ASME STP 381 (1965).
[2] Zienkiewicz, O. C. and Cheung, T. K., *The Finite Element Method in Structural and Continuum Mechanics.* McGraw-Hill (1968).
[3] Swedlow, J. L., Williams, M. L. and Yang, W. H., *Proc. First Int. Conf. Frac.*, 1, p. 259 (1966).
[4] Chan, S. K., Tuba, I. S. and Wilson, W. K., *Engrg. Fracture Mechanics*, 2, No. 1, p. 1 (1970).
[5] Kobayashi, A. S., Maiden, D. E. and Simon, B. J., ASME Paper No. 69-WA/PVP-12 (1969).
[6] Wilson, W. K., *On Combined Mode Fracture Mechanics.* University of Pittsburgh, Ph.D. Dissertation (1969).
[7] Wilson, W. K., Crack Tip Finite Elements for Plane Elasticity, Westinghouse Report 71-1E7-FMPWR-P2 (1971).
[8] Oglesby, J. J. and Lamacky, O., An Evaluation of Finite Element Methods for the Computation of Elastic Stress Intensity Factors. NSRDS, Report No. 3751 (1972).

[9] Williams, M. L., *J. of Applied Mechanics*, 24, No. 1, p. 109 (1957).
[10] Kantoravich, L. V. and Krylov, V. I., *Approximate Methods of Higher Analysis*. Noordhoff Ltd., Groningen, The Netherlands (1964).
[11] Hilton, P. D. and Hutchinson, J. W., *Journal of Engineering Fracture Mechanics* (to appear).
[12] Sih, G. C. and Chen, E. P., *Journal of the Franklin Institute*, 290, No. 1, p. 25 (1970).
[13] Sih, G. C. and Liebowitz, H., *Fracture*. Vol. II, edited by H. Liebowitz, Academic Press, New York, p. 67 (1968).
[14] Tracey, D. M., Finite Elements for Determination of Crack Tip Elastic Stress Intensity Factors. Army Materials and Mechanics Research Center TR 71-27, Watertown, Massachusetts (1971).
[15] Bowie, O. L., *J. of Applied Mechanics*, 31, p. 208 (1968).
[16] Rice, J., *Fracture*. Vol. II, edited by H. Liebowitz, Academic Press, New York, p. 191 (1968).
[17] McClintock, F., *Fracture*. Vol. III, edited by H. Liebowitz, Academic Press, New York, p. 47 (1968).
[18] Hutchinson, J. W., *J. of Mechs. and Phys. of Solids*. 16, p. 13 (1968).
[19] Neuber, H., ASME Trans., Series E, p. 544 (1961).
[20] Hutchinson, J. W., *J. of Mech. Phys. Solids*. 16, p. 337 (1968).
[21] Rice, J. R. and Rosengren, G., *J. of Mech. Phys. of Solids*. 16, p. 13 (1968).
[22] Hilton, P. D., *Int. J. Fracture Mechs.* (to appear).
[23] Kibler, J. J. and Roberts, R., *J. of Engrg. for Industry* (to appear).
[24] Budiansky, B., Trans. ASME (Series E), Vol. 81, p. 259 (1959).
[25] Gross, G. J., Srawley, J. and Brown, W., Stress Intensity Factors for a Single-Edge-Notch Tension Specimen by Boundary Collocation of a Stress Function. NASA TN-D-2395 (1964).
[26] Isida, M., *Proc. Fourth Congress on Applied Mechanics*. Vol. II, p. 955 (1962).
[27] Bueckner, H. F., Coefficients for Computation of the Stress Intensity Factor $K$ for a Notched Round Bar. ASTM STP 381, p. 82 (1965).

*W. K. Wilson*

# 9 *Finite element methods for elastic bodies containing cracks*

## 9.1 Introduction

One of the most powerful methods of numerical stress analysis presently available is the finite element method [1, 2]. In the past decade, the developments and refinements in this field of analysis have been very impressive. By means of the finite element method approximate solutions can be calculated for a wide range of structures of complex geometry and loading conditions and rather general material properties. The method has been applied to three dimensional bodies, plane bodies, axi-symmetric bodies, plates, shells, etc. Linear and non-linear materials, time independent and time dependent materials, and non-homogeneous and anisotropic bodies can be analyzed with almost equal ease.

When applying finite element methods of solution to cracked bodies nearly all of this versatility remains available. Thus, by use of the finite element method, the engineer can calculate elastic crack tip stress intensity factors within acceptable accuracy for a wide range of practical problems.

There are two general approaches which can be used in the application of finite element method to cracked elastic bodies. In the first, the cracked body is represented geometrically by the more conventional finite elements such as constant strain elements. While such elements cannot adequately represent the stress singularity condition at the crack tip, a number of techniques for extracting estimates of the stress intensity factor from the finite element solution are available.

The second approach involves the use of specially constructed finite elements at the crack tip. The stress singularity condition at the crack tip is built into these crack tip elements by means of the imposed displacement pattern. The use of crack tip elements of this type relieve the need for a very fine element mesh near the crack tip, allows for a direct calculation of the stress intensity factor, and are mathematically more elegant.

Before these two approaches are described in more detail and applied to a number of examples, the mathematical basis for the finite element method will be outlined. More detailed consideration of the general method is readily available in the literature [1]. In this chapter the displacement method of finite element analysis is used.

The finite element method considered here is based on the stationary total potential energy principle of elasticity, using a modified Rayleigh–Ritz procedure for obtaining approximate solutions of the stationary value problem. In brief, the body of interest is divided into a finite number of closed subregions (e.g. Figure 9.1) referred to as elements. For each subregion, a displacement function in terms of discrete parameters is prescribed. These discrete parameters $\delta_i$ $(i=1, \ldots, n)$ are normally displacements at specific points (nodal points) on the boundary of the element.

If the prescribed displacement patterns satisfy certain requirements [3, 4] with respect to completeness and conformity (continuity of displacements at element boundaries), then the best solution that can be obtained for the prescribed displacement representation is that which minimizes the total potential energy $\Pi$ of the system under the constraints imposed. That is, the best values of $\delta_i$ $(i=1, \ldots, n)$ are those that satisfy the system of equations

$$\frac{\partial \Pi}{\partial \{\delta\}} = \begin{Bmatrix} \dfrac{\delta \Pi}{\partial \delta_1} \\ \vdots \\ \dfrac{\partial \Pi}{\partial \delta_n} \end{Bmatrix} = [\mathbf{K}]\{\delta\} - \{\boldsymbol{F}\} = 0 \tag{9.1}$$

where $[\mathbf{K}]$ is the total stiffness matrix of the body and $\{\boldsymbol{F}\}$ is the generalized force vector. As the size of the elements decrease indefinitely the solutions obtained from the minimization of potential energy converge to the exact solution.

The system of equations (9.1) can be formed by summing over all elements of the body. Therefore, it is convenient to have an expression for the partial derivatives of the potential energy $\Pi_e$ of each element, $e$, with respect to the $m$ discrete parameters $\{\delta_e\} = \{\delta_1, \ldots, \delta_m\}$ which characterize that element's displacement function

$$\frac{\partial \Pi_e}{\partial \{\delta_e\}} = \left\{\frac{\partial \Pi_e}{\partial \delta_1} \cdots \frac{\partial \Pi_e}{\partial \delta_m}\right\}^T . \tag{9.2}$$

From the equation for the total potential energy of an elastic body in the absence of thermal stresses and body forces

$$\Pi = \tfrac{1}{2} \iiint_V \{\boldsymbol{\sigma}\}^T \{\boldsymbol{\varepsilon}\} \mathrm{d}V - \iint_{S_\sigma} \{\boldsymbol{u}\}^T \{\boldsymbol{X}\} \mathrm{d}S \tag{9.3}$$

where $\{\boldsymbol{\sigma}\}$, $\{\boldsymbol{\varepsilon}\}$, $\{\boldsymbol{u}\}$, and $\{\boldsymbol{X}\}$ represent the stress, strain, displacement, and surface traction vectors respectively, equation (9.2) can be expressed in terms of the element stiffness matrix $[\mathbf{k}^e]$ and the element generalized force vector $\{\boldsymbol{f}^e\}$.

$$\frac{\partial \Pi_e}{\partial \{\delta_e\}} = [\mathbf{k}^e]\{\delta_e\} - \{\boldsymbol{f}^e\} \tag{9.4}$$

where

$$[\mathbf{k}^e] = \iiint_{V_e} [\mathbf{N}]^T [\mathbf{C}]^T [\mathbf{D}]^T [\mathbf{C}] [\mathbf{N}] \mathrm{d}V \tag{9.5}$$

$$\{\boldsymbol{f}^e\} = \iint_{S_{\sigma e}} [\mathbf{N}]^T \{\boldsymbol{X}\} \mathrm{d}S \tag{9.6}$$

and the matrices $[\mathbf{N}]$, $[\mathbf{C}]$, and $[\mathbf{D}]$ are defined by the equations

$$\{\boldsymbol{u}\} = [\mathbf{N}]\{\delta_e\} \tag{9.7}$$

$$\{\boldsymbol{\varepsilon}\} = [\mathbf{C}]\{\boldsymbol{u}\} \tag{9.8}$$

$$\{\boldsymbol{\sigma}\} = [\mathbf{D}]\{\boldsymbol{\varepsilon}\} \tag{9.9}$$

The matrix $[\mathbf{N}]$ is a function of spatial coordinates and describes the defined displacement pattern of the element in terms of its discrete parameters. The matrix $[\mathbf{C}]$ is a partial differential operator and $[\mathbf{D}]$ is the elasticity matrix.

From the submatrices (9.5) and (9.6) obtained for each of the $l$ elements, the total stiffness matrix and the generalized force vector can be constructed by the assembly rules

$$[\mathbf{K}_{ij}] = \sum_{e=1}^{l} [\mathbf{k}_{ij}^e] \tag{9.10}$$

$$\{\boldsymbol{F}_i\} = \sum_{e=1}^{l} \{\boldsymbol{f}_i^e\} . \tag{9.11}$$

Equation (9.1) is then solved for $\{\delta\}$ and subsequently the strains and stresses can be calculated by means of equations (9.8) and (9.9) respectively.

## 9.2 Analysis with conventional finite elements

Since the development of the finite element concept an extensive number of different types of elements have been presented in the literature. These elements were intended for the analysis of non-cracked structures, but most of them can be used to obtain approximate solutions for cracked structures. While conventional elements cannot adequately represent the singular stress state at the crack tip, it is possible to obtain rather accurate estimates of the stress intensity factor by proper interpretation of the finite element solution. Basically, there are two different techniques for calculating stress intensity factors from conventional finite element solutions.

The first technique requires the use of an extremely fine element grid in the vicinity of the crack tip. The grid is to be fine enough such that a correlation of the finite element solution at a small distance from the crack tip with the classical crack tip stress and displacement equations will allow an acceptable estimate of the intensity factor to be made.

For the second technique finite element solutions are obtained for two slightly different crack lengths. From the calculated strain energy at each crack length the strain energy release rate for the crack can be estimated, and

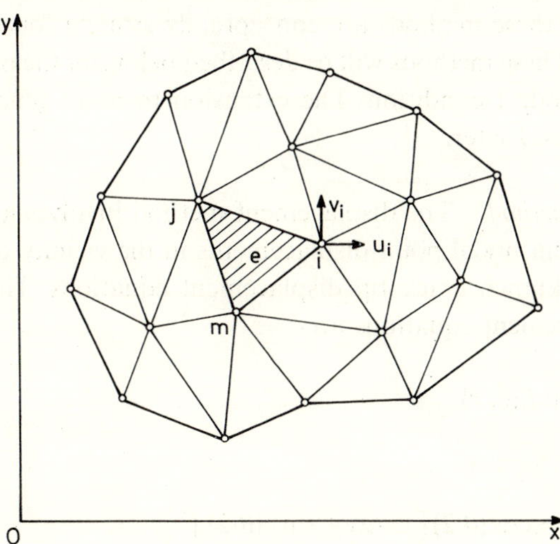


Figure 9.1. Finite element idealization of plane region.

in turn, the stress intensity factors can be calculated. The accuracy of this techniques has been found to be quite good for relatively coarse meshes.

In the following sections both of these techniques will be explored in some detail. Nearly all of the examples considered will involve the use of the constant strain triangular (CST) element (Figure 9.1) for plane strain conditions. The discrete parameters used for each element displacement pattern are displacements at the three nodal points located at the verticles of each triangle.

$$\{\delta_t\} = \{u_i v_i u_j v_j u_m v_m\}^T . \tag{9.12}$$

The well known stiffness matrix for this element is given in [1].

## 9.3 Estimates based on near tip solution

The cracked body of concern is geometrically represented by a finite element grid. The mesh away from the crack tip must be fine enough to adequately represent the overall flexibility of the body while the mesh in the vicinity of the crack tip must be fine enough to allow adequate representation of the crack tip singularity condition at a short distance from the tip. Once the numerical solution has been obtained for a particular finite element representation, crack tip stress intensity factors can be estimated by the use of established crack tip relations. The three methods of estimating considered here are: the displacement method, the stress method, and the line integral method. All of these methods are conceptually straight forward and easily implemented. These methods will be described below for the plane strain and plane stress mode I condition. The extension to more general conditions will be considered later.

*Displacement method.* The displacement method involves a correlation of the finite element nodal point displacements in the vicinity of the crack tip with the well known crack tip displacement equations. The plane strain mode I displacement equations are

$$u_i = \frac{k_1}{\mu} [r/2]^{\frac{1}{2}} f_i(\theta, \nu) \tag{9.13}$$

where $u_1 = u \quad u_2 = v$

$$f_1(\theta, \nu) = \cos(\theta/2)[1 - 2\nu + \sin^2(\theta/2)]$$
$$f_2(\theta, \nu) = \sin(\theta/2)[2 - 2\nu - \cos^2(\theta/2)]$$

and $\mu$ is the shear modulus. Similar equations can be written for plane stress. By substituting a nodal point displacement $u_i$ at some point $(\bar{r}, \bar{\theta})$ near the crack tip into equation (9.1), an estimate of the stress intensity factor

$$k_{\mathrm{I}}^* = [2/r]^{\frac{1}{2}} \mu u_i^* / [f_i(\theta, \nu)] \tag{9.14}$$

can be calculated. Calculations from the displacements at nodal points too close to the crack tip can give poor estimates due to the inability of the conventional elements to adequately represent the crack tip singularity condition. Whereas, $k_{\mathrm{I}}^*$ calculations based on displacements at nodal points distant from the crack tip give poor estimates because the crack tip displacement equations (9.13) are only accurate in the limit as $r \rightarrow 0$.

As shown by Chan *et al.* [5] a good estimate of $k_1$ can be made from a plot of $k_{\mathrm{I}}^*$ as a function of $r$, where $k_{\mathrm{I}}^*$ is calculated from the $v$ displacements on the crack surface ($\theta = \pi$). Such plots were obtained for a fracture test specimen model (Figure 9.2) for various element mesh (Figure 9.3) refinements. These plots are shown in Figure 9.4 for five cases. A measure of the element size near to and also away from the crack tip are listed in Table I for each case. These finite element curves are compared with a $k_{\mathrm{I}}^*$ curve calculated from displacements obtained by boundary collocation on the Williams stress function [6]. The collocation solution is considered to be accurate within 0.50%.

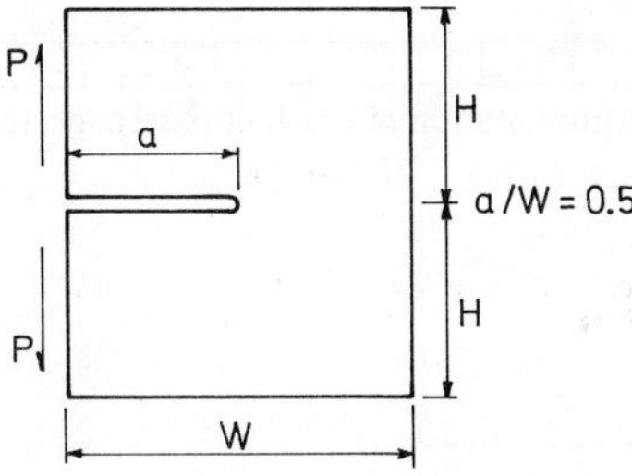


Figure 9.2. Geometry and loading condition of fracture toughness test specimen model.

A study of the $k_{\mathrm{I}}^*$ curves of Figure 9.3 shows a breakdown in the finite element solution very near the crack tip ($r \rightarrow 0$). The finer the element mesh near the crack tip, the smaller the region over which this breakdown occurs. The finer the element mesh away from the crack tip, the closer the constant slope portion of the $k^*$ curve away from the crack tip is to the exact solution.

As discussed in [5] by Chan *et al.*, a good estimate of the stress intensity

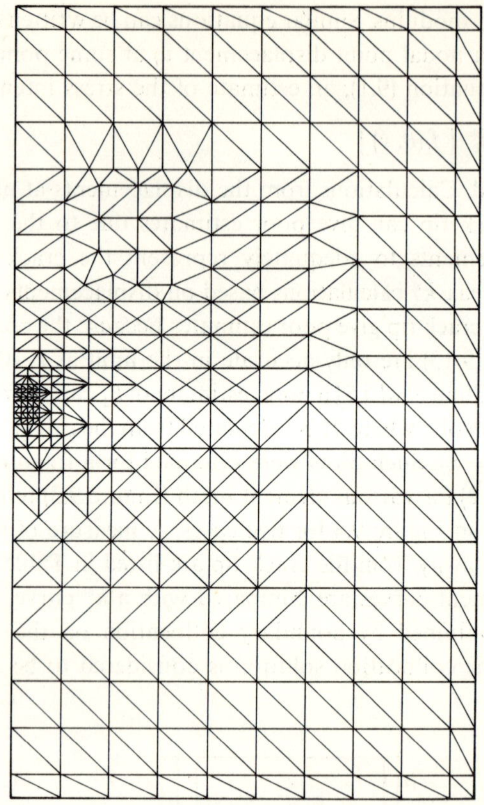

Figure 9.3. Finite element representation of one half of test specimen model (Figure 9.2) with $H/W = 0.6$.

TABLE I

*Element sizes for cases shown in Figure* 9.4

| *Case* | *Element size*⋆ $A/a^2$ | |
|---|---|---|
| | Inner ($10^{-6}$) | Outer ($10^{-2}$) |
| 1 | 312 | 2 |
| 2 | 78 | 2 |
| 3 | 20 | 2 |
| 4 | 1.20 | 2 |
| 5 | 1.20 | 1 |

⋆ $A$ = Element area
$a$ = Crack length

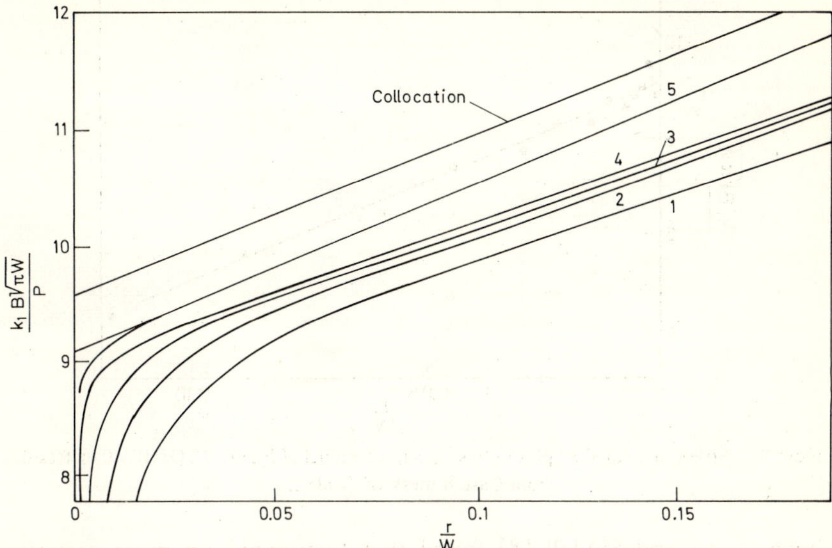


Figure 9.4. Effect of mesh size on displacement calculated $k_1^*$ for test specimen model of Figure 9.2 ($H/W = 0.6$). Mesh sizes for five cases given in Table I.

factor can be obtained from $k_1^*$ curves by extrapolating the straight portion of the curve beyond crack tip breakdown back to the vertical axis ($r = 0$) and taking the intercept value as the estimate. On the other hand, Kobayashi *et al.* [7] find that the $k_1^*$ value at a fixed number of nodes from the crack tip on the crack surface gives a good estimate of the intensity factor.

*Stress method.* The stress method for determining stress intensity factors is similar to the displacement method. The nodal point stresses are correlated with the well known crack tip stress equations. Now $k_1^*$ curves are calculated from nodal point stresses near the crack tip, and estimates of $k_1$ can be made by extrapolating back to the crack tip.

Chan [5] found that good results could be obtained from the $\sigma_y$ stress component on the $\theta = 0$ plane.

$$k_1^* = \sigma_y^* (2r)^{\frac{1}{2}} . \tag{9.15}$$

In Figure 9.5, such $k_1^*$ values are shown for mesh 4 of the example considered

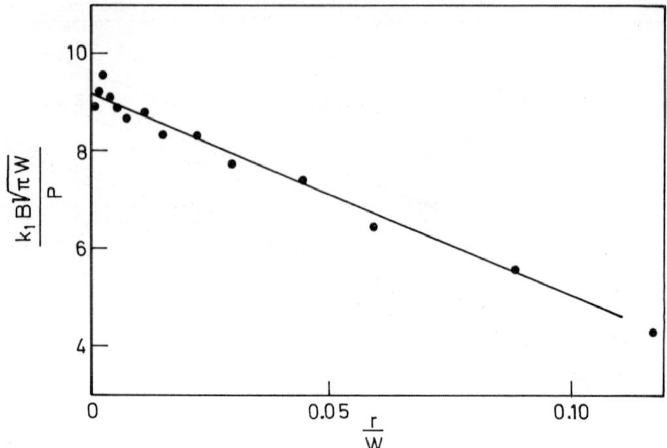


Figure 9.5. Stress calculated $k_1^*$ plot for test specimen model of Figure 9.2 ($H/W=0.6$). Results from Case 4 mesh of Table I.

above. Ayres and Siddall [8] found that they could get more accurate estimates from $k_1^*$ plots calculated from the $\sigma_y$ stress on the $\theta=90°$ plane for a number of examples.

Since it is well known that the finite element displacements are more accurate than the stresses when using the method of direct stiffness, it is best to construct $k_1^*$ plots from nodal point displacements.

*Line integral method.* Rice [9] has shown that the value of the line integral

$$J = \int_{\Gamma} \left( W\,\mathrm{d}y - \mathbf{T} \cdot \frac{\partial \mathbf{u}}{\partial x}\,\mathrm{d}s \right) \tag{9.16}$$

where $\Gamma$ is an arbitrary contour surrounding the crack tip, as shown in Figure 9.6, is proportional to the square of the crack tip stress intensity factor for plane linear elastic bodies free of body forces. The crack surface must also be traction free between the crack tip and the contour intercepts. For a mode I condition the relationship is

$$J = (1-\nu^2)\frac{\pi k_1^2}{E} \qquad \text{(plane strain)}$$

$$J = \frac{\pi k_1^2}{E} \qquad \text{(plane stress)} \tag{9.17}$$

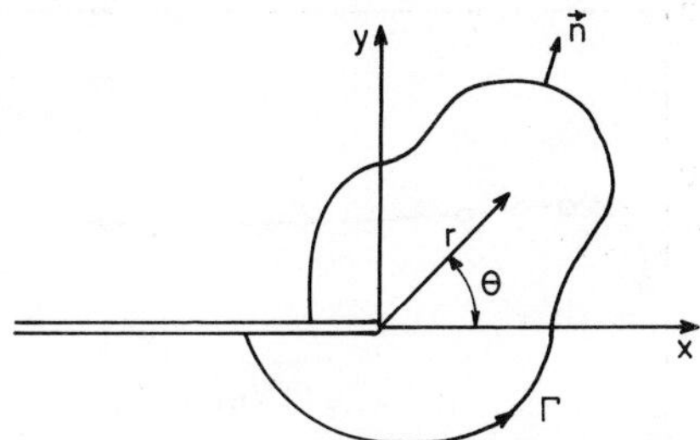


Figure 9.6. Crack tip coordinates and typical contour $\Gamma$.

In equation (9.16) $W$ is defined as the strain energy density, $\mathbf{T}$ is the traction vector defined according to the outward normal along $\Gamma$, $T_i = \sigma_{ij} n_j$, and $\mathbf{u}$ is the displacement vector. The line integral is evaluated in a counter-clockwise sense starting from the lower crack surface and continuing along the path $\Gamma$ to the upper flat surface and d$s$ is an element of arc length along $\Gamma$.

By numerically evaluating the $J$ integral for the finite element solution over a path surrounding the crack tip, an estimate of the crack tip stress intensity factor can be made by use of equation (9.17). Since the finite element solution is approximate the values of $J^*$ calculated over various paths surrounding the crack tip do vary somewhat.

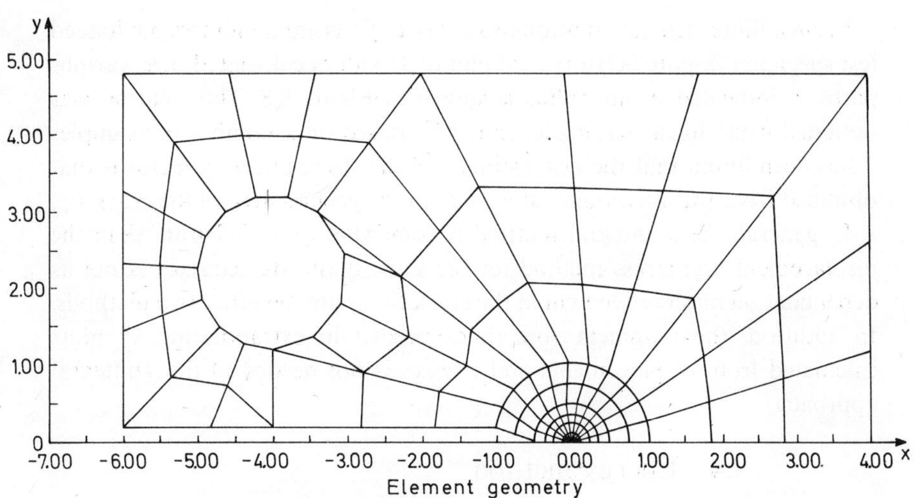


Figure 9.7. Finite element representation of one half of a pin loaded compact fracture toughness specimen.

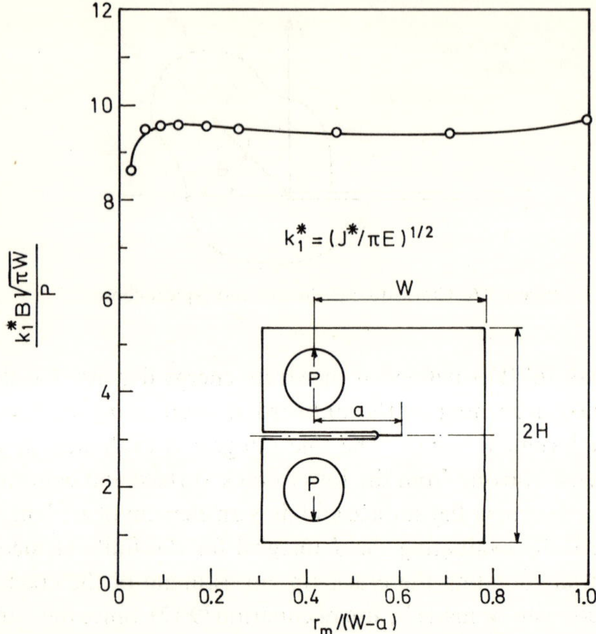


Figure 9.8. $k_{\mathrm{I}}^*$ calculated from $J$-integral over paths of mean radius $r_m$ from the crack tip. The $J$ is calculated from the finite element results obtained for the mesh of Figure 9.7.

From a finite element solution of a compact fracture toughness pin loaded test specimen (Figure 9.7) a typical plot of $J^*$ values calculated over various paths of indicated mean radius is shown in Figure 9.8. The solution was obtained using linear strain elements [7]. Based on a number of examples it has been found that the best estimate of the stress intensity factor is that obtained from the maximum value of $J^*$ in the general area of the crack tip.

In general, the $J$ integral method is somewhat more accurate than the displacement and stress method for the same mesh. Its accuracy is not as dependent on mesh refinement at the crack tip as are the other two methods. In addition, the judgment sometimes needed in extrapolating $k_{\mathrm{I}}^*$ plots calculated from displacements and stresses is not needed in the $J$ integral approach.

## 9.4 Energy method

As shown by Irwin [11] the crack tip stress intensity factor can be related to the strain energy release rate, $G$, by the following relation for a mode I loading condition.

$$G = (1-\nu^2)\frac{\pi k_1^2}{E} \quad \text{(plane strain)}$$
$$G = \frac{\pi k_1^2}{E} \quad \text{(plane stress)} \tag{9.18}$$

The strain energy release rate is the amount of energy made available at the crack tip for the crack extension process per unit area extension of the crack. For a crack extending under a fixed load condition

$$G = \frac{\mathrm{d}U}{\mathrm{d}A} \tag{9.19}$$

where $U$ is the elastic strain energy of the body containing the crack and $A$ is the crack area. For fixed displacement condition $G = -\mathrm{d}U/\mathrm{d}A$.

By obtaining finite element solutions for the cracked structure of interest at two or more slightly different crack lengths, $G$, and subsequently the crack tip stress intensity factor, can be estimated by means of equation (9.19). The strain energy stored in the body for each solution is determined by summing over the elements and $G$ is calculated by a finite difference representation of equation (9.19).

This method has been developed and successfully used by Watwood [12], Kobayashi *et al.* [7], and Anderson *et al.* [13]. Good accuracy has been obtained by the use of relatively coarse meshes. While the strain energy calculated for each crack length may not be very accurate for a coarse mesh, there is a cancelling of errors when taking the difference in strain energy at two different crack lengths. For this reason, as pointed out by Watwood, it is important that the coarseness of the mesh be approximately the same for each crack length. Watwood has considered possible bounds for stress intensity factors calculated by this method.

As shown by Paris and Sih [14] the energy release rate can also be related to the rate of change of compliance $\lambda$, the inverse spring constant, with crack extension by the equation

$$G = \frac{P^2}{2}\frac{\partial \lambda}{\partial A} \tag{9.20}$$

where $P$ is applied load. Dixon and Pook [15] calculated the compliance of

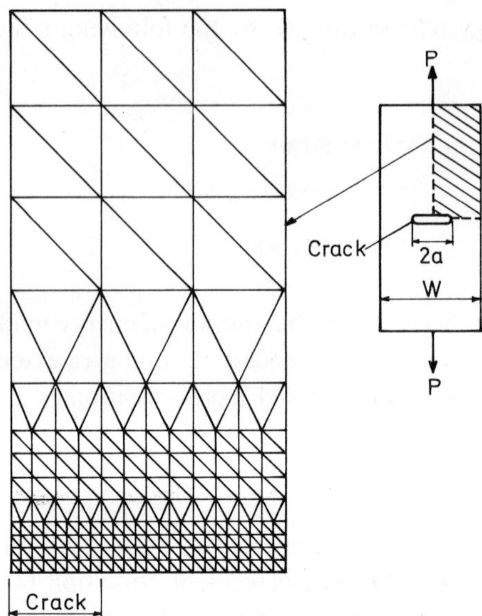


Figure 9.9. Finite element idealization for center-crack specimen. Dixon [15].

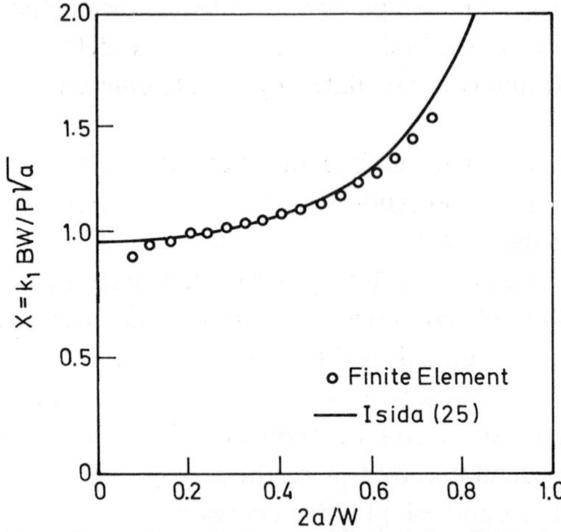


Figure 9.10. Variation of stress intensity factor with crack length/sheet width. Dixon [15].

a crack structure for a number of different crack lengths by the finite element technique and then by means of equation (9.20) calculated the energy release rate. This approach is equivalent to the strain energy approach described above except that now the strain energy is calculated as one-half the summation of the product of applied forces times their respective displacements.

The accuracy obtained by Dixon for a center cracked plate loaded in tension with the element mesh of Figure 9.9 is shown in Figure 9.10.

## 9.5 Extension of method

In discussing the above methods for determining stress intensity factor by use of conventional elements the mode I plane strain and plane stress conditions were specifically considered. Of course, many non-plane strain crack problems are of concern. In addition, the cracks of interest can also be subject to mode II loading, mode III loading, separately or simultaneously. The methods described above are extendable to these more complex situations in varying degrees.

The displacement and stress methods are in principle applicable to any two-dimensional or three-dimensional crack problems provided that the near field character of the stresses is known. For combined mode loading the magnitude of the respective stress intensity factors can be determined. For example, $k_1$, $k_2$, and $k_3$ can be estimated by means of the $v$, $u$, and $w$ displacements respectively, on the crack surface ($\theta=\pi$) or by means of the $\sigma_y$, $\tau_{xy}$, and $\tau_{yz}$ stresses ahead of the crack ($\theta=0$). The limitation of these methods is the availability of necessary computer facilities to allow sufficient element mesh refinement where needed.

The path independent $J$-integral is valid for plane problems. For a crack tip subject to combine modes I and II loading the relation between $J$ and the stress intensity factors is

$$
\begin{aligned}
J &= \frac{\pi(1-\nu^2)}{E}\,[k_1^2+k_2^2] \quad \text{(plane strain)} \\
J &= \frac{\pi}{E}\,[k_1^2+k_2^2] \quad \text{(plane stress)}
\end{aligned}
\tag{9.21}
$$

Therefore, by calculating the line integral $J$ (equation (9.16)) from the finite element results an estimate of the sum of the squares of the stress intensity factors can be made. An estimate of the magnitude of the separate intensity components cannot be made in the most general situation.

The energy (or compliance) method is most applicable to problems in which the stress intensity factors do not vary over the leading edge of the crack, i.e. plane problems, axisymmetric problems, etc. For combined mode conditions the magnitude of the separate mode intensity factors cannot be calculated. For plane problems, as with the $J$ method, only a sum of their squares can be calculated.

## 9.6 Special crack tip finite elements

Since the form of the elastic crack tip singularity is well known the construction of special crack tip finite elements which have the singularity condition built into the imposed displacement pattern is possible. Elements of this type eliminate any need for a fine element grid near the crack tip and are more appealing from a mathematical point of view than the use of more conventional elements at the crack tip. Highly accurate solutions can be obtained by placing these specially constructed elements at the crack tip and surrounding them with conventional elements which represent the rest of the structure.

There are many possibilities for both the geometric shape and the imposed displacement pattern which can be selected for a crack tip element. A number of these special finite elements have been presented in the literature. They can be classified into two groups. In the first group the crack tip is embedded in the interior of the element and in the second group the crack tip is enclosed by a number of special elements all of which have one of their nodal points at the crack tip.

The first elements of this type were developed by Wilson [16] and by Hilton and Hutchinson [17]. These elements were circular in shape, centered at the crack tip, and contained a crack extending radially from the center to the outer edge (Figure 9.11). The prescribed displacements for Wilson's elements were those associated with the mode III elastic singularity. Hilton and Hutchinson's element displacements were those associated with

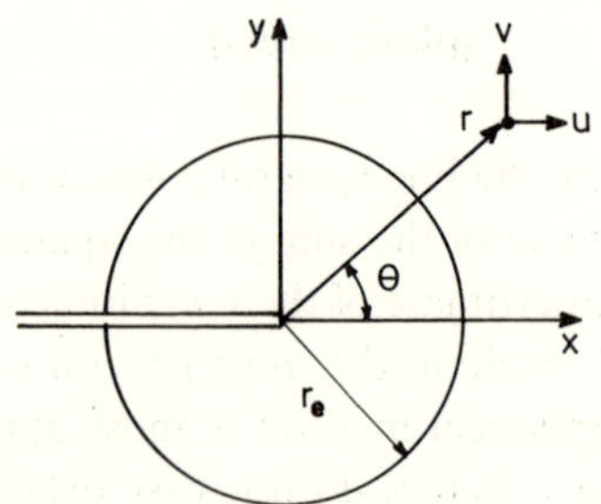


Figure 9.11. Circular crack tip element and coordinates.

the singularity for an elastic-plastic power hardening material under mode I plane stress [18] and mode III anti-plane [19] conditions.

The crack tip singularity stress distributions and associated displacement equations are exact in the limit as the crack tip is approached. Normally, these displacement equations give a good estimate of the exact displacements over a region local to the crack tip. But this region is small compared to the crack length. For this reason crack tip elements whose displacement fields are those corresponding to the singularity condition must still be restricted to a somewhat limited region near the crack tip. It is possible to relieve this limitation by adding more terms to the displacement pattern of the element, that is, by constructing high order crack tip elements. Elements of this type have been developed by Byskov [20] and Wilson [21] for plane bodies. These elements are described in the following sections.

*High order circular element.* The element constructed by Wilson is circular in shape (Figure 9.11) and the displacement terms used are those associated with the leading terms of the Williams stress function [6]. Only the symmetric terms are used, but the extension to skew-symmetric terms is clear.

The symmetric terms of the Williams stress function are:

$$\phi(r,\theta) = \sum_{n=1,2,3}^{\infty} \Big\{(-1)^{n-1} d_{2n-1} r^{n+\frac{1}{2}} \left[-\cos(n-\tfrac{3}{2})\theta + \frac{2n-3}{2n+1}\cos(n+\tfrac{1}{2})\theta\right] + (-1)^n d_{2n} r^{n+1} [-\cos(n-1)\theta + \cos(n+1)\theta]\Big\}. \tag{9.22}$$

The function is an eigen-function expansion of the Airy stress function about the crack tip of a semi-infinite crack whose surfaces are traction free. The displacements for this stress function are given in [22] with an errata in [23]. For this high order circular tip element, designated as SSC-4, the displacements corresponding to the first four terms of the Williams stress function are used along with a term for rigid body translation of the element in the $x$-direction. The displacement function is expressed in convenient form below where the notation is defined in Figure 9.11.

$$\begin{aligned} u &= u_0 + \sum_{k=1}^{4} \delta_k \left(\frac{r}{r_e}\right)^{k/2} f_k(\theta) \\ v &= \sum_{k=1}^{4} \delta_k \left(\frac{r}{r_e}\right)^{k/2} g_k(\theta) \end{aligned} \tag{9.23}$$

where

$$
\begin{aligned}
f_k(\theta) &= F_k(\theta)\cos\theta - G_k(\theta)\sin\theta \\
g_k(\theta) &= F_k(\theta)\sin\theta + G_k(\theta)\cos\theta
\end{aligned}
\tag{9.24}
$$

and

$$
\begin{aligned}
F_1(\theta) &= -(\tfrac{5}{2}-4\sigma)\cos(\tfrac{1}{2}\theta)+\tfrac{1}{2}\cos(\tfrac{3}{2}\theta) \\
F_2(\theta) &= (2-4\sigma)+2\cos(2\theta) \\
F_3(\theta) &= (\tfrac{3}{2}-4\sigma)\cos(\tfrac{1}{2}\theta)+(\tfrac{1}{2})\cos(\tfrac{5}{2}\theta) \\
F_4(\theta) &= -(1-4\sigma)\cos\theta-3\cos(3\theta) \\
G_1(\theta) &= (\tfrac{7}{2}-4\sigma)\sin(\tfrac{1}{2}\theta)-\tfrac{1}{2}\sin(\tfrac{3}{2}\theta) \\
G_2(\theta) &= -2\sin(2\theta) \\
G_3(\theta) &= (\tfrac{9}{2}-4\sigma)\sin(\tfrac{1}{2}\theta)-\tfrac{1}{2}\sin(\tfrac{5}{2}\theta) \\
G_4(\theta) &= -(5-4\sigma)\sin\theta+3\sin(3\theta)
\end{aligned}
$$

where

$$
\begin{aligned}
\sigma &= \nu \qquad \text{(plane strain)} \\
&= \frac{\nu}{1+\nu} \qquad \text{(plane stress)}
\end{aligned}
$$

and $r_e$ is the radius of the element. The discrete parameters for this element are

$$
\{\delta_{\text{SSC-4}}\} = \{\delta_1\,\delta_2\,\delta_3\,\delta_4\,u_0\}^T . \tag{9.25}
$$

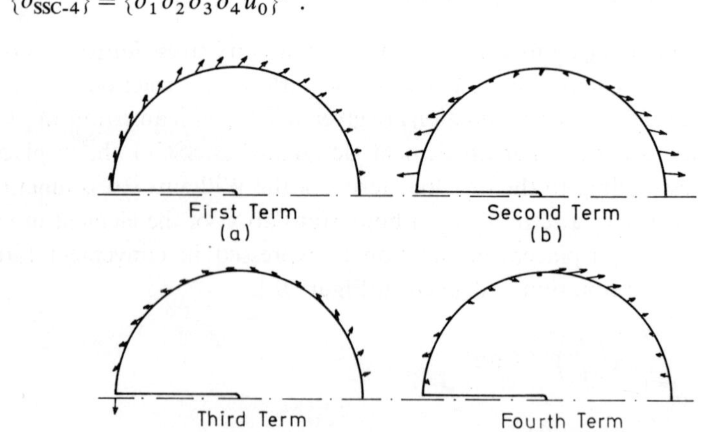


Figure 9.12. Normalized boundary displacement vectors for one half of circular crack tip element. Displacements corresponding to each term of series in equation (9.23).

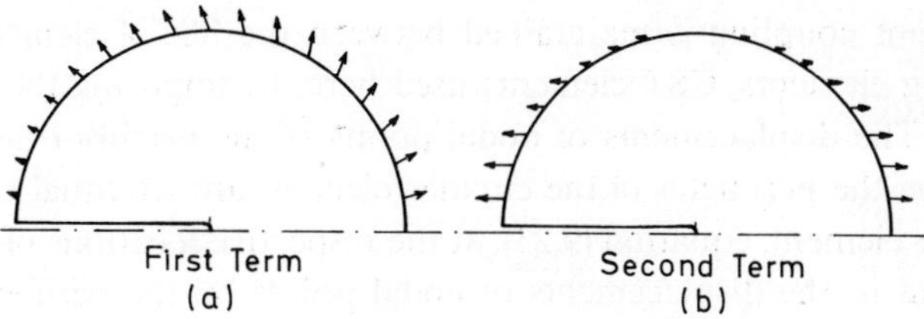


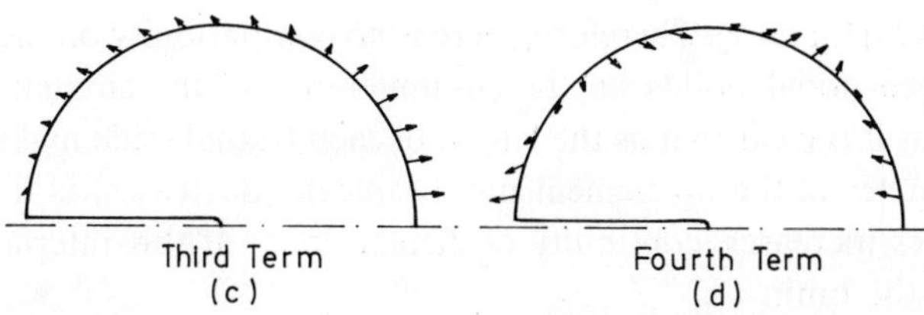


Figure 9.13. Normalized boundary traction vectors for one half of circular crack tip elements. Tractions corresponding to each term of series in displacement equation (9.23).

The displacement patterns at the outer radius of the element for each of the $\delta_i$ terms are shown in Figure 9.12. The surface tractions associated with each of these displacement terms are shown in Figure 9.13. The relation between $\delta_1$ and the crack tip stress intensity factor is

$$k_1 = -\mu \left(\frac{8}{r_e}\right)^{\frac{1}{2}} \delta_1 . \tag{9.26}$$

By means of equation (9.5) the stiffness matrix of this element can be determined. The rigid body displacement term, $u_0$, only contributes zero elements to the matrix. Therefore, for convenience only the elements of the four by four symmetric stiffness matrix $[\mathbf{k}^e] = [\mathbf{a}]$ corresponding to $\{\delta_1 \delta_2 \delta_3 \delta_4\}$ are listed below.

$$\begin{aligned}
&a_{11} = \mu(\pi/2)(5-8\sigma) && a_{12} = \mu(64/15)(-2+5\sigma) \\
&a_{13} = \mu(3\pi)(-1+2\sigma) && a_{14} = \mu(128/105)(6-7\sigma) \\
&a_{22} = \mu(16\pi)(1-\sigma) && a_{23} = \mu(64/5)(2-3\sigma) \\
&a_{24} = 0 && a_{33} = \mu(\pi/2)(15-24\sigma) \\
&a_{34} = \mu(128/35)(-6+5\sigma) && a_{44} = \mu(16\pi)(3-\sigma) .
\end{aligned} \tag{9.27}$$

Displacement coupling is maintained between the SSC-4 element and its neighboring elements, CST elements used here, by imposing the following condition. The displacements of nodal points of the neighboring elements which lie on the perimeter of the circular element are set equal to those of the circular element, equation (9.23), at the respective locations of the nodal points. That is, the displacements of nodal points on the perimeter of the SSC-4 element are a function of $\{\delta_{\text{SSC-4}}\}$.

Displacements between two neighboring nodal points on the circumference of the SSC-4 element vary linearly in the bounding constant strain triangular element and vary in the tip element as a function of $\theta$ as given by equation (9.23) for $r=r_e$. Therefore, there is no compatibility on the common edge between nodal points on the circumference of the circular crack tip element. But it is clear that as the length of each triangle side making up the outer perimeter of the tip element monotonically decreases, as the number of such sides increases, continuity of displacement at this interface will be reached in the limit.

This incompatability of displacements between nodal points at $r=r_e$ means that monotonic convergence of solutions with increasing numbers of finite elements cannot be guaranteed by means of existing proofs. But a rough numerical demonstration of convergence has been provided by Wilson [21].

A simple example which demonstrates the influence of the higher order displacement terms used in the circular crack tip element is now considered. The geometry is a circular cylinder with a radial crack extending from the center to the outer radius ($r=r_c$). Displacement boundary conditions are applied at the outer radius. The boundary displacements are those given by the equations

$$\begin{aligned} u &= -(r/r_c)^{\frac{1}{2}} f_1(\theta) - (r/r_c)^{\frac{3}{2}} f_3(\theta) \\ v &= -(r/r_c)^{\frac{1}{2}} g_1(\theta) - (r/r_c)^{\frac{3}{2}} g_3(\theta) \end{aligned} \tag{9.28}$$

with $r=r_c$, where the $f(\theta)$ and $g(\theta)$ functions are those of equations (9.24). It can easily be shown that the solution for this geometry and boundary condition is equations (9.28) and the stress intensity factor is $k_1=\mu(8/r_c)^{\frac{1}{2}}$.

Using the SSC element and CST elements (Figure 9.14), finite element solutions to the problem were obtained for various size circular tip elements (Figure 9.15). For one set of results the displacement pattern of the tip element (SSC-1) was restricted to the first term of the series of equation (9.23). For the second set of results the displacement pattern of the tip

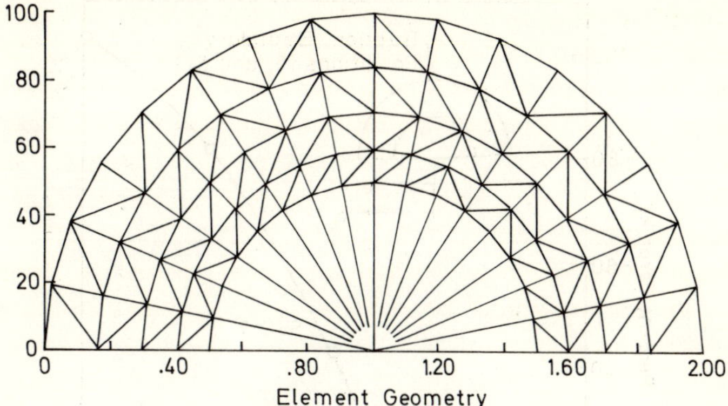


Figure 9.14. Finite element representation of one half of a cracked circular cylinder of radius $r_c$. Radius of circular crack tip element is $r_e = r_c/2$.

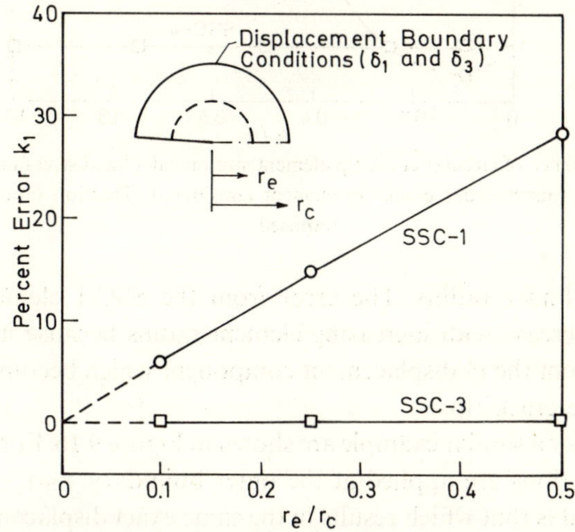


Figure 9.15. Influence of circular crack tip element size on calculated stress intensity factor. A first order and a third order crack tip element considered. Displacement boundary condition imposed.

element (SSC-3) included the first three terms of the series. In all cases the tip elements were coupled with enough triangular elements such that a further increase in their numbers would have an insignificant influence on the results shown. As observed in Figure 9.15, the use of the SSC-3 element results in a very accurate solution even for an element radius equal to one

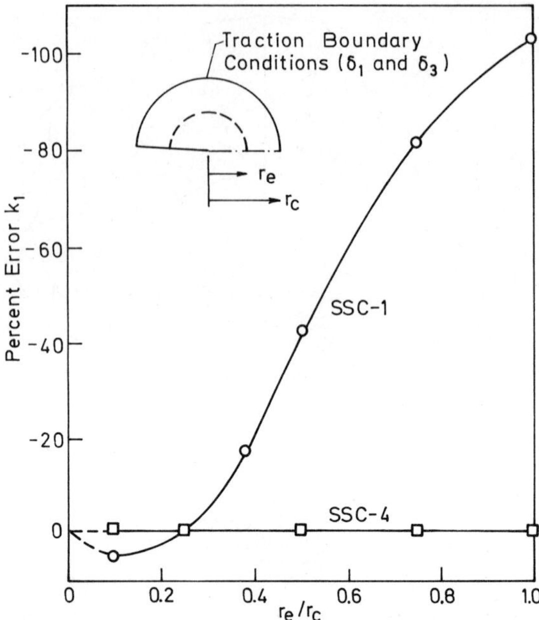


Figure 9.16. Influence of circular crack tip element size on calculated stress intensity factor. A first order and a fourth order crack tip element considered. Traction boundary condition imposed.

half of the cylinder radius. The error from the SSC-1 element solutions continually increase with increasing element radius because it cannot adequately represent the $r^{\frac{3}{2}}$ displacement component which becomes influential away from the crack tip.

The results for a similar example are shown in Figure 9.16. For this example boundary tractions are applied at the outer boundary, $r=r_c$. The traction pattern applied is that which results in the same exact displacement solution as that of the previous example. Again two sets of finite element solutions were obtained. One set was obtained for a single term ($\delta_1$) tip element (SSC-1) and the other for the four term tip element (SSC-4). General trends similar to those in Figure 9.14 are shown as the radius of the tip element is decreased. As would be expected, the error for the SSC-1 element is generally greater under the traction boundary condition than under the displacement boundary condition.

*High order triangular element.* A high order crack tip element similar to the one described in detail above has been presented by Byskov [20]. He proposed a cracked element which is a polygonal area containing a linear crack. The specific element he developed and applied is triangular in shape with a straight crack extending from one vertex of the triangle into its interior (Figure 9.17). The displacement terms used for the element are obtained from a complex stress function of Muskhelishvili which is equivalent to the Williams stress function. The specific displacement terms used for the element are essentially the same as the rigid body translation and first three terms of the series of the displacement function (equation (9.23)) of the SSC-4 element described above, except that the corresponding skew-symmetric components are also included.

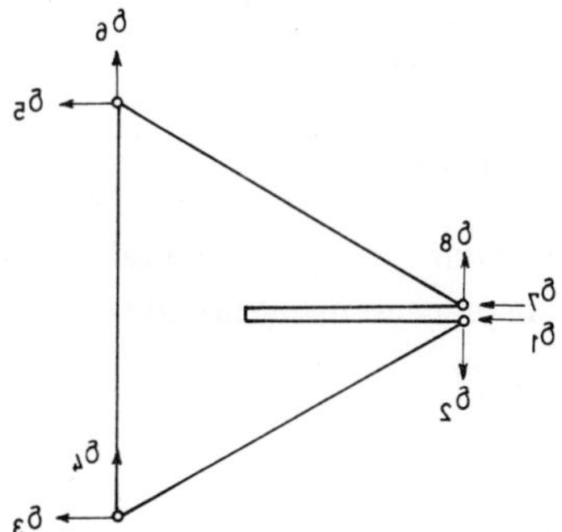

Figure 9.17. Cracked triangle. Byskov [20].

Incompatibility of displacements with neighboring elements also exists between nodal points for this triangular crack tip element. The stiffness matrix of the element is determined numerically for different triangular shapes. Byskov obtained finite element solutions for a number of examples using the cracked triangular element and obtained reasonable accuracy.

Besides shape, the primary difference between the circular crack tip element and the triangular crack tip element is in the technique of coupling with the neighboring conventional elements. In the circular element technique the number of conventional elements in contact with the perimeter of the circular element can be varied, thus allowing some control over the degree of incompatibility which occurs over the common inter-surface. With the triangular tip element technique of Byskov only three conventional constant strain triangular elements bound the tip element and the influence of the incompatibility cannot be controlled.

## 9.7 Special elements with nodal point at crack tip

Crack tip finite elements which have one nodal point at the crack tip can be constructed. These elements are assigned a displacement pattern compatible with the elastic singularity condition. The crack tip is then enclosed by a group of these special elements. In Figure 9.18 a group of such triangular elements is shown. Triangular elements of this type have been developed by Tracey [24] and Wilson [21] for plane bodies subject to in-plane loading.

The rather simple displacement pattern used by Wilson for each triangular element at the crack tip is

$$u = u_0 + \left[\frac{\theta_j - \theta}{\theta_j - \theta_i} u_i + \frac{\theta - \theta_i}{\theta_j - \theta_i} u_j\right]\left(\frac{r}{r_e}\right)^{\frac{1}{2}}$$
$$v = v_0 + \left[\frac{\theta_j - \theta}{\theta_j - \theta_i} u_i + \frac{\theta - \theta_i}{\theta_j - \theta_i} u_j\right]\left(\frac{r}{r_e}\right)^{\frac{1}{2}} \qquad (9.29)$$

where $\theta_i$ and $\theta_j$ are angular positions of nodal points located at $r_e$ and $(u_i\, v_i\, u_j\, v_j\, u_0\, v_0)$ are the displacements at nodal points $i$, $j$, and 0 (Figure 9.19), respectively.

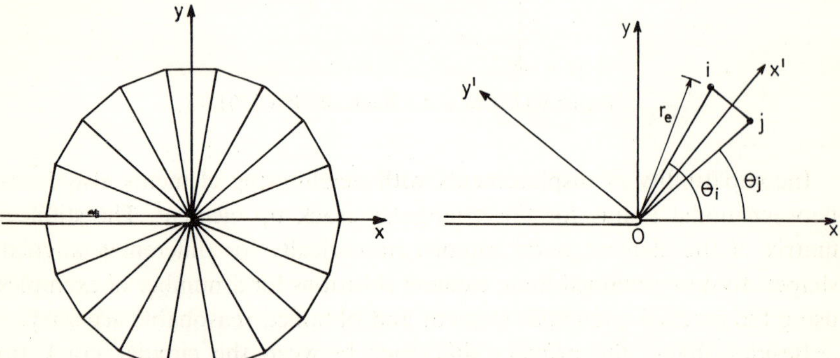


Figure 9.18. Crack tip enclosed by triangular crack tip elements.

Figure 9.19. Triangular crack tip element coordinates and nodal points.

These displacements vary linearly with the angular coordinate $\theta$ and are proportional to the square root of the radius $r$. Therefore, these elements are capable of representing the displacement field associated with the plane crack tip elastic singularity condition to any degree of accuracy needed.

The displacement patterns of equations (9.29) ensure that displacements are continuous at the interface of neighboring crack tip elements. Displacements at the interface of one of these crack tip elements and the conventional linear displacement element which bounds it in the radial direction (i.e. Figure 9.20) are only continuous at their two common nodal points. This is so because the displacements on the radial side of this crack tip triangular element do not vary quite linearly. But as the wedge angle of the tip element decreases indefinitely, the displacement over the radial edge of the element approaches a linear relation and continuity at this interface is approached in the limit.

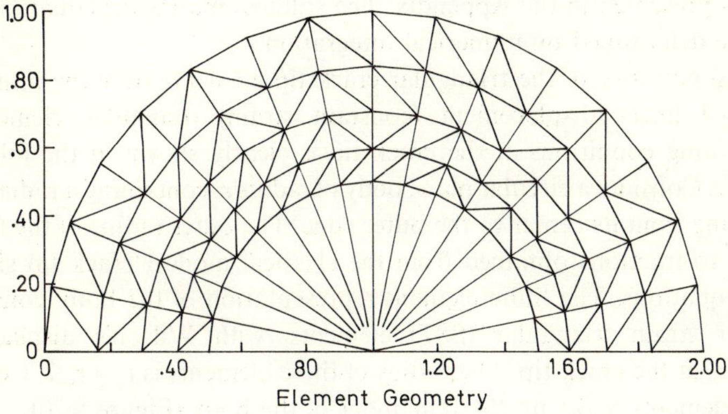


Figure 9.20. Finite element representation of one half of a cracked circular cylinder of radius $r_c$. Radius of outer nodal points of crack tip triangular elements is $r_e = r_c/2$.

Tracey has developed a triangular crack tip element which allows the displacements to vary linearly over the radial edge of the triangle and thus ensures compatibility at the radial edge with a bounding finite element whose displacements also vary linearly over this common edge. In terms of the notation of Figure 9.19 Tracey's crack tip displacement functions are

$$u = u_0 + \left[ \frac{(\cos\theta_j \cos\theta)^{\frac{1}{2}}(\tan\theta - \tan\theta_j)}{(\cos\theta_j \cos\theta_i)^{\frac{1}{2}}(\tan\theta_i - \tan\theta_j)} u_i + \frac{(\cos\theta \cos\theta_i)^{\frac{1}{2}}(\tan\theta_i - \tan\theta)}{(\cos\theta_j \cos\theta_i)^{\frac{1}{2}}(\tan\theta_i - \tan\theta_j)} u_j \right] \left(\frac{r}{r_e}\right)^{\frac{1}{2}} \tag{9.30}$$

and a similar expression for the $v$ displacement. Tracey originally expressed his displacement functions in terms of natural coordinates and obtained the functions by means of mapping the triangle into a square. These crack tip elements were coupled with isoparametric elements in examples considered by Tracey.

As the included angle at the crack tip nodal point decreases Wilson's displacement function (equations (9.29)) becomes equivalent to that of Tracey. In fact, for the case of 24 elements at the crack tip, as used by Tracey in one of his examples, the displacement patterns for the two functions differ by about 0.1%. The difference decreases linearly with the square of the included angle.

The element stiffness matrix for the displacement function of equations (4.29) is presented in the Appendix. The stiffness matrix for equation (4.30) must be determined by numerical integration.

The superiority of the triangular crack tip elements over the more conventional linear displacement (constant strain) triangular elements in representing conditions at the crack tip is clearly shown in the following example. Consider a circular plane body of radius $r_c$ containing a radial crack projecting from its center to the outer edge. The outer radius of the body is subject to tractions obtained from the classical mode I crack tip singular stress equations. The finite element representation of the body consists of singular strain triangular (SST) elements, with Wilson's displacement function, at the crack tip. The radius of these elements is $r_e = r_c/2$. Constant strain elements make up the remainder of the body (Figure 9.20).

The relative error in the nodal point displacements of the tip elements at $r = r_e$ are shown in Figure 9.21 for the case of 18 SST elements at the crack tip. The error is defined as the difference between finite element nodal displacement and the exact displacement divided by the maximum exact $v$-displacement at $r = r_e$ which occurs at $\theta = \pi$. These results are compared with those obtained when the SST elements were replaced with constant strain triangular (CST) elements. As expected, the SST results results are very much better.

In Figure 9.22 the influence of the number of elements at the crack tip on the root mean square of nodal point displacement error at $r = r_e$ where $r_e = r_c/2$ is shown both for SST elements and for CST elements at the crack tip. The percent error of the SST element solution is approaching zero with increasing number of crack tip elements, whereas the error of the CST element solutions appears to be leveling off at about 22%.

There are a number of situations in which the triangular crack tip elements are an advantage over the elements having the crack tip embedded in them. These are cases in which the material near the crack tip is non-homogeneous

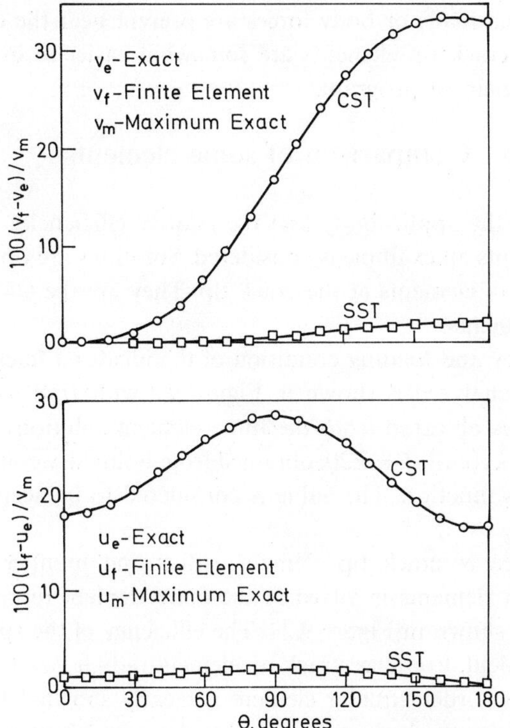


Figure 9.21. Percent error of finite element nodal point displacements at $r=r_e$. Results for cracked cylinder subject to $k_1$ tractions. Eighteen elements at crack tip ($r_e=r_c/2$).

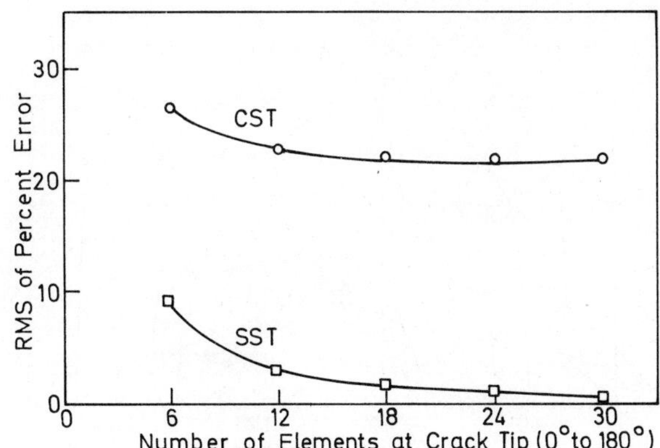


Figure 9.22. Root mean square of finite element nodal point displacement errors at $r=r_e$. Results for cracked cylinder subjected to $k_1$ tractions ($r_e=r_c/2$).

or when thermal loads or body forces are present near the crack tip. Also the triangular crack tip elements are somewhat easier to incorporate into existing finite element programs.

## 9.8 Comparison of some elements

To investigate the applicability and the relative efficiencies of some crack tip finite elements an example is considered. Solutions are obtained for four different types of elements at the crack tip. They are the CST, SST, SSC-1 and SSC-4 elements.

The geometry and loading condition of the idealized fracture toughness test specimen analyzed is shown in Figure 9.2 with $H/W=0.5$. The stress intensity factors obtained from the finite element solutions are compared with the value, $k_1(\pi a)^{\frac{1}{2}}/P=7.20$, obtained from boundary collocation on the Williams stress function. The value is considered to be accurate to within 0.2 percent.

The influence of crack tip element radius and number of additional constant strain elements involved in the finite element representation (e.g. Figure 9.23) is shown in Figure 9.24. The efficiency of the special crack tip elements is evident. For large crack tip element radii ($r_e/a=0.25$) the advantage of the high order circular element is clearly shown. The influence of element radius for the high order circular element is further demonstrated in Figure 9.25. High accuracy can be obtained with the element radius equal to one-half of the crack length.

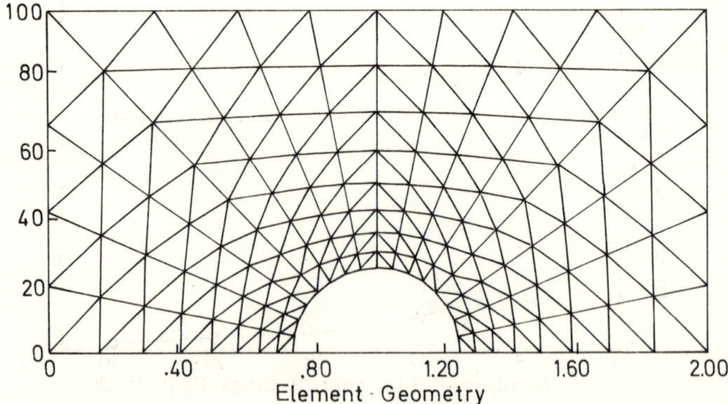


Figure 9.23. Finite element representation of one half of a test specimen model (Figure 9.2, $H/W=0.5$). Radius of circular crack tip element is $r_e=a/4$.

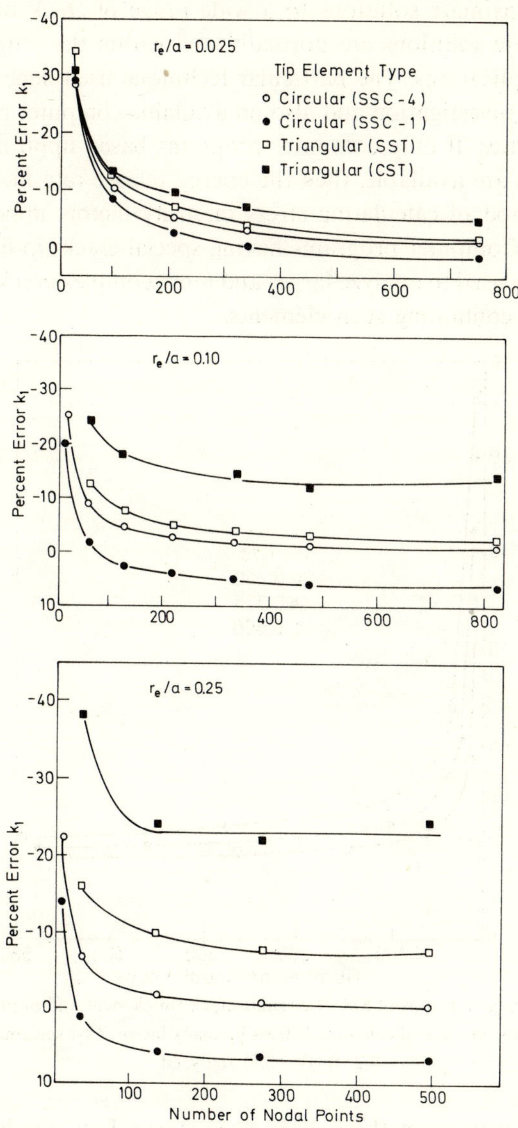


Figure 9.24. Influence of number of triangular element nodal points on error of calculated stress intensity factor for four types of crack tip elements. Test specimen model (Figure 9.2, $H/W = 0.5$) analyzed. Data shown for three crack tip element radii.

By the techniques considered above, finite element methods can be used to obtain approximate solutions to a wide range of crack problems. The accuracy of these solutions are normally well within the range needed for engineering applications. The particular technique used depends upon the problem under investigation and also on available computer programs and computer facilities. If only computer programs based upon more conventional elements are available, then the energy release rate method and the $J$-integral method of calculating stress intensity factors appear to be the most effective. Computer programs having special crack tip finite elements available can be used to analyze larger and more complex cracked structures than those not containing such elements.

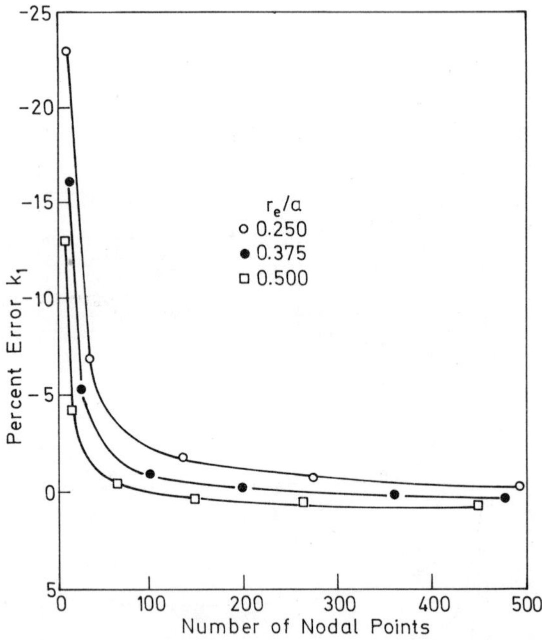


Figure 9.25. Influence of radius of SSC-4 circular crack tip element and number of triangular element nodal points on error of calculated stress intensity factor. Test specimen model (Figure 9.2, $H/W=0.5$) analyzed.

While finite element methods have been successfully applied to cracked bodies by many investigators, there is still a need for more rigorous mathematical consideration of the convergence criteria for these methods. Little work has been done in the area of convergence criteria for finite elements in the presence of a singularity, nor for element displacement functions as complex as those used with crack tip finite elements.

Future work in the area of the finite element analysis of cracked bodies will involve refinements of crack tip finite elements and extensions of these elements to other classes of problems such as plates [26] and shells. Another area of crack analysis in which finite elements offer promise is that of dynamic crack problems.

## 9.9 Appendix: Stiffness matrix of triangular crack tip (SST) element

The stiffness matrix of the triangular crack tip element whose displacement functions are equations (9.29) is given here in terms of the local $x'$–$y'$ coordinates (Figure 9.19) and associated nodal point displacement vector

$$\{\delta_t'\} = \{u_i'\ v_i'\ u_j'\ v_j'\}^T \tag{9.31}$$

where

$$\begin{Bmatrix} u' \\ v' \end{Bmatrix} = \begin{bmatrix} \cos\theta_c & \sin\theta_c \\ -\sin\theta_c & \cos\theta_c \end{bmatrix} \begin{Bmatrix} u \\ v \end{Bmatrix} \tag{9.32}$$

and $\theta_c = (\theta_i + \theta_j)/2$. The rigid body translation $(u_0, v_0)$ is not considered here since it adds only zero terms to the matrix. The displacement functions of equations (9.29) transform to

$$\begin{aligned} u' &= \tfrac{1}{2}\left[\left(1 - \frac{\theta'}{\theta_0}\right)u_i' + \left(1 + \frac{\theta'}{\theta_0}\right)u_j'\right]\left(\frac{r}{r_e}\right)^{\frac{1}{2}} \\ v' &= \tfrac{1}{2}\left[\left(1 - \frac{\theta'}{\theta_0}\right)v_i' + \left(1 + \frac{\theta'}{\theta_0}\right)v_j'\right]\left(\frac{r}{r_e}\right)^{\frac{1}{2}} \end{aligned} \tag{9.33}$$

where $\theta' = \theta - \theta_c$ and $\theta_0 = (\theta_i - \theta_j)/2$.

By means of equation (9.5) the stiffness matrix

$$[\mathbf{k}_t'] = \frac{\cos\theta_0}{4}[\mathbf{s}']$$

corresponding to the vector of equation (9.31) can be evaluated. The elements of the symmetric matrix $[\mathbf{s}']$ are

$$s'_{11} = 2B_1 + 2B_2/\theta_0^2 \qquad s'_{12} = -(B_6 + B_5)/\theta_0$$
$$s'_{13} = 2B_1 - 2B_2/\theta_0^2 \qquad s'_{14} = (B_6 - B_5)/\theta_0$$
$$s'_{22} = 2B_3 + 2B_4/\theta_0^2 \qquad s'_{23} = -s'_{14}$$
$$s'_{24} = 2B_3 - 2B_4/\theta_0^2 \qquad s'_{33} = s'_{11}$$
$$s'_{34} = -s'_{12} \qquad s'_{44} = s'_{22}$$

where

$$B_1 = \tfrac{1}{8}[(\lambda + 2\mu)I_1 + I_3]$$
$$B_2 = \tfrac{1}{8}(\lambda + 2\mu)(I_4 - 4I_2 + 4I_3) + \tfrac{1}{8}(\mu)(I_5 + 4I_2 + 4I_1)$$
$$B_3 = \tfrac{1}{8}[(\lambda + 2\mu)I_3 + \mu I_1]$$
$$B_4 = \tfrac{1}{8}(\lambda + 2\mu)(I_5 + 4I_2 + 4I_1) + \tfrac{1}{8}(\mu)(I_4 - 4I_2 + 4I_3)$$
$$B_5 = \tfrac{1}{4}(\lambda)(I_2 - 2I_3) + \tfrac{1}{4}(\mu)(I_2 + 2I_1)$$
$$B_6 = \tfrac{1}{4}(\lambda)(I_2 + 2I_1) + \tfrac{1}{4}(\mu)(I_2 - 2I_3)$$

with

$$I_1 = 2 \sin \theta_0$$
$$I_2 = 2(\sin \theta_0 - \theta_0 \cos \theta_0)$$
$$I_3 = -2 \sin \theta_0 + 2 \ln \tan \left( \frac{\pi}{4} + \frac{\theta_0}{2} \right)$$
$$I_4 = 4\theta_0 \cos \theta_0 + 2(\theta_0^2 - 2) \sin \theta_0$$
$$I_5 = 2 \int_0^{\theta_0} \theta^2 \sin \theta \tan \theta \, d\theta$$

and

$$\lambda = Ev/[(1 + v)(1 - 2v)] \quad \text{(plane strain)}$$
$$= Ev/[(1 - v)(1 + v)] \quad \text{(plane stress)}$$

Integral $I_5$ cannot be determined in closed form but a highly accurate approximation is

$$I_5 = 2[\tfrac{1}{5}\theta_0^5 + \tfrac{1}{42}\theta_0^7 + \tfrac{31}{3240}\theta_0^9 + \tfrac{173}{55440}\theta_0^{11}].$$

## References

[1] Zienkiewicz, O. C. and Cheung, Y. K., *The Finite Element Method in Structural and Continuum Mechanics*, McGraw-Hill Book Company, Inc., New York (1967).

[2] Zienkiewicz, O. C., *Applied Mechanics Review*, 23, 3, p. 249 (1970).

[3] DeArantes e Oliveira, E. R., *International Journal of Solids and Structures*, 4, 10, p. 929 (1968).

[4] Pian, T. H. H. and Ping Tong, *International Journal of Solids and Structures*, 3, 5, p. 865 (1967).

[5] Chan, S. K., Tuba, I. S. and Wilson, W. K., *Engineering Fracture Mechanics*, 2, 1, p. 1 (1970).

[6] Williams, M. L., *Journal of Applied Mechanics*, 24, 1, p. 109 (1957).

[7] Kobayashi, A. S., Maiden, D. E. and Simon, B. J., ASME Paper No. 69-WA/PVP-12, 1969, not published in ASME Transactions but available through ASME.

[8] Ayres, D. J. and Siddall, Jr., W. F., ASME Paper No. 70-PVP-23, 1970, not published in ASME Transactions but available through ASME.

[9] Rice, J. R., *Journal of Applied Mechanics*, 35, 2, p. 379 (1968).

[10] Visser, C., Gabrielse, S. E. and VanBuren, W., HSST Technical Report No. 4, Oak Ridge National Laboratory (1970).

[11] Irwin, G. R., in *Encyclopedia of Physics*, Vol. 7, p. 551, Springer, Berlin (1958).

[12] Watwood, V. B., *Nuclear Engineering and Design*, 11, 4, p. 323 (1969).

[13] Anderson, G. P., Ruggles, V. L. and Stibor, G. S., *International Journal of Fracture Mechanics*, 7, 1, p. 63 (1971).

[14] Paris, P. C. and Sih, G. C., in *Fracture Toughness Testing and Its Applications*. ASTM Special Technical Publications No. 381, p. 30 (1965).

[15] Dixon, J. R. and Pook, L. P., *Nature*, 224, 5215, p. 166 (1969).

[16] Wilson, W. K., *Combined Mode Fracture Mechanics*. Ph.D. Dissertation, University of Pittsburgh (1969).

[17] Hilton, P. D. and Hutchinson, J. W., Plastic Intensity Factor for Cracked Plates. *Engineering Fracture Mechanics*, in press.

[18] Hutchinson, J. W., *Journal of the Mechanics and Physics of Solids*, 16, 5, p. 337 (1968).

[19] Rice, J. R., *Journal of Applied Mechanics*, 34, 2, p. 287 (1967).

[20] Byskov, E., *International Journal of Fracture Mechanics*, 6, 2, p. 159 (1970).

[21] Wilson, W. K., Some Crack Tip Finite Elements for Plate Elasticity, presented at the Fifth National Symposium on Fracture Mechanics, to be published in ASTM Special Technical Publication, p. 513, Part I.

[22] Gross, B., Roberts, E., Jr. and Srawley, J. E., *International Journal of Fracture Mechanics*, 4, 3, p. 267 (1968).

[23] Gross, B., Roberts, E., Jr. and Srawley, J. E., *International Journal of Fracture Mechanics*, 6, 1, p. 87 (1970).

[24] Tracey, D. M., *Engineering Fracture Mechanics*, 3, 3, p. 255 (1971).

[25] Isida, M., *Transactions of Japan Society of Mechanical Engineers*, 21, 107, p. 511 (1955).

# Subject index